EIGHTH EDITION

BUSINESS MATHEMATICS

A COLLEGIATE APPROACH

NELDA W. ROUECHE

VIRGINIA H. GRAVES

Upper Saddle River, New Jersey
Columbus, Ohio

Library of Congress Cataloging-in-Publication Data
Roueche, Nelda W.
Business mathematics: a collegiate approach / Nelda W. Roueche, Virginia H. Graves.—8th ed.
p. cm.
Includes index.
ISBN 0-13-084730-5
1. Business mathematics. I. Graves, Virginia H., 1943- II. Title.

HF5691 .R68 2001
650'.01'513—dc21 99-088768

Vice President and Publisher: Dave Garza
Editor in Chief: Stephen Helba
Editor: Elizabeth Sugg
Associate Editor: Michelle Churma
Production Editor: Louise N. Sette
Design Coordinator: Robin G. Chukes
Text Designer: Rosemarie Votta
Cover Designer: Tom Mack
Cover photo: © Albright-Knox Gallery
Production Manager: Brian Fox
Marketing Manager: Shannon Simonson

This book was set in Times Roman by Carlisle Communications, Ltd. It was printed and bound by Von Hoffmann Press, Inc. The cover was printed by Von Hoffmann Press, Inc.

10 9 8 7 6 5 4 3 2
ISBN 0-13-084730-5

Chapter/Page	Formula	Description
Chapter 13 page 345	_____% × Cost = Markup	% markup based on cost
Chapter 13 page 345	_____% × Selling price = Markup	% markup based on selling price
Chapter 14 page 371	Total handling cost = Cost + Overhead	Total handling cost
Chapter 14 page 371	Operating profit (or loss) = Actual selling price − Total handling cost	Operating profit or operating loss
Chapter 14 page 371	Actual selling price = Total handling cost	Breakeven point
Chapter 14 page 371	If Actual selling price < Cost, then Cost − Actual selling price = Absolute loss (or Gross loss)	Absolute loss or gross loss
Chapter 15 page 384	$I = Prt$	Simple interest
Chapter 15 page 384	$M = P + I$ or $M = P(1 + rt)$	Maturity value at simple interest
Chapter 15 page 401	$P = \frac{M}{1 + rt}$	Present value at simple interest
Chapter 16 page 410	$D = Mdt$	Simple discount
Chapter 16 pages 410 and 413	$p = M - D$ or $p = M(1 - dt)$	Proceeds
Chapter 18 page 473	$M = P(1 + i)^n$	Amount at compound interest
Chapter 18 page 486	$P = M(1 + i)^{-n}$	Present value at compound interest
Chapter 19 page 495	$M = \text{Pmt.} \times \text{Amt. of ann. tab.}_{\overline{n}\rvert i}$	Amount of annuity
Chapter 19 page 498	$\text{P.V.} = \text{Pmt.} \times \text{P.V. ann. tab.}_{\overline{n}\rvert i}$	Present value of annuity
Chapter 20 page 507	$\text{Pmt.} = M \times \text{S.F. tab.}_{\overline{n}\rvert i}$	Sinking fund payment
Chapter 20 page 511	$\text{C.Y.} = \frac{\text{Annual interest}}{\text{Bond mkt. or purchase price}}$	Current yield of bond
Chapter 20 page 517	$\text{Pmt.} = \text{P.V.} \times \text{Amtz. tab.}_{\overline{n}\rvert i}$	Amortization payment

PREFACE

In its 31st year of publication, *Business Mathematics: A Collegiate Approach* continues to meet the needs of business students throughout the nation. Over the years, the text has evolved into its current highly accepted position. The success of the book confirms that it has accomplished its primary objective—to familiarize students with a wide range of business procedures that require the use of mathematics.

Business Mathematics is a comprehensive textbook. A wide range of mathematical procedures exposes students to various business applications. The teaching methodology used in *Business Mathematics* has proved to be successful with many types of students and in varied academic environments. The text is used extensively in the business programs of two-year community and technical colleges, as well as in the lower-level programs at four-year colleges and universities. The content is sufficient for a full year's course, or selected topics/chapters may be used for a one-term course.

The authors' overriding concern has always been for the student. Extreme care is taken to present each topic in a clear and logical manner, with all steps included to facilitate understanding. In addition, concise discussions describe the business applications of each topic, so that the student can appreciate its relevance. Undoubtedly, this emphasis on the student's needs has contributed to the enduring acceptance of the text.

Business Mathematics sharpens the mathematical skills of students preparing to enter business employment, as it simultaneously provides an introduction to accounting, finance, insurance, statistics, taxation, and other math-related subjects. If anything, the widespread use of calculators and computers in business has raised the expectation that employees will be knowledgeable in mathematical procedures. The fact that machines perform the final calculations in no way lessens the necessity that employees understand what needs to be done and what sequence of operations is required. Thus, students anticipating a career in business should see this course as valuable preparation for a wide range of career opportunities.

Even those students who plan nonbusiness careers will receive valuable consumer education. The text provides an understanding of consumer issues such as bank reconciliations, discounts, markups and markdowns, as well as installment purchases and simple and compound interest.

STUDENT-ORIENTED FEATURES

Among the aspects that make *Business Mathematics* especially student-oriented are the following features:

- **Precision:** The terminology and business applications of each topic are presented in a precise, readable discussion that is as concise as possible without sacrificing content. The first use and definition of each important term are highlighted through the use of **boldface.** Moreover, each chapter concludes with a glossary that defines all the boldface key terms of that chapter.
- **Thoroughness:** An unusually large number of examples cover every type of problem that the student may be assigned. The thorough explanations include every step, to ensure that the student will understand the entire solution. In addition, illustrations of many business forms, notes, and statements are featured.
- **Extensive Practice:** Over 1,000 problems are included in this text. Most exercise sets contain two types of problems: tabular problems designed for quick mastery of technique, followed by written "word" problems to ensure full understanding of the topic under discussion. All odd/even problems form a pair—providing one problem for classwork and a similar one for homework—and are arranged in order of increasing difficulty. Special care has been taken to use compatible number combinations, so as to emphasize the inherent business procedures and avoid tedious exercises in arithmetic.
- **Workplace Skills:** Student learning objectives describe the workplace skills to be developed. Each skill is completely cross-referenced to the text, identifying the section and examples where the related information is explained, as well as which problems should be studied to develop the skill.
- **Reinforcement:** Brief reviews of arithmetic, equations, and percent are included for the benefit of students whose math skills need refreshing. This essential review is made more relevant by using review problems that are also business oriented, which contributes to the overall goals of the course.
- **Retention:** Elementary equation-solving is used throughout the text in all topics where it is logical and natural to do so. Students can thus apply the same techniques to many topics and avoid trying to memorize several variations of each formula (which inexperienced math students find nearly impossible to do).
- **Calculators:** The use of hand-held LCD calculators, which is allowed (or even required) at many institutions, is introduced in a topic that covers basic functionality and gives students general practice. Throughout the text, high-

lighted Calculator Techniques illustrate how to apply specific calculator operations to the current section's examples, in order to help students compute the problems correctly and efficiently.

SURVEY RESULTS

Prior to this eighth edition, business mathematics instructors at a number of colleges were surveyed to determine (a) whether any existing topics were considered obsolete and (b) whether any specific new topics were recommended. Consensus strongly reinforced the text's existing subject matter, with appropriate updating. Although not all suggestions could be incorporated into the eighth edition, the authors are grateful to those instructors who responded. Among the ideas recommended for this edition are the following:

- **New Student Problems:** Over 50% of the student problems are new in this edition. The new problems update the text to reflect business trends, while providing instructors with fresh assignment selections.
- **Decimals:** Because students now use calculators extensively, emphasis has been placed on decimals rather than on fractions in examples and problems. Where appropriate, fractional equivalents have been retained.
- **Spreadsheets:** Spreadsheet problems and instructions have been added to Chapters 4, 9, 10, and 20. A CD-ROM containing templates for the problems has been included with the text.
- **Payroll Regulations:** The payroll chapter (Chapter 8) reflects changes in Social Security, Medicare, and federal income tax deductions. Examples and problems incorporate 1999–2000 tables, rates, and forms. Included are illustrations of Form W-4, the Employee's Withholding Allowance Certificate; Form W-2, Wage and Tax Statement; and a quarterly individual earnings summary, as well as illustrations and problems for Form 941, the Employer Quarterly Tax Return.
- **Global Economy:** To facilitate international transactions in the global market, metrics and foreign exchange conversions have been included in Appendix B.
- **Reference Tables:** Extensive tables with 10 decimal places have been retained, as requested by a majority of survey participants. Thus, instructors and students can determine the appropriate number of decimals for text problems and personal computations. For the eighth edition, the tables have been placed in a supplement, Tables Booklet, which will facilitate test taking.
- **Currency:** Interest rates, prices, and wages in the examples and problems have been adjusted to reflect current trends.
- **Topic Sequence:** The same sequence of topics as presented in previous editions has been maintained. The survey respondents recommended the retention of the first three chapters of basic mathematics review, preferring it in the body of the text.

SUPPLEMENTARY SUPPORT

An extensive *Instructor's Manual* is available upon adoption of the text. For the instructor's convenience, complete solutions are included for all problems in the text. In addition, suggestions for organizing the course and teaching each chapter are offered, as well as sample quizzes and a final examination. Reflecting the wide acceptance of *Business Mathematics,* an extensive array of supplementary materials is available for instructors and/or students, as follows:

- Test Item File
- Test Manager (computerized version)
- Student Solutions Manual (available upon approval of the instructor or institution)
- Spreadsheet files to accompany exercises in chapters 4, 9, 10, and 20 (included with the text)
- Achievement Tests
- Instructor's Solutions Manual with Transparency Masters
- Tables Booklet
- Companion Website

The content of this eighth edition is due in large measure to the contributions of the following people, to whom we express special appreciation:

- The instructors who participated in the prerevision survey for this eighth edition. The time you took to respond is greatly appreciated, for your suggestions ensure a comprehensiveness and consensus that would be impossible to achieve if the modifications were derived only from the authors. Our thanks to the following survey participants: Professor Mark Bambach, Ed.D., Community College of Philadelphia (PA); Professor Bobbie Corbett, Northern Virginia Community College (VA); Professor Robert E. Foss, Madison Area Technical College (WI); and Professor Luther Guynes, Los Angeles City College (CA).
- The following professors at the Alexandria campus of Northern Virginia Community College for sharing their expertise in various technical areas of business: Professor Douglas Bracy; Professor James Gale; Professor Ann Marie Klinko; Professor James Lock; Professor William T. McDaniel; Professor Michael Palguta; Professor Yoland Smith, J.D.; Professor Joyce Wood; Professor Dale Wurzer; and Mr. William Dudley for assistance with the manuscript.
- Professor James V. Gray provided federal tax information and forms for Chapter 8.
- Professor Lynn Pape served as the quantitative editor, offering helpful suggestions in procedures and steps for examples and problems, as well as checking the mathematical solutions in the *Instructor's Manual.* Our special

thanks to Mr. Pape and Mr. Gray, both also of the Alexandria campus of Northern Virginia Community College.

- Ms. Susanne Stevenson, a former business mathematics student, colleague, and friend, did the word processing of the *Instructor's Manual,* a project of myriad complexities in itself.
- Many experts outside the academic community provided important information concerning their area of expertise: Mr. John Bukovinsky of International Business Machines Corporation; Dr. James Duggan of the U.S. Department of the Treasury; Ms. Kathryn D. Bryan of Sears, Roebuck and Co.; Ms. Gesila H. Mathison of Hunt Wesson, Inc; Ms. Ellen Zimny of the Coca-Cola Company, and Ms. Barbara Krebs of First Virginia Bank.
- Mr. Steve Helba, editor-in-chief at Prentice Hall, who coordinated activities with many different people.
- Ms. Michelle Churma, associate editor at Prentice Hall, who cheerfully assisted the authors in numerous ways.
- Louise Sette, production editor at Prentice Hall.
- Special thanks to our families. Their encouragement and understanding provided the supportive environment to complete this revision.

It is our sincere hope that all who teach or study *Business Mathematics* will find it to be a very worthwhile experience.

Nelda W. Roueche
Austin, Texas

Virginia H. Graves
Alexandria, Virginia

How to Study Business Math

The authors remember their time on "the other side of the desk" as students. We clearly remember wishing that a math book had more examples, or trying to determine what happened in the steps that were left out, or wondering what some vague explanation was supposed to mean. So when we began *Business Mathematics: A Collegiate Approach,* we designed it not for the instructor but especially for you, the student.

Successful completion of business math requires only two things—time and effort. If you are willing to supply these, you can be assured that your efforts will be rewarded. And you will find that it is well worth the effort required, for business math is a very practical course. Regardless of what position you may later accept, almost any mathematics required by the job will have been introduced in business math. (The text will also serve as a good on-the-job reference book after you finish the course.)

We think that by following these suggestions, you will be successful and enjoy the course:

- The first step seems so easy, and yet we are amazed at how many students do not do it—***read the book.*** Read it carefully and thoroughly. If something seems unclear when you first read it, read it again. The explanations will familiarize you with the business applications of each topic and the special terminology associated with it. For every problem in the text, there is an example that includes the steps to be followed to find the solution. Try working out the example itself with paper and pencil. Make a practice of reading each section before it is discussed in class; this will enable you to better understand the instructor's discussions and examples.

- ***Keep a notebook*** for taking notes in class. Try to take very complete notes on both the instructor's lecture and the examples given. Copy instructions and examples written on the board or overhead projector. Also, write down questions you encounter in reading the text, in your instructor's lecture, and in your attempts at solving problems.
- ***Ask questions***—never hesitate to ask questions in class or after class. If you go on to the next section with a question still unanswered, you may find yourself unable to understand the topic discussed there. (Besides, when one student has a question, others usually have the same question.)
- ***Work assigned problems*** using a pencil (everybody makes a few careless mistakes), and don't skip steps, unless you are quite adept at math. If a set of problems requires a formula, write it with each problem. Always start by listing the information that is given and indicating what must be found, the unknown. Allow plenty of space for each problem; many mistakes are made because the problem becomes so crowded that the student cannot read his/her numbers or determine what the sequence of steps has been. If you absolutely cannot solve a problem, ask for assistance from your instructor.
- Learn to ***use a calculator.*** Section 2 of Chapter 1 provides general instructions for calculator usage. Throughout the text, Calculator Techniques show steps to be taken to solve numerous examples. Work through these exercises with your calculator. Also, read the instruction manual that comes with your calculator to fine-tune your skills.
- Before working a problem, ***estimate the answer.*** It may be difficult to determine in your head the answer to 25% of 496, but you should be able to estimate the answer to be approximately 125 (1/4 of 500). ***Compare your computed answer*** to your estimate for reasonableness.
- ***Keep homework papers organized*** to study before quizzes and exams. Test yourself on the problems, and briefly write down the steps or formulas that would be required, or simply set up the problem. Then check what you have written against a similar problem that you have worked previously. ***Review*** the chapter learning objectives as well as the glossaries.
- During a quiz or exam, ***read the problem through*** completely before attempting to solve it. Work as quickly as possible without being careless. ***Double-check your calculations.*** When quizzes are returned to you, ***correct your errors and save the papers*** as study material for the final exam.

As indicated above, this study plan will require substantial time and effort, but you will learn much useful information in the process. Best wishes for success in business math, and good luck in the business world.

NELDA W. ROUECHE
VIRGINIA H. GRAVES

Discover the Companion Website Accompanying This Book

THE PRENTICE HALL COMPANION WEBSITE: A VIRTUAL LEARNING ENVIRONMENT

Technology is a constantly growing and changing aspect of our field that is creating a need for content and resources. To address this emerging need, Prentice Hall has developed an online learning environment for students and professors alike—Companion Websites—to support our textbooks.

In creating a Companion Website, our goal is to build on and enhance what the textbook already offers. For this reason, the content for each user-friendly website is organized by chapter and provides the professor and student with a variety of meaningful resources. Common features of a Companion Website include:

For the Professor—

Every Companion Website integrates Syllabus Manager™, an online syllabus creation and management utility.

- Syllabus Manager™ provides you, the instructor, with an easy, step-by-step process to create and revise syllabi, with direct links into Companion Website and other online content without having to learn HTML.

- Students may logon to your syllabus during any study session. All they need to know is the web address for the Companion Website and the password you've assigned to your syllabus.
- After you have created a syllabus using Syllabus Manager™, students may enter the syllabus for their course section from any point in the Companion Website.
- Clicking on a date, the student is shown the list of activities for the assignment. The activities for each assignment are linked directly to actual content, saving time for students.
- Adding assignments consists of clicking on the desired due date, then filling in the details of the assignment—name of the assignment, instructions, and whether or not it is a one-time or repeating assignment.
- In addition, links to other activities can be created easily. If the activity is online, a URL can be entered in the space provided, and it will be linked automatically in the final syllabus.
- Your completed syllabus is hosted on our servers, allowing convenient updates from any computer on the Internet. Changes you make to your syllabus are immediately available to your students at their next logon.

For the Student—

- Chapter Objectives—outline key concepts from the text
- Interactive self-quizzes—complete with hints and automatic grading that provide immediate feedback for students

After students submit their answers for the interactive self-quizzes, the Companion Website Results Reporter computes a percentage grade, provides a graphic representation of how many questions were answered correctly and incorrectly, and gives a question by question analysis of the quiz. Students are given the option to send their quiz to up to four email addresses (professor, teaching assistant, study partner, etc.).

- Message Board—serves as a virtual bulletin board to post—or respond to—questions or comments to/from a national audience
- Web Destinations—links to www sites that relate to chapter content

To take advantage of these and other resources, please visit the *Business Mathematics,* Eighth Edition, Companion Website at

www.prenhall.com/roueche

CONTENTS

Basics

CHAPTER 1

REVIEW OF OPERATIONS

OBJECTIVES

Upon completion of Chapter 1, you will be able to:

1. Define and use correctly the terminology associated with each topic.
2. Multiply efficiently using (Section 1):
   a. Numbers containing zero (example: $307 \times 1{,}400$) (Examples 1, 2; Problems 1, 2)
   b. Decimal numbers with fractions (example: $15 \times 3.5\frac{1}{4}$) (Example 3; Problems 1, 2)
   c. Numbers with exponents (example: 3^4) (Example 4; Problems 1, 2).
3. Simplify fractions containing (Section 1: Example 5; Problems 3, 4):
   a. Decimal parts $\left(\text{example: } \frac{6}{1.5}\right)$ b. Fractional parts $\left(\text{example: } \frac{6}{1\frac{1}{2}}\right)$.
4. Find negative differences (deficits) (Section 1: Example 6; Problems 3, 4).
5. Simplify expressions containing parentheses of the type $M(1 - dt)$ or $P(1 + rt)$ (Section 1: Examples 7–9; Problems 5, 6).
6. Round off numbers to a specified decimal place (Section 1: Example 10; Problems 7, 8).
7. Round a multiplication product to an accurate number of digits, based on the digits of the original numbers (Section 1: Example 11; Problems 9–12).
8. Using a hand-held LCD calculator (battery or solar-powered), perform computations for (Section 2):
   a. Basic arithmetic (Example 1; Problems 1, 2)
   b. Percent calculations (Example 2; Problems 3, 4)
   c. Memory operations (Example 3; Problems 5, 6).

The normal day-to-day operations of most businesses require frequent computations with numbers. Many of these computations are done in modern business by calculators and computers. Some computations, however, are still performed manually. And even those processes whose final calculations will be done automatically must first be set up correctly, so that the proper numbers will be fed into the machines and the proper operation performed. Section 1 covers these basic arithmetic techniques. In Section 2, you will have an opportunity to practice using a calculator for basic operations.

SECTION 1

ARITHMETIC TECHNIQUES

The following topics are presented to eliminate some of the weaknesses that many students have in working with specific types of numbers. A more comprehensive review is included as Appendix A, Arithmetic, for those students who need further practice.

MULTIPLICATION

Whenever a number is to be multiplied by some *number ending in zeros,* the zeros should be written to the right of the actual problem; the zeros are then brought down and affixed to the right of the product without actually entering into the **operation.** If either number contains a decimal, this also does not affect the problem until the operation has been completed.

Example 1 (a) 125 × 40 (b) 2.13 × 1,500 (c) 13,000 × 18

$$\begin{array}{rr|l} & 125 & \\ \times & 4 & 0 \\ \hline & 5{,}00 & 0 \end{array} \qquad \begin{array}{rr|l} & 2.13 & \\ \times & 15 & 00 \\ \hline & 1\ 065 & \\ & 2\ 13 & \\ \hline & 3{,}195. & 00 \end{array} \qquad \begin{array}{rr|l} & 18 & \\ \times & 13, & 000 \\ \hline & 54 & \\ & 18 & \\ \hline & 234, & 000 \end{array}$$

When multiplying by a *number containing inner zeros,* students often write whole rows of zeros unnecessarily to assure themselves that the other digits will be aligned correctly. The useless zeros can be eliminated if you will remember the following rule: On each line of multiplication, the first digit written down goes directly underneath the digit that was used for multiplying.

Example 2 (a) $2{,}145 \times 307$ (b) $1{,}005 \times 7{,}208$

Right

$$\begin{array}{rr} & 2{,}145 \\ \times & 307 \\ \hline & 15\ 015 \\ & 643\ 5 \\ \hline & 658{,}515 \end{array}$$

Inefficient

$$\begin{array}{rr} & 2{,}145 \\ \times & 307 \\ \hline & 15\ 015 \\ & 00\ 00 \\ & 643\ 5 \\ \hline & 658{,}515 \end{array}$$

(b)

$$\begin{array}{rr} & 7{,}208 \\ \times & 1{,}005 \\ \hline & 36\ 040 \\ & 7\ 208 \\ \hline & 7{,}244{,}040 \end{array}$$

When multiplying a **whole number** by a **mixed number** (a mixed number is a whole number plus a fraction), you should first multiply by the fraction and then multiply by the other numbers in the usual manner. If a decimal is involved, the decimal point is marked off in the usual way—the fraction does not increase the number of decimal places.

Example 3 (a) $24 \times 15\frac{3}{4}$ (b) $35 \times 1.3\frac{3}{7}$

$$\begin{array}{rrl} & 24 & \\ \times & 15\frac{3}{4} & \\ \hline & 18 & (\frac{3}{4} \times 24 = 18) \\ & 120 & \\ & 24 & \\ \hline & 378 & \end{array}$$

$$\begin{array}{rr} & 3\ 5 \\ \times & 1.3\frac{3}{7} \\ \hline & 1\ 5 \\ & 10\ 5 \\ & 35 \\ \hline & 47.0 \end{array}$$

There are several different ways to indicate that multiplication is required. The common symbols are the "times" sign ("×") and the raised dot ("·"). Many formulas contain **variables,** which are letters or symbols used to represent numerical values. Often variables are written together; this indicates that the numbers that these variables represent are to be multiplied. A number written beside a variable indicates multiplication of the number and variable. Numbers or variables within parentheses written together should also be multiplied. A number or variable written adjoining parentheses should be multiplied by the expression within the parentheses. Thus,

$$3 \cdot 5 = 3 \times 5$$

$$Prt = P \times r \times t$$

$$4k = 4 \times k$$

$$(2.5)(4)(6.8) = 2.5 \times 4 \times 6.8$$

$$7(12) = 7 \times 12$$

Some few problems require the use of exponents. An **exponent** is merely a number which, when written as a superscript to the right of another number, called the **base,** indicates how many times the base is to be written in repeated multiplication times itself.

Example 4

(a) $x^2 = x \cdot x$ (2 = exponent, x = base)　　(b) $5^3 = 5{\cdot}5{\cdot}5 = 125$ (3 = exponent, 5 = base)

(c) $2^4 = 2{\cdot}2{\cdot}2{\cdot}2 = 16$　　(d) $(1.02)^3 = (1.02)(1.02)(1.02) = 1.061208$

DIVISION

Recall that a fraction indicates division—the numerator of the fraction (above the line) is to be divided by the denominator (below the line). Thus, if the denominator contains a fraction, it must be inverted and multiplied by the numerator *of the entire fraction.* If the denominator contains a decimal, it must be moved to the end of the number and the decimal in the numerator moved a corresponding number of places.

Example 5

(a) $\dfrac{3}{\frac{3}{4}} = \dfrac{3}{1} \div \dfrac{3}{4} = \dfrac{\overset{1}{\cancel{3}}}{1} \times \dfrac{4}{\cancel{3}} = \dfrac{4}{1} = 4$

(b) $\dfrac{6}{1\frac{1}{2}} = \dfrac{6}{\frac{3}{2}} = \dfrac{6}{1} \div \dfrac{3}{2} = \dfrac{\overset{2}{\cancel{6}}}{1} \times \dfrac{2}{\cancel{3}} = \dfrac{4}{1} = 4$

(c) $\dfrac{4}{0.5} = \dfrac{4.0}{0.5} = \dfrac{40}{5} = 8$

(d) $\dfrac{5.2}{0.13} = \dfrac{5.20}{0.13} = \dfrac{520}{13} = 40$

(e) $\dfrac{2.31}{0.3} = \dfrac{2.31}{0.3} = \dfrac{23.1}{3} = 7.7$

SUBTRACTION

Unfortunately, business expenses sometimes exceed the funds budgeted for them. In this case, the **deficit** is called a **negative difference.** (It is also common to say that the account is "in the red.") A negative difference is found by taking the numerical difference between the larger and smaller numbers. The result is indicated as being a deficit either by placing a minus sign before the number or by placing the result in parentheses (the method used on most financial statements).

Example 6

(a)

Bank balance	$598.00
Checks written	− 650.00
Deficit	−$ 52.00

(b)

Profit earned	$3,500.00
Salary owed	− 4,700.00
Negative difference	($1,200.00)

PARENTHESES

Several formulas used in finding simple interest and discount contain parentheses. You should be familiar with the correct procedure to follow in working with parentheses: If parentheses contain both multiplication and addition, or multiplication and subtraction, the multiplication should be performed first and the addition or subtraction last. This procedure simply follows the standard rule for order of operations: Multiplication and division should always be performed before addition and subtraction. Thus, $3 \times 4 - 8 = 12 - 8 = 4$; and $9 + 16 \div 8 = 9 + 2 = 11$.

Example 7

(a) $$(1 + rt) = \left(1 + \frac{\overset{3}{\cancel{6}}}{100} \cdot \frac{1}{\cancel{2}}\right) \qquad \left(\text{where } r = \frac{6}{100} \text{ and } t = \frac{1}{2}\right)$$
$$= \left(1 + \frac{3}{100}\right)$$
$$= \left(\frac{100}{100} + \frac{3}{100}\right)$$
$$= \left(\frac{103}{100}\right)$$

(b) $$(1 - dt) = \left(1 - \frac{\overset{2}{\cancel{8}}}{100} \cdot \frac{1}{\cancel{4}}\right) \qquad \left(\text{where } d = \frac{8}{100} \text{ and } t = \frac{1}{4}\right)$$
$$= \left(1 - \frac{2}{100}\right)$$
$$= \left(\frac{100}{100} - \frac{2}{100}\right)$$
$$= \left(\frac{98}{100}\right)$$

(c) $$(1 + rt) = \left(1 + \frac{\overset{2}{\cancel{8}}}{100} \cdot \frac{7}{\underset{3}{\cancel{12}}}\right) \qquad \left(\text{where } r = \frac{8}{100} \text{ and } t = \frac{7}{12}\right)$$
$$= \left(1 + \frac{14}{300}\right)$$
$$= \left(\frac{300}{300} + \frac{14}{300}\right)$$
$$= \left(\frac{314}{300}\right)$$

Note. Parts (a) through (c) are left "unreduced" intentionally, since each fraction represents part of a larger problem.

If a parenthesis is preceded by a number, that number must be multiplied times the whole expression within the parentheses. That is, the terms within the parentheses should be consolidated into a single number or fraction, if possible, *before* multiplying. Consider the formula $P(1 + rt)$ when $P = \$600$, $r = \frac{5}{100}$, and $t = \frac{1}{3}$:

Example 8

Right

$$\begin{aligned} P(1 + rt) &= \$600\left(1 + \frac{5}{100}\cdot\frac{1}{3}\right) \\ &= 600\left(1 + \frac{5}{300}\right) \\ &= 600\left(\frac{300}{300} + \frac{5}{300}\right) \\ &= \overset{2}{\cancel{600}}\left(\frac{305}{\cancel{300}}\right) \\ &= \$610 \end{aligned}$$

Wrong

$$\begin{aligned} P(1 + rt) &= \$600\left(1 + \frac{5}{100}\cdot\frac{1}{3}\right) \\ &= 600\left(1 + \frac{5}{300}\right) \\ &= \overset{2}{\cancel{600}}\left(1 + \frac{5}{\cancel{300}}\right) \\ &= 2(6) \\ &= \$12 \end{aligned}$$

If it is impossible to consolidate the terms within the parentheses into a single number or fraction, then the number in front of the parentheses must be multiplied times *each* separate term within the parentheses. (Separate terms may be identified by the fact that a plus or a minus sign always appears between them.)

Example 9

(a) $P(1 + rt) = P\cdot 1 + P\cdot rt = P + Prt$

(b) $M(1 - dt) = M\cdot 1 - M\cdot dt = M - Mdt$

(c) $M(1 - dt)$ with $M = \$400$ and $d = \frac{6}{100}$ becomes

$$\begin{aligned} M(1 - dt) &= \$400\left(1 - \frac{6}{100}t\right) \\ &= \$400\cdot 1 - \overset{4}{\cancel{400}}\cdot\frac{6}{\cancel{100}}t \\ &= \$400 - 24t \end{aligned}$$

ROUNDING OFF DECIMALS

The general rule for rounding off decimals is as follows:

> *If the last decimal place that you wish to include is followed by any digit from 0 through 4, the digit in question remains unchanged. If followed by any digit from 5 through 9, the digit in question is increased by one.*

Example 10

(a) Rounding off to tenths (to one decimal place):
14.3)274 = 14.3

(b) Rounding off to hundredths (to two decimal places):
5.37)812 = 5.38

(c) Rounding off to thousandths (to three decimal places):
0.032)569 = 0.033

(d) Rounding off 24.14759:
to tenths: 24.1
to hundredths: 24.15
to thousandths: 24.148

Note. A special case sometimes arises when the last digit that you wish to use is followed by a single 5 (with no other succeeding digits). In this case, particularly when working with numbers that should total 100%, you may use the following rule:

If the last digit to be included is an odd number, round it off to the next higher number; if the digit in question is even, leave it unchanged.

The following example illustrates this case:

Example 10 (cont.)

(e) Round to the nearest percent.

	62.5%	=	62%
+	37.5%	=	38%
	100.0%	=	100%

Accuracy of Computation

Business, engineering, manufacturing, and the like all demand careful accuracy in their calculations, processes, and products. Computational accuracy is dependent on the number of significant digits used. **Significant digits** are digits obtained by precise measurement rather than by rough approximation. For example, 4,283′ has four significant digits; 4,300′ has only two significant digits if it was measured only to the nearest hundred feet. Similarly, 8.75 has three significant digits; 9 has one significant digit. There are two significant digits in 9.0 if a measurement made to the nearest tenth falls between 8.95 and 9.05.

The concept of significant digits requires that the following rule be observed: *The result obtained from a mathematical calculation can never be more accurate than the least accurate figure used in making the calculation.* Specifically, additional accuracy (more significant digits) cannot be created artificially by the simple use of a mathematical operation (such as multiplication).

This rule requires an initial calculation to be rounded off in many instances. The answer must at least be rounded off so as to contain no more than the number of decimal places contained in the original figure which had the *least number of decimal places.* To be more restrictive, the answer should contain no more significant digits than the original number which had the *fewest significant digits.*

Example 11 (a) Find the area of a room 17.27 feet in length and 13.6 feet in width.

$A = lw$

$= 17.27 \times 13.6$

$= 234.872$

$A = 234.9$ square feet (since 13.6 had only one decimal place)

or

$A = 235$ square feet (since 13.6 had three significant digits)

This rule is of particular importance in computing amounts invested at compound interest, since tables containing many decimal places are used. Students will want to minimize work by using no more of these decimal places than are necessary—but at the same time, obtain an answer that is correct to the nearest cent.

To do this, you should first estimate the answer to determine the number of digits it will contain (including the cent's place). This number plus one more (to ensure absolute accuracy) will determine the number of digits you need to copy from the table.

It should be noted that any known exact amount (such as $350) is considered accurate for any number of decimal places desired.

Example 11 (cont.) (b) Suppose an interest problem requires that $200 be multiplied by the table value 1.48594740. We wish to use only enough digits from the table to ensure that our answer is correct to the nearest cent.

First, estimate the value of the tabular number: It is approximately 1.5. Therefore, the answer we obtain will be approximately $\$200 \times 1.5 = \$300$.

The number $300.00 contains 5 digits; thus we must copy $5 + 1 = 6$ digits from the table. (The sixth digit from the table will be rounded off using the previously discussed rules for rounding off decimals.) The number 1.48594740, rounded to 6 digits, equals 1.48595.

Thus, $\$200 \times 1.48595 = \$297.19000 = \$297.19$ is the solution correct to the nearest cent.

(c) Find $1,500 times 1.86102237 correct to the nearest cent, using no more digits than necessary.

1.86102237 equals approximately 2.
$2 \times \$1,500 = \$3,000$, which contains 6 digits (including the cents digits).
$6 + 1 = 7$ digits are required from the table.
$\$1,500 \times 1.861022 = \$2,791.533000 = \$2,791.53$, correct to cents.

(d) Find $\$800 \times 0.50752126$ correct to cents, using no more digits than necessary.

0.50752126 equals approximately 0.5.
$\$800 \times 0.5 = \$400$, which has 5 digits (to the nearest cent).
$5 + 1 = 6$ digits are required from the table.
$\$800 \times 0.507521 = \$406.016800 = \$406.02$.

Note. The zero before the decimal in the number 0.50752126 does not qualify as a significant digit (that is, as a digit which should be counted), because the zero could have been omitted without changing the value of the number.

Another suggestion that minimizes work is to substitute fractional equivalents for percents (such as $\frac{1}{3}$ for $33\frac{1}{3}\%$, or $\frac{3}{7}$ for $42\frac{6}{7}\%$) in computations. Such equivalents usually simplify calculations done manually and often provide more accurate results than the more tedious computations performed with rounded decimal equivalents. For this reason, even students using calculators are often advised to use fraction equivalents (a two-step process). Thus, all students should familiarize themselves with the percents having convenient fractional equivalents. A table of these equivalents appears on page 51.

PROBLEM SOLVING

Students often read a problem, decide that they do not know how to solve it, and simply "give up" without writing anything at all. By attacking problems in an organized manner, however, you can succeed with many problems for which the solution process was not at all obvious on first reading.

1. Make a list of all information that was given. Name the term that each given amount represents, and/or associate each with a variable that can be used in an equation or formula. (For instance, overhead = \$3,000, or $r = 5\%$.)
2. Based on the question asked in the problem, indicate what needs to be found. (For example, markup = ?% of selling price, or $P = ?\$$.)
3. Ask yourself how this unknown quantity could logically be found. (A formula, such as $I = Prt$? An algebraic equation based on the question in the problem? Some process, such as using a table to determine annual percentage rate for monthly payments?)
4. After deciding on a logical method, reexamine your list of given information. Do you already have everything required to use this method? If not, could your given information be used to compute the additional information required by this method?
5. Attempt some calculations using the method you selected. Examine your result to decide whether it really answers the question that was asked and whether that answer is reasonable. (If necessary, go back to step 3 and consider another method. For instance, the unknown quantity may be included in several different formulas and could perhaps be found using a different formula than was first tried.)

The foregoing approach may not be foolproof, of course, but it certainly increases your chances for success. You have no doubt heard the expression "If all else fails, read the directions." In this case, the "directions" are the examples in the text, and the authors assume that students will study them before attempting any problems!

SECTION 1 PROBLEMS

Find the product.

1. a. 83×30 b. 50×256 c. 183×600 d. $1,300 \times 44$
e. $3,600 \times 118$ f. $1,641 \times 302$ g. $405 \times 1,765$ h. $2,009 \times 13,202$
i. $1,070 \times 423$ j. $34 \times 6.4\frac{1}{2}$ k. $5.2 \times 18\frac{1}{4}$ l. $28\frac{1}{3} \times 0.48$
m. $5.4 \times 32\frac{1}{6}$ n. $2.4 \times 3.5\frac{3}{4}$ o. 5^4 p. 4^8
q. 7^3 r. 2.01^2

2. a. 54×60 b. 80×362 c. 300×119 d. $57 \times 1,500$
e. $2,400 \times 353$ f. $3,154 \times 206$ g. $107 \times 6,332$ h. $3,002 \times 20,442$
i. $7,060 \times 814$ j. $82 \times 5.3\frac{1}{2}$ k. $6.3 \times 24\frac{1}{6}$ l. $45\frac{2}{5} \times 0.60$
m. $7.2 \times 91\frac{5}{6}$ n. $6.8 \times 2.8\frac{3}{4}$ o. 6^4 p. 8^5
q. 10^3 r. 3.06^2

Divide or subtract, as indicated.

3. a. $\frac{48}{\frac{6}{7}}$ b. $\frac{72}{4\frac{1}{2}}$ c. $\frac{20}{0.5}$ d. $\frac{3.6}{0.12}$ e. $\frac{1.36}{0.8}$

f. Net sales $66,708
Cost of goods sold − 68,134

g. Travel allowance $300
Travel expenses − 475

h. Net income $82,500
Partners' salaries −100,000

i. Escrow for taxes $2,575
Taxes assessed − 3,100

4. a. $\frac{111}{\frac{3}{5}}$ b. $\frac{85}{2\frac{1}{8}}$ c. $\frac{50}{0.4}$ d. $\frac{4.5}{0.15}$ e. $\frac{6.21}{0.3}$

f. Checkbook balance $356.44
Checks written − 372.60

g. Gross profit $51,382
Operating expenses −56,309

h. Advertising budget $45,000
Advertising expenses − 49,810

i. Account balance $1,620
Payment to balance −1,682

Find the value of each expression.

5. a. $\left(1 + \frac{16}{100}\cdot\frac{1}{4}\right)$ b. $\left(1 + \frac{7}{100}\cdot\frac{1}{3}\right)$ c. $\left(1 - \frac{3}{100}\cdot\frac{1}{6}\right)$
d. $\left(1 - \frac{4}{100}\cdot\frac{1}{2}\right)$ e. $5,000\left(1 + \frac{18}{100}\cdot\frac{4}{9}\right)$ f. $1,500\left(1 + \frac{28}{100}\cdot\frac{2}{7}\right)$
g. $1,000\left(1 - \frac{6}{100}\cdot\frac{2}{5}\right)$ h. $500\left(1 - \frac{12g}{100}\right)$ i. $300\left(1 + \frac{9b}{100}\right)$
j. $j(kl + m)$ k. $w(1 - xy)$

6. a. $\left(1 + \frac{42}{100}\cdot\frac{1}{6}\right)$ b. $\left(1 + \frac{3}{100}\cdot\frac{1}{5}\right)$ c. $\left(1 - \frac{8}{100}\cdot\frac{1}{2}\right)$

d. $\left(1 - \frac{21}{100}\cdot\frac{1}{3}\right)$ e. $900\left(1 + \frac{18}{100}\cdot\frac{2}{9}\right)$ f. $720\left(1 + \frac{3}{100}\cdot\frac{5}{27}\right)$

g. $1{,}000\left(1 - \frac{12}{100}\cdot\frac{10}{15}\right)$ h. $600\left(1 - \frac{3b}{100}\right)$ i. $400\left(1 + \frac{6x}{100}\right)$

j. $a(bc + d)$ k. $e(1 - wh)$

Round off each number as indicated.

7. a. To tenths: 43.258, 156.643, 1,680.952
b. To hundredths: 8.9426, 26.4453, 160.0639
c. To thousandths: 18.92453, 0.56641, 337.00894
d. To tenths, hundredths, and thousandths: 5.08473, 23.67521

8. a. To tenths: 51.174, 204.526, 1,875.973
b. To hundredths 9.3345, 81.6666, 104.1748
c. To thousandths: 15.04525, 0.94368, 257.00172
d. To tenths, hundredths, and thousandths: 8.06481, 16.54546

Compute each product. Round first according to the least accurate decimal, and then round according to the least number of significant digits.

9. a. 14.2 × 12.35 b. 4.56 × 7.3 c. 5.8 × 7.83 d. 1.111 × 3.85
10. a. 23.6 × 34.47 b. 8.11 × 4.7 c. 9.2 × 5.62 d. 7.634 × 8.25

Compute each product, correct to the nearest cent. Use no more digits than are necessary.

11. a. \$300 × 1.91301845 b. \$500 × 1.52161826
c. \$4,000 × 0.37440925 d. \$20 × 19.08162643
12. a. \$400 × 1.44354605 b. \$600 × 1.81246018
c. \$5,000 × 0.26423817 d. \$10 × 16.01964522

SECTION 2

USING A CALCULATOR

Calculators are allowed (or required) for the business mathematics courses at many colleges, usually beginning with Chapter 4 after the Basics unit is complete. This current section teaches some basic techniques for using a calculator. Then, beginning with Chapter 4, various topics throughout the text will contain boxes that illustrate calculator techniques applicable to those corresponding problems.

Different types of calculators require slightly different methods of operation. However, the techniques described here apply for the basic calculators used by most students: hand-held LCD (liquid crystal display) calculators allowing a maximum of eight digits (or seven digits after the decimal point). Only basic key-functions are assumed; thus, if your calculator contains special keys not described here, you should consult the instructions that came with your calculator to learn their uses.

If needed, press the [ON/C] key to activate your calculator. (Some light/solar-powered calculators have no "on" key but activate as soon as they are uncovered.) After each problem, press [ON/C] again to "clear" the calculator, resetting the display to zero.

ARITHMETIC OPERATIONS

Arithmetic operations compute in the order you enter them into the calculator. The first number is "known" by the calculator as soon as you press the digits; thus, the first operation (or function) symbol that you press determines what process happens with the second number. In the examples, square boxes such as "[+]" indicate that you should press the key containing that symbol. An arrow "→" points to the solution that should display on your calculator.

Example 1 This series of additions and subtractions is typical of calculations you would use for a checkbook, where you make deposits (add) and write checks (subtract).

(a) Find $550 - 75 - 80 + 100 - 125 = 370$.

550 [−] 75 [−] 80 [+] 100 [−] 125 [=] ⟶ 370

Notice that as you press the operation symbol before the third number (and similarly before all subsequent numbers), the current subtotal temporarily displays, just before you enter the number itself. This subtotal could be used to update your checkbook balance after each transaction.

Hints. If you accidentally press the wrong *operation* key (such as [+] when you meant [−]), simply press the correct key immediately. It will cancel the previous operation and take effect itself. That is, 550 [+] [−] 75 computes as 550 [−] 75. If you enter a *number* incorrectly, immediately press the [CE] (cancel entry) key; then enter the correct number and continue with the calculation.

Multiplication and division are required for many business mathematics computations. Problems containing fractions also make use of the same techniques.

Example 1 (cont.)

(b) Find $3{,}600 \div 90 \times 15 = 600$.

3,600 [÷] 90 [×] 15 [=] ⟶ 600

(c) Compute $\frac{3}{5}(750) = 450$.

3 [×] 750 [÷] 5 [=] ⟶ 450

Observe in this case that 3 [÷] 5 [×] 750 [=] 450, as previously calculated. On a calculator, however, reversing the order for a fraction does not always produce exactly the same result, as demonstrated by the next example.

(d) Multiply $\frac{3}{7} \times 280 = 120$.

3 [×] 280 [÷] 7 [=] ⟶ 120 (preferred method)

3 [÷] 7 [×] 280 [=] ⟶ 119.99999

Thus, you should make a habit of multiplying by the numerator first and then dividing by the denominator. (Most problems in this text are designed for convenient calculations by students without calculators; thus, denominators usually "cancel" into the second number, but they may not divide evenly into the numerator, as shown in this example.)

PERCENT CALCULATIONS

A large portion of business mathematics calculations involve percent. The [%] key produces an immediate result on your calculator; it is not necessary to press the [=] key for multiplication problems that involve percent. However, this dictates the order in which numbers must be entered, as follows.

Example 2 (a) Find $720 \times 35\% = 252$.

720 [×] 35 [%] ⟶ 252

(b) Multiply $48\% \times 250 = 120$.

250 [×] 48 [%] ⟶ 120

Notice from Part (b) that the percent (here, 48%) must be entered *last,* although it appears first in the problem.

MEMORY OPERATIONS

The memory capability of calculators allows you to save values that you will use again—either for repeated, similar calculations or else for a later portion of the current computation. For instance, you might store a tax rate or a Social Security rate and use that rate in several similar calculations. Or, given a complicated fraction, you would determine the denominator first and save it in memory, to use after the numerator has been computed.

The keys [M+] and [M−] are used to add or subtract a value into memory, combining it with any value already existing in memory. If a multiple-step computation is in progress, the [M+] or [M−] key can be used without pressing [=]; that is, the memory key will simultaneously determine the solution and store it into memory. An "M" will appear at one side of your display, indicating that memory currently contains a nonzero value. You can use the memory value simply by pressing the memory recall key, [MR] (sometimes labeled [M^R_C]), at the corresponding point where that value is used in a calculation.

Example 3 (a) Place into memory the sales tax rate 6.25%. Then find the sales tax on purchases of $20 and $50.

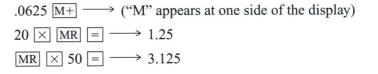

.0625 [M+] ⟶ ("M" appears at one side of the display)

20 [×] [MR] [=] ⟶ 1.25

[MR] [×] 50 [=] ⟶ 3.125

Notice that the [MR] key can be used in either order in a multiplication problem; for instance, the last line could just as easily be computed as 50 [×] [MR].

Hint. If you are unsure what decimal value is equivalent to a percent, multiply 1 times the percent to enter its decimal value. For instance, the first line above could be: 1 [×] 6.25 [%] [M+] → 0.0625.

Before proceeding, you need to "clear" the current value from memory. This procedure varies according to the type of key your calculator has: (1) If your calculator has an [MC] (or memory clear) key, press that key. The "M" will then disappear, although the last solution remains in the display. (2) If there is no [MC] key, you could clear the memory simply by pressing [OFF] and then again pressing [ON/C]. Alternatively, you can press [MR] to recall the memory value and then press [M−] to subtract it from itself, which leaves the memory value at zero. As before, the "M" immediately disappears from your display, indicating that memory has cleared, although the number itself remains as the displayed value. (3) If you have an [AC] (or all clear) key, this will clear both memory and the current display simultaneously.

(b) Upon completing Part (a), clear 6.25% from the calculator memory.

If you have an "MC" key:	[MC] ⟶	("M" disappears; 3.125 remains)
Without an "MC" key:	[MR] ⟶	0.0625
	[M−] ⟶	("M" disappears; 0.0625 remains)
If you have an "AC" key:	[AC] ⟶	0. ("M" disappears; zero appears)

When using fractions that have multiple parts in the denominator, evaluate the denominator first and save it into memory, before entering the numerator. Similarly, multiple values inside parentheses should be computed first and stored into memory, before you multiply by the value preceding the parentheses.

Example 3 (cont.)

(c) Calculate the value of $\frac{2{,}000 + 47.5}{1 - 0.025} = 2{,}100$.

1 [−] .025 [M+] ⟶ 0.975 ("M" appears and remains until you clear memory)

2000 [+] 47.5 [=] ⟶ 2047.5 [÷] [MR] [=] ⟶ 2100

Note. The last line can be computed correctly as 2000 [+] 47.5 [÷] [MR] [=] → 2100. However, most students feel more comfortable including the first [=], in order to see the 2047.5 total value of the numerator.

Hint. Remember to clear memory before each new problem. For assurance, you may also wish to clear the display by pressing [ON/C] (although that is not essential if the display contains the result of a previous calculation).

Business mathematics requires many financial formulas like the next example, where a "rate × time" calculation (the last two items in parentheses below) must be added/subtracted into memory as a separate step. The "1 [M+]" step can be done either before or after you place the "rate × time" value into memory; here, it is shown as the first step.

Example 3 (cont.)

(d) Calculate $1{,}200\left(1 - 0.06 \times \frac{5}{12}\right) = 1{,}170$.

1 [M+] ⟶ ("M" appears; 1 remains)

.06 [×] 5 [÷] 12 [M−] ⟶ 0.025

1200 [×] [MR] [=] ⟶ 1170

As you press [MR], notice that this briefly displays (recalls) the memory value 0.975, which is the computed value within parentheses.

Note. On some calculators, the second row above can also be computed as: 5 [×] 6 [%] [÷] 12 [M−], to give the same result in exact dollars. In most real-world calculations, however, your final result must be rounded to the nearest cent. Observe that the [M−] key is used here because a "−" appears before the "$0.06 \times \frac{5}{12}$." Had a "+" appeared there, [M+] would be used to add the "rate × time" value into memory.

SECTION 2 PROBLEMS

Perform the following operations using a calculator.

1. a. 3,120 − 48 + 188 − 251

b. 964 − 410 + 17 + 8

c. 76 × 9 ÷ 4

d. 12 ÷ 8 × 46

e. $\frac{3}{8} \times 400$

f. $\frac{4}{9}(108)$

g. $\frac{2}{7}(420)$

h. $\frac{608}{8} \times 12$

2. a. $4{,}261 - 36 + 216 - 378$ b. $883 - 520 + 23 - 5$ c. $66 \times 6 \div 9$

d. $24 \div 6 \times 5$ e. $\frac{4}{5} \times 55$ f. $\frac{3}{7}(560)$

g. $\frac{5}{8}(72)$ h. $\frac{468}{9} \times 18$

3. a. $136 \times 22\%$ b. $582 \times 30 \times 5\%$

c. $6.2\% \times 405$ d. $90\% \times 60\% \times 1{,}500$

4. a. $246 \times 37\%$ b. $332 \times 15 \times 9\%$

c. $4.8\% \times 308$ d. $40\% \times 80\% \times 2{,}700$

5. a. 12.25% of 32; of 60; of 180 b. $14\% \times 12$; $\times 25$; $\times 1{,}400$

c. $\dfrac{335 + 785}{1 - 0.80}$ d. $\dfrac{874 - 56.42 - 21.26}{12.34 + 4.25}$

e. $4{,}000(1 + 0.08 \times 24)$ f. $1{,}600\left(1 - \frac{3}{4} \times 9\%\right)$

g. $720\left(1 + \frac{5}{8} \times 5\%\right)$ h. $\dfrac{126.9}{47 \times 15\%}$

6. a. 28% of 15; of 48; of 206 b. $65\% \times 18$; $\times 34$; $\times 2{,}740$

c. $\dfrac{82 + 136}{1 - 0.75}$ d. $\dfrac{587 - 5.62 - 149.08}{24.9 + 3.92}$

e. $3{,}200(1 + 0.09 \times 47)$ f. $45\left(1 - \frac{2}{5} \times 8\%\right)$

g. $850\left(1 + \frac{1}{2} \times 12\%\right)$ h. $\dfrac{1{,}029.6}{66 \times 20\%}$

CHAPTER  GLOSSARY

Base. A number that is to be used in multiplication times itself.

Deficit. The result when a larger number is subtracted from a smaller number.

Exponent. A superscript that indicates how many times a base is to be written in repeated multiplication times itself. (Example: The exponent 3 denotes that $5^3 = 5 \cdot 5 \cdot 5 = 125$.)

Mixed number. A number that combines a whole number and a fraction. (Example: $12\frac{1}{2}$)

Negative difference. (See "Deficit.")

Operation. Any of the arithmetic processes of addition, subtraction, multiplication, and division.

Significant digits. Digits obtained by precise measurement rather than by rough approximation (or artificially by computation).

Variable. A letter or symbol used to represent a number.

Whole number. A number from the set

$$\{0, 1, 2, 3, \ldots\}$$

which has no decimal or fractional part.

CHAPTER 2

USING EQUATIONS

OBJECTIVES

Upon completion of Chapter 2, you will be able to:

1. Define and use correctly the terminology associated with each topic.
2. Solve basic equations that require only (Section 1: Examples 1–11; Problems 1–36):
   - **a.** Combining the similar terms of the equation; and/or
   - **b.** Using the operations of addition, subtraction, multiplication, and division.
   - **c.** Examples:

$$x - 12 = 16, \qquad 5y - 4 = 6 - 3y, \qquad \text{or} \qquad \frac{3x}{4} + 5 = 23$$

3. Express any written problem in a concise sentence which provides the structure for the equation that solves the problem (Section 2: Examples 1–8; Problems 1–36).
4.
   - **a.** Express numbers in ratios (Section 3: Example 1; Problems 1, 2, 5–8).
   - **b.** Use proportions to find numerical amounts (Section 3: Examples 2–4; Problems 3, 4, 9–28).

The procedure required to solve many problems in business mathematics is much more obvious if you have a basic knowledge of equation-solving techniques. The equations we will consider, such as $x - 7 = 5$ or $\frac{2}{3}y + 5 = y + 2$, are **first-degree equations in one variable.** "First-degree" means that the variable, or letter, in each equation has an exponent of 1 (although this exponent is not usually written; that is, x means x^1). Notice that the "one variable" may appear more than one time, but there will be only one distinct variable in any given equation.

It is important to note the difference between an expression and an equation. An **expression** is any indicated mathematical operation(s) written with no equal sign. An **equation** is a mathematical statement that two expressions (or an expression and a number) are equal. For instance, "$2x - 4$" is an expression, whereas "$2x - 4 = 10$" is an equation.

Before you start to solve any actual equations, a discussion of some characteristics of equations may prove helpful.

SECTION 1

BASIC EQUATIONS

An equation may be compared to an old-fashioned balancing scale. The "equals" sign is the center post of the scale, and the two sides of the equation balance each other as do the pans of the scale.

We know that you can either add weights to or remove them from the pans of a scale, and as long as you make the same changes in both pans, the scales remain in balance. The same is true of equations: In solving an equation, you may perform any operation (addition, subtraction, multiplication, or division), and as long as the *same operation* is performed with the *same numbers* on *both sides* of the equation, the balance of the equation will not be upset.

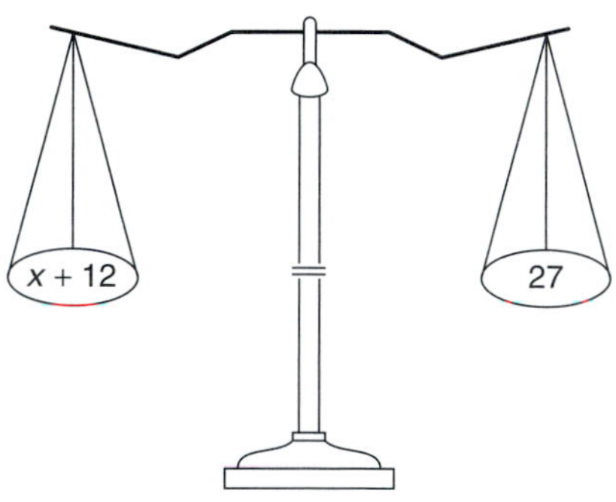

The object in solving any equation with unknowns in it is to determine what value of the unknown quantity (the *variable*) will make the equation a true statement. Basically, this is accomplished by isolating the variable **term** or terms on one side of the equation, with the ordinary numbers on the other side, as follows:

$$\text{Variables} = \text{Numbers}$$

or

$$\text{Numbers} = \text{Variables}$$

For instance, given $5p - 3 = 2p + 15$, we isolate the variable terms on one side and the numbers on the other to obtain $5p - 2p = 15 + 3$. The **similar terms** (those that include the same variable) are isolated by performing exactly the same operation(s) on both sides of the equation, as demonstrated in the examples that follow.

There are only four basic equation forms—those whose solution requires a single operation of either addition, subtraction, multiplication, or division. All other equations are variations or combinations of these four forms. We shall now consider several elementary equations, beginning with the four basic forms.

In the basic equation forms to follow, there will be only one number associated with the unknown (variable) on the same side of the equation. You should first determine which operation is involved with the number. The solution of the equation will then require that the *opposite* operation be performed on both sides of the equation.

Example 1 Solve for x in the equation: $x + 5 = 38$

To isolate the variable x on the left side of the equation, we need to remove the 5. In the original equation, the 5 is added to x; therefore, we must perform the opposite of addition, or subtract 5 from both sides.

$$x + 5 = 38$$

(Subtract 5 from both sides) $$x + 5 - 5 = 38 - 5$$

($5 - 5 = 0$, and $x + 0 = x$) $$\underbrace{x + \not{5} - \not{5}} = \underbrace{38 - 5}$$

$$x = 33$$

Note. Numbers or variables in a given equation may be added or subtracted only if they are on the *same side* of the equation; thus, in Example 1, equation-solving techniques were used to get 38 and 5 on the same side of the equation before the numbers could be combined. Also, students who have had little experience in solving equations should develop the following habit: On each succeeding line of the problem, first copy the adjusted equation resulting from the preceding step; then perform the next step.

Example 2 Solve for r: $r - 26 = 47$

The 26 is subtracted from r; therefore, we must *add* 26 to both sides of the equation.

$$r - 26 = 47$$

(Add 26 to both sides) $$r - 26 + 26 = 47 + 26$$

$(-26 + 26 = 0)$ $$\underbrace{r - \cancel{26} + \cancel{26}} = \underbrace{47 + 26}$$

$$r = 73$$

Example 3 Solve for y: $6y = 72$

The expression $6y$ means "6 times y" (or $6 \times y$). Any number that multiplies a variable is called a **coefficient;** in this equation, 6 is the coefficient of y.

Because 6 multiplies y, we must perform the opposite of multiplication, or divide both sides of the equation by 6.

$$6y = 72$$

(Divide both sides by 6) $$\frac{6y}{6} = \frac{72}{6}$$

$\left(\frac{6}{6} = 1;\ 1y = y\right)$ $$\frac{\overset{1}{\cancel{6}}y}{\cancel{6}} = \frac{72}{6}$$

$$y = 12$$

Example 4 Solve for k: $\frac{k}{3} = 17$

Since k is divided by 3, we *multiply* both sides of the equation by 3.

$$\frac{k}{3} = 17$$

(Multiply both sides by 3) $$\frac{3}{1} \cdot \frac{k}{3} = 17 \cdot 3$$

$$\frac{\overset{1}{\cancel{3}}}{1} \cdot \frac{k}{\cancel{3}} = 17 \cdot 3$$

$$k = 51$$

Now let us consider some variations and combinations of the basic equation forms.

Example 5 Solve for t: $\frac{2t}{3} = 36$

Just as $\frac{2 \times 12}{3}$ gives the same result as $\frac{2}{3} \times 12$, so $\frac{2t}{3}$ is the same as $\frac{2}{3}t$. In order to solve the equation for t, we need to remove the coefficient $\frac{2}{3}$ and obtain a coefficient of 1. This can be accomplished easily by multiplying by the reciprocal of $\frac{2}{3}$. The **reciprocal** of any number is the result obtained when the number is inverted (that is, the numerator and denominator are interchanged). Thus, the reciprocal of $\frac{2}{3}$ is $\frac{3}{2}$.

$$\frac{2t}{3} = 36$$

$\left(\text{Multiply by } \frac{3}{2}, \text{ the reciprocal of } \frac{2}{3}\right)$ $$\frac{\cancel{3}^{1}}{\cancel{2}} \cdot \frac{\cancel{2}^{1}}{\cancel{3}} t = \cancel{36}^{18} \cdot \frac{3}{\cancel{2}}$$

$\left(\frac{3}{2} \cdot \frac{2}{3} = 1;\ 1t = t\right)$ $$t = 54$$

Example 6 Solve for p: $6p + p - 3p = 48$

We must first combine the three p terms into a single p term. This is done by adding and/or subtracting the coefficients, whichever the signs indicate. (Recall that $p = 1p$.)

$$6p + p - 3p = 48$$

(Combine the coefficients: $6 + 1 - 3 = 4$) $$4p = 48$$

(Divide both sides by 4) $$\frac{\cancel{4}^{1}p}{\cancel{4}} = \frac{48}{4}$$

$$p = 12$$

Example 7 Solve for m: $4m + 5 = 33$

When one side of an equation contains both a variable term and a number term, it is customary to work with the number term first, in order to obtain an altered equation of the type

$$\text{Variable term} = \text{Numbers}$$

$$4m + 5 = 33$$

(Subtract 5 from both sides) $$4m + \cancel{5} - \cancel{5} = 33 - 5$$

$$4m = 28$$

(Divide both sides by 4) $$\frac{\overset{1}{\cancel{4}}m}{\cancel{4}} = \frac{28}{4}$$

$$m = 7$$

Example 8 Solve for n: $6 = \frac{n}{7} - 3$

We must first isolate the variable term so that

$$\text{Numbers} = \text{Variable term}$$

$$6 = \frac{n}{7} - 3$$

(Add 3 to both sides) $$6 + 3 = \frac{n}{7} - \cancel{3} + \cancel{3}$$

$$9 = \frac{n}{7}$$

(Multiply both sides by 7) $$7 \cdot 9 = \frac{n}{\cancel{7}} \cdot \overset{1}{\cancel{7}}$$

$$63 = n$$

Example 9 Solve for s: $\frac{3s}{5} - 4 = 14$

$$\frac{3s}{5} - 4 = 14$$

(Add 4) $$\frac{3s}{5} - \cancel{4} + \cancel{4} = 14 + 4$$

$$\frac{3s}{5} = 18$$

$$\left(\text{Multiply by } \frac{5}{3}\right) \qquad \frac{\overset{1}{\cancel{5}}}{\cancel{3}} \cdot \frac{\overset{1}{\cancel{3}}s}{\cancel{5}} = \overset{6}{\cancel{18}} \cdot \frac{5}{\cancel{3}}$$

$$s = 30$$

Example 10 Solve for z: $9(z - 3) = 45$

Recall that the coefficient 9 must multiply times each term within the parentheses.

$$9(z - 3) = 45$$

$$9z - 27 = 45$$

$$(\text{Add } 27) \qquad 9z - \cancel{27 + 27} = 45 + 27$$

$$9z = 72$$

$$(\text{Divide by } 9) \qquad \frac{\overset{1}{\cancel{9}}z}{\cancel{9}} = \frac{72}{9}$$

$$z = 8$$

Example 11 Solve for d: $8d + 2 = 5d + 17$

Remember that variable terms or number terms can be added and/or subtracted only when they appear on the same side of the equation. Therefore, the first steps will involve getting all the variable terms on one side of the equation and all the number terms on the other. When the solution of an equation requires that both of these steps be done, it is customary—although not essential—to work with the variable terms first.

$$8d + 2 = 5d + 17$$

$$(\text{Subtract } 5d \text{ from both sides}) \qquad \underbrace{8d - 5d} + 2 = \cancel{5d} - \cancel{5d} + 17$$

$$3d + 2 = 17$$

$$(\text{Subtract 2 from both sides}) \qquad 3d + \cancel{2} - \cancel{2} = 17 - 2$$

$$3d = 15$$

$$(\text{Divide both sides by } 3) \qquad \frac{\overset{1}{\cancel{3}}d}{\cancel{3}} = \frac{15}{3}$$

$$d = 5$$

SECTION 1 PROBLEMS

Solve the following equations.

1. $x + 23 = 88$

2. $y + 15 = 68$

3. $x - 9 = 21$

4. $d - 8 = 52$

5. $4y = 112$

6. $7w = 434$

7. $12b = 108$

8. $6c = 246$

9. $8x - 6 = 18$

10. $18a - 4 = 86$

11. $7c + 48 = 125$

12. $4t + 12 = 236$

13. $26n + 3 = 601$

14. $17d + 13 = 438$

15. $8z - 8 = 2z + 52$

16. $15q - 4 = 6q + 158$

17. $7q = 5q + 16$

18. $12v = 5v + 84$

19. $6(y + 3) = y + 128$

20. $8(x + 6) = x + 195$

21. $8f + 22 = f + 57$

22. $10r + 9 = r + 81$

23. $15x - 6 = 7x + 10$

24. $27k - 12 = 6k + 282$

25. $2h - 10 = 32 - 4h$

26. $3t - 6 = 78 - 9t$

27. $6(x - 2) = 4x - 2$

28. $8(m - 3) = 3m - 4$

29. $\frac{3a}{8} = 48$

30. $\frac{4e}{5} = 56$

31. $\frac{r}{7} - 6 = 12$

32. $\frac{x}{8} - 15 = 16$

33. $\frac{3x}{5} + 8 = 98$

34. $\frac{6d}{7} + 9 = 81$

35. $47 = \frac{3d}{4} - 22$

36. $66 = \frac{2z}{5} - 8$

SECTION 2
WRITTEN PROBLEMS

Many business problems of a mathematical nature can have no predetermined formula applied. In these cases, the business person must be able to set the facts of the situation into an original equation to obtain the solution.

An equation is simply a mathematical sentence (a statement of equality). If you can express the facts of a mathematical problem in a clear, concise, English sentence, the mathematical sentence (equation) will follow in exactly the same pattern. (By the same token, a student who cannot express the facts in a clear, English sentence probably does not understand the situation well enough to be able to obtain a mathematical solution.)

One fact that helps immensely in converting English sentences into mathematical sentences is that the *verb* of the English sentence corresponds to the *equal sign* of the equation. The mathematical equivalents of several other words will be pointed out in the following examples.

Example 1 What number increased by 15 gives 68?

What number	increased by	15	gives	68?
↓	↓	↓	↓	↓
n	$+$	15	$=$	68?

Thus,

$$n + 15 = 68$$

$$n + 15 - 15 = 68 - 15$$

$$n = 53$$

Example 2 Advertising expense last month was $300 less than utilities expense. If $745 was spent for advertising, how much were utilities?

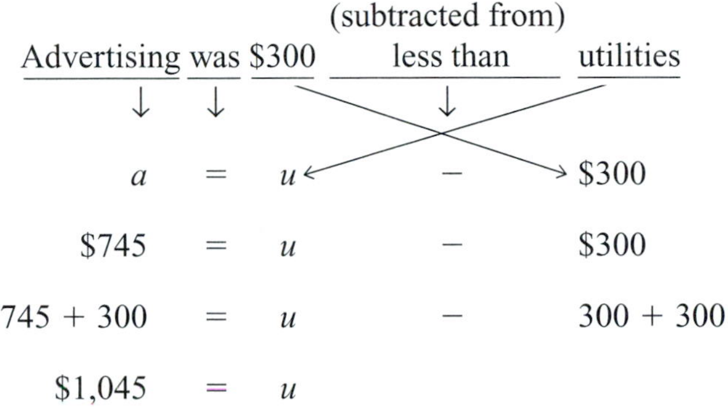

(subtracted from)
Advertising was $300 less than utilities
a = u − $300
$745 = u − $300
745 + 300 = u − 300 + 300
$1,045 = u

Example 3 About $\frac{1}{6}$ of a family's net (after-tax) monthly income is budgeted for food. What is their net monthly income if food bills average $105 per week? (Assume that 4 weeks equal 1 month.)

One-sixth	of	net income	is	food expense
↓	↓	↓	↓	↓
$\frac{1}{6}$	$\times$	I	$=$	$(4 \times \$105)$

$$\frac{I}{6} = 420$$

$$(\not{6})\frac{I}{\not{6}} = 420(6)$$

$$I = \$2{,}520$$

Example 4 A training workshop held by Field Sales Corp. cost the firm \$4,800. If 75 salespeople attended the workshop, what was the cost per participant?

Number of participants	times	cost per participant	equals	total
75	$\times$	c	=	\$4,800

$$75c = 4{,}800$$

$$\frac{\cancel{75}c}{\cancel{75}} = \frac{4{,}800}{75}$$

$$c = \$64$$

Example 5 The state welfare department pays $\frac{2}{3}$ of the cost of day care for indigent children. The remaining balance is financed by Whitworth County. If the state paid \$1,400 for day care, what was the total cost of the day-care project?

Two-thirds	of	total cost	was	state's share
$\frac{2}{3}$	$\times$	c	=	\$1,400

$$\frac{2c}{3} = 1{,}400$$

$$\frac{3}{2} \times \frac{2c}{3} = \overset{700}{\cancel{1{,}400}} \times \frac{3}{\cancel{2}}$$

$$c = \$2{,}100$$

Example 6 The number of television sets sold at Entertainment Associates was three times the combined total of compact disc players and radios. If their sales included 72 TV sets and 8 radios, how many CD players were sold?

TVs	were	three	times	CDs and radios combined
t	=	3	$\times$	$(c + r)$

$$72 = 3 \times (c + 8)$$

$$72 = 3(c + 8)$$

$$72 = 3c + 24$$

$$72 - 24 = 3c + 24 - 24$$

$$48 = 3c$$

$$\frac{48}{3} = \frac{\cancel{3}c}{\cancel{3}}$$

$$16 = c \quad \text{(CD players)}$$

Example 7 Harris and Smith together sold 36 new insurance policies. If Harris sold three times as many policies as Smith, how many sales did each make?

Note. When a problem involves two amounts, it is usually easier to let the variable of your equation represent the smaller quantity.

Since Harris sold three times as many policies as Smith, then Smith is the smaller quantity:

$$\underline{\text{Harris}}\ \underline{\text{sold}}\ \underline{\text{three}}\ \underline{\text{times}}\ \underline{\text{Smith}}$$
$$H = 3 \times S$$

Then

$$\underline{\text{Harris}}\ \underline{\text{and}}\ \underline{\text{Smith}}\ \text{together}\quad \underline{\text{sold}}\quad \underline{\text{36 policies}}$$
$$H + S = 36$$
$$\downarrow$$
$$3S + S = 36$$
$$\frac{\cancel{4}S}{\cancel{4}} = \frac{36}{4}$$

(Smith's sales) $S = 9$ policies

(Harris's sales) $H = 3S$

$= 3 \times 9$

$H = 27$ policies

Example 8 The Texas Barbeque charges \$9 for a barbeque dinner and \$15 for a steak dinner. A recent dinner for 250 people totaled \$2,850. How many of each type of dinner were served?

We know that a total of 250 dinners were served, some barbeque and some steak. So, suppose we knew there were 50 barbeque dinners; how would we find the number of steak dinners? (Steak $= 250 - 50$.) Regardless of the actual number of each type of dinners,

$$\text{Barbeque} + \text{Steak} = 250$$
$$b + s = 250$$
$$b = 250 - s$$

Now,

Cost of barbeque	plus	cost of steak	totals	\$2,850
$\$9 \times b$	+	$\$15 \times s$	=	\$2,850
↓				
$9(250 - s)$	+	$15s$	=	2,850
$2{,}250 - 9s$	+	$15s$	=	2,850
2,250	+	$6s$	=	2,850
$2{,}250 - 2{,}250$	+	$6s$	=	$2{,}850 - 2{,}250$
		$6s$	=	600
		$\frac{\not{6}s}{\not{6}}$	=	$\frac{600}{6}$
(Steak dinners)		s	=	100 dinners
(Barbeque dinners)		b	=	$250 - s$
			=	$250 - 100$
		b	=	150 dinners

The following list is a review of words and their mathematical equivalents. Knowing the mathematical equivalents of these words will be helpful in solving equations.

ENGLISH WORD	SYMBOL
Is, are, was, were, gives, sold	=
Increased by, more than, combined, together, sum	+
Decreased by, less than, fewer than	−
Of, times, product of	×
Per, out of	÷

SECTION 2 PROBLEMS

"Translate" the following expressions into algebraic symbols.

1. a. 6 times x
b. Apples and oranges combined

2. a. 10 times y
b. Samuel and José together

c. A number increased by 10
d. A number decreased by 18
e. $\frac{2}{3}$ of c
f. 5 more than $\frac{1}{4}$ of p
g. Twice the sum of r and s
h. g costs \$4 less than h
i. d totaled 2 times a and b together
j. b costs 8.5 times as much as f
k. m equals 9 less than $\frac{1}{3}$ of n
l. \$10 per barrel

c. A number increased by 12
d. A number decreased by 44
e. $\frac{1}{6}$ of k
f. 17 more than $\frac{1}{3}$ of j
g. Three times the sum of x and y
h. h made \$24 less than g
i. c totaled 6 times m and n together
j. p costs 5.5 times as much as t
k. l equals 4 less than $\frac{2}{5}$ of k
l. \$25 per ticket

Express each of the following as a math equation, and solve.

3. What number decreased by 26 yields 56?

4. What number increased by 23 yields 84?

5. The Corner Market charges \$1.50 less than Wilson's Mart for the same size package of dog food. If the Corner Market's price is \$12, what does Wilson's Mart charge?

6. In July, a patio chair was priced at \$4.99 less than the April price. If the July price was \$10.99, what was the April price?

7. The Scott Shop charges \$15 more than the Garcia Co. for a winter jacket. If the Scott Shop's price is \$77, how much does the Garcia Co. charge?

8. Quality Lumber Co. charges \$1.50 per decking board more than Handy Carpenter charges. If Quality Lumber's price is \$4.15 per board, what is Handy Carpenter's price?

9. Three-fifths of a sports shop's sales were charge sales. What were its charge sales, if the total sales were \$3,000?

10. Three-fifths of a firm's monthly expenditures are related to payroll expenses. If the total monthly expenses are \$6,000, how much are the payroll expenses?

11. One-fourth of Happy Ed's Auto sales last year were repeat customers. If 800 cars were sold to repeat customers, how many cars were sold altogether?

12. Two-fifths of the people enrolled in a seminar held master's degrees. If 18 people had master's degrees, how many people enrolled in the course?

13. Eight less than $\frac{2}{5}$ of employees at a manufacturing plant took no sick leave during January. If there were 40 employees without an absence, how many people does the plant employ?

14. At a back-to-school sale, single-subject and multiple-subject theme books were sold. If 15 less than $\frac{1}{2}$ of the total sales were for single-subject books, and 50 single-subject theme books were sold, how many total books were sold?

15. Utility expenses for February were 1.2 times March utility expenses for the Dawson family. If the February utilities totaled \$192, how much were the expenses for March?

16. Before a car engine was overhauled, it got 0.75 times the number of miles per gallon as it did afterwards. The car got 21 miles per gallon before the overhaul. How many miles per gallon did it get after the overhaul was completed?

17. Depreciation amounting to \$1,176 was claimed last year on robotic equipment used by an auto manufacturer. The equipment was used 4,900 hours last year. What was the depreciation per hour?

18. Last week, Jeff worked 40 hours and earned \$384. What was his wage per hour?

19. Salaries for managers are 1.8 times the salaries for staff. If the total salaries are $280,000, how much is each salary?

20. The $2.90 selling price for a bag of chips is obtained by adding the cost and the markup. If the markup is 0.45 times the cost, find the cost and the markup.

21. An electronic store's September sales of cellular phones and pagers totaled $1,900. Cellular phones sold for $50, and pagers sold for $60. If 35 items were sold, how many of each were sold?

22. Sales of denim shorts and denim jeans for May totaled $1,955. The shorts sold for $15 each, and the jeans sold for $28 per pair. If 100 items were sold, how many of each were sold?

23. A homeowner pays $1,005 per month for mortgage, taxes, and home insurance combined. The mortgage payment is 9 times the insurance payment, and the taxes are $15 more than the insurance. How much is spent for each item?

24. Carol, Debbie, and Frieda practice the piano for a total of 23 hours per week. Carol practices twice as much as Frieda, and Debbie practices 3 hours more than Frieda every week. How many hours does each woman practice?

25. A store sells three brands of watches. Last week, the number of Brand A watches sold was 3 times as many as Brands B and C together. Find the number of Brand C watches sold, if 129 Brand A watches and 25 Brand B watches were sold.

26. At a Labor Day sale, a beach shop ran a special sale on beach towels, umbrellas, and chairs. Beach towel sales totaled 3 times as many as umbrella and chair sales together. Find the number of chairs sold, if 54 towels and 12 umbrellas were sold.

27. A man purchased a jacket, a pair of slacks, and a flannel shirt for $137. The jacket cost 2.5 times as much as the slacks, and the flannel shirt cost $7 less than the slacks. How much did each item cost?

28. The combined ages of a woman, her mother, and her sister equal 130 years. The mother is 1.5 times the age of the woman. The sister is 10 years younger than the woman. What is the age of each?

29. A gym shop sold 54 leotards for a total of $900. The cotton leotards sold for $15 each, and the Lycra sold for $18 each. How many of each were sold?

30. Yesterday, a mail-order firm's sales of sleeping bags totaled $3,840. A child's sleeping bag sold for $40, while the adult sleeping bag sold for $100. If 60 sleeping bags were sold altogether, how many of each were sold?

31. One style of athletic shoes costs $20 wholesale in canvas and $36 in leather. An invoice for $776 accompanies an order for 26 pairs of shoes. How many pairs of each were bought?

32. A best-seller novel sells for $28 in a hardback edition and $12 in a paperback edition. A book store's sales of both editions totaled $1,356. If 65 copies of this novel were sold, how many of each edition were sold?

33. A theater sold tickets for a total of $7,050 on Friday and Saturday. Friday night tickets were priced at $30 each, while Saturday matinee tickets were $22 each. If 275 tickets were sold, how many of each were sold?

34. At a close-out sale, an appliance store sold microwave ovens and toaster ovens, with total sales revenue amounting to $1,616. The microwave ovens sold for $80 each, while the toaster ovens sold for $32 each. If 25 total ovens were sold, how many of each were sold?

35. The Lamplighter sold 200 lamps during a two-week special sale. Desk lamps sold for $19 each, and floor lamps sold for $40 each. If $5,165 was received from sale of the 200 lamps, how many of each were sold?

36. At KBC Co., the nationally advertised brand refrigerator sold for $990, while the store brand sold for $560. If 50 refrigerators were sold for a total of $36,170, how many of each brand were sold?

SECTION 3

RATIO AND PROPORTION

A **ratio** is a way of using division or fractions to compare numbers. When two numbers are being compared, the ratio may be written in any of three ways: for example, (1) 3 to 5, (2) 3:5, or (3) $\frac{3}{5}$. Method 3 indicates that common fractions (which are **rational numbers**) are also ratios. (Notice that "rational" and "ratio" are variations of the same word.) If more than two numbers are being compared, the ratio is written in either of the first two ways: as 3 to 2 to 4, or 3:2:4.

Ratios are often used instead of percent in order to compare items of expense, particularly when the percent would have exceeded 100%. In this case it is customary to reduce the ratio to a comparison to 1, such as 1.3 to 1 or as $\frac{2.47}{1}$. (A ratio is "reduced" by expressing it as a common fraction and reducing the fraction. A ratio is reduced to "a comparison to 1" by dividing the denominator into the numerator.) Ratios may also be used as a basis for dividing expenses among several categories, which we will consider when apportioning overhead in a later chapter.

Example 1 Determine the following ratios:

(a) There are 20 doctors and 48 registered nurses on the staff of Neuman County Hospital. What is the ratio of doctors to nurses?

$$\begin{aligned}\text{Doctors to registered nurses} &= \frac{\text{Doctors}}{\text{Nurses}} \\ &= \frac{20}{48} \\ &= \frac{5}{12}\end{aligned}$$

The ratio of doctors to registered nurses is 5 to 12, or 5:12, or $\frac{5}{12}$.

(b) Net sales of Warner Corp. were $105,000 this year, compared to $84,000 last year. What is the ratio of this year's sales to last year's?

$$\begin{aligned}\text{Current sales to previous} &= \frac{\text{Current sales}}{\text{Previous sales}} \\ &= \frac{\$105{,}000}{\$84{,}000} \\ &= \frac{5}{4} \\ &= 1.25 \quad \text{or} \quad \frac{1.25}{1}\end{aligned}$$

The ratio may be given correctly in any of the following forms: 5 to 4 or 1.25 to 1; 5:4 or 1.25:1; $\frac{5}{4}$ or $\frac{1.25}{1}$. Since most actual business figures would not reduce to a common fraction like $\frac{5}{4}$, business ratios are usually computed by ordinary division, without attempting to reduce the fraction. Thus,

$$\frac{\$105{,}000}{\$84{,}000} = 1.25 = \frac{1.25}{1} \quad \text{or} \quad 1.25 \text{ to } 1 \quad \text{or} \quad 1.25:1$$

The term *proportion* is frequently used in connection with ratio. A **proportion** is simply a mathematical statement that two ratios are equal. Thus, $\frac{6}{10} = \frac{3}{5}$ is a proportion. The same proportion could also be indicated as 6:10 :: 3:5, which is read "six is to ten as three is to five."

A proportion may be used to find an unknown amount if we first know the ratio that exists between this unknown and another, known amount. This method is particularly useful when relationships (ratios) are given as percent. Such proportions are then solved like ordinary equations.

Example 2 The net profit of Shaw Hardware was 7% of its net sales. What was the net profit if the firm had $60,000 in net sales?

Since $7\% = \frac{7}{100}$, then

$$\frac{\text{Profit}}{\text{Net sales}} = \frac{7}{100}$$

$$\frac{P}{\$60{,}000} = \frac{7}{100}$$

$$\overset{1}{(\cancel{60{,}000})}\frac{P}{\cancel{60{,}000}} = \frac{7}{\cancel{100}}\overset{600}{(\cancel{60{,}000})}$$

$$P = \$4{,}200$$

For any given proportion, the **cross products** are always equal. Given $\frac{6}{10} = \frac{3}{5}$, the cross products are computed as

$$\frac{6}{10} \times \frac{3}{5}$$

$$6 \cdot 5 = 3 \cdot 10$$

$$30 = 30$$

Cross products* are useful in solving any proportion where the unknown appears in the denominator, as follows.

*Cross products are a shortcut which produces the same result that is obtained when both sides of an equation are multiplied by the product of the two given denominators. Given $\frac{6}{10} = \frac{3}{5}$, then $50(\frac{6}{10}) = 50(\frac{3}{5})$, which gives $30 = 30$, as before.

Example 3 Lien and Pham, who are sales representatives for Nguyen & Co., received orders last week in the ratio of $\frac{2}{3}$. How many orders did Pham sell, if Lien sold 56 orders?

$$\frac{\text{Lien sales}}{\text{Pham sales}} = \frac{2}{3}$$

$$\frac{56}{P} = \frac{2}{3}$$

$$(3)56 = 2P$$

$$168 = 2P$$

$$\frac{168}{2} = \frac{\cancel{2}P}{\cancel{2}}$$

$$84 = P$$

Example 4 Citizens of San Rio paid \$39 for 700 kilowatt-hours (kWh) of electricity. What was the bill for a family that used 1,400 kWh?

$$\frac{\text{Cost}}{\text{kWh}} = \frac{\$39}{700}$$

$$\frac{c}{1{,}400} = \frac{39}{700}$$

$$(\cancel{1{,}400})\frac{c}{\cancel{1{,}400}} = \frac{39}{\cancel{700}}(\overset{2}{\cancel{1{,}400}})$$

$$c = \$78$$

SECTION 3 PROBLEMS

Reduce the following ratios and express in each of the three forms.

1. a. 8 to 21 b. 150 to 450 c. \$1,000 to \$2,500
d. \$400 to \$640 e. 36,000 to 16,000

2. a. 7 to 24 b. 6 to 54 c. 90 to 150
d. \$2,100 to \$2,700 e. 72,000 to 9,000

Find the missing element in each proportion.

3. a. $\frac{c}{14} = \frac{2}{7}$ b. $\frac{3}{8} = \frac{r}{16}$ c. $\frac{3}{g} = \frac{9}{24}$ d. $\frac{15}{10} = \frac{9}{z}$

4. a. $\frac{a}{18} = \frac{2}{9}$ b. $\frac{6}{21} = \frac{t}{14}$ c. $\frac{15}{k} = \frac{5}{8}$ d. $\frac{3}{2} = \frac{12}{d}$

Use ratio or proportion to complete the following problems. Reduce ratios or proportions to lowest terms.

5. Carlita read 8 books over spring break, while Susanne read only 2. What was the ratio of books read by Carlita to books read by Susanne?

6. Karl made 6 client calls out of state last week, whereas Brian made 3. What was the ratio of calls made by Karl to those made by Brian?

7. In a beginning word processing class, 22 people had completed a previous keyboarding class, whereas 4 had not. What was the ratio of students with previous keyboarding skills to those who had none?

8. Carter made 20 deliveries yesterday, while Doug made 25. What was the ratio of Carter's deliveries to Doug's deliveries?

9. The operating expenses of Randolph Motors were 28% of their net sales last month. Find the operating expenses if net sales were $66,000.

10. This year, the production department at Clinton Race Cars received 80% of the number of specialized orders that it received last year. Last year, the company received 105 specialized orders. How many specialized orders were received this year?

11. Fifteen percent of Pam's gross wages is deducted for an annuity. If her annuity deduction was $33, what was the amount of her gross wages?

12. Gary pledged 6% of his gross wages this month to a soup kitchen. If his deduction for the soup kitchen was $240, what was the amount of his gross wages?

13. The ratio of men to women in an aerobics program was 3:8. If 6 men were enrolled, how many women were enrolled?

14. The ratio of successful to unsuccessful calls by a telemarketing firm was 3 to 7. If 1,008 calls yielded successful sales, how many were unsuccessful?

15. The ratio of business administration majors to marketing majors at a local community college was 5 to 2. If there were 240 business administration majors, how many marketing majors were there?

16. The ratio of situps completed by Omar and Paul was 2:3. Omar completed 76 situps. How many situps did Paul complete?

17. The ratio of team sales to individual sales for warm-up suits was 2:7 for Gray Sports Shop. If sales to individuals totaled 1,260, how many warm-ups were sold to teams?

18. The ratio of family-home sales to single-person home sales was 5:2. If 26 single-person home sales were made last month, how many sales were made to families?

19. A computer store sold 54 packages of Brand X software. The ratio of sales for Brand X to Brand Y was 3:2. How many Brand Y packages were sold?

20. The Collins Co. has a ratio of 2:1 for its current assets compared to its current liabilities. If current assets total $88,500, how much are current liabilities?

21. A secretary transcribes 900 words in 30 minutes. How many words can she transcribe in 3½ hours?

22. A check processor can process 1,500 checks in 20 minutes. How many checks can the equipment process in 2 hours?

23. George packs 10 boxes in 45 minutes. How many boxes can he pack in 3 hours?

24. A homeowner paid $112 for 1,400 kilowatt-hours (kWh) of electricity. At this rate, how much would 1,850 kWh of electricity cost?

25. The laser printer just purchased by Wood & Associates will print 50 pages in 5 minutes. How many pages can it print in 1 hour?

26. A nurse counted 20 heartbeats in 15 seconds. At this rate, how many times will the heart beat in 1 minute?

27. After 6 hours, a truck driver had driven 330 miles. At this rate, how long will it take to drive 550 miles?

28. On a vacation, a family covered a distance of 144 miles in 3 hours. At this rate, how long would it take them to drive a total of 1,200 miles?

CHAPTER 2 GLOSSARY

Coefficient. A number that multiplies a variable or a quantity in a set of parentheses. [Example: 3 is a coefficient in $3x$ or in $3(x + z)$.]

Cross products. The equal products obtained when each numerator of a proportion is multiplied by the opposite denominator. (Example: Given $\frac{3}{4} = \frac{9}{12}$, the cross products 3?12 and 4?9 both equal 36.)

Equation. A mathematical statement that two expressions (or an expression and a number) are equal.

Expression. Any indicated mathematical operation(s) written without an equal sign.

$$\left(\text{Examples: } 2x + 3; \frac{3t}{5} - 1\right)$$

First-degree equation in one variable. An equation with one distinct variable whose highest exponent (degree) is 1. (Example: $2x + 7 = x + 12$.)

Proportion. A mathematical statement that two ratios are equal. (Example: $\frac{3}{4} = \frac{9}{12}$)

Ratio. A comparison of two (or more) numbers, frequently indicated by a common fraction. (Examples: 4 to 7; 4:7; $\frac{4}{7}$)

Rational number. A number that can be expressed as a common fraction (the quotient of two integers).

Reciprocal. The result obtained when a number is inverted by interchanging the numerator and the denominator. Or, that value which, when multiplied by the original value, yields 1. (Example: The reciprocal of $\frac{3}{4}$ is $\frac{4}{3}$.)

Similar terms. Terms that include the same variable. (Example: The expression $5x + x - 2x$ contains three similar terms.)

Terms. The products, quotients, or numbers in a mathematical expression that are separated by plus or minus signs.

$$\left(\text{Example: The expression } 4x - \frac{3x}{5} + 7 \text{ contains three terms.}\right)$$

CHAPTER 3

REVIEW OF PERCENT

OBJECTIVES

Upon completion of Chapter 3, you will be able to:

1. Define and use correctly the terminology associated with each topic.
2. a. Change a percent to its equivalent decimal or fraction (Section 1: Examples 1, 2; Problems 1–30).
   b. Change a decimal or fraction to its equivalent percent (Section 1: Examples 3, 4; Problems 31–60).
3. Use an equation to find the missing element in a percentage relationship (examples: What percent of 30 is 25? 60 is 120% of what number?) (Section 2: Example 1; Problems 1–36).
4. Use the basic equation form "___% of Original = Change?" to find the percent of change (increase or decrease) (example: $33 is what percent more than $27?) (Section 2: Example 2; Problems 37–46).
5. Use an equation to find the original number when the percent of change (increase or decrease) and the result are both known (example: What number increased by 25% of itself gives 30?) (Section 2: Example 3; Problems 47–56).
6. Given a word problem containing percents, express the problem in a concise sentence which translates into an equation that solves the problem (Section 3: Examples 1–5; Problems 1–48).

Percent is a fundamental topic with which everyone has some familiarity. We use a percent like a ratio to make a comparison. A ratio is often expressed as a fraction, such as $\frac{18}{24}$ or $\frac{3}{4}$, whereas a percent would express a relationship using 100 as the denominator of the fraction, such as $\frac{75}{100}$. Percent compares the number of parts out of 100 to which the fraction is equivalent.

Because percent is one of the most frequently applied mathematical concepts in all areas of business, it is extremely important for anyone entering business to be capable of doing accurate percent calculations.

For a further discussion of percents, fractions, and decimals, see Appendix A.

SECTION 1

BASIC PERCENT

The use of percent will be easier if you remember that **percent** means **hundredths.** That is, the expression "46%" may just as correctly be read "46 hundredths." The word "hundredths" denotes either a common fraction with 100 as denominator or a decimal fraction of two decimal places. Thus,

$$46\% = 46 \text{ hundredths} = \frac{46}{100} \quad \text{or} \quad 0.46$$

A percent must first be changed to either a fractional or a decimal form before that percent of any number can be found. These conversions are simplified by applying the fact that "percent" means "hundredths."

A. CHANGING A PERCENT TO A DECIMAL

When people think of percent, they normally think of the most common percents, those between 1% and 99%. These are the percents that occupy the first two places in the decimal representation of a percent. (For example, 99% = 99 hundredths = 0.99, and 1% = 1 hundredth = 0.01.)

A.1 *When converting percents to decimals, write the whole percents between 1% and 99% in the first two decimal places.*

When this rule is followed, the other digits will then naturally fall into their correct places, as demonstrated next.

Example 1 **% to Decimal**

Express each of the following percents as a decimal.

(a) 5% = 5 hundredths = 0.05
(b) 86% = 86 hundredths = 0.86
(c) 16.3% = 0.163 (since 16% = 0.16)
(d) 131% = 1.31 (since 31% = 0.31)

(e) $122.75\% = 1.2275$ (since $22\% = 0.22$)
(f) $6.09\% = 0.0609$ (since $6\% = 0.06$)
(g) $0.4\% = 0.004$ (since $0\% = 0.00$)
(h) $0.03\% = 0.0003$ (since $0\% = 0.00$)

Some percents are written using common fractions. To change a fraction to a decimal, you must divide the numerator by the denominator.

$$\frac{n}{d} \quad \text{or} \quad d\overline{)n}$$

A.2 *When a fractional percent is to be changed to a decimal, first convert the fractional percent to a decimal percent, and then convert the decimal percent to an ordinary decimal.*

(Machine calculations of the decimal percent may be rounded to the third decimal place.)

Example 1 (cont.) **Fractional % to Decimal**

Express each of the following percents as a decimal.

(i) $\frac{7}{8}\%$: Recall that $\frac{7}{8}$ means $\begin{array}{r} .875 \\ 8\overline{)7.000} \end{array} = 0.875$

And since $\frac{7}{8} = 0.875$

then $\frac{7}{8}\% = 0.875\% = 0.00875$

(j) $\frac{4}{7}\%$: First, $\frac{4}{7}$ means $\begin{array}{r} .57\frac{1}{7} \\ 7\overline{)4.00} \end{array} = 0.57\frac{1}{7}$ (or 0.571)

Then $\frac{4}{7} = 0.57\frac{1}{7}$

implies that $\frac{4}{7}\% = 0.57\frac{1}{7}\% = 0.0057\frac{1}{7}$ (or 0.00571)

A fractional percent always indicates a percent less than 1% (or it indicates that fractional part of 1%). Thus, $\frac{1}{4}\% = \frac{1}{4}$ of 1%; $\frac{3}{5}\% = \frac{3}{5}$ of 1%.

When the given fraction is a common one for which the decimal equivalent is known, the process of *converting a fractional percent to a decimal* may be shortened.

A.3 *Write two zeros to the right of the decimal point* (0.00) *to indicate that the percent is less than* 1%, *and then write the digits that are normally used to denote the decimal equivalent of the fraction.*

Example 1 (cont.)

(k) $\frac{1}{4}\%$: $\left(\frac{1}{4}\% = \frac{1}{4} \text{ of } 1\%\right)$

To show that the percent is less than 1%, write: 0.00

To indicate $\frac{1}{4}$, affix the digits "25" $\left(\frac{1}{4} = 25 \text{ hundredths}\right)$: 0.0025

Thus, $\frac{1}{4}\% = 0.0025$.

(This may be easily verified, since $4 \times 0.0025 = 0.0100 = 1\%$.)

(l) $\frac{2}{3}\%$: $\left(\frac{2}{3}\% = \frac{2}{3} \text{ of } 1\%\right)$

Indicate a percent less than 1%: 0.00

Affix the digits $66\frac{2}{3}$: $0.0066\frac{2}{3}$

B. CHANGING A PERCENT TO A FRACTION

The procedure for changing a percent to a fraction is summarized as follows:

B.1 *After an ordinary percent has been changed to hundredths (by dropping the percent sign and placing the number over a denominator of 100), the resulting fraction should be reduced to lowest terms.*

An improper fraction—one in which the numerator is greater than the denominator—is considered to be in lowest terms provided that it cannot be further reduced.

Example 2 **% to Fraction**

Convert each percent to its fractional equivalent in lowest terms.

(a) $45\% = 45 \text{ hundredths} = \frac{45}{100} = \frac{9}{20}$

(b) $8\% = 8 \text{ hundredths} = \frac{8}{100} = \frac{2}{25}$

(c) $175\% = 175 \text{ hundredths} = \frac{175}{100} = \frac{7}{4}$

Percents containing fractions may be converted to their ordinary fractional equivalents by altering the above procedure slightly. The fact that "percent" means "hundredths" may also be shown as

$$\% = \text{hundredth} = \frac{1}{100}$$

B.2 *Percents containing fractions are changed to ordinary fractions by substituting $\frac{1}{100}$ for the % sign and multiplying.*

Now let us apply this fact to change some fractional percents to common fractions.

Example 2 (cont.) **% with Fraction to Fraction**

(d) $12\frac{1}{2}\% = 12\frac{1}{2} \times \frac{1}{100} = \frac{25}{2} \times \frac{1}{100} = \frac{25}{200} = \frac{1}{8}$

(e) $55\frac{5}{9}\% = 55\frac{5}{9} \times \frac{1}{100} = \frac{500}{9} \times \frac{1}{100} = \frac{500}{900} = \frac{5}{9}$

(f) $\frac{1}{3}\% = \frac{1}{3} \times \frac{1}{100} = \frac{1}{300}$

B.3 *Decimal percents may be converted to fractions by applying the following procedure:*

1. *Change the percent to its decimal equivalent.*
2. *Pronounce the value of the decimal; then write this number as a fraction and reduce it.*

Example 2 (cont.) **Decimal % to Fraction**

(g) 22.5%:
1. Since 22% = 0.22, then 22.5% = 0.225.
2. 0.225 = 225 thousandths $= \frac{225}{1{,}000} = \frac{9}{40}$.

(h) 0.56%: Since 0% = 0.00, then

$$
\begin{aligned}
0.56\% &= 0.0056 \\
&= 56 \text{ ten-thousandths} \\
&= \frac{56}{10{,}000} \\
&= \frac{7}{1{,}250}
\end{aligned}
$$

(i) 0.6%:

$$0.6\% = 0.006$$
$$= 6 \text{ thousandths}$$
$$= \frac{6}{1{,}000}$$
$$= \frac{3}{500}$$

C. CHANGING A DECIMAL TO A PERCENT

Just as "percent" means "hundredths," so is the opposite also true—that is, *hundredths = percent.* Rule C states an important concept that should be remembered when working with decimals and percents.

> C. *When a decimal number is to be changed to a percent, the hundredths places of the decimal indicate the whole "percents" between* 1% *and* 99% (0.01 = 1%, *and* 0.99 = 99%).

By first isolating the whole percents between 1% and 99%, you will have no difficulty in placing the other digits correctly.

If there is to be a decimal point in the percent, it will come after the hundredth's place of the decimal number. (This is why it is sometimes said: "To change a decimal to a percent, move the decimal point two places to the right and add a percent sign.")

Example 3 **Decimal to %**

Express each of these decimals as a percent.

(a) 0.67 = 67 hundredths = 67%
(b) $0.82\frac{1}{2} = 82\frac{1}{2}$ hundredths $= 82\frac{1}{2}\%$
(c) 0.08 = 8 hundredths = 8%
(d) 0.721: Since 0.72 = 72%, then 0.721 = 72.1%.

It may prove helpful to circle the first two decimal places, as an aid in identifying the percents between 1% and 99%.

Example 3 (cont.)

(e) 1. (45)9: Given 1.459, the .45 indicates 45%.
Thus, 1.459 = 145.9%.

(f) 0.(9): Given 0.9, the .9 or .90 denotes 90%.

So, 0.9 = 90%.

(g) 0.(00)26: Given 0.0026, the 0.00 represents 0%.

Hence, 0.0026 = 0.26%.

D. CHANGING A FRACTION TO A PERCENT

The procedure for changing a fraction to a percent is summarized in Rule D.

D. *To change a fraction to a percent, apply the following steps:*

1. *Convert the fraction to its decimal equivalent by dividing the numerator by the denominator:*

   $$\frac{n}{d}\Big) \quad \text{or} \quad d\overline{)n}$$

   (Machine calculations may be rounded to the third decimal place.)
2. *Then change the decimal to a percent, as illustrated in the preceding example.*

Example 4 **Fraction to %**

Change each fraction to its equivalent percent.

(a) $\frac{5}{8} = 8\overline{)5.000}^{\,.625} = 0.625 = 62.5\%$

(b) $\frac{7}{18} = 18\overline{)7.00}^{\,.38\frac{16}{18}} = 0.38\frac{8}{9} = 38\frac{8}{9}\%$ (or 38.9%)

(c) $\frac{5}{4} = 4\overline{)5.00}^{\,1.25} = 1.25 = 125\%$

(d) $1\frac{4}{5}\left(\text{since } \frac{4}{5} = 5\overline{)4.00}^{\,.80} = 0.80\right) = 1.80 = 180\%$

or $1\frac{4}{5} = \frac{9}{5} = 5\overline{)9.00}^{\,1.80} = 1.80 = 180\%$

(e) $2\frac{3}{4}\left(\text{since } \frac{3}{4} = 0.75\right) = 2.75 = 275\%$

or $2\frac{3}{4} = \frac{11}{4} = 2.75 = 275\%$

SECTION 1 PROBLEMS

Express each percent as (a) a decimal and (b) a fraction in lowest terms.

1. 11% **2.** 23% **3.** 32%
4. 18% **5.** 2% **6.** 6%
7. 30.5% **8.** 44.2% **9.** 6.25%
10. 3.35% **11.** 174% **12.** 450%
13. 128.4% **14.** 130.6% **15.** 0.80%
16. 0.48% **17.** 1.75% **18.** 1.30%
19. 105% **20.** 260% **21.** $\frac{3}{4}$%
22. $\frac{1}{8}$% **23.** $\frac{2}{5}$% **24.** $\frac{1}{20}$%
25. $\frac{3}{8}$% **26.** $\frac{4}{5}$% **27.** $1\frac{3}{5}$%
28. $87\frac{1}{2}$% **29.** $12\frac{1}{2}$% **30.** $8\frac{1}{5}$%

Express each of the following as a percent.

31. 0.09 **32.** 0.04 **33.** 0.40
34. 0.52 **35.** 0.674 **36.** 0.232
37. 2.11 **38.** 1.45 **39.** 1.28
40. 2.05 **41.** 0.001 **42.** 0.022
43. 0.5 **44.** 0.8 **45.** 0.005
46. 0.008 **47.** 3.1 **48.** 4.6
49. 0.03 **50.** 0.02 **51.** $\frac{1}{2}$
52. $\frac{3}{5}$ **53.** $\frac{1}{200}$ **54.** $\frac{3}{500}$
55. $\frac{5}{12}$ **56.** $\frac{4}{9}$ **57.** $1\frac{3}{4}$
58. $3\frac{1}{2}$ **59.** $4\frac{4}{5}$ **60.** $6\frac{1}{8}$

SECTION 2

PERCENT EQUATION FORMS

All problems that involve the use of percent are some variation of the basic percent form: "Some percent of one number equals another number." This basic **percent equation form** may be abbreviated as

_____% of _____ = _____

Given in reverse order, the equation form is

_____ = _____% of _____

Another way of expressing the preceding formula is

$$\text{rate} \times \text{base} = \text{percentage} \qquad (r \cdot b = p)$$

Because there can be only one unknown in an elementary equation, there are only three variations of the basic equation form: The unknown can be either (1) the percent (or **rate**), (2) the first number (or **base**), or (3) the second number (the **percentage**). For instance, consider

$$70\% \text{ of } 52 = 36.4$$

Here, 70% is the rate, 52 is the base, and 36.4 is the percentage. (This is somewhat confusing because the percentage is not a percent; the rate is the percent. This confusion, however, is avoided by using the percent equation form above.)

Equation-solving procedures that were studied earlier will be applied to solve percentage problems. *Recall that before a percent of any number can be computed, the percent must first be changed to either a fraction or a decimal.* Conversely, if the unknown represents a percent, the solution to the equation will be a decimal or fraction that must then be converted to the percent.

Example 1 (a) 16% of 45 is what number?

$$0.16 \times 45 = n$$

$$7.2 = n$$

(b) What percent of 30 is 6?

$$r \times 30 = 6$$

$$30r = 6$$

$$\frac{\cancel{30}r}{\cancel{30}} = \frac{6}{30}$$

$$r = \frac{1}{5}$$

$$r = 20\%$$

(c) $66\frac{2}{3}\%$ of what number is 12?

Because $66\frac{2}{3}\%$ is exactly $\frac{2}{3}$, we will use $\frac{2}{3}$:

$$\frac{2}{3} \times n = 12$$

$$\frac{2n}{3} = 12$$

$$\frac{\cancel{3}}{\cancel{2}} \times \frac{\cancel{2}n}{\cancel{3}} = 12 \times \frac{3}{2}$$

$$n = 18$$

Many of the formulas used to solve mathematical problems in business are nothing more than the basic percent equation form with different words (or variables) substituted for the percent and the first and second numbers. The methods used to solve the formulas are identical to those of Example 1.

One such formula is used to find **percent of change** (that is, percent of increase or decrease). In words, the formula can be stated, "What percent of the original number is the change?" In more abbreviated form it is

$$\_\_\_\_\%\text{ of Original} = \text{Change?}$$

Example 2

(a) What percent more than 24 is 33?

1. *Original number.* To have "more than 24" implies that we originally had 24; thus, 24 is the original number. In general, it can be said that the "original number" is the number that follows the words "more than" or "less than" in the stated problem.
2. *Change.* The "change" is the amount of increase or decrease that occurred. That is, the change is the numerical difference between the two numbers in the problem. (The fact that it may be a negative difference is not important to this formula.) Thus, the change in this example is $33 - 24 = 9$.

$$\_\_\_\_\%\text{ of Original} = \text{Change}$$

$$\_\_\_\_\%\text{ of } 24 = 9$$

$$24x = 9$$

$$\frac{\cancel{24}x}{\cancel{24}} = \frac{9}{24}$$

$$x = \frac{3}{8}$$

$$x = 37\tfrac{1}{2}\%$$

Thus, 24 increased by 37½% of itself gives 33.

(b) 12 is what percent less than 18?

Original = 18

Change = 18 − 12

= 6

$$\_\_\_\_\%\text{ of Original} = \text{Change}$$

$$18r = 6$$

$$\frac{\cancel{18}r}{\cancel{18}} = \frac{6}{18}$$

$$r = \frac{1}{3}$$

$$r = 33\tfrac{1}{3}\%$$

Thus, 18 decreased by $33\frac{1}{3}\%$ of itself is 12.

The "percent of change" formula is often applied to changes in the prices of merchandise or stocks and to changes in volume of business from one year to the next. Other applications include increase in cost of living or unemployment and decrease in expenses or net profit.

A third type of problem is a variation of the "percent of change" type. In this application we know the percent of change (percent increase or decrease) and the result, and we compute the amount of the original number. The original number is either increased or decreased by a percent of itself. The number is expressed as 100%n or $1n$ so that the variables can be combined.

Example 3 (a) What number decreased by 12% of itself gives 66?

What number decreased by (12% of itself) gives 66?

$$n \quad - \quad (0.12 \times n) \quad = \quad 66$$

$$n^* \quad - \quad 0.12n \quad = \quad 66$$

$$0.88n = 66$$

$$\frac{\cancel{0.88}n}{\cancel{0.88}} = \frac{66.00}{0.88}$$

$$n = 75$$

(b) What number increased by $33\frac{1}{3}$% of itself gives 28?

When the percent contains a repeating decimal, it is easier to work with the fraction equivalent instead of the decimal equivalent.

$$\frac{3}{3}n + \frac{1n}{3} = 28$$

$$\frac{4n}{3} = 28$$

$$\frac{\cancel{3}}{\cancel{4}} \times \frac{\cancel{4}n}{\cancel{3}} = \overset{7}{\cancel{28}} \times \frac{3}{\cancel{4}}$$

$$n = 21$$

*Recall that $n = 1n$; thus,

$$n - 0.12n = 1n - 0.12n$$

$$= 1.00n - 0.12n$$

$$= 0.88n$$

SECTION 2 PROBLEMS

Use equations to obtain the following solutions. Express percent remainders either as fractions or rounded to tenths (example: $28\frac{4}{7}\%$ or 28.6%).

Part One

1. 8% of 700 is what number?
2. 15% of 460 is what number?
3. 20% of 640 is how much?
4. 40% of 520 is how much?
5. 25% of 78 is how much?
6. 42% of 90 is what amount?
7. What is ⅗% of 37,000?
8. What is ¾% of 6,000?
9. What is 12½% of 560?
10. What is 50½% of 900?
11. What is 4½% of 600?
12. What is 9¾% of 400?
13. ¼% of 14,000 is what amount?
14. ⅕% of 12,000 is how much?
15. What percent of 90 is 13?
16. What percent of 100 is 12?
17. 28 is what percent of 80?
18. 450 is what percent of 3,000?
19. 5.4 is what percent of 180?
20. 45 is what percent of 900?
21. What percent of 18 is 4.5?
22. What percent of 140 is 42?
23. 15% of what number is 48?
24. 120% of what amount is 78?
25. 44% of what number is 33?
26. 11½% of what number is 115?
27. 240 is 40% of what amount?
28. 196 is 28% of what amount?
29. 90 is 33⅓% of what number?
30. 50 is 33⅓% of what amount?
31. 100 is 1¼% of what amount?
32. 243 is 4½% of what amount?
33. 3.5 is 0.7% of what number?
34. 27 is 0.9% of what number?
35. 48 is ½% of what amount?
36. 9 is ¾% of what amount?

Part Two

37. What percent more than 5 is 7?
38. What percent more than 150 is 180?
39. 44 is what percent less than 55?
40. 105 is what percent less than 150?
41. 130 is what percent less than 325?
42. 26 is what percent less than 260?
43. 135 is what percent more than 108?
44. 300 is what percent more than 200?
45. What percent less than $600 is $597?
46. What percent less than $800 is $794?

Part Three

47. What number increased by 25% of itself gives 90?
48. What number increased by 40% of itself gives 77?
49. What amount decreased by 25% of itself gives 36?
50. What amount decreased by 5% of itself gives 76?
51. What number increased by 50% of itself gives 108?
52. What number increased by 20% of itself gives 480?
53. What amount decreased by 33⅓% of itself gives 150?
54. What amount decreased by 45% of itself gives 550?
55. What number increased by 25% of itself yields 2,500?
56. What amount increased by 37½% of itself yields 8,250?

SECTION 3

WORD PROBLEMS USING PERCENTS

Three primary types of percent applications are illustrated by the following five examples. These types are as follows:

1. The basic percent equation form: _____% of _____ = _____ (Examples 1–3).
2. Finding *percent* of change (increase or decrease), using _____% of Original = Change (Example 4).
3. Finding an *amount* that has been increased or decreased by a percent of itself (Example 5).

Example 1 A direct-mail campaign by a magazine produced \$16,000 in renewals and new subscriptions. If renewals totaled \$10,000, what percent of the subscriptions were renewals?

What percent of subscriptions were renewals?

$$\_\_\_\_\% \times \$16{,}000 = \$10{,}000$$

$$16{,}000r = 10{,}000$$

$$\frac{16{,}000r}{16{,}000} = \frac{10{,}000}{16{,}000}$$

$$r = 0.625$$

$$r = 62.5\%$$

Example 2 Thirty-five percent of a payment made to a partnership land purchase was tax deductible. If a partner receives a notice that he qualifies for a \$700 tax deduction, how much was his total payment?

35% of payment was tax deduction

$$35\% \times p = \$700$$

$$0.35p = 700$$

$$\frac{0.35p}{0.35} = \frac{700.00}{0.35}$$

$$p = \$2{,}000$$

Example 3 A quarterback completed 40% of his attempted passes in a professional football game. If he attempted 65 passes, how many did he complete?

$$\underline{40\%}\ \underline{\text{of}}\ \underline{\text{attempted passes}}\ \underline{\text{were}}\ \underline{\text{completed passes}}$$

$$\begin{aligned} 0.4 \times 65 &= c \\ 0.4(65) &= c \\ 26 &= c \end{aligned}$$

Example 4 Last year's sales were $45,000; sales for this year totaled $48,600. What was the percent of increase in sales?

$$\begin{aligned} \text{Original} &= \$45{,}000 \\ \text{Change} &= \$48{,}600 - \$45{,}000 \\ &= \$3{,}600 \end{aligned}$$

$$\begin{aligned} \_\_\_\_\%\text{ of Original} &= \text{Change?} \\ \_\_\_\_\%\text{ of }\$45{,}000 &= \$3{,}600 \\ 45{,}000r &= 3{,}600 \\ \frac{\cancel{45{,}000}r}{\cancel{45{,}000}} &= \frac{3{,}600}{45{,}000} \\ r &= 0.08 \\ r &= 8\% \end{aligned}$$

Example 5 Labor expenses (wages) at a construction project increased by 10% this month. If wages totaled $13,200 this month, how much were last month's wages?

$$\underline{\text{Old wages}}\ \underline{\text{increased by}}\ \underline{10\%\text{ (of itself)}}\ \underline{\text{equals new wages}}$$

$$\begin{aligned} w + (10\% \times w) &= \$13{,}200 \\ w + 0.1w &= 13{,}200 \\ \frac{\cancel{1.1}w}{\cancel{1.1}} &= \frac{13{,}200.0}{1.1} \\ w &= \$12{,}000 \end{aligned}$$

Note. Percents are never used just by themselves. In an equation, any percent that is included must be a percent *of* ("times") some other number or variable.

For convenience, the following problems are divided into two parts, each of which covers all the given examples. Part One covers the three types of problems presented: (1) Examples 1–3, see Problems 1–16; (2) Example 4, see Problems 17–26; (3) Example 5, see Problems 27–30. Additional practice is provided in Part Two, but the order is scrambled.

The following percents (many with fractional remainders) will be given in many problems so that students without calculators can conveniently use the exact fractional equivalents. (Students using calculators may also use the fractional equivalents by entering the numerators and denominators separately.) A summary of conversions is given in Table 3–1 on page 56.

PERCENT EQUIVALENTS OF COMMON FRACTIONS			
$\frac{1}{2} = 50\%$			
$\frac{1}{3} = 33\frac{1}{3}\%$	$\frac{2}{3} = 66\frac{2}{3}\%$		
$\frac{1}{4} = 25\%$	$\frac{3}{4} = 75\%$		
$\frac{1}{5} = 20\%$	$\frac{2}{5} = 40\%$	$\frac{3}{5} = 60\%$	$\frac{4}{5} = 80\%$
$\frac{1}{6} = 16\frac{2}{3}\%$	$\frac{5}{6} = 83\frac{1}{3}\%$		
$\frac{1}{7} = 14\frac{2}{7}\%$	$\frac{2}{7} = 28\frac{4}{7}\%$	$\frac{3}{7} = 42\frac{6}{7}\%$	
$\frac{1}{8} = 12\frac{1}{2}\%$	$\frac{3}{8} = 37\frac{1}{2}\%$	$\frac{5}{8} = 62\frac{1}{2}\%$	$\frac{7}{8} = 87\frac{1}{2}\%$
$\frac{1}{9} = 11\frac{1}{9}\%$	$\frac{2}{9} = 22\frac{2}{9}\%$	$\frac{4}{9} = 44\frac{4}{9}\%$	$\frac{5}{9} = 55\frac{5}{9}\%$
$\frac{1}{10} = 10\%$	$\frac{3}{10} = 30\%$	$\frac{7}{10} = 70\%$	$\frac{9}{10} = 90\%$
$\frac{1}{12} = 8\frac{1}{3}\%$			

SECTION 3 PROBLEMS

Part One

1. A yogurt store's overhead last quarter was $6,750, while sales were $15,000. The store's overhead was what percent of its sales?
2. A real estate agent earned a commission of $11,200 on a sale of $160,000. What percent of the sale was the commission?
3. Out of its total budget of $60,000, the human resources department spent $51,600 on wages. What percent of the budget was spent on wages?

4. Depreciation expenses totaled $33,600 last year at Fairmont Industries. If total operating expenses were $280,000, what percent of the total operating expenses were the advertising costs?

5. Income from advertisements placed in the *Montana Ledger* newspaper amounted to 75% of total revenue. What were the receipts from advertising if revenues totaled $150,000 last month?

6. Mary Ann earned a 15% return on her investment of $4,600. What was the dollar amount of her investment return?

7. Eight percent of a stock's cost was paid in dividends last quarter. If the stock cost $85, what dividend did it yield?

8. Today, 13% of insurance enrollees in this country are in managed-care plans. If there are 40 million people covered by insurance, how many of these people are in managed-care plans?

9. Charles receives 20% of his income from investments. What was his total income if his investment income had been $10,200?

10. A candidate in a campus senate race won 60% of the total votes cast. If the candidate received 492 votes, how many votes were cast for all candidates in this race?

11. Inventory represents 60.25% of a hardware store's total assets. What are the total assets if the inventory is valued at $180,750?

12. In a recent survey of television shows, "Prime Events" was watched by 48% of households polled. If there were 240 households watching this show, how many households were surveyed?

13. A collection agency charges a fee of 33⅓% of the accounts receivable it collects. The agency collected $186,000 last month. What fee did it earn for these collections?

14. Electricity costs for a restaurant amount to 16⅔% of its monthly expenses. If monthly expenses total $8,700, how much does electricity cost?

15. The raw material cost to manufacture a pair of safety boots is $10. The cost of raw materials equals 25% of the total production cost. What is the total production cost?

16. The gross profit made on a music system was 8% of the sales price. What was the sales price if the gross profit was $18.80?

17. Monthly cable charges last year were $36, while the same charges this year amount to $37.98. What percent increase does this represent?

18. Construction costs have risen so that a building that could have been built for $940,000 five years ago would now cost $1,015,200 to construct. What percent increase in construction costs does this represent?

19. The cost-of-living index reveals that a standard market basket of food that cost $50 last year now costs $52.80 for the same items. What is the percent of increase?

20. Hannah's bonus last year amounted to $246, while this year's bonus was $263.22. What percent of increase does this represent?

21. Jack made 12 account calls on Monday and 8 calls on Tuesday. What percent decrease does this represent?

22. After a firm installed new equipment, its cost of producing a business letter dropped from $24 to $20. What percent decrease does this represent?

23. The price of a mutual fund dropped from $30 per share to $28.20 per share. What percent decrease is this?
24. After an efficiency study, factory expenses were reduced from $8,800 per month to $7,876. What percent decrease is this?
25. Phillip's travel expense vouchers totaled $1,500 last month compared to $1,312.50 this month. What is the percent decrease in his travel expenses?
26. Accidents at McCorkle Co. fell from 18 last year to 15 this year. What percent decrease in accidents is this?
27. What number increased by 40% of itself yields 4,200?
28. What amount increased by 37.5% of itself yields 1,925?
29. What amount decreased by 20% of itself leaves $36?
30. What amount decreased by 14% of itself leaves 86?

Part Two

31. At its grand opening, Rossbach Co. estimated that $16\frac{2}{3}\%$ of its customers were teenagers. Nine hundred thirty customers visited the store during this time. How many were teenagers?
32. Zeigler Professional Services estimates that $1\frac{3}{4}\%$ of its credit sales of $6,500 will be uncollectible. How much is its estimated uncollectible accounts?
33. The cost of coffee to a grocery store has increased from $200 per case to $206 per case over the past six months. What is the percent increase in cost?
34. The price of a prescription medication has increased from $32 to $48. By what percent has the price increased?
35. Seventeen percent of recent graduates of Parker College had secured jobs before graduation. If 68 people had jobs before graduation, how many graduates were there?
36. A store had 16% of its total sales returned. If $450 worth of goods was returned, how much were the total sales?
37. The liabilities of Quinn and Pfister Co. decreased from $23,000 last year to $17,480 this year. What was the percent of change?
38. The number of students taking weekend classes dropped from 1,820 students during the winter term to 1,729 in the spring term at a college. What percent decrease was this?
39. A gross profit of $160 was made on a sale of $480. What percent of the selling price was the gross profit?
40. Griffin Cabinet Co. employs 150 sales representatives. If 84 of these employees reached their sales quota last year, what percent of the sales representatives is this?
41. The price of the latest edition of a software package increased by 18% over the previous edition. The new software was priced at $122.72. What was the previous edition's price?
42. Within the past five years, the average price of a haircut has increased by 20%. If a haircut costs $24 today, what was the cost five years ago?
43. During a special promotion, the price of a multimedia system was reduced by 30%. What was the price before the reduction if the sale price was $2,100?
44. The owner of a desktop publishing firm has experienced a 10% drop in her net worth over the last three years. If her current net worth is $225,000, what was her net worth three years ago?

TABLE 3-1 SUMMARY OF CONVERSIONS

GIVEN:	TO CONVERT TO:	
PERCENT	A. DECIMAL	B. FRACTION
Basic % (1% to 99%)	Basic percents become the first two (hundredths) decimal places. (Other digits align accordingly.) (Example: 76% = 76 hundredths = 0.76)	Percent sign is dropped and given number is placed over a denominator of 100; reduce. (Example: 76% = 76 hundredths = $\frac{76}{100} = \frac{19}{25}$)
Decimal % (125.73%; 0.45%)	Digits representing percents greater than 99% or less than 1% are aligned around the first two (hundredths) decimal places (see also Section 1, "Basic Percent"). (Example: 0.45% = 0.0045)	Convert the percent to a decimal (as indicated at left). Pronounce the decimal value; then write this as a fraction and reduce. (Example: 0.45% = 0.0045 = 45 ten-thousandths = $\frac{45}{10{,}000} = \frac{9}{2{,}000}$)
Fractional % ($\frac{1}{4}$%; $\frac{7}{5}$%; $1\frac{2}{3}$%)	Convert the fractional percent to a decimal percent (by dividing the numerator by the denominator). Then convert this decimal percent to an ordinary decimal (as shown above). (Example: $\frac{7}{5}$% = 1.4% = 0.014)	Percent sign is replaced by $\frac{1}{100}$; multiply and reduce. (Example: $\frac{7}{5}\% = \frac{7}{5} \times \frac{1}{100} = \frac{7}{500}$)

	TO CONVERT TO PERCENT
C. *Decimals* (0.473; 1.2; 0.0015)	Isolate the first two (hundredths) decimal places to identify the basic percent (1% to 99%). Other digits will then be aligned correctly. (Example: 0.473 = 0.(47)3 = 47.3%)
D. *Fractions* ($\frac{2}{9}$; $\frac{7}{3}$; $1\frac{1}{2}$)	Convert the fraction to a decimal (by dividing the numerator by the denominator). Then follow the above procedure for converting a decimal to percent. (Example: $\frac{2}{9} = 0.22\frac{2}{9} = 0.(22)\frac{2}{9} = 22\frac{2}{9}\%$)

45. The value of a store's inventory increased from $60,000 to $63,000. By what percent has the value increased?

46. Orders for a math textbook increased from 5,000 to 5,750 at a publishing company this year. What was the percent increase in textbooks ordered?

47. Find a newspaper or magazine article that includes percents, and bring it to class for discussion. Does the use of percents clarify or enhance the information in the article? Why or why not? Are the percents presented in a way that would enable you to compute additional data not specifically stated in the article?

48. Continue with the newspaper or magazine article from Problem 47. Create a word problem using information from the article, and solve the problem using techniques you have used for earlier problems in this assignment. Repeat this for a second word problem.

CHAPTER 3 GLOSSARY

Base. The number that a percent multiplies: the first number in the basic percent equation form. (The base represents 100%.)

Hundredths. Equals percent; a value associated with the first two decimal places or with a common fraction having a denominator of 100.

Percent. Means hundredths; indicates a common fraction with a denominator of 100 or a decimal fraction of two decimal places.

Percentage. A part of the base (a number or an amount) that is determined by multiplying the given percent (rate) times the base. (See the accompanying figure.)

Percent equation form. The basic equation _____% of _____ = _____; in words, what % of a first number (base) is a second number (percentage)?

Percent of change. The equation _____% of original = change; a variation of the basic equation form, used to find a percent of increase or decrease.

Rate. The percent that multiplies a base. (See the accompanying figure.)

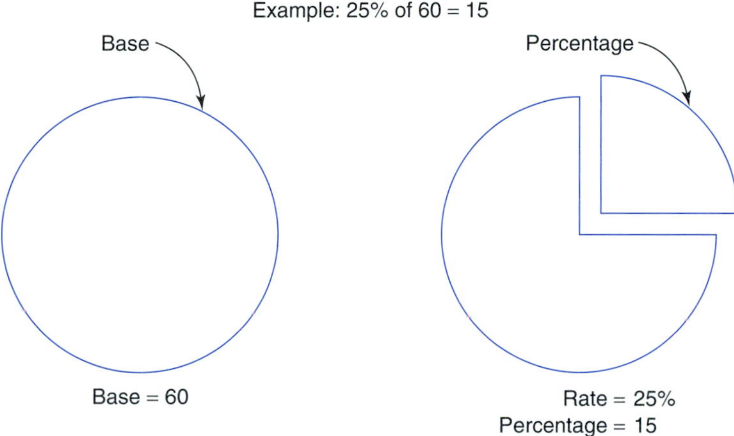

CHAPTER 4

BASIC STATISTICS AND GRAPHS

OBJECTIVES

Upon completion of Chapter 4, you will be able to:

1. Define and use correctly the terminology associated with each topic.
2. Compute the three simple averages: mean, median, and mode (Section 1: Examples 1–7; Problems 1–18).
3. Use a frequency distribution to determine mean, median, and modal class for grouped data (Section 2: Examples 1, 2; Problems 1–8).
4. Using data that assume a normal curve distribution (Section 3):
   a. Determine the percents associated with various combinations of intervals $\pm 1\sigma$, $\pm 2\sigma$, and $\pm 3\sigma$ (Example 1; Problems 1–14)
   b. Compute standard deviation for ungrouped data, using the formula

   $$\sigma = \sqrt{\frac{\Sigma d^2}{n}}$$

   (Example 2; Problems 5–14).
5. Interpret index numbers and compute simple price relatives (Section 4: Examples 1, 2; Problems 1–6).
6. Construct the common types of graphs: bar, line, and circle (Section 5: Examples 1, 2; Problems 1–12).

Decisions about what course a business may take are rarely made without first consulting records indicating what past experience has been. When numerical records are involved, these data must be organized before they can present meaningful information. The process of collecting, organizing, tabulating, and interpreting numerical data is called **statistics.** The information presented as a result of this process is also known, collectively, as **statistics.** Statistics are invaluable to business in making comparisons, indicating trends, and analyzing facts affecting operations, production, profits, and so forth. However, the significance of statistics can often be realized more easily when the information is displayed in graph or chart format. Thus, graphs/charts are an integral part of statistics.

SECTION 1

AVERAGES

If anyone attempted to analyze a large group of numbers simultaneously, it would usually be impossible to obtain much useful information. Thus, it is customary to condense such a group into a single number representative of the entire group. This representative number is generally some central value around which the other numbers seem to cluster; for this reason, it is called a **measure of central tendency** or an **average.** The three most common averages are the **arithmetic mean** (or just the "mean"), the **median,** and the **mode,** all of which we will consider briefly.

MEAN

The **arithmetic mean** is the common average with which you are already familiar. It is the average referred to in everyday expressions about average temperature, average cost, the average grade on a quiz, average age, and so on. The mean is found by adding the various numbers and dividing this sum by the number of items.

Example 1 The five departments at Smith Sporting Goods made sales of the following amounts: A, \$1,050; B, \$1,200; C, \$1,350; D, \$1,800; E, \$1,500. What were the average (mean) sales per department?

$$\frac{\$1{,}050 + \$1{,}200 + \$1{,}350 + \$1{,}800 + \$1{,}500}{5} = \frac{\$6{,}900}{5} = \$1{,}380$$

The mean sales per department were \$1,380. Notice that there was not a single department that had sales of exactly \$1,380. It is quite common for the "average" of a group of numbers to be a value not actually contained in the group. However, this mean is still representative of the entire group in that it is similar to most of the values in the group—some are somewhat larger and others somewhat smaller.

Further observe that if each department had made sales of $1,380, the total sales would also have been $6,900. This kind of observation applies to any situation in which an arithmetic mean is utilized.

WEIGHTED MEAN

The ordinary mean will present a true picture only if the data involved are of equal importance. If they are not, a weighted mean is needed. A **weighted arithmetic mean** is found by "weighting" or multiplying each number according to its importance. These products are then added and that sum is divided by the total number of weights.

A weighted average of particular importance to students is the **grade-point average.**

Example 2 Listed below are Carol Dickerson's course credit hours and term grades for the spring quarter at West Ridge Community College. Each A carries 4 quality points; a B, 3 points; a C, 2 points; and a D, 1 point. Compute Carol's grade-point average (average per credit hour) for the quarter.

CREDIT HOURS	FINAL GRADE	QUALITY POINTS
4	A	4
3	C	2
3	B	3
3	C	2
2	C	2

If we simply computed the average of Carol's quality points,

$$\frac{4 + 2 + 3 + 2 + 2}{5} = \frac{13}{5} = 2.6$$

the result would be misleading, because her courses are not of equal importance from a credit standpoint. We must therefore weight the quality points for each course according to the number of credit hours each course carries, as follows:

CREDIT HOURS		QUALITY POINTS		
4	×	4	=	16
3	×	2	=	6
3	×	3	=	9
3	×	2	=	6
+ 2	×	2	=	4
15				41

$$\frac{41}{15} = 2.733\ldots = 2.73$$

Carol's grade-point average for the quarter is thus 2.73. This is somewhat better than the 2.6 obtained by the ordinary averaging method.

Example 3 The same model of microwave oven is available from several appliance stores. Prices and number of ovens sold are shown below. What are (a) the average price per store and (b) the average price per oven sold?

STORE	NUMBER SOLD	PRICE
1	40	$450
2	60	400
3	120	350
4	80	375

It may help to first analyze the question. The word "per" indicates the division line of a fraction. The word following "per" tells which total number of objects to divide by. Hence, average price "per store" is found by dividing by the total number of stores, and average price "per oven" is found by dividing by the total number of ovens.

(a) $$\text{Average price per store} = \frac{\text{Sum of prices at all stores}}{\text{Total number of stores}}$$

$$\frac{\$450 + \$400 + \$350 + \$375}{4} = \frac{\$1{,}575}{4}$$

$$= \$393.75 \qquad \text{Average price per store}$$

(b) $$\text{Average price per oven} = \frac{\text{Sum of prices of all ovens}}{\text{Total number of ovens}}$$

For this average, the various prices must be weighted according to the number of ovens sold at each price.

NUMBER		PRICE		
40	×	$450	=	$ 18,000
60	×	400	=	24,000
120	×	350	=	42,000
+ 80	×	375	=	30,000
300				$114,000

$$\frac{\$114{,}000}{300} = \$380 \qquad \text{Average price per oven}$$

Example 4 David Cohen invested $5,000 in a business on January 1. On May 1, he withdrew $200; on July 1, he reinvested $1,000; and on August 1, he invested another $1,400. What was his average investment during the year?

$$\text{Average investment} = \frac{\text{Sum of investments of all months}}{\text{Total number of months}}$$

Before we can find the average investment, we must first determine the new balance after each change. Each succeeding balance must then be multiplied (weighted) by the number of months it remained invested.

DATE	CHANGE	AMOUNT OF INVESTMENT		MONTHS INVESTED		
January 1		$5,000	×	4	=	$20,000
May 1	− $200	4,800	×	2	=	9,600
July 1	+1,000	5,800	×	1	=	5,800
August 1	+1,400	7,200	×	5	=	36,000
				12		$71,400

$$\frac{\$71{,}400}{12} = \$5{,}950 \qquad \text{Average investment}$$

MEDIAN

Even when a group of numbers are of equal importance, the arithmetic mean may not convey an accurate description if the group contains a few values which differ greatly from the others. In such a case, the median may be the better average (or measure of central tendency). The **median** is the midpoint (or middle number) of a group of numbers. It is the central value of an array (or listing of data).

In order to find the median, we must arrange the numbers in order of size (such an arrangement is called an **ordered array**). If the group contains an odd number of entries, the median is the exact middle number. Its position can be determined quickly by first dividing the number of entries by 2. Then the next whole number larger than the quotient indicates the position held by the median when the numbers are ordered in a series from smallest to largest.

When a group contains an even number of values, the median is again found by first dividing the number of entries by 2. The median is then the arithmetic mean (average) of the number in this position and the next number, when the numbers are ordered from smallest to largest.

Example 5 The Wingate Insurance Agency has seven agents with the following respective years of experience: 5, 3, 22, 8, 6, 10, and 30. What is the median years of experience of these agents?

Notice that in this case the mean years of experience,

$$\frac{5 + 3 + 22 + 8 + 6 + 10 + 30}{7} = \frac{84}{7} = 12 \text{ years}$$

does not give a very good indication of central tendency, since only two of the seven agents have had at least that much experience. The median is more descriptive:

Since there are seven numbers in the group, divide 7 by 2: $\frac{7}{2} = 3+$

The median is the next larger (or fourth) number when the numbers are arranged in increasing order of size. Hence, in the series, 3, 5, 6, 8, 10, 22, 30, the median is 8 years of experience.

Example 6 Wilcox & Co., a clothing store, employs six people. Their monthly salaries are $1,240, $1,350, $1,160, $1,400, $1,450, and $2,400. What is the median salary at the store?

$2,400
1,450
1,400
1,350
1,240
1,160

Since there are six employees, there is no exact middle number. The median is the arithmetic mean of the third ($\frac{6}{2} = 3$) and fourth salaries:

$$\frac{\$1,400 + \$1,350}{2} = \frac{\$2,750}{2} = \$1,375$$

The median salary is $1,375. (The mean salary, however, is $1,500.)

A major difference between the mean and the median is that all the given values have an influence on the mean, whereas the median is determined basically by position. The median is often an appropriate measure of central tendency where the items can be ranked. However, if the number of items is quite large, it may be difficult to prepare the array necessary for determining the median. An alternative measure sometimes used in such circumstances is the **modified mean,** in which the mean is computed in the usual manner after the extremely high and/or low scores have been deleted. If each highest and each lowest score were deleted in Examples 5 and 6, the modified means would be 10.2 years of experience and $1,360 per month, both of which are much nearer to the median scores than were the ordinary means.

MODE

The **mode** of a group of numbers is the value (or values) that occurs most frequently. If all the values in a group are different, there is no mode. A set of numbers may have two or more modes when these all occur an equal number of times (and more often than the other values).

The mode may be quite different from the mean and the median, and it does not necessarily indicate anything about the quality of the group. Like the median, the mode is not influenced by all of the given values, and it may be very erratic. Unlike the other averages, however, the mode always represents actual items from the given data. The mode's principal application to business, from a production or sales standpoint, is in indicating the most popular items.

Example 7 The employees of Noland Construction Co. worked the following numbers of hours last week: 40, 48, 25, 36, 43, 48, 45, 60, 52, 43, 56, 54, 48, 33, and 53. What is the modal number of hours worked?

The numbers of hours worked, arranged in numerical order, are as follows:

Hours	
60	
56	
54	
53	
52	There are three employees who worked exactly 48 hours last week; hence, 48 is the mode. If there were only two employees with 48-hour weeks, then 48 and 43 would both be modes, since these numbers would each be worked by two employees.
48 }	
48 }	
48 }	
45	
43	
43	
40	
36	
33	
25	

SECTION 1 PROBLEMS

Find (1) the arithmetic mean, (2) the median, and (3) the mode for each of the following groups of numbers.

1. a. 10, 22, 29, 22, 45, 30, 43, 18, 39, 28, 59, 41, 42, 40, 12
   b. 170, 154, 120, 133, 142, 161, 115, 121, 130, 154

2. a. 20, 18, 20, 16, 22, 19, 21, 23, 17, 20, 24
   b. 342, 332, 351, 343, 347, 348, 340, 352, 344, 351, 349, 341

3. The credit hours and final grades for Julio's four courses are listed below. He receives 4 quality points for each A, 3 points for each B, 2 points for each C, and 1 point for a D.
   a. What are Julio's average (mean) quality points per class?
   b. What is his quality-point average for the semester (his average per credit hour)?

CREDIT HOURS	GRADES
3	D
3	C
4	A
5	B

4. By the end of the semester, Genene had earned the following grades. She earns 4 quality points for each A, 3 points for each B, 2 points for each C, and 1 point for a D.
   a. What were her average (mean) quality points per class?
   b. What was her quality-point average for the semester (her average per credit hour)?

CREDIT HOURS	GRADES
3	A
3	B
3	A
2	C
1	C

5. Scott's Heating and Air Conditioning Service employs 20 people. Their jobs and wages are shown below.
   a. What was the mean wage per job?
   b. What was the mean wage per employee?

JOB	NUMBER IN JOB	WEEKLY WAGES
Manager	1	$1,000
Sales representative	4	450
Technician	10	550
Office support	5	400

6. Gloria's Hair Cuttery employs 10 people. Their jobs and wages are listed below.
   a. What was the mean wage per job?
   b. What was the mean wage per employee?

JOB	NUMBER IN JOB	WEEKLY WAGES
Manager	1	$1,200
Stylist	6	600
Receptionist	2	450
Custodian	1	350

7. Six stores in a city sell the same brand and model of vacuum cleaner. The price and number sold at each store are listed below.
   a. What is the average price per store?
   b. What is the mean price per vacuum cleaner sold?

STORE	PRICE	NUMBER SOLD
A	$ 85	12
B	135	8
C	96	22
D	89	20
E	112	9
F	107	9

8. Computer stores in a city sold the following laser printer cartridges last week. Determine the following:
   a. The mean price per store
   b. The mean price per cartridge

STORE	PRICE	NUMBER SOLD
Compuland	$ 70	50
ComputerWorld	85	42
Computer Connection	103	30

9. Hope Powers invested $5,000 in the Aerobic Connection Partnership on January 1. On March 1, she withdrew $200. She reinvested $400 on August 1 and withdrew $600 on September 1. Find Powers's average (mean) investment during the year.
10. Graham Grimes invested $20,000 on January 1 in an office supply store. He invested another $15,000 on May 1. He withdrew $2,000 on August 1 and reinvested $5,000 on November 1. What was his average (mean) investment during the year?
11. Determine the median salary earned by 7 college presidents throughout the state last year if the salaries earned were $151,000, $133,000, $108,000, $130,000, $138,000, $110,000, $140,000.
12. Determine the median-priced executive office chair sold at Kelley's Office Furniture Co. last month if the selling prices were $155, $194, $400, $350, $300, $290, $180, $375, $265.
13. Determine the most popularly priced (modal) CD/cassette player sold at Johnson Electronics last week if the selling prices were $130, $86, $82, $94, $82, $79, $124, $82, $79, $86, $158.
14. What was the most popularly priced coffeemaker sold (the modal price) at Flemming Kitchens, Inc. last week if the selling prices were $58, $99, $18, $25, $18, $33, $58, $18, $42, $18, $25, $33, $61.
15. The junior partners and law clerks in Jones, Hecht, and Isaac worked the following hours last week: 48, 50, 54, 60, 48, 54, 56, 60, 59, 68, 52, 60, 51, 60, 45, 47. Find the following:
    a. Mean number of hours worked
    b. Median number of hours worked
    c. Most frequent (mode) number of hours worked
16. The years of experience at Ratner Realty were as follows: 3, 5, 1, 2, 8, 6, 3, 7, 9, 12. Determine the following:
    a. Mean number of years of experience
    b. Median number of years of experience
    c. Most frequent (modal) number of years of experience
17. Employees at Armez Textiles worked on a production basis. They produced the following number of units last month: 1,600, 1,630, 1,640, 1,630, 1,620, 1,640, 1,560, 1,640, and 1,620.
    a. What was the mean number of units?
    b. What was the median number of units?
    c. What was the modal number of units?
18. During the last month, the sales representatives at Harrison Auto sold the following number of cars: 26, 15, 9, 12, 22, 20, 15.
    a. What was the mean number of cars sold?
    b. What was the median number of cars sold?
    c. What was the modal number of cars sold?

SECTION 2

AVERAGES (GROUPED DATA)

Business statistics often involve hundreds or thousands of values. It would be extremely time-consuming to arrange each value in an array. It is easier to handle large numbers of values by organizing them into groups (or **classes**). The difference between the lower limit of one class and the lower limit of the next class is the **interval** of the class. (All classes of a problem usually have the same-sized interval.) Each class is handled by first determining the number of values within the class (the **frequency**) and then using the **midpoint** of the class as a basis for calculation. The end result of this classification and tabulation is called a **frequency distribution.**

The first step in making a frequency distribution is to decide what the class intervals shall be. For the sake of convenience, the intervals are usually round numbers. A tally is then made as each given value is classified; the total tally in each class is the frequency f of the class. After the midpoint of each class is computed, the frequency is multiplied by the midpoint value to obtain the final column of the frequency distribution. Example 1 shows how to compute the interval midpoints and the arithmetic mean of the frequency distribution.

Example 1 Lady Fair Mills is a textile firm producing ladies' lingerie. Using a frequency distribution, determine (a) the mean and (b) the median number of garments processed per employee last week in the sewing department. Production per employee (in dozens) was as follows:

368	363	352	333
340	347	380	321
386	342	308	357
320	382	341	374
355	335	364	391
372	325	378	348

The smallest number is 308, and the largest is 391; thus, a table from 300 to 400 will include all entries. The difference between the upper and lower limits (400 − 300) gives a **range** of 100. We can conveniently divide this range into five classes.* Since $\frac{100}{5} = 20$, each class interval will be 20.

CLASS INTERVALS	TALLY	f	MIDPOINT	$f \times$ MIDPOINT
380–399	||||	4	390	1,560
360–379	~~||||~~ |	6	370	2,220
340–359	~~||||~~ |||	8	350	2,800
320–339	~~||||~~	5	330	1,650
300–319	|	+ 1	310	+ 310
		24		8,540

*The choice of 5 classes is arbitrary. Four classes with an interval of $\frac{100}{4} = 25$ could be chosen, as could 10 classes with an interval of $\frac{100}{10} = 10$.

The midpoint of each interval is the arithmetic mean of the lower limit of that class and the lower limit of the next class.[†]

$$\text{Example: } \frac{320 + 340}{2} = \frac{660}{2} = 330$$

(a) The arithmetic mean of the frequency distribution is found by dividing the $f \times$ midpoint grand total by the total number of items (the frequency total). Thus,

$$\frac{8{,}540}{24} = 355\frac{5}{6} \quad \text{or} \quad 356$$

The mean production per employee in the sewing department was 356 dozen garments, to the nearest whole dozen.

MEDIAN FOR GROUPED DATA

The *median for grouped data* is found as follows:

1. Divide the total number of items (total frequency) by 2. This quotient represents the position which the median holds in the tally.
2. Counting from the smallest number, find the class where this item is located, and determine how many values within this class must be counted to reach this position.
3. Using the number of values that must be counted (step 2) as a numerator and the *total* number within the class as a denominator, form a fraction to be multiplied times the class interval.
4. Add this result to the lower limit of the interval to determine the median.

Example 1 (cont.)

(b) The median for this example is thus determined as follows:

(1) $\frac{24}{2} = 12$; we are concerned with the 12th item from the bottom.

(2) The bottom classes contain only $1 + 5 = 6$ items. We must therefore include 6 more items from the class 340–359.

(3) There are 8 values in the class, and the class interval is 20. Hence,

$$\frac{6}{8} \times 20 = \frac{3}{\not{4}} \times \overset{5}{\not{20}} = 15$$

(4) $340 + 15 = 355$

The median of 355 is thus the dividing point where half of the employees obtained higher production and half had lower.

[†]Adherence to strict statistical methods requires a midpoint that is the mean of the lower and upper boundaries of the interval. For instance, $\frac{320 + 339}{2} = 329.5$, rather than the 330 midpoint determined above. The simplicity of the method in Example 1 seems preferable for introductory business mathematics; however, the instructor may specify the standard statistical midpoint if desired. [The group mean of part (a) would then be $355\frac{1}{3}$ rather than $355\frac{5}{6}$.] Further discussion appears in the Instructor's Manual.

As a point of interest, compare the mean and median for Example 1 when the data are computed individually (without a frequency distribution). The mean is then 353.4 (compared to the grouped data mean of 356), and the median is 353.5 (compared to the 355 obtained using a frequency distribution). The differences between these means and medians are small, even though we used a small number of items (24). For large numbers of items, the results obtained by the individual and the grouped-data methods are almost identical.

Since a frequency distribution does not contain individual values, there is no single mode. However, the distribution may contain a *modal class* (the class containing the largest number of frequencies). The class interval 340–359 is the modal class of Example 1. This indicates that the most common production per employee was between 340 and 359 dozen garments.

Example 2 What is the median production if a frequency distribution contains 25 items and is distributed as follows?

CLASS INTERVALS	*f*
2,000–2,399	3
1,600–1,999	4
1,200–1,599	10
800–1,199	7
400–799	+ 1
	25

The steps for solution (see the preceding rules) are as follows:

(1) $\frac{25}{2} = 12.5$; we must locate the class containing the imaginary item numbered 12.5.

(2) The two bottom classes contain only 1 + 7 = 8 entries. Thus, 12.5 is in the next class, 1,200–1,599, and 4.5 more values from this class complete the required 12.5.

(3) The class 1,200–1,599 contains 10 items, and the interval is 400. Thus,

$$\frac{4.5}{\not{1}\not{0}} \times \overset{40}{4\not{0}\not{0}} = 4.5 \times 40 = 180$$

(4) The median production is 1,200 + 180 = 1,380 pieces.

Note. The median can also be found by counting *down from the top* to the required item, multiplying the obtained fraction times the interval in the usual manner, and then *subtracting* this product from the *lower limit* of the adjacent class. For example, since the top two classes in Example 2 contain 3 + 4 = 7 items, 5.5 more values from the class 1,200–1,599 must be included. Now, $\frac{5.5}{10} \times 400 = 220$; and 1,600 − 220 gives 1,380, as before.

SECTION 2 PROBLEMS

Use a frequency distribution to determine (a) the mean, (b) the median, and (c) the modal class for the following sets of data.

1. Use intervals of 0–19, 20–39, etc.

82	22	28	36	39
68	23	61	38	12
96	44	10	21	45
73	78	51	40	26
14	30	17	65	56

2. Use intervals of 0–9, 10–19, etc.

4	15	66	8	2	61
14	25	38	12	10	21
22	34	55	68	32	11
45	27	26	39	48	50
29	13	56	24	30	62

3. Use intervals of $100, starting with $500.

$ 570	$ 640	$590	$ 960	$980	$ 800
670	940	660	730	820	1,050
710	860	850	740	770	680
970	1,020	890	1,080	820	760
1,000	780	950	870	500	840

4. Use intervals of $50, starting with $200.

$260	$390	$540	$460	$263
500	445	312	435	421
210	430	442	324	505
236	248	515	403	452

5. During a Presidents' Day Sale, the 20 sales associates at Stephen's registered sales in the amounts shown below. Make a frequency distribution using $100 intervals (starting with $50), and determine (a) the mean, (b) the median, and (c) the most typical sales interval (modal class).

$155	$ 95	$ 75	$385	$190
420	160	50	260	330
170	50	180	175	240
140	250	300	80	200

6. Island InterCon received the following number of help-desk calls during the last half of the year (26 weeks). Construct a frequency distribution to determine (a) the mean, (b) the median, and (c) the mode. Use intervals of 100 beginning with 400.

600	742	554	645	1,000	904
680	591	812	725	1,050	880
540	804	770	780	910	605
490	860	620	900	924	594
750	700				

7. Wexford Laboratories employs 26 lab technicians whose years of experience are listed below. Use 3-year intervals in completing a frequency distribution, and compute (a) the mean, (b) the median, and (c) the most common interval (modal class) of experience.

15	13	6	3	3
12	16	4	8	6
8	7	10	0	6
7	18	6	22	14
6	1	12	20	5
3				

8. The net weekly sales of the branches of Duggan Software are listed below. Prepare a frequency distribution with $2,000 intervals, starting with the smallest sales figure. Find (a) the mean, (b) the median, and (c) the modal class size (most common sales interval).

$ 9,200	$ 6,600	$ 8,000	$ 5,400
5,400	5,550	3,990	7,000
2,000	4,900	6,080	5,900
3,500	10,200	8,750	9,900
10,000	8,800	2,700	10,101
9,600	8,500	3,650	8,200

SECTION 3

STANDARD DEVIATION

Measures of central tendency—the arithmetic mean, median, and mode—are used to identify some typical value(s) around which the other values cluster. It is frequently also helpful to know something about how spread out the data are: Are many (or most) of the given amounts close in value to the measure of central tendency? Or do the amounts scatter out without concentrating near any particular value? To provide some indication of this scatteredness (or **variability**), we compute **measures of dispersion.** One simple measure of dispersion which has already been mentioned is the **range**—the difference between the upper and lower limits of a frequency distribution or between the highest and lowest values in a group of numbers. Probably the most widely used measure of dispersion is the **standard deviation,** which we will briefly consider.

The tally column of a frequency distribution provides a visual illustration of the central tendency and of the dispersion of a set of data. If a tally column were turned sideways and a smooth line drawn outlining the tallies, we could obtain a simple graph of the data, as shown below. This example—with a heavy concentration of values in the middle and quickly tapering off symmetrically on both sides—is typical of the distribution for many sets of data. For example, the graphs of many human characteristics (such as height, weight, intelligence, athletic ability) would form such a graph, with most people obtaining some average, central score and those with higher and lower scores distributed evenly on both sides.

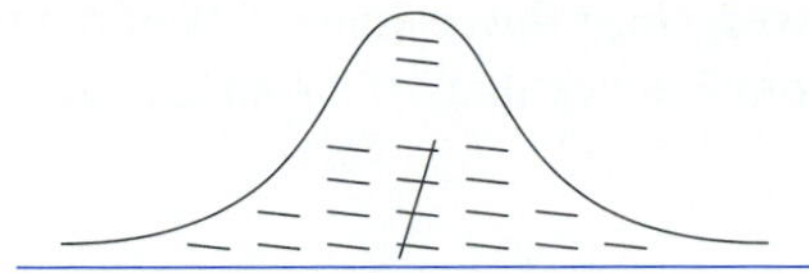

Many sets of business-related data would also form a similar graph. For example, on quality-control tests of the chemical content of a medicine or on tests of the sizing accuracy of a mechanical part, most test results would match the desired standard, and variations higher and lower should taper off quickly. In business and industry, however, it is vital to know exactly how much variation exists.

When statistical data are collected on large groups of people who are typical of the entire population (rather than just some specialized group), the graph of these data would approach the perfect mathematical model known as the **normal curve** (or **bell curve**), shown in the following diagram. In this normal distribution, the mean, median, and mode all have the same value (*M*): the value (on the horizontal line) which corresponds to the highest point on the curve. The range of values is evenly divided into six equal intervals, each of which represents one standard deviation, σ. The six intervals are marked off symmetrically around the mean at $\pm 1\sigma$, $\pm 2\sigma$, and $\pm 3\sigma$.*

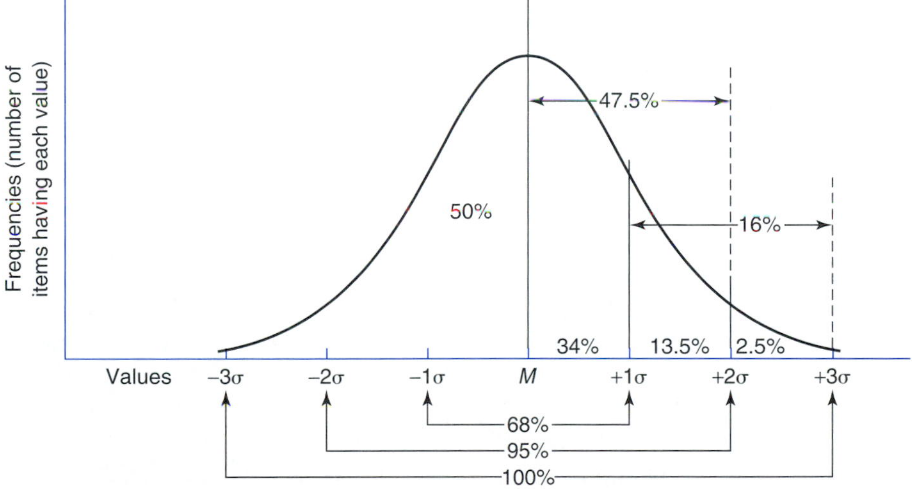

The standard deviation is significant because we know what percent of the total items (people) fall within each interval or combination of intervals. These percent categories are indicated on the normal curve above (and apply symmetrically on each side of the mean). Thus, as shown below, if the mean height of American males is 5′10″ and one standard deviation $\sigma = 2''$, then 68% of all men have heights between 5′8″ and 6′0″ tall; 95% of the men fall between 5′6″ and 6′2″; and 100% of the men fall between 5′4″ and 6′4″. (Obviously, there are men shorter than 5′4″ and taller than

*Technically, the normal curve continues infinitely in both directions, so that additional higher and lower values are possible. However, we assume for practical purposes that the given range of values lies entirely within $\pm 3\sigma$, since the area within $\pm 3\sigma$ includes 99.73% of all items depicted by the normal curve.

6′4″. However, since fewer than 0.3% of the men would fall in either category, we assume for convenience that 100% have been included.)

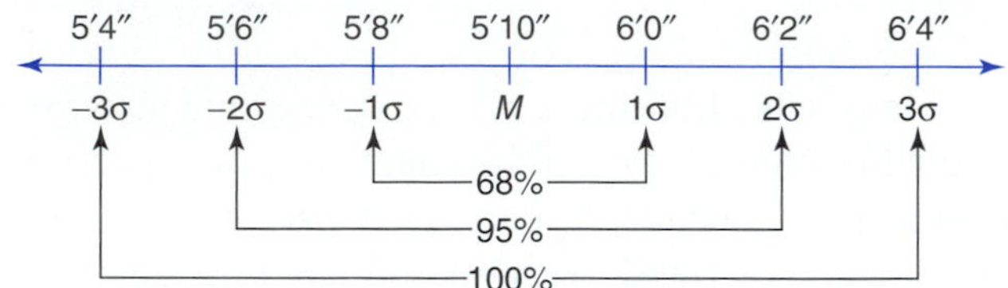

Since the standard deviation for a set of data can be computed, industries can test their products to determine how accurate their manufacturing processes are and what percent of their products fall into various tolerance categories. Clothing manufacturers or other firms whose products are sized may use the percents associated with standard deviation as a guide in deciding how many items of each size to produce.

Example 1 Assume that the mean size of women's dresses is size 12, with a standard deviation of 4, as shown below. Use the percent categories of the normal curve to answer each question.

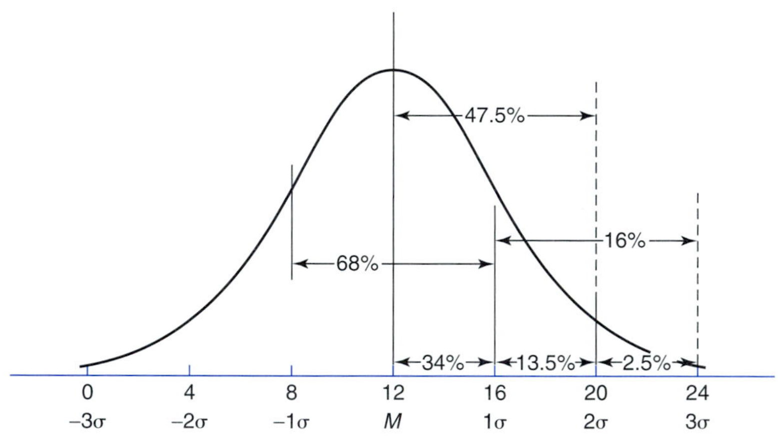

(a) What percent of the women are size 12 or larger? 50%
(b) What percent of the women are size 16 or larger? 16%
(c) What percent are 24 or smaller? 100%
(d) What percent are smaller than size 4? 2.5%
(e) What percent are smaller than size 4 *or* larger than size 20? 2.5% + 2.5% = 5.0%
(f) If a dress company plans to manufacture 5,000 dresses, how many of those should be in sizes 8 through 16? 68% × 5,000 = 3,400

The computation of standard deviation from a set of grouped data in a frequency distribution requires more time and explanation than can be included in introductory business math. However, we can compute the standard deviation for a set of ungrouped values. Using this type of data, the formula for standard deviation is

$$\sigma = \sqrt{\frac{\Sigma d^2}{n}}$$

where the symbols are as follows:

σ = standard deviation
Σ = "the sum of"
d^2 = the square of the deviations from the mean (for each given value)
n = the number of values given
$\sqrt{\ }$ = "the square root of"

This procedure for finding standard deviation is outlined as follows.

STANDARD DEVIATION PROCEDURE

1. Find the mean: Add the given data, and divide by n.
2. Find each deviation, d: Subtract the mean from each given value, obtaining either a positive or a negative difference. (The sum of these deviations must be 0.)
3. Square each deviation, and find their sum, Σd^2.
4. Apply the standard deviation formula, $\sigma = \sqrt{\frac{\Sigma d^2}{n}}$: Divide Σd^2 by n, and take the square root. (If the square root is not a whole number, let the standard deviation remain in $\sqrt{\ }$ form.)

Example 2 Find the standard deviation of the following values: 61, 62, 65, 67, 68, and 73.

MEAN	DEVIATION, d	d^2	STANDARD DEVIATION
61	$61 - 66 = -5$	25	$\sigma = \sqrt{\frac{\Sigma d^2}{n}}$
62	$62 - 66 = -4$	16	
65	$65 - 66 = -1$	1	
67	$67 - 66 = 1$	1	$= \sqrt{\frac{96}{6}}$
68	$68 - 66 = 2$	4	
+ 73	$73 - 66 = 7$	49	$= \sqrt{16}$
$6\overline{)396} = 66$	0	$\Sigma d^2 = 96$	$\sigma = 4$

The standard deviation is 4. Thus, the values associated with each interval would appear on a line graph as shown below. The dots represent the six values given above. For a normal curve distribution, 68% of the values should fall within $\pm 1\sigma$; that is, 68% of the 6 numbers (or 4 numbers) should fall between 62 and 70. Four of our given numbers (62, 65, 67, and 68) fall within $\pm 1\sigma$. Within $\pm 2\sigma$ should be 95% of the 6 numbers, which equals 6 values to the nearest whole unit. All 6 of our given values do fall within $\pm 2\sigma$, which includes values from 58 through 74.

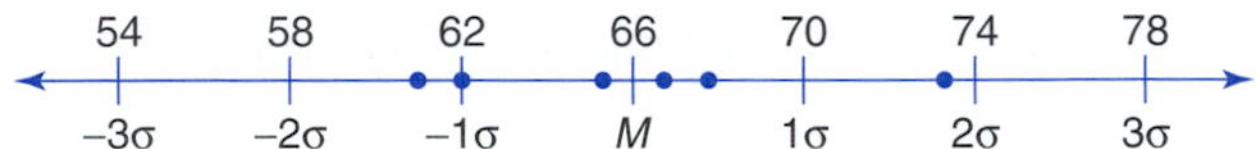

Our given set of data above contains too few numbers to include values in the two outside intervals under the normal curve. Sets of data containing many more values, however, would closely match the normal distribution. Industries using standard deviation for quality control would thus use larger groups of data in order to obtain dependable statistical results.

Calculator techniques . . . FOR EXAMPLE 2

Set up the problem using the format shown in the example. Then

61 [+] 62 [+] 65 [+] 67 [+] 68 [+] 73 [=] ⟶ 396 [÷] 6 ⟶ 66

Compute the Deviation and d^2 columns mentally; then calculate as follows to find their sum and take the square root (using the [$\sqrt{x}$] key):

25 [+] 16 [+] 1 [+] 1 [+] 4 [+] 49 [=] ⟶ 96 [÷] 6 ⟶ 16 [$\sqrt{x}$] ⟶ 4

Complete the line graph by hand.

SECTION 3 PROBLEMS

Use the percent categories associated with the normal curve in order to answer the following questions.

1. Assume that 75 was the mean test score on a finance exam, with a standard deviation of 5 points as shown below.

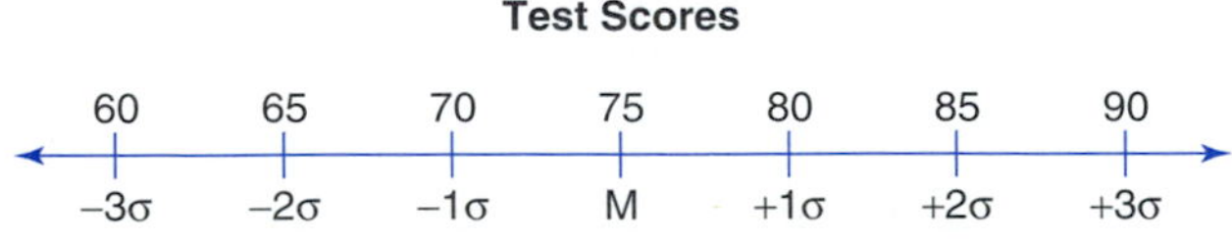

a. What percent of students had scores of 75 or higher?
b. What percent had 85 or higher?
c. What percent had scores between 65 and 75?
d. What percent had 65 or higher?
e. If 50 people took the exam, how many had scores between 60 and 70?
f. How many of the 50 people scored over 70?

2. Employees at Sparkman Electronics had the following number of engines fail inspection last month:

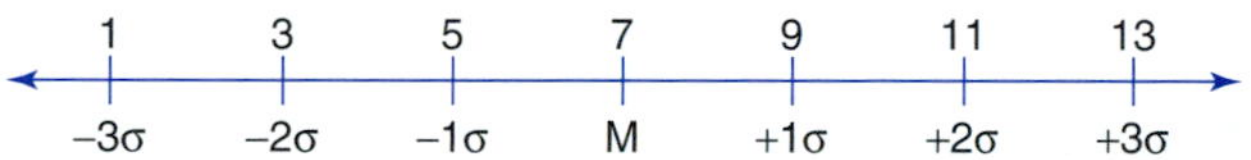

Assume that 7 was the mean and the standard deviation was 2. Then determine the following:
a. What percent of the employees had 7 or more failures?
b. What percent had 3 or fewer failures?
c. What percent had between 7 and 9 failures?
d. What percent had between 1 and 5 failures?
e. If there were 100 employees, how many had between 5 and 9 failures?
f. How many of the 100 had 7 failures or fewer?

3. Suppose that the mean temperature for Jacksonville for the month of August was 92° with a standard deviation of 2°. Prepare a line graph to show the temperatures associated with each deviation $\pm 1\sigma$, $\pm 2\sigma$, and $\pm 3\sigma$.
a. What percent of the days had temperatures of 88° or above?
b. What percent of the days had temperatures of 92° or below?
c. What percent of the days had temperatures between 90° and 94°?
d. What percent of the days had temperatures of 94° or above?
e. For the 31 days in August, how many days were 94° or above?
f. For the 31 days in August, how many days were 92° or below?
g. For the 31 days in August, how many days were between 90° and 94°?

4. Assume that the average individual frozen lasagna dinner contains 22 grams of fat with a standard deviation of 4 grams. Prepare a line graph to show the grams of fat associated with each deviation $\pm 1\sigma$, $\pm 2\sigma$, $\pm 3\sigma$.
a. What percent of the dinners have between 14 and 26 grams of fat?
b. What percent of the dinners have 14 grams or fewer?
c. What percent have 26 grams or more?
d. What percent have 18 grams or more?
e. If 6,000 dinners were manufactured, how many had 22 grams of fat or less?
f. How many of the 6,000 dinners have between 18 and 26 grams?
g. How many of the 6,000 dinners had 30 grams or above?

For the following sets of data:

(a) Compute each standard deviation.
(b) Prepare a line graph showing the value associated with each deviation $\pm 1\sigma$, $\pm 2\sigma$, and $\pm 3\sigma$.

(c) *Check each interval (between $\pm 1\sigma$, $\pm 2\sigma$, and $\pm 3\sigma$) to determine whether it contains the expected number of given values. If not,what does this mean?*

5. 8, 10, 12, 16, 19

6. 41, 45, 53, 54, 57

7. 28, 34, 35, 35, 40, 44

8. 26, 29, 29, 30, 30, 31, 32, 33

9. 31, 32, 35, 41, 44, 44, 46, 47

10. 55, 57, 58, 59, 59, 60, 60, 63, 64, 65

11. 23, 24, 25, 31, 31, 34, 36, 36

12. 68, 72, 72, 73, 73, 73, 74, 75, 75, 75

13. 14, 15, 16, 18, 20, 20, 23, 23, 25, 26

14. 74, 79, 80, 80, 81, 81, 82, 84, 84, 85

SECTION 4

INDEX NUMBERS

An **index number** is similar to a percent in both appearance and purpose. Like percent, index numbers use a scale of 100 in order to show the relative size between certain costs or to indicate change in costs over a period of time, rather than showing the actual, numerical differences. Whereas the actual dollars-and-cents changes in costs often could not be readily analyzed, both the index number and the percent allow the reader to make comparisons between items more easily. Index numbers are often used in business to depict long-range trends in prices or volume (of sales, production, and so forth).

The simplest index number that can be computed involves a single item. Such a simple price index is sometimes called a **price relative.** This index is simply the ratio of a given year's price to the base year's price, multiplied by 100:

$$\begin{matrix}\text{Index number} \\ \text{(or Price relative)}\end{matrix} = \frac{\text{Given year's price}}{\text{Base year's price}} \times 100$$

Example 1 Suppose that a half-gallon of milk cost \$2.50 in 1998, \$2.40 in 2000, and \$2.80 in 2003. Compute the price relatives (or index numbers) for the milk using 1998 as the base year.

YEAR	PRICE		PRICE INDEX (1998 = 100.0%)
1998	\$2.50	Base year	100.0
2000	2.40	$\frac{2.40}{2.50} \times 100 =$	96.0
2003	2.80	$\frac{2.80}{2.50} \times 100 =$	112.0

The student should realize that this index is nothing more than a simple percent, but written without the percent sign. The question "What percent of the 1998 price was the 2003 price?" would be answered "112%."

An excellent example of the use of index numbers is the Consumer Price Index, published monthly by the U.S. Department of Labor (Bureau of Labor Statistics). The costs of a large number of goods and services are indexed for the United States as a whole and for selected cities and urban areas individually.

TABLE 4-1 CONSUMER PRICE INDEX OF SELECTED EXPENSES IN SELECTED URBAN AREAS

(1982–84 = 100)

EXPENSE	U.S. URBAN AVERAGE	ATLANTA	CHICAGO	DALLAS	LOS ANGELES	NEW YORK	SEATTLE
Food	160.0	167.1	164.4	157.0	165.2	164.5	164.0
Housing	156.7	158.2	154.2	142.3	153.7	169.7	161.9
Utilities	128.4	142.8	119.9	127.4	145.2	114.3	121.7
Transportation	140.5	125.8	135.3	143.2	137.7	151.4	147.8
Medical care	241.4	241.9	249.3	233.4	234.1	255.3	232.8

Source: U.S. Department of Labor.

Example 2 **(a) Indicating Rise of Price in Each City Separately**

The index numbers in Table 4-1 show the "current" year-end costs of the major consumer expenses compared with the costs of these same expenses in *that* locality during the base years, 1982–84. These three years, 1982, 1983, and 1984, were averaged to give a composite base year. The same food that cost \$100 in Dallas in 1982–84 would cost \$157 at the present time. Similarly, housing that cost \$100 in Seattle in 1982–84 has now increased to \$161.90. (These indexes also may be read "Transportation that cost \$10 in Chicago in 1982–84 would now cost \$13.53.")

Various index numbers cannot be used for comparison among themselves unless they refer to the same base amount. For instance, Table 4-1 gives the price index for medical care in Atlanta as 241.9 and in New York as 255.3. This means that medical care that cost \$100 in Atlanta in 1982–84 would now cost \$241.90 and that medical care that cost \$100 in New York in 1982–84 would now cost \$255.30. It does *not* mean that \$100 in 1982–84 would have purchased the same medical services and drugs in both Atlanta and New York.

Thus, Table 4-1 does not necessarily indicate that medical care in New York costs more today than similar care in Atlanta; it simply indicates that the cost of medical care in New York has *increased* more rapidly since 1982–84 than have medical expenses in Atlanta (255.3 versus 241.9). If a medical procedure in New York had cost substantially less in 1982–84 than did the same procedure in Atlanta, then the New York medical procedure could still be less expensive than the Atlanta procedure (although New York medical fees have increased faster than have Atlanta fees).

Example 2 (cont.) **(b) Comparing Actual \$ Prices between Cities**

As Part (a) implies, the actual cost of living in Los Angeles compared with the cost of living in Chicago (or any other locality) cannot be determined from the

TABLE 4-2 COMPARATIVE INDEX OF SELECTED EXPENSES IN SELECTED URBAN AREAS AT INTERMEDIATE STANDARD OF LIVING

U.S. Urban Average = 100

EXPENSE	ATLANTA	CHICAGO	DALLAS	LOS ANGELES	NEW YORK	SEATTLE
Food	104.2	102.6	95.3	115.5	143.9	108.4
Housing	103.5	106.7	96.4	168.8	460.5	103.6
Utilities	94.8	98.0	96.7	103.4	176.6	72.2
Transportation	101.9	104.5	105.6	125.2	119.3	105.5
Medical care	105.6	98.3	102.7	121.1	184.3	140.3
Combined	102.2	105.1	98.6	127.8	231.3	101.8

Source: American Chamber of Commerce Research Association, 1998.

Consumer Price Index, but would have to be obtained from some other statistical source that contains an index of *comparative* living costs. Such an index, released periodically by the American Chamber of Commerce Researchers Association, measures price levels for consumer goods and services. The national average is set at 100, and each city's index is read as a percentage of the national average.

A typical index of this type is given in Table 4-2, based on the U.S. urban average of 100. This comparative index enables us to determine that, for every $100 that the average urban family spends, the Seattle family spends $101.80, the Dallas family spends $98.60, and the Chicago family $105.10—all to maintain the same intermediate standard of living. As a matter of interest, notice on the comparative index of transportation expenses that transportation in Chicago (104.5) is more expensive than in Atlanta (101.9), despite the fact that the Consumer Price Index showed transportation costs have increased more rapidly in Atlanta (140.5) than in Chicago (135.3).

Example 1 illustrated the computation of price relatives for only a single item—milk. Actually, the term "index number" is often reserved for a number derived by combining several related items, each weighted according to its relative importance. For instance, the index numbers for food expenses in Example 2 are a composite of many different grocery items, each weighted according to its portion of the total food budget in a typical family (as one consideration). There are many methods for constructing such composite index numbers, all of which require more statistical knowledge than can be acquired in basic business math. Even for the same data, these composite index numbers vary significantly according to the method used; hence, the best method to use depends on careful analysis of the purpose that the index is intended to serve. Obtaining an accurate and meaningful composite index number is not a simple matter. Our index number problems will therefore be limited to simple price relatives.

SECTION 4 PROBLEMS

Base your answers for Problems 1 and 2 on the Consumer Price Index in Table 4-1 and the Comparative Index in Table 4-2. (Include the appropriate index number with each answer.)

1. a. How much would a Seattle resident pay now for housing that cost $1,000 in the composite base period, 1982–84?

b. How much would a Chicago resident pay now for food that cost $100 in 1982–84?

c. How much does the average U.S. citizen pay today for transportation that cost $10 in 1982–84?

d. In which city are utility costs the highest?

e. In which city have utility costs risen fastest?

f. In which city have housing costs increased most slowly?

g. In which city have medical costs increased fastest?

h. In which city are food costs the most expensive?

i. In which city have transportation costs increased most slowly?

2. a. How much would a Los Angeles resident spend today for housing that cost $1,000 in the composite base period, 1982–84?

b. If a medical fee was $10 in 1982–84, how much should a Dallas resident expect to pay now for the same service?

c. How much does the average U.S. citizen pay today for utilities that cost $100 in 1982–84?

d. In which area have utility costs risen most slowly?

e. Where are food prices least expensive?

f. In which area are housing costs most expensive?

g. In which area are transportation costs least expensive?

h. In which area have medical costs risen most slowly?

i. In which area is medical care least expensive?

3. Calculate the price relative for a popular magazine, using 1998 as the base year.

YEAR	PRICE	INDEX (1998 = 100.0)
1998	$1.50	
1999	1.47	
2000	1.65	
2001	1.80	
2002	1.95	

4. Determine the price index for a package of gum, using 1997 as the base year.

YEAR	PRICE	INDEX (1997 = 100.0)
1997	$0.50	
1998	0.45	
1999	0.55	
2000	0.60	
2001	0.70	

5. Determine the price index for a can of soft drink, using 1990 as the base year.

YEAR	PRICE	INDEX (1990 = 100.0)
1990	$0.60	
1995	0.42	
1999	0.75	
2001	0.84	

6. Determine the price index for a candy bar, using 1996 as the base year.

YEAR	PRICE	INDEX (1996 = 100.0)
1996	$0.80	
1998	0.84	
2000	0.72	
2002	0.76	

SECTION 5

GRAPHS

Many people read business statistics without fully realizing their significance. Statistics usually make a much stronger impression if they can be visualized. This is particularly true when the data are used to make comparisons among several items or when the statistics give the various parts that comprise a whole. **Graphs** or **charts** are visual representations of numerical data. The purpose of business graphs is to present data in a manner that will be both accurate and easy to interpret. Today, businesses assemble, tabulate, and graph information through computer programs. These computer-generated graphs make a presentation visually appealing and interesting. For instance, a spreadsheet is used to assemble and tabulate a firm's cash flow over several months. From the same spreadsheet, the results can be presented in a variety of graph formats. The most common graphs are *bar graphs, line graphs,* and *circle graphs.*

All graphs have certain elements in common, whether they are computer generated or hand drawn. Every graph should have a title so that the reader will know what facts are presented. Every graph should show the origin (zero) and also some low value on the scale to help establish the scale for the reader. Very large figures may be rounded off before graphing.

Breaks in the scale, to eliminate unused space on the graph, should be used with care. The difference between amounts may appear much greater proportionally than they actually are; thus, a break can cause a distortion in the graph. Both of the accompanying bar graphs illustrate the same information. The break in the left bar graph creates an overemphasis on the difference between the net sales for the two years. The difference appears greater than it actually is; thus, a distortion is created. The vertical scale in the bar graph on the right corrects this distortion by eliminating the break and by using a lower value as the beginning number.

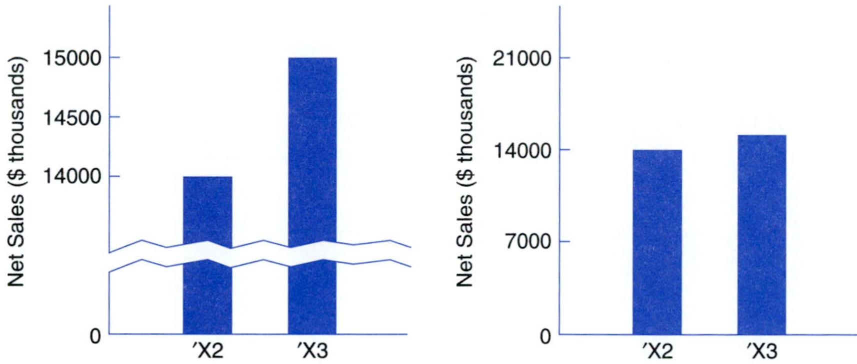

You are undoubtedly familiar with bar graphs, since they are so widely used. For this topic, we will therefore consider in detail only line graphs and circle graphs.

LINE GRAPHS

The line graph usually records change (in sales, production, income, and so forth) over a period of time. The only basic difference between it and the vertical bar graph is that, instead of having bars that come up to the various values, the values are connected by a line. The line may be broken or unbroken, straight or curved. Two or more related statistics are often presented on the same line graph, using additional lines.

Line graphs should ordinarily be drawn on graph paper. After the scale of values has been labeled (time should be on the horizontal scale), locate the appropriate values and mark each with a dot. Finally, connect the dots with a line. When the graph contains more than one line, be sure to make each line distinct and to identify each.

Example 1 Koenig Building Distributors, Inc. experienced the following totals of net sales, cost of goods sold, and net profit during a 5-year period. Depict these data on a line graph.

	20X1	20X2	20X3	20X4	20X5
Net sales	$180,000	$210,000	$160,000	$220,000	$237,000
Cost of goods sold	105,000	110,000	80,000	112,000	95,000
Net profit	25,000	32,000	8,000	10,000	42,000

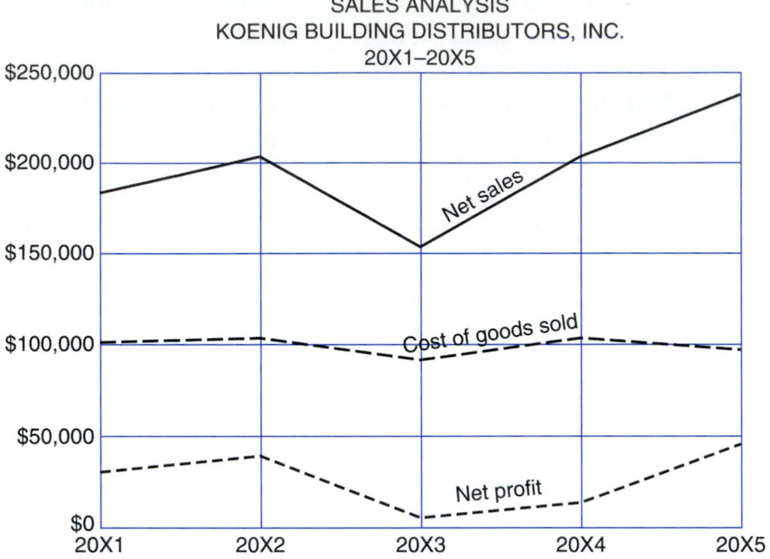

CIRCLE GRAPHS

The circle graph (also called a "pie chart") presents the breakdown of a whole into its various parts. For example, it may show what portion of a budget is spent for each item. Thus, the entire circle represents 100%, and each item should occupy an area proportional to its percentage of the total.

Before preparing the graph, convert each item to its corresponding percent of the total (if this is not given). Multiply each percent times 360° to determine the number of degrees required for each item. Use a compass to draw the circle, and use a protractor to measure the number of degrees needed for each part of the graph.

Example 2 The $200,000 in total assets of Straiton Parts Co. are distributed on December 31, 20XX, as listed below. Represent these assets on a circle graph.

CURRENT ASSETS		PLANT ASSETS	
Cash	$14,000	Plant	$40,000
Accounts receivable	36,000	Land	12,000
Inventory	70,000	Equipment	28,000

To prepare a circle graph, you must first convert each item to a rate (percent) of the total assets by answering the question, "What percent of the total assets is each separate item?"

$$\_\_\_\_\%\text{ of total assets} = \text{Each item}$$
$$\_\_\_\_\%\text{ of }\$200{,}000 = \text{Each item}$$
$$200{,}000r = \text{Each item}$$
$$r = \frac{\text{Each item}}{\$200{,}000}$$

Thus,

$$\text{Cash} = \frac{\$14{,}000}{\$200{,}000} = 7\%; \quad \text{Plant} = \frac{\$40{,}000}{\$200{,}000} = 20\%; \quad \text{etc.}$$

Each asset would require the following number of degrees:

Cash	7%	of	360°	=	25.2°	or	25°
Accounts receivable	18%	of	360°	=	64.8°	or	65°
Inventory	35%	of	360°	=			126°
Plant	20%	of	360°	=			72°
Land	6%	of	360°	=	21.6°	or	22°
Equipment	+ 14%	of	360°	=	50.4°	or	+ 50°
	100%						360°

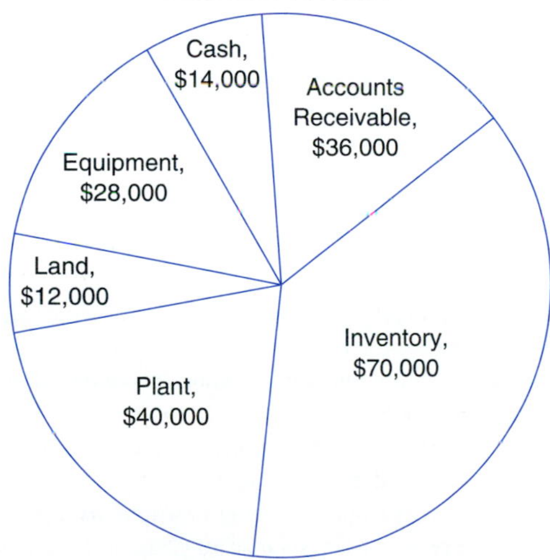

SPREADSHEETS AND FINANCIAL DECISIONS

Businesses frequently use computer **spreadsheets** to organize and calculate data to help them make financial decisions. Just as the multicolumn worksheet is a valuable tool for the accountant at the end of the fiscal period, the spreadsheet is used to gather, calculate, and analyze data. A spreadsheet is composed of columns (vertical) and rows (horizontal). Columns are labeled by letters (A, B, C or a, b, c, and so on), and rows are labeled by numbers (1, 2, 3, and so on). The intersection of a column and a row is referred to as a **cell.** The first cell on a spreadsheet is labeled A1 or a1 (cell location, address, or reference). In the following illustration, Sales is located in A1, and the amount ($100,000) is located in B1.

	A	B	C
1	Sales	$100,000	
2	Sales returns	20,000	
3	Net sales		$80,000

After the data have been assembled on the spreadsheet, a graph or chart can be generated from the spreadsheet to present the information visually. The problems in Section 5 can be solved manually or with an Excel® spreadsheet. The computer disk that accompanies this textbook contains files to be used in solving the problems. Directions for spreadsheets and charts are provided in each problem if you wish to use the spreadsheet approach. First, we should define some terms.

TERM	ACTION TO BE TAKEN OR DEFINITION
Chart Wizard	A shortcut method of creating charts from a spreadsheet.
Click	Press the left mouse button.
Click and drag	Hold down the left mouse button while moving the cursor through the text, which will highlight or select the text.
Close a folder or file	Click the *Close* button on the File Menu. The folder or file is removed from the screen.
Desktop	The opening screen or window containing icons of or shortcuts to application software. It is the background for your computer work.
Dialog Box	A box or window containing commands or menu options such as choosing a filename for saving a document.
Disk	A storage device. Save your documents on a 3½″ disk that is normally inserted in *Drive A* of the computer.
File	A document such as a business letter or a spreadsheet saved on a disk.
Folder	A place where related files, such as all Chapter 4 problems, can be stored together.
Icon	A button or an image on a tool bar or the desktop.
Open a folder or file	Click the folder or file; then click the *Open* button. The folder or file is then active, and you can see what it contains. You can also double-click on a file or folder name to open it.
Print	Click the *Print* button on the File Menu. A hard copy of the file will be printed.
Save	Click the *Save* button on the File Menu. You must specify on which drive you wish to save the data.
Tool Bar	A vertical or horizontal bar containing buttons or icons that a user clicks with the mouse to initialize a computer application or a task.
Tool Tip	A message that appears when the mouse pointer is placed on top of an icon or a button. The Tool Tip identifies each icon or button.

SECTION 5 PROBLEMS

1. Use a bar graph to depict the mean salaries for office managers in the various cities shown below:

Athens	$36,000	Florence	$41,000
Calhoun	32,000	Greenville	40,000
Clifton	48,000	Hoover	46,000
Dothan	38,000		

The *Open* Dialog Box for Excel® is shown in Figure 4-1. This box will appear on the computer monitor when the *Open* button from the File drop-down menu on the Menu Bar is clicked with the mouse (Instructions 1, 2, and 3 below). When the *Chapter 4* folder is opened (Instruction 4), the computer monitor will display the *Open* Dialog Box shown in Figure 4-2.

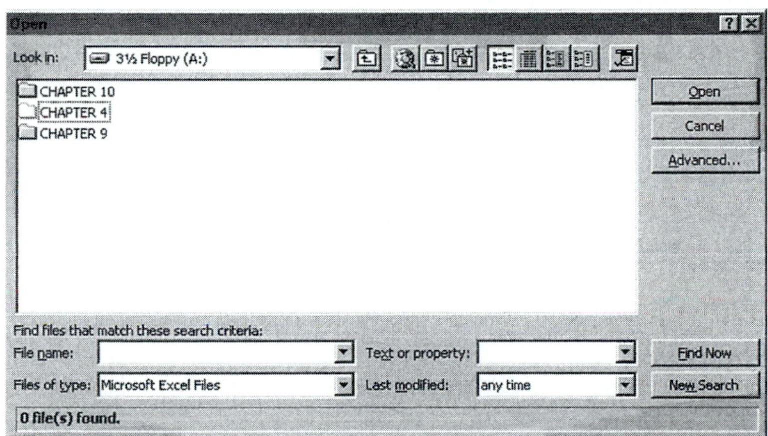

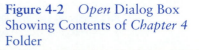

Figure 4-1 *Open* Dialog Box for Excel Showing Three Folders

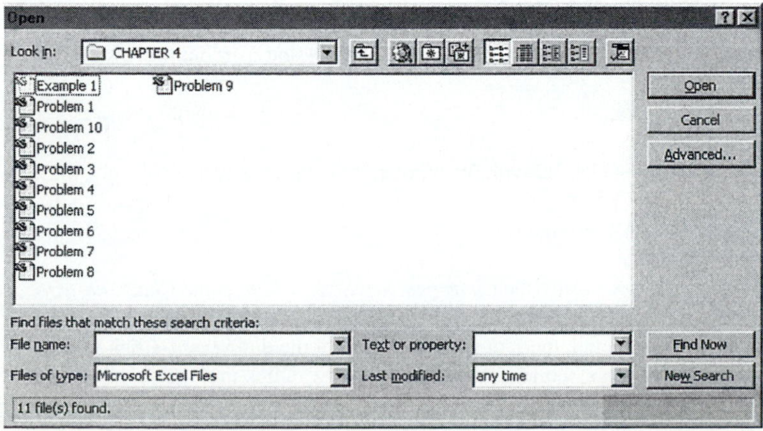

Figure 4-2 *Open* Dialog Box Showing Contents of *Chapter 4* Folder

EXCEL® SPREADSHEET DIRECTIONS

1. Open the Excel program on your computer.
2. From the File Menu, click the *Open* button.
3. At the *Open* Dialog Box, change the *Look-in* Box to *Drive A.*
4. Click the *Chapter 4* folder. Click the *Open* button.
5. Click the *Problem 1* file. Click the *Open* button.
6. Key in the Athens salary in Cell B2. Do not key in the dollar sign or comma. The cells have been formatted to display the amounts with dollar signs and commas. Continue keying the remaining salaries in Column B. Use the *Enter* key to move to B3.
7. Select Cells A2 through B8 by clicking the left mouse button and dragging from Cell A2 through Cell B8. Cell A2 will not be highlighted, but the other cells will be.
8. On the Standard Tool Bar, click the *Chart Wizard* icon. Move the mouse pointer over the icons on the Standard Tool Bar until the tool tip displays "Chart Wizard."
9. In the *Chart Wizard—Step 1 of 4—Chart type* Dialog Box, click *Bar* for the *Chart type,* and click *Stacked Bar* (second column, first row). Click the *Next* button. Click the *Next* button in the *Step 2 of 4* Dialog Box.
10. In the *Chart Wizard—Step 3 of 4—Chart Options* Dialog Box, click the *Titles* tab at the top of the box. Click in the *Chart title* Box, and key in "Mean Salaries for Office Managers." Click the *Legend* tab at the top of the box; click the *Show legend* Box to remove the check mark. Click the *Next* button.
11. In the *Chart Wizard—Step 4 of 4—Chart location* Dialog Box, click the *As new sheet.* Click the *Finish* button.
12. Save the spreadsheet and chart.
13. At the bottom of the screen are located several tabs. *Chart 1* is currently highlighted and active. Print the chart by clicking the *Print* icon on the Standard Tool Bar. Click *Sheet 1* tab at the bottom of the window; click *Print.*
14. Close the file. Do not close Excel if you plan to continue working problems.

2. Use a bar graph to depict the number of seniors in Charles County high schools who were accepted for college admission in the year 200X.

Dooley	115	Madison	220
Jefferson	110	Washington	160
Livingston	180	Woodlawn	200

EXCEL® SPREADSHEET DIRECTIONS

1. Open the Excel program (if not already open) on your computer.
2. From the File Menu, click the *Open* button.
3. At the *Open* Dialog Box, change the *Look-in* Box to *Drive A* if necessary. Steps 4 and 5 are not necessary if you have not exited from Excel.
4. Click the *Chapter 4* folder. Click the *Open* button.
5. Click the *Problem 2* file. Click the *Open* button.
6. Key in the numbers for each school, beginning in Cell B2 with the number 115. Move between cells in Column B with the *Enter* key.
7. Select the cells by clicking and dragging Cells A2 through B7.
8. Click the *Chart Wizard* on the Standard Tool Bar.
9. In the *Chart Wizard—Step 1 of 4—Chart type,* click *Bar* for the *Chart type* and *Stacked Bar* as the *Chart sub-type* (Column 2, Row 1). Click the *Next* button in this dialog box and again in the *Step 2 of 4* Dialog Box.
10. From *Chart Wizard—Step 3 of 4—Chart options,* click the *Titles* tab at the top of the dialog box. Click in the *Chart title* Box, and key in "Charles County Seniors, College Acceptance, 200X." Click the *Legend* tab, and click the *Show legend* Box to remove the legend from the chart. Click the *Next* button.

(*continued*)

EXCEL® SPREADSHEET DIRECTIONS

11. From *Chart Wizard—Step 4 of 4—Chart location,* click *As new sheet.* Click the *Finish* button.
12. Save. Print the chart and the spreadsheet. See Problem 1, Step 13.
13. Close the file.

3. Create a line graph for Coleman Interiors showing net sales, cost of goods sold, and gross profit for 3 consecutive years.

	1999	2000	2001
Net sales	$180,000	$200,000	$220,000
Cost of goods sold	85,000	90,000	75,000
Gross profit	95,000	110,000	145,000

EXCEL® SPREADSHEET DIRECTIONS

1. Open Excel; open the *Chapter 4* folder; open *Problem 3.* See Problems 1 and 2 if necessary.
2. Key the 1999 amounts in Column B, beginning with 180000 in Cell B3, 85000 in B4, and 95000 in B5. Key the 2000 amounts in Column C and the 2001 amounts in Column D. These cells have been formatted with dollar symbols and commas.
3. Click and drag Cells A2 through D5 to select the cells.
4. Click the *Chart Wizard* on the Standard Tool Bar.
5. In the *Chart Wizard—Step 1 of 4—Chart type* Dialog Box, click *Line* as the *Chart type,* and click *Line with markers* (Column 1, Row 2). Click the *Next* button in this dialog box and the *Step 2 of 4* Dialog Box.
6. In the *Chart Wizard—Step 3 of 4—Chart options* Dialog Box, click the *Titles* tab; click the *Chart title;* and key in "Coleman Interiors, Inc." Click the *Legend* tab; be sure that a check mark appears beside the *Show legend* button so that the legend will show. Click the *Next* button.
7. From the *Chart Wizard—Step 4 of 4—Chart location* Dialog Box, click *As new sheet.* Click the *Finish* button.
8. Save and print. See Problem 1 directions.
9. Close the file.

4. Construct a line graph for Solomon Repair Services showing its current assets, current liabilities, and working capital for 3 years.

	1999	2000	2001
Current assets	$200,000	$206,000	$210,000
Current liabilities	150,000	178,000	170,000
Working capital	50,000	28,000	40,000

EXCEL® SPREADSHEET DIRECTIONS

1. Open Excel if necessary. Open the *Chapter 4* folder. Open *Problem 4.* (See Problem 1, Steps 1–5.)
2. Key the 1999 figures in Column B, beginning with 200000 in Cell B3. Key the 2000 figures in Column C, beginning with 206000 in Cell C3. Key the 2001 figures in Column D, beginning with 210000 in Cell D3. Do not key in the dollars and commas. The cells have been formatted.

(*continued*)

EXCEL® SPREADSHEET DIRECTIONS

3. Click and drag Cells A2 through D5 to select the cells for the chart. Click the *Chart Wizard* on the Standard Tool Bar.
4. From the *Chart Wizard—Step 1 of 4—Chart type* Dialog Box, click *Line* as the *Chart type* and *Line with markers* (Column 1, Row 2) as the *Chart sub-type.* Click the *Next* button in this dialog box and the *Step 2 of 4* Dialog Box.
5. From the *Chart Wizard—Step 3 of 4—Chart options* Dialog Box, click the *Titles* tab; click the *Chart title* Box, and key "Solomon Repair Services." Click the *Legend* tab; *Show legend* should be checked to show the legend in the chart. Click the *Next* button.
6. From the *Chart Wizard—Step 4 of 4—Chart location* Dialog Box, click *As new sheet.* Click the *Finish* button.
7. Save, print, and close.

5. Prepare a line graph to show the number of new policies issued by American Insurance Co. each year during a 10-year period. Plot these policies on the same line graph.

YEAR	TERM	UNIVERSAL	LIMITED PAYMENT
1993	27,000	70,000	42,000
1994	29,500	72,000	45,000
1995	30,000	76,500	38,000
1996	30,000	78,000	39,500
1997	35,500	82,500	39,500
1998	33,000	96,000	40,500
1999	32,000	88,000	43,000
2000	31,000	88,500	41,000
2001	34,500	90,000	40,500
2002	36,500	92,500	40,000

EXCEL® SPREADSHEET DIRECTIONS

1. Open Excel if necessary. Open the *Chapter 4* folder. Open *Problem 5.*
2. Key the numbers of policies in the appropriate cells for the types of policies sold during each of the 10 years.
3. Click and drag Cells A2 through D12 to select the cells.
4. Click *Chart Wizard* on the Standard Tool Bar.
5. At the *Step 1 of 4—Chart type* Dialog Box, click *Line* for the *Chart type.* Click *Line with markers* (Column 1, Row 2) for the *Chart sub-type.* Click the *Next* button.
6. At the *Step 2 of 4—Chart source data* Dialog Box, click the *Series* tab at the top of the box. Under *Series,* click *Year;* click the *Remove* button. In the *Category (X) axis* labels Box, key in exactly "={1993,1994,1995,1996,1997,1998,1999,2000,2001,2002}." Put commas but no spaces between the years. Click the *Next* button.
7. At the *Step 3 of 4—Chart options* Dialog Box, click the *Titles* tab; click the *Chart title* Box; and key "American Insurance Co." Click the *Legend* tab; there should be a check mark beside the *Show legend* Box. Click the *Next* button.
8. At the *Step 4 of 4—Chart location* Dialog Box, click *As new sheet.* Click the *Finish* button.
9. Save, print, and close.

6. Prepare a line graph to show the July expenses, 20X1 for Brown Company for the various departments shown below.

	SALARIES	DEPRECIATION	UTILITIES	SUPPLIES
Department A	$60,000	$15,000	$3,500	$5,000
Department B	70,000	25,000	3,000	3,000
Department C	50,000	12,000	2,600	4,000

EXCEL® SPREADSHEET DIRECTIONS

1. Open Excel if necessary. Open the *Chapter 4* folder. Open *Problem 6.*
2. Key the salaries in Column B, beginning with 60000 in Cell B2. Key the remaining amounts for depreciation in Column C, utilities in Column D, and supplies in Column E. Remember, the dollars and commas should not be keyed.
3. Select Cells A2 through E5. Click the *Chart Wizard.*
4. At the *Step 1 of 4—Chart type* Dialog Box, click *Line* as *Type* and *Line with marker* (Column 1, Row 2) for *Chart sub-type.* Click the *Next* button. Click the *Next* button in the *Step 2 of 4* Dialog Box.
5. At the *Step 3 of 4—Chart options* Dialog Box, click the *Title* tab; key "Brown Company, July Expenses, 20X1" in the *Chart title* Box. Click the *Legend* tab; be sure that a check mark appears beside *Show legend.* Click the *Next* button.
6. At the *Step 4 of 4—Chart location* Dialog Box, click *As new sheet.* Click *Finish.*
7. Save, print, and close.

7. Present a circle graph or a pie chart from a spreadsheet showing the federal budget income sources. A federal dollar is derived from the following sources:

Individual income taxes	$0.48
Social Security/Medicare	0.34
Corporate income taxes	0.10
Excise taxes	0.04
Other fees	0.04

EXCEL® SPREADSHEET DIRECTIONS

1. Open Excel. Open the *Chapter 4* folder. Open *Problem 7.*
2. Key in 0.48 in Cell B2; continue keying the amounts in Column B.
3. Select Cells A2 through B6. Click *Chart Wizard.*
4. At the *Step 1 of 4—Chart type* Dialog Box, click *Pie* for the *Chart type* and *Pie with a 3-D visual effect* (Column 2, Row 1) under the *Chart sub-type.* Click the *Next* button. At the *Step 2 of 4* Box, click the *Next* button.
5. At the *Step 3 of 4—Chart options* Dialog Box, click the *Title* tab; click the *Title* Box; and key "Federal Budget Income Sources." Click the *Legend* tab; click *Show legend* to remove the check mark. Click the *Data labels* tab; click *Show label and percent.* Be sure that there is a check mark beside the *Show leader lines* Box at the bottom of the box. Click the *Next* button.
6. At the *Step 4 of 4—Chart location* Dialog Box, click *As new sheet.* Click the *Finish* button.
7. When the chart displays, click one data label. Click it a second time, holding the mouse button down, and drag it away from the pie chart. *Note:* This is not a double-click; there should be a 1-second pause between clicks. Click each of the remaining labels only once, holding the mouse button down, and drag each label away from the chart.
8. Save, print, and close.

8. Present a circle graph or a pie chart from a spreadsheet showing the federal government spending. A representative federal dollar is spent on the following items:

Direct benefits	$0.51
Miscellaneous programs	0.16
National defense	0.15
Interest on the debt	0.12
Surplus	0.06

EXCEL® SPREADSHEET DIRECTIONS

1. Open Excel if necessary. Open the *Chapter 4* folder. Open *Problem 8.*
2. Key the amounts, beginning with 0.51 in Cell B2.
3. Select Cells A2 through B6. Click the *Chart Wizard.*
4. At the *Step 1 of 4—Chart type* Dialog Box, click *Pie* for *Type* and *Pie with 3-D visual effect* (Column 2, Row 1) under *Chart sub-type.* Click the *Next* button. Click the *Next* button at *Step 2 of 4* Dialog Box.
5. At the *Step 3 of 4—Chart options* Dialog Box, click the *Title* tab and key "Federal Budget Expenditures" as the *Title.* Click the *Legend* tab; click *Show legend* to remove the legend from the chart. Click the *Data labels* tab; click *Show label and percent.* Be sure that *Show leader lines* is checked. Click the *Next* button.
6. At the *Step 4 of 4—Chart location* Dialog Box, click *As new sheet.* Click *Finish.*
7. When the chart displays, click each data label to select, and drag the labels away from the pie. See Step 7 of Problem 7.
8. Save, print, and close.

9. Wadge Industries had the following operating expenses last year. Construct a circle graph or a pie chart from a spreadsheet to compare these expenses.

Wages	$126,000
Utilities	32,000
Depreciation	20,000
Supplies	18,000
Taxes	5,000
Other	3,000

EXCEL® SPREADSHEET DIRECTIONS

1. Open Excel if necessary. Open the *Chapter 4* folder. Open *Problem 9.*
2. Key the amounts in Column B, beginning with 126000 in Cell B2. Do not key the dollar signs or commas.
3. Select Cells A2 through B7. Click the *Chart Wizard.*
4. At the *Step 1 of 4—Chart type* Dialog Box, click *Pie;* click *Pie with a 3-D visual effect* (Column 2, Row 1). Click the *Next* button. Click the *Next* button at the *Step 2 of 4* Dialog Box.
5. At the *Step 3 of 4—Chart options* Dialog Box, click *Chart title* and key "Wadge Industries." Click the *Legend* tab, and click *Show legend* to remove the legend. Click the *Data labels* tab, and click *Show label and percent.* Be sure that *Show leader lines* is checked. Click *Next.*
6. At the *Step 4 of 4—Chart location* Dialog Box, Click *As new sheet.* Click *Finish.*
7. When the chart displays, click each data label to select and drag the labels away from the chart. See Step 7 of Problem 7.
8. Save, print, and close.

10. The current assets of Bostick Technologies for the first quarter of the year are shown below. Construct a circle graph or a pie chart from a spreadsheet to compare these assets.

Cash	$150,000
Notes receivable	100,000
Accounts receivable	75,000
Supplies	4,500
Merchandise inventory	320,000

EXCEL® SPREADSHEET DIRECTIONS

1. Open Excel. Open the *Chapter 4* folder. Open *Problem 10.*
2. Key the amounts in Column B, beginning with 150000 in Cell B2. Do not key the dollar symbols or commas.
3. Select Cells A2 through B6. Click the *Chart Wizard.*
4. At the *Step 1 of 4—Chart type* Dialog Box, click *Pie* and *Pie with a 3-D visual effect* (Column 2, Row 1). Click the *Next* button. Click the *Next* button at the *Step 2 of 4* Dialog Box.
5. At the *Step 3 of 4—Chart options* Dialog Box, key "Bostick Technologies" in the *Chart title* Box. Click the *Legend* tab, and click *Show legend* to remove the legend. At the *Data label* Box, click *Show label and percent. Show leader lines* should be checked. Click the *Next* button.
6. At the *Step 4 of 4—Chart location* Dialog Box, click *As new sheet.* Click the *Finish* button.
7. When the chart displays, drag each data label away from the pie. See Step 7 of Problem 7.
8. Save, print, and close.

11. Find examples of bar, line, and circle graphs in recent periodicals. What information is depicted in the graphs? Using the data in one of the graphs, describe in narrative form the information that is shown.

12. Choose a graph from a periodical. Using the same data, construct a graph different from that depicted in the periodical. In other words, if the periodical depicts a bar graph, use the same data to construct a line or circle graph.

CHAPTER  GLOSSARY

Arithmetic mean. (See "Mean.")

Array, ordered. A listing of numerical data in order of size.

Average. A typical value or measure of central tendency for a group of numbers. (Examples: Mean, median, and mode.)

Central tendency, measure of. A representative number around which the other numbers in a group seem to cluster; an average.

Chart. (See "Graph.")

Classes. The groups (or intervals) into which given values are categorized in a frequency distribution.

Dispersion, measure of. A value that gives some idea of how spread out (or how scattered from the mean) a group of numbers are. (Examples: Range and standard deviation.)

Frequency. The number of values within a class interval of a frequency distribution.

Frequency distribution. A process for classifying large numbers of values and tabulating the data in groups (classes) rather than individually; used to compute measures of central tendency and/or dispersion.

Grade-point average. A weighted scholastic average where each letter grade is assigned a quality-point value that is multiplied by the number of credit hours that the course earns. This total, divided by the total hours, gives the grade-point average (or average per credit hour).

Graph (or chart). A visual representation of numerical data.

Index number. A number based on a scale of 100 which shows the relative size between certain costs or indicates change in costs over a period of time. (Example: If 1997 = 100 and the current index is 115, then \$115 is now required to purchase an item that cost \$100 in 1997.)

Interval. On a frequency distribution, the difference between the lower limit of one class and the lower limit of the next class.

Mean, arithmetic. The most common average; the sum of all given values divided by the number of values.

(Example: $\frac{7 + 5 + 6}{3} = \frac{18}{3} = 6$)

Median. The midpoint (or middle value) of a group of numbers; the central value of an array. (Examples: Given 1, 2, 3, 7, and 9, the median is 3; given 1, 2, 3, 7, 9, and 10, the median is

$\frac{3 + 7}{2} = \frac{10}{2} = 5$.)

Midpoint. Of a class in a frequency distribution, the arithmetic mean of the lower limit of that class and the lower limit of the next class.

Mode. The value in a set of data that occurs most frequently. (Example: Given 1, 3, 3, 5, 5, 5, 7, 9, and 10, the mode is 5.)

Modified mean. An arithmetic mean computed after the extreme high and low values have been deleted.

Normal (or bell) curve. A symmetrical, bell-shaped graph associated with the expected distribution for many types of data; the graph for which standard deviation applies.

Price relative. A simple price index for one item:

$$= \frac{\text{current price}}{\text{base price}} \times 100$$

Range. The difference between the highest and lowest values in a group of numbers (or between the upper and lower limits in a frequency distribution); a simple measure of dispersion.

Standard deviation (σ). For a normal curve distribution, the deviation (difference) between the mean and two crucial points symmetrical to the mean, within which interval fall 68% of all the given values.

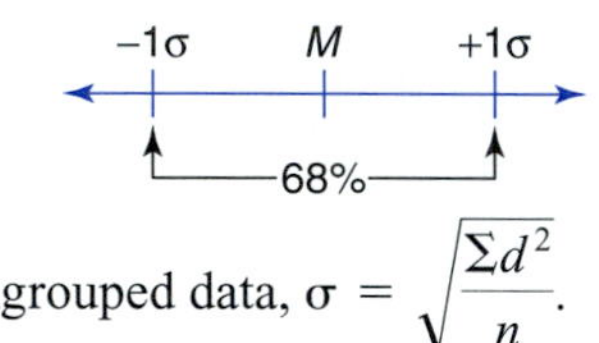

For ungrouped data, $\sigma = \sqrt{\frac{\Sigma d^2}{n}}$.

Statistics. Informally, the process of collecting, organizing, tabulating, and interpreting numerical data; also, the information presented as a result of this process.

Variability. Dispersion (or scatteredness) of data.

Weighted mean. A mean computed for data whose importance differs and where each value is weighted (or multiplied) according to its relative importance.

CHAPTER 5

TAXES

OBJECTIVES

Upon completion of Chapter 5, you will be able to:

1. Define and use correctly the terminology associated with each topic.
2. Compute sales tax based on a percentage rate (Section 1: Examples 1, 2; Problems 1–6, 11–24).
3. Find the marked price of an item, given the sales (or other) tax rate and the total tax or the total price (Section 1: Examples 2–4; Problems 1, 2, 7–22).
4. Express a given (decimal) property tax rate (Section 2: Examples 1, 2; Problems 1–6):
   a. As a percent
   b. As a rate per $100
   c. As a rate per $1,000
   d. In mills.
5. Calculate the property tax rate of a tax district (Section 2: Example 1; Problems 5, 6).
6. Using the property tax formula $R \cdot V = T$ and given two of the values, determine the unknown (Section 2: Examples 3–7; Problems 7–22):
   a. Property tax
   b. Tax rate
   c. Assessed value of property.
7. Determine the change in the property tax rate or the assessed value from one year to another (Section 2: Example 7; Problems 23–26).

Increased government activity (at the federal, state, and local levels) has made taxes—personal as well as business—an integral part of modern life. Two of the most common taxes are the sales tax and the property tax. The sales tax has become an everyday part of commerce. Property owners, including business organizations, must pay a property tax on their possessions. The following sections present the calculation of sales tax and property tax. (A third important tax is income tax. However, any significant study of income tax would require an entire course by itself and could not be accomplished in one unit of a business math course. Chapter 8, "Wages and Payrolls," discusses the taxes that must be deducted from payrolls.)

SECTION 1

SALES TAX

A large majority of state legislatures have approved **taxes on retail sales** as a means of securing revenue. Retail merchants are responsible for collecting these taxes at the time of each sale and relaying them to the state. In some states the legislatures have also allowed their cities the option of adopting a municipal sales tax in addition to the state tax in order to obtain local funds. Sales tax is computed as a specified percent of the selling price, with the percent usually varying from 4% to 8%.

Sales taxes historically have applied only to sales made within the jurisdiction of the taxing body. Thus, merchandise sold outside a state (such as by mail or by telephone) has not usually been subject to that state's sales tax. However, merchandise brought into another state for use in the state is usually subject to a **use tax.** For instance, merchandise sold in Maryland by telephone to a Virginia resident is not subject to the Maryland state sales tax but is subject to the Virginia use tax. All states imposing sales taxes also impose use taxes. Credit is given for sales tax paid to other

Illinois Department of Revenue **Use Tax Return**

Do not use this form for motor vehicle or aircraft purchases.

Please type or print using black ink. **Make remittance payable to:** Illinois Department of Revenue **Mail to:** Illinois Department of Revenue, Retailers' Occupation Tax, Springfield IL 62796-0001

Purchaser's name: James S. Thomas
Street address: 106 Quincy Street
City: Cairo | State: IL | ZIP code: 62914
County: Alexander | (area code) Telephone number: 618) 466-2123

Item(s) description	Purchase price	Purchase date
Sofa/chair	$1,500.00	6 mo. / x6 year

Col. 1) Tax Due	Col. 2) Tax Paid at Purchase	Col. 3) Net Tax Due (710)
$75.00	0	$75.00

Official Use
I.B.T. No.
0775-0005

Complete (A) if you have an FEIN.
Complete (B) if (A) is not applicable.

(A) Federal Employer Identification Number

(B) Social Security Number: 1 2 3 4 5 6 7 8 9

This form is authorized as outlined by the Illinois Use Tax Act. Disclosure of this information is REQUIRED. Failure to provide information could result in a penalty. This form has been approved by the Forms Management Center. IL-492-1880

Under penalties of perjury and other penalties as provided by law which include a fine, or imprisonment, or both, I declare that this return and statements, to the best of my knowledge and belief are true, correct and complete.

RUT-44 (R-11/86) Signature: X James S. Thomas Date: 4/1/x7

Official Use

Figure 5-1

states, although there are a few states that do not provide credit for sales tax on motor vehicle purchases in other states. Many consumers are unaware of the use tax—they think their out-of-state purchase simply is not taxed. The state collects the use tax from the out-of-state vendor or from the purchaser. Because the purchase is difficult to track—and thus the use tax is difficult to collect—many states do lose tax revenue. That is one of the reasons many state legislatures are considering passage of sales taxes on all retail purchases regardless of the state of residence of the buyer. The Urban Institute estimates that only ten percent of the combined sales and use tax revenue comes from the use tax. Figure 5-1 shows a state's use tax return.

Example 1 Watts Plumbing Equipment Inc. made the following sales, subject to a 6% sales tax rate: (a) $40.10 and (b) $405.00. Determine the sales tax and the total amount collected for each sale.

(a) To find the sales tax, use the basic percent formula,
_____% of _____ = _____.

$$6\% \times \text{Price} = \text{Tax}$$

$$0.06 \times \$40.10 = t$$

$$\$2.41 = t$$

The total amount due on the first sale is

$$\begin{aligned} \text{Marked price} &= \ \ \$40.10 \\ \text{Sales tax} &= \underline{+\ \ 2.41} \\ \text{Total price} &= \$\ 42.41 \end{aligned}$$

(b) To find the sales tax for the second sale,

$$6\% \times \text{Price} = \text{Tax}$$

$$0.06 \times \$405 = t$$

$$\$24.30 = t$$

The total amount due on the sale is

$$\begin{aligned} \text{Marked price} &= \ \ \$405.00 \\ \text{Sales tax} &= \underline{+\ \ \ 24.30} \\ \text{Total price} &= \ \ \$429.30 \end{aligned}$$

Example 2 A 5% sales tax is charged on all retail sales made by Sun Country Pool Supplies. (a) How much tax must be charged on a $34.85 sale? (b) The sales tax on a case of pool chlorine was $2.85. What was the price of the case of chlorine to the nearest dollar?

(a) We will use the basic percent formula, _____% of _____ = _____.

$$5\% \times \text{Price} = \text{Tax}$$

$$0.05 \times \$34.85 = t$$

$$\$1.74 = t$$

The sales tax due is $1.74.

(b) The same formula is used to find the price when the sales tax is known. Since the tax was $2.85,

$$5\% \times \text{Price} = \text{Tax}$$

$$5\% \times p = \$2.85$$

$$0.05p = 2.85$$

$$p = \frac{2.85}{0.05}$$

$$p = \$57$$

The case of chlorine was priced at $57; thus, the total price, including the sales tax, was $59.85.

Example 3 The total price of a camera, including a 6% sales tax, was $137.80. What was the marked price of the camera?

$$\text{Marked price} + \text{Sales tax} = \text{Total price}$$

$$\text{Price} + 6\% \text{ of price} = \text{Total price}$$

$$p + 0.06p = \$137.80$$

$$1.06p = 137.80$$

$$p = \frac{137.80}{1.06}$$

$$p = \$130$$

The marked price of the camera was $130.

Note. Observe that the sales tax is 6% of the marked price, *not* 6% of the total price. Therefore, problems of this type *cannot* be solved by taking 6% of the total price and subtracting it to find the marked price. The results would be close, but they would differ by a few cents; hence, they would be wrong. (You are invited to try a few examples if you are still skeptical! The only exception is for small amounts under $2.00.)

Excise taxes are levied on items to help limit the purchase of potentially harmful products such as firearms, or to help pay for products or services used only by certain people, such as highways financed by gasoline taxes. Other examples of excise taxes include taxes on automobiles, air transportation, telephone service, sporting and recreational equipment, alcoholic beverages, tobacco, entertainment, and business licenses.

A similar tax levied by the federal government is a **customs duty, import tax,** or **tariff.** These are taxes levied on products brought into the United States from outside the country. The purpose of these taxes is to protect American business against unfair foreign competition. Over the years, bilateral and multilateral agreements such as NAFTA (Northern American Free Trade Agreement) and GATT (General Agreement on Tariffs and Trade) have eliminated or reduced many of these taxes.

Another federal tax is the **luxury tax.** This 10% tax is levied on automobiles whose market price exceeds a certain amount. The threshold amount, which began at $32,000 in 1993, is adjusted annually for inflation. This tax also had been applied to the purchase of furs, jewelry, boats, and airplanes, but Congress repealed the tax on these items in 1993. Luxury taxes will not be included in the problems in this text.

Example 4 The $52.65 total price of a professional hockey ticket includes a 7% sales tax and a 10% federal excise tax. What was the marked price of the ticket?

$$\text{Marked price} + \text{Sales tax} + \text{Excise tax} = \text{Total price}$$

$$\text{Price} + 7\% \text{ of price} + 10\% \text{ of price} = \$52.65$$

$$p + 0.07p + 0.10p = 52.65$$

$$1.17p = 52.65$$

$$p = \frac{52.65}{1.17}$$

$$p = \$45$$

SECTION 1 PROBLEMS

Compute the missing items.

	MARKED PRICE	SALES TAX RATE	SALES TAX	TOTAL PRICE
1.	a. $15.75	4%		
	b. 66.00	8		
	c.	6	$1.95	
	d.	5	2.05	
	e.	6	3.48	
	f.	7		$25.68
	g.	8		38.88
	h.	5		91.35
2.	a. $83.00	5		
	b. 12.00	4½		
	c.	6	$3.90	
	d.	7	1.82	
	e.	5½	1.87	
	f.	8		$46.44
	g.	6		99.64
	h.	9		51.23

3. Determine the sales tax and the total price to be charged for an answering machine marked $29, if a 6% sales tax is required.

4. A curio shelf has a list price of $98. Determine the sales tax and the total price if the tax rate is 5%.

5. If a 9% sales tax is to be collected, determine the sales tax and the total price on a five-piece dining set marked $310.

6. What are (a) the amount of sales tax and (b) the total price for a storage chest marked $220 if a 7% sales tax is to be collected?

7. A 6% sales tax on a humidifier was $5.34.

a. What was the marked price?

b. What was the total price including the sales tax?

8. If a sales tax of 7% amounted to $5.53,

a. What was the marked price?

b. What was the total price including sales tax?

9. The import duty on a crate of citrus fruit was $7.50. If a 15% duty was charged,

a. What was the crate's declared value?

b. What was the total price including the duty?

10. The excise tax on a strand of pearls was $30. If a 12% excise tax was levied on the jewelry,
    a. What was the marked price of the pearls?
    b. What was the total price including the excise tax?
11. The total cost of a tent was $198.55. If the tax rate was 4½%,
    a. What was the marked price of the tent?
    b. What was the sales tax?
12. The total cost of a vacuum cleaner was $178.69. If the tax rate was 7%,
    a. What was the marked price?
    b. What was the sales tax?
13. The total cost of a hammock was $93.45. If the tax rate was 5%,
    a. What was the marked price?
    b. What was the sales tax?
14. The total cost of a sewing machine was $172.53. If the tax rate was 6½%,
    a. What was the marked price?
    b. What was the sales tax?
15. After a 6½% sales tax was added, the total cost of a braided rug was $66.03.
    a. What was the marked price of the rug?
    b. What was the sales tax paid?
16. A mulching mower had a total price of $156.60 after an 8% sales tax was added.
    a. What was the marked price of the mower?
    b. What was the sales tax?
17. The price of a camera was $93.96, which included an 8% sales tax.
    a. What was the marked price of the camera?
    b. How much was the sales tax?
18. Patio furniture had a total price of $284.08, which included a 6% sales tax.
    a. What was the marked price of the furniture?
    b. How much sales tax was added?
19. The gasoline bill for a car was $15.60, which included a 5% sales tax and a 15% excise tax.
    a. What was the basic charge for the gasoline?
    b. How much sales tax was paid?
    c. How much excise tax was paid?
20. A recent telephone bill totaled $58, which included a 6% sales tax and a 10% excise tax.
    a. What was the basic charge for telephone usage?
    b. What was the amount of sales tax?
    c. What was the amount of excise tax?
21. The price of a five-piece place setting of imported china was $84. This price included a 7% city/state sales tax and a 13% import duty.
    a. What was the marked price of the china?
    b. What was the amount of sales tax?
    c. How much import tax was included?

22. A gold ring had a total price of $495.60. This price included a 6% city/state sales tax and a 12% federal excise tax.

a. What was the marked price of the ring?

b. What was the amount of city/state sales tax?

c. What was the amount of federal excise tax?

23. Imported photographic equipment had a price tag of $25,000. This price was subject to a 6% state sales tax and a 12% import tax.

a. What was the total price of the equipment including the taxes?

b. What was the sales tax?

c. What was the import tax?

24. An imported car had a price tag of $37,000. In addition, the buyer must pay 7% state sales tax and a 10% import duty.

a. What was the total price of the car including the taxes?

b. What was the sales tax?

c. What was the import tax?

SECTION 2

PROPERTY TAX

The typical means by which counties, cities, and independent public school boards secure revenue is the **property tax.** This tax applies to **real estate** (land and the building improvements on it), as well as to **personal property** (cash, cars, household furnishings, appliances, jewelry, and so on). Often, the real-property tax and personal-property tax are combined into a single assessment. Two things affect the amount of tax that will be paid: (1) the assessed value of the property and (2) the tax rate.

Property tax is normally paid on the **assessed value** of property rather than on its market value. Assessed value in any given locality is usually determined by taking a percent of the estimated fair market value of property. The percent used varies greatly from locality to locality, although assessed valuations between 50% and 100% are probably most common.

Thus, the tax rate alone is not a dependable indication of how high taxes are in a community; one must also know how the assessed value of property is determined. Even though two towns have the same tax rate, taxes in one of the towns might be much higher if the property there is assessed at a higher rate. Similarly, a community with a higher tax rate than surrounding areas might actually have lower taxes if property there is assessed at a much lower value. The recent trend has been for property to be assessed at higher percents of market value and for tax rates to be reduced somewhat—with the net result that property taxes continue to increase. Assessed valuations supposedly equal to market value are becoming more common, although even then the property would probably bring somewhat more if actually sold.

The tax *rate* of an area is usually expressed in one of three ways: as a percent, as an amount per $100, or as an amount per $1,000. The latter method is less common than percent or per $100 rates. A variation of the amount per $1,000 is to express the rate in mills. **Mills** indicates "thousandths."

The county commissioners, city council members, or other board responsible for establishing the tax rate use the following procedure: After the budget (total tax needed) for the coming year has been determined and the total assessed value of property in the area is known, the rate is computed:

$$\text{Rate} = \frac{\text{Total tax needed}}{\text{Total assessed valuation}}$$

Example 1 Markly County has an annual budget of $4,800,000. If the taxable property in the county is assessed as $300,000,000, what will be the tax rate for next year? Express this rate (a) as a percent, (b) as an amount per $100, (c) as an amount per $1,000, and (d) in mills.

$$\text{Rate} = \frac{\text{Total tax needed}}{\text{Total assessed value}}$$

$$= \frac{\$4,800,000}{\$300,000,000}$$

$$\text{Rate} = 0.016$$

(a) The property tax rate is 1.6% of the assessed value of property.
(b) When the rate is expressed per $100, the decimal point is marked off after the hundredths place. Thus, the rate is $1.60 per C (or per $100).
(c) The rate per $1,000 is found by marking off the decimal point after the thousandths place. Hence, the rate is $16.00 per M (or per $1,000).
(d) The rate in mills is found the same way as in part (c), but the $ sign is not used (16 mills).

In most instances, the tax rate does not divide out to an even quotient as in Example 1. When a tax rate does not come out even, the final digit to be used is always rounded up to the next higher digit, regardless of the size of the next digit in the remainder.

Example 2 The tax rate initially computed for the Richland Independent School District is 0.0152437+. Express this rate (a) as a percent to two decimal places, (b) per $100 of valuation (correct to cents), (c) per $1,000 of valuation (correct to cents), and (d) in whole mills.

(a) 1.53% (b) $1.53 per C (c) $15.25 per M (d) 16 mills

Property tax is found using the formula

$$\text{Rate} \times \text{Valuation} = \text{Tax}$$

The same basic formula is also used to find the rate or the valuation (assessed value) of property when either one is unknown.

Example 3 How much property tax will be due on a home assessed at $75,600 if the tax rate is 2%?

$R = 2\%$	$R \times V = T$
$V = \$75{,}600$	$2\% \times \$75{,}600 = T$
$T = ?$	$\$1{,}512 = T$

The real property tax on this home is $1,512.

Example 4 A duplex apartment in Orange County is assessed at $79,500 and is subject to tax at the rate of $1.60 per C. Calculate the real property tax.

$R = \$1.60$ per C	$R \times V = T$
$V = \$79{,}500$	$1.60 \times \$795 = T$
$= \$795$ hundred	$\$1{,}272 = T$
$T = ?$	

A property tax of $1,272 is due on the duplex.

Note. When the tax formula is solved for valuation (V), the quotient obtained represents that many hundreds or thousands of dollars of valuation, according to whether the rate is per $100 or per $1,000. The original quotient must be converted to a complete dollar valuation. (For example, a solution of "$721" means "$721 hundred" when the rate is per $100; hence, the assessed value must be rewritten $72,100.)

Example 5 A property tax of $1,120 was paid in Brazos City, where the tax rate is $17.50 per M. What is the assessed value of the property?

$R = \$17.50$ per M	$R \times V = T$
$V = ?$	$17.50V = \$1{,}120$
$T = \$1{,}120$	$V = \dfrac{1{,}120}{17.50}$
	$= 64$ thousand
	$V = \$64{,}000$

The property is assessed at $64,000.

Example 6 The office building of Wesson & White, Inc., which is assessed at \$86,500, was taxed \$1,557. Determine the property tax rate per \$100.

$$R = ?\text{ per C} \qquad R \times V = T$$

$$V = \$86{,}500 \qquad R \times \$865 = \$1{,}557$$

$$= \$865\text{ hundred} \qquad 865R = 1{,}557$$

$$T = \$1{,}557 \qquad R = \frac{1{,}557}{865}$$

$$R = \$1.80\text{ per C}$$

The tax rate is \$1.80 per \$100 of assessed valuation.

Example 7 The property tax was \$792 on a home when the rate was 1.65%. During the succeeding year, the assessed value of the home was increased by \$3,000, and the property tax amounted to \$867. Determine the amount of increase in the tax rate.

$$R = 1.65\% \qquad R \times V = T$$

$$V = ? \qquad 0.0165V = \$792$$

$$T = \$792 \qquad V = \frac{792}{0.0165}$$

$$V = \$48{,}000$$

The assessed value of the home for the first year was \$48,000. The assessed value of the home for the second year was \$48,000 + \$3,000 = \$51,000.

$$R = ? \qquad R \times V = T$$

$$V = \$51{,}000 \qquad R(\$51{,}000) = \$867$$

$$T = \$867 \qquad R = \frac{867}{51{,}000}$$

$$= 0.017$$

$$R = 1.7\%$$

$$\begin{array}{r} 1.70\% \\ -1.65 \\ \hline 0.05\% \end{array} \quad \text{Increase in rate}$$

SECTION 2 PROBLEMS

In the following problems, express each tax rate as indicated.

1. 0.0184402+
- a. As a percent with two decimal places
- b. As an amount per $100, correct to cents
- c. As an amount per $1,000, correct to cents
- d. In whole mills

2. 0.0224615+
- a. As a percent with two decimal places
- b. As an amount per $100, correct to cents
- c. As an amount per $1,000, correct to cents
- d. In whole mills

3. 0.0192356+
- a. As a percent with two decimal places
- b. As an amount per $100, correct to cents
- c. As an amount per $1,000, correct to cents
- d. In whole mills

4. 0.0140671+
- a. As a percent with two decimal places
- b. As an amount per $100, correct to cents
- c. As an amount per $1,000, correct to cents
- d. In whole mills

For Problems 5 and 6, determine the tax rate (a) as a percent with one decimal place, (b) as an amount per $100, correct to cents, (c) as an amount per $1,000, correct to cents, and (d) in whole mills.

5. In Queens County, the county commissioners voted a budget of $24,000,000. The total assessed value of all real estate in the county amounted to $980,000,000. Express the tax rate in the four variations.

6. The total assessed valuation of all real property in Hudson Hills was $660,000,000. The city council voted an annual budget of $23,000,000. Express the tax rate in the four variations.

For Problems 7 and 8, find each missing factor relating to property tax.

	Rate	Assessed Value	Tax
7.	a. 2.8%	$94,000	
	b. $1.25 per C	59,200	
	c. $12.40 per M	75,000	
	d. 2.1%		$1,050
	e. $2.60 per C		1,768
	f. $18.80 per M		1,222
	g. ?%	82,000	1,148
	h. ? per C	74,500	2,384
	i. ? per M	95,000	912

8.	a. 1.8%	$75,000	
	b. $2.10 per C	88,000	
	c. $11.20 per M	65,000	
	d. 2.4%		$2,256
	e. $1.15 per C		1,380
	f. $15.30 per M		3,060
	g. ?%	600,000	10,200
	h. ? per C	45,000	675
	i. ? per M	540,000	7,020

9. Determine the tax due on real estate in Charles County assessed at $35,000 if the tax rate is 1.8%.
10. Property owners in Boise pay a tax rate of 2.8%. Determine the tax due on real property assessed at $85,000.
11. The property tax rate in Marshall City is $2.05 per C. A commercial building has been assessed at $90,000. How much tax is due?
12. The property tax rate in Salzburg is $1.12 per C. Determine the tax due on a residence assessed at $130,000.
13. A property tax of $429 was paid on undeveloped land in Denver. What was the assessed value of the property if the tax rate was $1.95 per C?
14. A property tax of $1,702 was paid on a commercial property. What was the assessed value of the property if the tax rate was $1.85 per C?
15. Determine the assessed value of a home in Wadley if the tax rate is $12.75 per M and the property tax was $918.
16. What was the assessed value of real property if the tax rate was $9.45 per M and the tax due was $4,725?
17. A resident paid $2,016 in property taxes on a home assessed at $84,000. What percent is the tax rate?
18. What percent is the tax rate in Prince William County if a property owner paid $4,806 in property taxes on a home assessed at $267,000?
19. A homeowner in Hope paid $525 in real estate taxes on a home assessed at $35,000. What was the tax rate expressed in whole mills?
20. Arvan Construction Co. paid $10,800 in property taxes on its office building assessed at $600,000. What was the tax rate expressed in whole mills?
21. Property taxes cost a small business $3,960. If the real property was assessed at $180,000, what was the rate per $100?
22. Property tax in Harrisburg cost a homeowner $1,666. If the home was assessed at $98,000, what was the tax rate per $100?
23. The property tax was $1,349 on commercial property when the assessed value was $95,000. During the succeeding year, the tax rate was increased by $0.08 per $100, and the property tax amounted to $1,434. Determine the amount of increase in the assessed value. (Hint: Solve each year separately.)
24. The property tax on real property was $10,560 when the assessed value was $800,000. During the succeeding year, the tax rate increased by $0.10 per $100 and the property tax amounted to $11,644. Determine the amount of increase in assessed value. (Hint: Solve each year separately.)
25. The first-year property tax on a new home was $390, and the property tax rate was 1.3%. During the second year, the assessed value of the home increased by $4,000, and the property tax amounted to $544. What was the tax rate for the second year?

26. The property tax rate in Carlisle was 1.6% and the property tax on a home was $5,440. During the next year, the assessed value of the home was increased by $1,000 and the property tax amounted to $5,797. Determine the tax rate for the second year.

CHAPTER 5 GLOSSARY

Assessed value. A property value (usually computed as some percent of true market value) which is used as a basis for levying property taxes.

Customs duty (import tax, tariff). A tax levied on products brought into this country.

Excise tax. A tax levied on the manufacture or sale of certain products and services, usually luxury items, or to finance projects related to the item taxed; paid in addition to sales tax.

Luxury tax. A tax on automobiles imposed on that portion of the marked price that exceeds a certain cutoff point.

Mills. A property tax rate that denotes "thousandths." (Example: 16 mills indicates a rate of 0.016.)

Personal property. Possessions (such as cars, cash, household furnishings, appliances, jewelry, etc.) which together with real property are subject to property tax.

Property tax. A tax levied on real property and personal property; the chief means by which counties, cities, and independent school districts obtain revenue.

Real property. Real estate (land and buildings) which together with personal property is subject to property tax.

Sales tax. Tax levied by states (and sometimes by cities and counties) on the sale of merchandise and services.

Use tax. A tax levied on merchandise brought into a state after its purchase in another state.

CHAPTER 6

INSURANCE

OBJECTIVES

Upon completion of Chapter 6, you will be able to:

1. Define and use correctly the terminology associated with each topic.
2. **a.** Compute fire insurance premiums (Section 1: Example 1; Problems 1–4, 11, 12)
   b. Compute the premium refund due when a policy is canceled (1) by the insured or (2) by the carrier (Examples 2, 3; Problems 3, 4, 13, 14)
3. **a.** Determine compensation due following a fire loss if a coinsurance clause requirement is not met (Section 1: Example 4; Problems 5, 6, 9, 10, 15–20)
   b. Find each company's share of a loss when full coverage is divided among several companies (Example 5; Problems 7, 8, 21–24)
   c. Find each company's share of a loss when a coinsurance requirement is not met (Example 6; Problems 9, 10, 25–28).
4. **a.** Compute the annual premium for (1) automobile liability and medical payment insurance and (2) comprehensive/collision insurance (Section 2: Examples 1, 3; Problems 1–12)
   b. Find the compensation due following an accident (Examples 2, 4; Problems 13–20).
5. Determine premiums for the following life insurance policies: (a) term, (b) whole life, (c) 20-payment life, and (d) variable universal life (Section 3: Examples 1–3; Problems 1, 2, 7–14).
6. Calculate the following nonforfeiture values: (a) cash value, (b) paid-up insurance, and (c) extended term insurance (Section 3: Example 4; Problems 3, 4, 13–18).
7. Compare the monthly (and total) annuities that a beneficiary could receive under various settlement options (Section 3: Examples 5–7; Problems 5, 6, 19–26).

All insurance is carried to provide financial compensation should some undesirable event occur. Insurance operates on the principle of shared risk. Many persons purchase insurance protection, and their payments are pooled in order to provide funds with which to pay those who experience losses. Thus, rates are lower when many different people purchase the same protection and divide the risk; rates would be much higher if only a few people had to pay in enough money to provide reimbursement for a major loss.

The risk that an insurance company assumes by insuring a person, organization, or event always influences how high the rates will be. The company determines the mathematical probability that the insured entity will incur losses and then sets its rates accordingly. The insurance company does not know which of its clients will suffer a loss, but it knows from experience how many losses to expect. The insurance rates must be high enough to pay the expected losses, to pay the expenses of operating the company, and to provide a reserve fund should losses exceed the expected amounts during any period.

The payment made to purchase insurance coverage is called a **premium.** Insurance premiums are usually paid either annually, semiannually, quarterly, or monthly. The total cost of insurance is slightly higher when payments are made more frequently, in order to cover the cost of increased bookkeeping. An insurance contract, or **policy,** defines in detail the provisions and limitations of the coverage. The amount of insurance specified by the policy is the **face value** of the policy. Insurance premiums are paid in advance, and the **term** of the policy is the time period for which the policy will remain in effect. The insurance company is often referred to as the **insurer,** the **underwriter,** or the **carrier.** The person or business that purchases insurance is called the **insured** or the **policyholder.** When an insured loss occurs, the payment that the insurance company makes to reimburse the policyholder (or another designated person) is often called an **indemnity.**

As indicated above, insurance is sold in order to provide financial protection to the policyholder; it is not intended to be a profit-making proposition when a loss occurs. Thus, insurance settlements are always limited to either the actual amount of loss or the face amount of the policy, whichever is smaller.

Individual states have insurance commissions that regulate premium structure and insurance requirements (such as required automobile insurance) passed by their legislatures. Thus, insurance practices and premium computation vary widely from state to state.

SECTION 1

BUSINESS INSURANCE

Certain standard forms of insurance are carried by almost all businesses; these types of insurance are discussed on the following pages. Because of its importance to private citizens as well as to business organizations, fire insurance receives particular attention.

FIRE INSURANCE

Basic fire insurance provides financial protection against property damage resulting from fire or lightning. (Insured property may be buildings and their contents, vehicles, building sites, agricultural crops, forest crops, and so forth.) Further damage resulting from smoke or from attempts to extinguish or contain a fire may be as costly as the fire damage itself. Therefore, **extended coverage** is usually included to provide protection against smoke, water, and chemical damage, as well as physical destruction caused by firefighters having to break into the property or taking measures to keep a fire from spreading. Extended coverage also covers riot, civil commotion, hurricane, wind, and hail damage.

Homeowners often purchase a **comprehensive** "package" **policy,** which, in addition to the coverage above, also provides protection against loss due to theft, vandalism, and malicious mischief; protection against accidental loss or damage; and liability coverage in case of personal injury or property damage to visitors on the property. Similar protection may also be purchased by business firms, although it is sometimes obtained in separate policies.

PREMIUMS

Historically, fire insurance policies were written for several years in advance, and reduced long-term rates were given for premium prepayments. (For instance, a prepaid 3-year policy might have cost only 2.7 times as much as a 1-year policy.) In recent years, however, spiraling costs for reimbursing fire losses have forced insurance companies to discontinue these practices. Most fire insurance policies are now written for either 1 or 3 years, and the prepaid cost of a 3-year policy is three times the 1-year premium. If the 3-year premium is paid in three annual payments, interest may also be charged for the latter 2 years. It is probable that this total cost would still be cheaper than three 1-year policies, however, since rates have been increasing so rapidly.

Each state has a fire-rating bureau that establishes fire insurance rates. A representative of the bureau inspects every commercial property and determines the rate to be paid by each. These rates vary greatly, because they are influenced by many different factors. The primary factors that the inspector must consider are (1) the construction of the building, (2) the contents stored or manufactured therein, (3) the quality of fire protection available, (4) the location of the buildings, and (5) past experience regarding the probability of a fire.

Thus, a building of fireproof construction (all steel and concrete, and with a fire-resistive roof) would have a much lower insurance rate than a frame building with a flammable roof. Similarly, property located near fire hydrants and served by a well-trained fire department in an area that receives abundant rainfall would require much lower premiums than property located in an arid region and some distance from a fire station. (The installation of fire extinguishers or a sprinkler system would reduce the rate somewhat.) A dress shop would normally have a lower rate than a paint store, because of the combustible nature of painting supplies; if the

TABLE 6-1 ANNUAL FIRE INSURANCE PREMIUMS PER $100 OF FACE VALUE

	STRUCTURE CLASS				CONTENTS CLASS			
TERRITORY	A	B	C	D	A	B	C	D
1	$0.45	$0.56	$0.66	$0.75	$0.51	$0.79	$0.79	$0.98
2	0.58	0.64	0.78	0.80	0.68	0.86	0.91	1.03
3	0.63	0.72	0.85	0.93	0.71	0.89	0.95	1.12

dress shop relocated to a site beside a gasoline storage area, however, its fire insurance rate would increase greatly.

Fire insurance rates are quoted as an amount per $100 of insurance coverage. Separate rates are charged for insurance on the building itself, on the furniture and fixtures, and on the stock or merchandise—normally in that order of increasing premium expense. Table 6-1 and Example 1 illustrate the way in which the premiums may be determined.

Example 1 Using Table 6-1, determine the fire insurance premium due for a $125,000 home policy in Territory 1, Class B category, with contents valued at $30,000.

Fire insurance premiums are computed using the procedure

$$\text{Premium} = \text{Rate} \times \text{Value of policy}$$

or simply

$$P = R \times V$$

Thus, the premium for the preceding home would be computed as follows:

$$\begin{aligned} R &= \$0.56 \text{ per } \$100 & P &= R \times V \\ V &= \$125{,}000 & &= \$0.56 \times \$1{,}250 \\ &= \$1{,}250 \text{ hundreds} & P &= \$700 \text{ (annual premium)} \\ P &= ? \end{aligned}$$

The annual premium on the building itself is $700. The premium to insure the contents would be separately computed in a similar manner.

$$\begin{aligned} R &= \$0.79 \text{ per } \$100 & P &= R \times V \\ V &= \$30{,}000 & &= \$0.79 \times \$300 \\ &= \$300 \text{ hundreds} & P &= \$237 \\ P &= ? \end{aligned}$$

The total premium for the building and the contents would be

$$P = \$700 + \$237$$

$$P = \$937$$

Note. Fire insurance premiums are rounded to the nearest whole dollar.

SHORT-TERM POLICIES

Sometimes a business may wish to purchase fire insurance for only a limited time in order to obtain coverage on merchandise that will soon be sold. Also, policies are sometimes canceled by the policyholders if they sell the insured property, if they move, and so on. If, for any reason, a policy is in effect for less than 1 year, it is considered a **short-term policy.** The premium to be charged for such a policy is computed according to a short-rate table, such as Table 6-2. A very short policy is relatively expensive because of the cost to the insurance company of selling, writing, and processing the policy.

TABLE 6-2 SHORT-TERM RATES INCLUDING CANCELLATION BY THE INSURED

MONTHS OF COVERAGE	PERCENT OF ANNUAL PREMIUM CHARGED	MONTHS OF COVERAGE	PERCENT OF ANNUAL PREMIUM CHARGED
1	20%	7	75%
2	30	8	80
3	40	9	85
4	50	10	90
5	60	11	95
6	70	12	100

Example 2 Grant Avery owned a home on which he paid $800 for fire insurance. He moved after 8 months and canceled the insurance policy. How much refund will Avery receive?

For an 8-month premium, 80% of the annual premium will be charged:

Annual premium	$800	Annual premium	$800
	× .80		− 640
8-month premium	$640	Refund due	$160

Occasionally, it becomes necessary for an insurance company to cancel a policy. This might happen because of excessive claims or a refusal of the policyholder to accept higher rates when the fire risk has increased. The company ordinarily must give advance notice that the policy will be canceled, thus allowing the policyholder time to purchase another policy before the coverage ceases. When a company cancels, it may keep only that fraction of the annual premium equivalent to the fraction of the year which has elapsed. Many companies compute this time precisely, using $\frac{\text{exact days}}{365}$; however, we shall use approximate time for our calculations. The insurance company is not entitled to short-term rates when it cancels the policy.

Example 3 A $640,000 commercial fire policy, sold at the rate of $0.75 per $100, is canceled by the underwriter after 5 months. How much refund will the policyholder receive?

The annual premium is determined in the usual manner:

$$R = \$0.75 \text{ per } \$100 \qquad \text{Annual premium} = R \times V$$
$$V = \$640{,}000 \qquad = \$0.75 \times \$6{,}400$$
$$= \$6{,}400 \text{ hundreds} \qquad P = \$4{,}800$$
$$P = ?$$

Because the underwriter canceled after 5 months, only $\frac{5}{12}$ of the annual premium may be retained:

$$\text{5-month premium} = \frac{5}{12} \times \$4{,}800$$
$$= \$2{,}000$$

Annual premium	$4,800
5-month premium	− 2,000
Refund due	$2,800

COMPENSATION FOR LOSS; COINSURANCE CLAUSE

Experience indicates that most fires result in the loss of only a small portion of the total value of the property. Realizing that the total destruction is relatively rare, many people would be tempted to economize by purchasing only partial coverage. However, to pay the full amount of these damages would be extremely expensive to an insurance company if most policyholders were paying premiums on only the partial value of their property. Therefore, to maintain an economically sound business operation, as well as to encourage property owners to purchase full insurance coverage, many policies contain a coinsurance clause. A **coinsurance clause** guarantees full payment after a loss (up to the face amount of the policy) only if the policyholder purchases coverage which is at least a specified percent of the total value of the property at the time the fire occurs. For example, if a house would cost $125,000 to replace, then this figure represents the total value, regardless of how much less money the house may have cost originally.

Many policies contain coinsurance clauses requiring 80% coverage, although lower coverage is sometimes allowed, and higher coverage (particularly for extended coverage policies) may often be necessary. If the required coverage is not carried, the company will pay damages equivalent to the ratio of the required coverage that is carried; the policyholder must then bear the remaining loss. (In this respect, the policyholder "coinsures" the property. The term "coinsurance" implies that the insurance company and the policyholder each assume part of the total risk.) Most policies on homes do provide full coverage; businesses are more likely than homeowners to partially coinsure themselves.

A property owner may purchase more insurance than the minimum requirement, of course. However, in no case will payment for damages exceed the actual value of the property destroyed, no matter how much insurance is carried. The maximum payment is always the amount of damages or the face value of the policy, whichever is smaller. When a fire occurs, the policyholder must contact the insurance company, which sends out an *insurance adjuster* to inspect the damage and determine the indemnity to be paid. (The policyholder may also have to obtain repair or reconstruction estimates from several companies.)

Example 4 Centex Distributors, Inc. operates in a building valued at \$750,000. The building is insured for \$500,000 under a fire policy containing an 80% coinsurance clause. (a) What part of a fire loss will the company pay? (b) How much will the company pay on a \$120,000 loss? (c) On a \$420,000 loss? (d) On a \$720,000 loss?

(a) According to the coinsurance clause, 80% of the property value must be insured:

$$\$750{,}000 \times 80\% = \$600{,}000 \text{ insurance required}$$

$$\frac{\text{Insurance carried}}{\text{Insurance required}} = \frac{\$500{,}000}{\$600{,}000} = \frac{5}{6}$$

Centex Distributors carries only $\frac{5}{6}$ of the insurance required by the coinsurance clause; therefore, the insurance company will pay only $\frac{5}{6}$ of any loss, up to the face value of the policy. By their failure to carry the required coverage, Centex has indirectly agreed to coinsure the remaining portion of any fire loss themselves.

(b) $\frac{5}{6} \times \$120{,}000 = \$100{,}000$ insurance reimbursement

(c) $\frac{5}{6} \times \$420{,}000 = \$350{,}000$ indemnity

(d) $\frac{5}{6} \times \$720{,}000 = \$600{,}000$

Preliminary calculations indicate the insurance company's share of a \$720,000 loss to be \$600,000. However, this exceeds the face value of Centex's policy; therefore, the insurance company's payment is limited to the full value of the policy, or \$500,000.

MULTIPLE CARRIERS

Insurance coverage on property may often be divided among several companies. This may happen because the value of the property is so high that no single insurance company can afford to assume the entire risk, because the owner wishes to distribute his business for public relations purposes, or simply because coverage on various portions of the property was purchased separately over a period of years.

When property is insured with multiple carriers, each company pays damages in the ratio that its policy bears to the total insurance coverage. As noted previously, the total amount paid by all carriers will never exceed the value of the damage; nor will any company's payment exceed the face value of its policy. Multiple policies may also contain coinsurance clauses.

Example 5 **Coinsurance Met**

Farnsworth Equipment suffered a $180,000 fire. Farnsworth was insured under the following policies: $150,000 with Company A, $100,000 with Company B, and $50,000 with Company C. What indemnity will each company pay? (Assume that Farnsworth meets the coinsurance requirement necessary to receive full coverage on a loss.)

Policies		
$150,000	A pays $\frac{\$150{,}000}{\$300{,}000}$	or $\frac{1}{2}$ of the loss
100,000	B pays $\frac{\$100{,}000}{\$300{,}000}$	or $\frac{1}{3}$ of the loss
50,000	C pays $\frac{\$\ 50{,}000}{\$300{,}000}$	or $\frac{1}{6}$ of the loss
$300,000		

The full $180,000 loss is insured; therefore, each company's payment is as follows:

$$\text{Company A:}\quad \frac{1}{2} \times \$180{,}000 = \$\ 90{,}000$$

$$\text{Company B:}\quad \frac{1}{3} \times \$180{,}000 = 60{,}000$$

$$\text{Company C:}\quad \frac{1}{6} \times \$180{,}000 = \underline{30{,}000}$$

$$\text{Total indemnity} = \$180{,}000$$

Example 6 **Coinsurance Not Met**

Centex Distributor's building (Example 4) is valued at $750,000 but is insured for only $500,000 under an 80% coinsurance requirement. Assume now that this $500,000 coverage is divided between Company A ($300,000) and Company B ($200,000). (a) What part of any loss is covered? (b) How much reimbursement will Centex receive after a $120,000 fire? (c) What is each insurance company's payment after the $120,000 loss?

(a) As in Example 4, the required coverage is 80% × $750,000, or $600,000. Thus, the part of any loss that is covered (up to the face value of the policy) as before is

$$\frac{\text{Insurance carried}}{\text{Insurance required}} = \frac{\$500{,}000}{\$600{,}000} = \frac{5}{6}$$

(b) Again, following a $120,000 fire,

$$\frac{5}{6} \times \$120{,}000 = \$100{,}000 \qquad \text{Total indemnity (to be shared by all insurance carriers)}$$

(c) Each company's share of this $100,000 indemnity (due following the $120,000 loss) is the ratio of its policy to the total insurance carried ($500,000):

$$\text{Company A:} \quad \frac{\$300{,}000}{\$500{,}000} \quad \text{or} \quad \frac{3}{5} \times \$100{,}000 = \$\ \ 60{,}000$$

$$\text{Company B:} \quad \frac{\$200{,}000}{\$500{,}000} \quad \text{or} \quad \frac{2}{5} \times \$100{,}000 = \underline{\ \ 40{,}000}$$

$$\text{Total indemnity} = \$100{,}000$$

Thus, with multiple carriers, the $100,000 will be reimbursed as $60,000 from Company A and $40,000 from Company B. Centex assumes the remaining $20,000 loss itself.

Several other types of insurance are carried by most businesses. The following paragraphs give a brief description of this additional insurance.

BUSINESS INTERRUPTION

Business interruption insurance is purchased as a supplement to fire and property insurance. Whereas fire and property insurance provides coverage for the actual physical damage resulting from a fire or natural disaster (earthquake, flood, hurricane, tornado, and so on), business interruption insurance provides protection against loss of income until the physical damage can be repaired and normal business resumed. This insurance provides money to meet business expenses that continue despite the lapse in operations: interest, loan payments, mortgage or rent payments, taxes, utilities, insurance, and advertising, as well as the salaries of key members of the firm and ordinarily even the anticipated net profit.

LIABILITY

Businesses may be held financially liable, or responsible, for many different occurrences; general **liability insurance** is a business essential in order to provide this protection. An accident causing personal injury to a customer often results in the customer's suing the business. When firms are declared negligent of their responsibility for the customer's safety, judgments of large sums of money are often granted by the courts. Hence, no well-managed company would operate without *public liability* insurance.

Another area in which court judgments have been quite large in recent years is bodily injury resulting from defective or dangerous products. Companies that manufacture mechanical products, food products, or pharmaceutical products that could conceivably be harmful can be insured by *product liability* policies.

Liability for the safety of employees is covered under **worker's compensation** policies. This insurance provides payments to an employee who loses work time because of sickness or accidental injury resulting directly from work responsibilities. The most common type of injury that leads to worker's compensation claims is back injury. State laws determine the amount of benefits paid under different situations. Disability income is paid to workers who are totally disabled temporarily, or it is paid permanently to those who are partially disabled on a permanent basis. The national average is about 75% of the worker's take-home pay. Some states set no limits on length of payments, while other states set a limit in years or a maximum dollar amount paid to the worker. Most worker's compensation policies also provide death benefits in case of a fatality.

The labor statutes in 47 states require employers to carry this insurance; some states have a state fund for worker's compensation, although in most cases it is purchased from commercial insurance companies. As would be expected, worker's compensation rates vary greatly according to the occupational hazards of different jobs. These rates are usually quoted as an amount per $100 in wages. The premiums are usually based only on regular wages; that is, premiums are not paid on overtime wages.

A business firm's liability for the operation of motor vehicles is similar to that of the general public; both are included in the following section of this chapter. Business policies, however, often contain special provisions covering fleets of cars, hired cars, or the cars of employees being used in company business.

MONEY AND SECURITIES

The theft of money and securities (such as checks, stocks, titles of ownership, and so forth) is a constant hazard in many businesses. A number of different policies cover the various conditions under which money and valuable securities may be stolen, such as burglary from an office safe or robbery of a messenger carrying money or securities to or from the business.

General theft policies do not cover the dishonesty of employees, which must be provided for by separate policies. *Fidelity bonds* protect against embezzlement or other loss by employees who handle large sums of money. These bonds are essential in banks, retail stores, and other offices where large amounts of cash or securities are involved.

The worst financial loss experienced by most retail stores results from shoplifting (by both customers and employees), a problem of major national proportion that is increasing annually. There is presently no insurance available that adequately protects against shoplifting losses; thus, the business's only means of recovery is to pass the loss along to consumers in the form of higher prices.

LIFE INSURANCE

Particularly in small businesses, the death of a partner or a key person in the firm may jeopardize the entire business operation. Many firms, therefore, purchase life insurance on critical personnel to protect against such financial loss. Such policies also enable the firm to buy from the estate the business interest of a deceased partner. Life insurance is discussed in more detail in a later section.

Disability

Disability insurance provides coverage for workers who become disabled and thus unable to work, due to a physical or mental reason. Monthly benefits are paid until the person is well enough to return to work. During the disability, personal expenses such as mortgage, utilities, medical bills, and food continue. The disability benefits help pay for these living expenses. Today, one in seven people becomes disabled for at least five years before reaching the age of 65. Approximately 40% of Americans have some disability insurance.

Group Insurance

Group insurance available at many companies provides life insurance and/or health and accident coverage for employees and their families. This coverage, unlike worker's compensation, provides coverage for illnesses or accidents not related to occupational duties. Most group life insurance plans offer term life insurance similar to the policies described later in this chapter. These group life policies are also available to many individuals through the official associations of their business or profession, alumni associations, or other special-interest groups.

Typical group hospitalization plans provide basic benefits for either accidental injuries or medical/surgical care. Each policy contains a long list of covered hospital services and surgical procedures, each specifying the maximum amount that will be paid to the hospital and surgeon. Maternity benefits are generally included in group policies, although a specified waiting period is often required for this coverage, as well as for coverage of certain preexisting medical conditions.

Some group policies also include a major medical supplement, which provides for certain expenses not covered by the basic benefits. For example, major medical coverage will pay a specified percent of the costs of physicians' fees, prescriptions, and many services or procedures not otherwise covered, as well as excess expenses above those reimbursed through the basic policy. All major medical plans and some basic policies specify a deductible amount that the patient must first pay before insurance coverage becomes effective.

The employer often pays half or more of the group insurance premiums as one of the organization's fringe benefits. Few companies pay the full amount of employees' premiums anymore, however. Even when the employee pays the entire cost, group insurance is cheaper than similar coverage purchased privately. Also, group policies may usually be purchased without a physical examination, which permits some people to receive insurance protection when they would not ordinarily qualify.

Group coverage usually terminates immediately when a worker leaves the firm, although legally the worker can remain on the group policy for a limited time by assuming the full cost. Under COBRA (Consolidated Omnibus Budget Reconciliation Act passed by Congress in 1986), coverage continues for employees under the former employer's group plan for as long as 18 months. Health insurance is guaranteed for employees terminated for any reason except for those terminated for "gross misconduct." The monthly premiums are paid in full by the individual instead of by the employer or employer/ employee. COBRA is also available for the widow(er) and dependent children of deceased employees for up to 36 months. As another alternative, a group policy may often be converted to a private policy (at an increased rate, of course).

In many cases, group insurance provides standard coverage for all employees. However, certain benefits may sometimes be increased according to the employee's salary or years of service to the business. A recent trend at many businesses, called "cafeteria coverage," lets employees choose from the various policies available, in order to meet the particular needs of themselves or their individual families.

SECTION 1 PROBLEMS

Compute the fire insurance premiums required for the following homes, using the premiums listed in Table 6-1.

	STRUCTURE VALUE	CONTENTS VALUE	CLASS	TERRITORY	TOTAL PREMIUM
1. a.	$350,000	$64,000	B	1	
b.	580,000	78,000	A	2	
2. a.	$575,000	$82,000	C	1	
b.	840,000	55,000	D	3	

Compute the premium and the refund due (if applicable) for the following amounts of insurance for the terms indicated. Assume an annual rate of $0.78 per $100. Use the short-term rates from Table 6-2 when applicable.

	AMOUNT OF INSURANCE	TERM	CANCELED BY	PREMIUM	REFUND DUE
3. a.	$ 290,000	1 year	X		X
b.	770,000	6 months	X		X
c.	830,000	9 months	Insured		
d.	1,500,000	4 months	Carrier		
4. a.	$ 260,000	1 year	X		X
b.	590,000	3 months	X		X
c.	720,000	7 months	Insured		
d.	2,800,000	10 months	Carrier		

Compute the compensation (indemnity) that will be paid under each of the following conditions.

	Property Value	Coinsurance Clause	Insurance Required	Insurance Carried	Amount of Loss	Indemnity
5. a.	$ 500,000	80%		$200,000	$ 250,000	
b.	440,000	80		352,000	90,000	
c.	600,000	90		450,000	420,000	
d.	900,000	80		600,000	780,000	
e.	1,000,000	80		700,000	950,000	
6. a.	$ 300,000	80%		$200,000	$ 180,000	
b.	750,000	90		500,000	135,000	
c.	600,000	90		360,000	450,000	
d.	400,000	80		300,000	380,000	
e.	1,500,000	80		900,000	1,500,000	

Determine the compensation to be paid by each carrier below (assume that coinsurance clause requirements are met).

	Company	Amount of Policy	Ratio of Coverage	Amount of Loss	Compensation
7. a.	N	$440,000		$477,000	
	O	620,000			
b.	I	$750,000		$840,000	
	N	400,000			
	S	250,000			
8. a.	A	$200,000		$140,000	
	T	500,000			
b.	R	$ 90,000		$120,000	
	I	62,000			
	S	50,000			
	K	38,000			

Calculate the settlement due from each carrier below under the given coinsurance terms.

	Prop. Value	Coins.	Insurance Required	Insurance Carried	Fire Loss	Total Indem-nity	Co.	Policy Value	Co. Ratio	Co. Pmt.
9. a.	$600,000	80%			$450,000		I	$200,000		
							J	120,000		
b.	800,000	70			770,000		K	280,000		
							L	200,000		

	Prop. Value	Coins.	Insurance Required	Insurance Carried	Fire Loss	Total Indem-nity	Co.	Policy Value	Co. Ratio	Co. Pmt.
10. a.	$900,000	80%			$450,000		M	$240,000		
							N	400,000		
b.	350,000	90			189,000		O	150,000		
							P	100,000		

11. The Southern Co. insures its merchandise that is warehoused for a 3-month period. The annual rate for this fire insurance policy is $0.60 per $100. How much does this short-term policy cost if the merchandise is insured for $180,000?

12. A manufacturer insures its merchandise that is in storage for 6 months. The annual rate for the fire insurance policy is $0.82 per $100. How much does this short-term policy cost if the merchandise is insured for $370,000?

13. An annual rate of $0.90 per $100 was charged on a fire insurance policy covering a $630,000 building. Determine the premium charged and the refund due if the policy is canceled after 5 months

a. By the insured

b. By the insurer

14. A fire policy covering a $1,800,000 warehouse cost $1.05 per $100. Determine the premium charged and the refund due if the policy is canceled after 8 months

a. By the insured

b. By the carrier

15. A fire insurance policy containing an 80% coinsurance clause is written for $600,000. The building covered by this policy is valued at $680,000. How much indemnity would be paid on

a. An $80,000 loss

b. A $500,000 loss

c. A $600,000 loss

16. A 90% coinsurance clause is written into the fire insurance policy covering a $2,000,000 plant. If the plant is insured for $1,900,000, how much will the policy pay after

a. A $50,000 fire

b. An $800,000 fire

c. A $1,900,000 fire

17. An office park valued at $5,000,000 was insured for $3,500,000. The fire insurance policy contained a 90% coinsurance clause. What settlement would be due following

a. A $36,000 loss

b. A $2,700,000 loss

c. A total loss

18. A dress shop valued at $1,000,000 is insured for $700,000. If the fire insurance policy includes an 80% coinsurance clause, determine the compensation due when a fire resulted in

a. An $180,000 loss

b. A $600,000 loss

c. A total loss

19. The owner of a small business housed in a $300,000 building purchased a $240,000 fire insurance policy to cover the building. The policy included a 90% coinsurance clause. How much indemnity would be paid when a fire caused

a. $81,000 damages

b. $189,000 damages

c. $288,000 damages

20. A shop valued at $620,000 is insured for $500,000. If the fire insurance policy included a 90% coinsurance clause, determine the compensation due when a fire resulted in

a. $11,160 damages

b. $223,200 damages

c. $600,000 damages

21. A store is insured for $240,000 with Company A, $340,000 with Company B, and $420,000 with Company C. The policies meet all coinsurance requirements. What settlement would each insurer make as a result of

a. A $50,000 fire

b. A $900,000 fire

c. A $1,000,000 fire

22. Atwell Drilling Co. is insured for $550,000 with Dee Company, $410,000 with Fee Company, and $640,000 with Gee Company. The policies meet coinsurance requirements. Determine each company's settlement if there is

a. A $256,000 fire loss

b. A $384,000 fire loss

c. A $1,600,000 fire loss

23. A meat-processing plant has fire protection with four underwriters: $300,000 with AA Company, $350,000 with BB Company, $250,000 with CC Company, and $100,000 with DD Company. Each policy meets coinsurance requirements. What is each company's share of

a. A $72,000 loss

b. A $600,000 loss

c. A $1,000,000 loss

24. A camera store carries the following policies, which meet all coinsurance clauses: $200,000 with Company U, $280,000 with Company V, $320,000 with Company W, and $350,000 with Company X. What is each company's share of

a. A $92,000 loss

b. A $207,000 loss

c. A $1,150,000 loss

25. A jewelry store is insured by American Insurance Co. for $700,000 and by Liberty Insurance Co. for $800,000. Its merchandise and building are valued at $2,500,000. If both companies have 80% coinsurance requirements,

a. What total amount would the store receive on fire damages of $2,200,000?

b. How much would each insurance company pay in indemnities?

26. Koslowski Realty Co. is insured by Company X for $1,000,000 and by Company Y for $1,100,000. Both policies require 90% coinsurance. If there is a fire loss of $2,880,000 on its $3,000,000 building,

a. What total amount will Koslowski Realty receive?

b. How much will each insurance company pay in settlement?

27. Mike's Computer Supply Co. is valued at $6,000,000. The store has fire insurance policies with Dickson Insurance Agency for $1,200,000, Eagle Insurance Company for $1,800,000, and Franklin Insurance Co. for $1,000,000. Each policy contains an 80% coinsurance clause. Determine
   a. The total indemnity paid for a fire loss of $4,500,000
   b. How much each company would pay toward this loss
28. Henley Hardware is valued at $5,000,000. The fire insurance policies are with Prentice Insurance Company for $2,000,000, with Randall Insurance Agency for $1,000,000, and with Quincy Insurance Company for $800,000. Each policy contains an 80% coinsurance clause. Determine
   a. The total compensation that would be due after a $3,500,000 fire loss
   b. How much each company would pay toward the damages

SECTION 2

MOTOR VEHICLE INSURANCE

Automobile and truck insurance is carried by responsible owners (private or industrial) of most vehicles. Insurance policies on motor vehicles are usually written for a maximum term of 1 year. The primary reason for the 1-year policy is the alarmingly high nationwide accident rate—each year, there are more than 25 million auto accidents in the United States. These accidents constantly increase rates for automobile insurance, because repair costs continue to spiral. Since insurance rates always reflect the risk taken by the insurance company, the driving record (and age) of the operator is another major factor in determining the cost of vehicle insurance. This further accounts for the 1-year policies, since a driver's record may fluctuate over a period of years.

Insurance on motor vehicles is divided into two basic categories: (1) liability coverage and (2) comprehensive/collision coverage.

LIABILITY INSURANCE

Liability insurance provides financial protection for damage inflicted by the **policyholder** to the person or property of other people. As a protection to its citizens, virtually all states require the owners of all vehicles registered in the state to purchase minimum amounts of liability insurance. Liability insurance covers **bodily injury liability** (liability for physical injury to others resulting from the policyholder's negligence, such as a pedestrian or occupants of another car being struck by the insured) and **property damage liability** (liability for damage caused by the policyholder to another vehicle or other property).

Also available on an optional basis is **medical payment** insurance (commonly called "medical pay"). Medical payment insurance pays for necessary medical and/or funeral services resulting from an auto accident. This insurance supplements other health and accident insurance the policyholder may carry, and it will begin paying medical costs immediately, without waiting for the courts to rule on a liability suit. Under liability coverage, the insurance company promises to pay for bodily injury only to the extent that the insured is legally responsible for having caused the injury.

Medical pay covers the insured driver, family members when they are either passengers or pedestrians struck by another motor vehicle, and any other person occupying the covered auto. Occupants of other cars struck by the insured or by a family member are not covered by this part of the policy.

Each state determines the minimum liability coverage for its drivers. For example, the state of Virginia has set "25/50/20" as the minimum coverage. This means that the policy will pay up to $25,000 for the bodily injury caused to a single person or, when more than one person is involved in the accident, a maximum total of $50,000 for the injuries inflicted to all victims. Also, a maximum of $20,000 will be paid for the property damage resulting from a single accident. On a nationwide basis, 25/50 for bodily injury and 10 to 30 for property damage is the most common coverage. Much higher liability coverage is available at rates only slightly higher than those for minimum coverage. Thus, many drivers purchase bodily injury coverage up to $300,000/$300,000 and property damage coverage up to $100,000.

If a court suit should result in a victim's being awarded a settlement that exceeds the policyholder's coverage, the additional sum would have to be paid by the insured. For example, suppose that a driver carried 15/30/10 liability insurance and a court awards $18,000 to one victim and $2,000 to each of three other victims. Even though the total awarded to all victims ($24,000) is less than the maximum total coverage ($30,000), the insurance company will not pay the total award: Since the maximum liability for a single victim is $15,000, the insurance company is responsible only for $21,000 ($15,000 to the first victim and $2,000 each to the other three victims). The remaining $3,000 of the $18,000 judgment must be paid by the policyholder.

Two factors determine the standard rates charged for liability insurance: the classification of the vehicle (according to how much it is driven, whether used for business, and the age of the driver) and the territory in which the vehicle is operated. As would be expected, rates are higher in populous areas, since traffic congestion results in more accidents. If a driver has a record of accidents and/or speeding tickets, insurance will be available only at greatly increased rates through an *assigned risk plan,* a joint underwriting association, or a reinsurance facility.

Table 6-3 contains excerpts of the driver classifications used to determine liability insurance premiums. As you can see from the table, drivers under 21 are intricately classified according to sex, age, and whether or not they have taken a driver training course. Males under 25 are classified according to marital status and whether they are the owners or principal operators of the automobiles. Unmarried males may continue to be so classified until they reach age 30. Most mature drivers would be classified "all others." (An automobile is classified according to the status of the youngest person who operates the car or according to the person whose driver classification requires the highest multiple.)

There is a positive correlation between scholastic records and driving records of full-time students. Full-time high school and college students who have a B average or better or are in the upper 20% of their class may often qualify for a *good student discount.* This discount is usually a 15% to 20% reduction in the annual premiums paid to the insurance company.

TABLE 6-3 DRIVER CLASSIFICATIONS

Multiples of Base Annual Automobile Insurance Premiums

		PLEASURE; LESS THAN 3 MILES TO WORK EACH WAY	DRIVES TO WORK 3 TO 9 MILES EACH WAY	DRIVES TO WORK 10 MILES OR MORE EACH WAY	USED IN BUSINESS
No young operators	Only operator is female, age 30–64	0.90	1.00	1.30	1.40
	One or more operators age 65 or over	1.00	1.10	1.40	1.50
	All others	1.00	1.10	1.40	1.50
Young females	Age 16 DT[a]	1.40	1.50	1.80	1.90
	Age 16 No DT	1.55	1.65	1.95	2.05
	Age 20 DT	1.05	1.15	1.45	1.55
	Age 20 No DT	1.10	1.20	1.50	1.60
Young males (married)	Age 16 DT	1.60	1.70	2.00	2.10
	Age 16 No DT	1.80	1.90	2.20	2.30
	Age 20 DT	1.45	1.55	1.85	1.95
	Age 20 No DT	1.50	1.60	1.90	2.00
	Age 21	1.40	1.50	1.80	1.90
	Age 24	1.10	1.20	1.50	1.60
Young unmarried males (not principal operator)	Age 16 DT	2.05	2.15	2.45	2.55
	Age 16 No DT	2.30	2.40	2.70	2.80
	Age 20 DT	1.60	1.70	2.00	2.10
	Age 20 No DT	1.70	1.80	2.10	2.20
	Age 21	1.55	1.65	1.95	2.05
	Age 24	1.10	1.20	1.50	1.60
Young unmarried males (owner or principal operator)	Age 16 DT	2.70	2.80	3.10	3.20
	Age 16 No DT	3.30	3.40	3.70	3.80
	Age 20 DT	2.55	2.65	2.95	3.05
	Age 20 No DT	2.70	2.80	3.10	3.20
	Age 21	2.50	2.60	2.90	3.00
	Age 24	1.90	2.00	2.30	2.40
	Age 26	1.50	1.60	1.90	2.00
	Age 29	1.10	1.20	1.50	1.60

[a]"DT" indicates completion of a certified driver training course.

TABLE 6-4 AUTOMOBILE LIABILITY AND MEDICAL PAYMENT INSURANCE

Base Annual Premiums

BODILY INJURY				PROPERTY DAMAGE			
COVERAGE	TERRITORY 1	TERRITORY 2	TERRITORY 3	COVERAGE	TERRITORY 1	TERRITORY 2	TERRITORY 3
15/30	$ 81	$ 91	$112	$ 5,000	$83	$ 95	$100
25/25	83	94	115	10,000	85	97	103
25/50	86	97	120	25,000	86	99	104
50/50	88	101	125	50,000	87	101	107
50/100	90	103	129	100,000	90	103	108
100/100	91	104	131	MEDICAL PAYMENT			
100/200	94	108	136				
100/300	95	110	139	$ 1,000	$62	$ 63	$ 64
200/300	98	112	141	2,500	65	66	67
300/300	100	115	144	5,000	67	68	69
				10,000	70	72	74

Table 6-4 presents some typical base annual rates for automobile liability insurance (bodily injury and property damage) and medical payment insurance in three different territories. The total premium cost for two or more vehicles would be reduced by a certain percent when the vehicles are covered under one insurance policy.

Example 1 Edwin Carter is 32 years old and drives his car to work each day, a one-way distance of 6 miles. The area in which he lives is classified Territory 1. (a) Determine the cost of 15/30/10 liability coverage on his automobile and $2,500 medical payment insurance. (b) What would Carter's premium be if he increased his insurance to 50/100 bodily injury and $25,000 property damage coverage (with the same $2,500 medical pay coverage)?

(a) Table 6-4 is used to compute the base annual premium of Carter's liability ($15,000 single and $30,000 total bodily injury; $10,000 property damage) and medical payment ($2,500) coverage in Territory 1:

$ 81	Bodily injury (15/30)
85	Property damage ($10,000)
+ 65	Medical payment ($2,500)
$231	Total base premium

The base annual premium in Territory 1 is $231 for 15/30/10 liability and $2,500 medical pay coverage. Because Carter is 32 years old, he falls into the category "all others" in the driver classification table. Table 6-3 indicates that because he drives 6 miles to work, the multiple 1.10 should be used to compute Carter's annual premium:

	$ 231	Base annual premium
×	1.10	
	$254.10	Actual annual premium

(b) Increased coverage would cost as follows:

	$ 90	50/100 Bodily injury
	86	$25,000 Property damage
+	65	$2,500 Medical payment
	$ 241	Base annual premium
×	1.10	
	$265.10	Total annual premium

Note. In an actual insurance office, the agent is not required to use the driver multiple (like 1.10 above) because extensive tables are available that show the rates in every category, with the appropriate multiple already applied. It is beneficial for students to compute rates using the multiples, however, since it enables you to see clearly how the significantly different premium rates for various ages and categories are obtained.

Example 2 Bill Hold of Hold Construction Co. was driving a company truck when it struck an automobile, injuring a mother and her son. Hold Construction was subsequently sued for $120,000 personal injuries and $4,500 property damage. The court awarded the victims a $60,000 personal injury judgment and the entire property damage suit. Hold Construction carries 25/50/10 liability coverage. (a) How much will the insurance company pay? (b) How much of the award will Hold Construction have to pay?

(a) The insurance company will pay the $50,000 maximum total bodily injury coverage ($25,000 maximum per person) and the total award for property damage ($4,500). Thus, the insurance company will pay a total of $54,500.

(b) Since the court awarded the claimants more than Hold Construction's bodily injury coverage, the firm will be responsible for the excess:

	$60,000	Bodily injury awarded
−	50,000	Liability coverage for bodily injury
	$10,000	Amount to be paid by Hold Construction

NO-FAULT INSURANCE

No-fault insurance has been tried, with varying degrees of success, in approximately half of the states. At one time Congress considered making a standard no-fault policy mandatory nationwide, but the proposed legislation failed to pass.

Under **no-fault insurance,** each motorist collects for *bodily injuries* from his or her own insurance company after an accident, regardless of who is at fault. Reimbursement is made for medical expenses and the value of lost wages and/or services (such as the cost of a housekeeper for an injured mother). The victim then forfeits any right to sue, unless the accident was so severe that the damages exceed a certain specified amount.

No-fault insurance was proposed as a means whereby motorists could save money on insurance but still be protected financially in case of an accident. However, the savings derived through no-fault insurance pertain only to reduced bodily injury premiums, since small suits are disallowed. Premiums for property damage liability, as well as comprehensive and collision insurance on the driver's own vehicle (described below), are not reduced by the adoption of no-fault insurance. Furthermore, considerably more than half of a motorist's premiums apply toward these latter types of insurance. Thus, rising automobile repair costs, in particular, plus the tendency of victims to sue for large amounts, as well as the increased operating costs of the insurance companies themselves, have combined to produce continually increasing premiums in most areas, despite the adoption of no-fault insurance. Since the originally high expectations for no-fault insurance have not been achieved, enthusiasm for its adoption has also lessened.

Comprehensive and Collision Insurance

Comprehensive and collision insurance protect against damage to the policyholder's own automobile. **Comprehensive insurance** covers damage to the car resulting from fire, theft, vandalism, acts of nature, falling objects, and so forth. Items such as coats left in a car are not covered. Equipment not built into the car, such as a CB radio, generally is not covered. Extra insurance can be purchased on these items.

Protection against collision or upset damage may be obtained through **collision insurance.** (That is, collision insurance also covers damage resulting from one-car accidents—including damage to runaway, driverless vehicles—where no collision of two vehicles occurred.) Collision insurance pays for repairs to the vehicle of the insured when the policyholder is responsible for an accident, when the insured's car was damaged by a hit-and-run driver, or when another driver was responsible for the collision but did not have liability insurance and was unable to pay for the property damage caused. Collision insurance does not pay for loss or damage to other vehicles or property. Because collision insurance (as well as comprehensive) will pay for damages regardless of who is at fault, it is in this respect already a type of no-fault insurance and thus is not affected whenever no-fault insurance is adopted in a state.

Also available is **uninsured motorists insurance,** which offers financial protection for the policyholder's own bodily injuries or those of a family member when hit by a driver who did not carry bodily injury liability insurance. It does not cover the uninsured motorist.

Since comprehensive and collision insurance provides compensation only to the insured, it is not required by state law. Under certain circumstances, however, this coverage is mandatory. If an automobile is purchased by monthly installment payments, the institution financing the purchase retains legal title to the automobile until all installments are paid. To protect their investment, these financial agencies ordinarily require the purchaser to carry comprehensive and collision insurance.

A vehicle's owner may purchase collision insurance which pays the entire cost of repairing damage; however, such insurance is quite expensive. Most collision insurance is sold with a "deductible" clause, which means that the policyholder pays a specified part of any repair cost (an amount specified in the deductible clause). For instance, if "$250 deductible" collision insurance is purchased, the insured must pay the first $250 in damage resulting from any one collision, and the carrier will pay the remaining cost, up to the value of the vehicle. (If damage is less than $250, the owner must pay the entire cost.) Collision policies are available with various deductible amounts specified; $250-deductible policies are most common, but deductibles of higher amounts are becoming more frequent. Whereas comprehensive policies formerly were usually written without deductible amounts, such policies now often have a deductible. Both collision and comprehensive coverage become considerably less expensive as the policyholder pays a higher amount of each repair cost.

Recall that standard liability rates are determined by the classification of vehicle use and the territory in which the vehicle is operated. In addition to the previous considerations, comprehensive/collision rates also depend on the make, model, and age of the vehicle (example: "Chevrolet, Camaro, 2 years old"). Each automobile model is assigned an identification letter from A to Z. (Less expensive models are identified by letters near the beginning of the alphabet, and successive letters identify increasingly expensive models.)

The age of the automobile influences comprehensive and collision costs, since newer automobiles are more expensive to repair. Automobiles of the current model year are classified as "Age Group 1"; cars in Group 2 (the first preceding model year) and Group 3 (the second preceding model year) both pay the same rate; and models 3 or more years old are classified in Group 4. Table 6-5 lists excerpts of base annual comprehensive and collision insurance rates in three territories for vehicles of various ages. As in the case of liability insurance, the base annual comprehensive/collision premiums must be multiplied by a multiple reflecting the driver's classification.

Example 3 Juan Lopez, who is 26 and single, uses his car in business in Territory 3. His car is a Model K, less than 1 year old. Find his annual premium for the following insurance: 50/50/10 liability insurance, $1,000 medical payment, full comprehensive, and $500-deductible collision.

Table 6-4 is used to determine the base annual liability (bodily injury and property damage) and medical pay premiums, as before. The comprehensive and collision base premiums are shown in Table 6-5. The total of all these base annual premiums must then be multiplied by the driver classification multiple from Table 6-3 to obtain the total annual premium.

TABLE 6-5 COMPREHENSIVE AND COLLISION INSURANCE

Base Annual Premiums

		TERRITORY 1			TERRITORY 2			TERRITORY 3		
MODEL CLASS	AGE GROUP	COMPRE-HENSIVE	$250-DEDUCTIBLE COLLISION	$500-DEDUCTIBLE COLLISION	COMPRE-HENSIVE	$250-DEDUCTIBLE COLLISION	$500-DEDUCTIBLE COLLISION	COMPRE-HENSIVE	$250-DEDUCTIBLE COLLISION	$500-DEDUCTIBLE COLLISION
(1)	1	$55	$ 82	$ 76	$59	$ 92	$ 80	$73	$100	$ 91
A-G	2,3	52	77	73	56	86	76	58	94	85
	4	49	71	67	51	79	70	54	85	78
(3)	1	63	111	101	69	128	108	75	141	127
J-K	2,3	59	103	95	64	118	101	68	131	118
	4	54	93	86	57	106	91	61	116	105
(4)	1	68	123	112	76	143	120	83	169	142
L-M	2,3	64	125	104	70	133	112	75	138	132
	4	57	102	94	62	117	100	66	129	117
(5)	1	77	140	126	86	164	136	95	183	162
N-O	2,3	70	130	117	77	151	126	85	168	150
	4	62	115	105	68	133	112	73	147	132

$ 125	50/50 Bodily injury
103	$10,000 Property damage
64	$1,000 Medical payment
75	Comprehensive (Model K, Age Group 1, Territory 3)
+ 127	$500-deductible collision (classified as above)
$494	Base annual premium
× 2.00	Driver classification (unmarried male, age 26, business use)
$988	Total annual premium

Example 4 Harold Granger carries 10/20/5 liability insurance and $250-deductible collision insurance. Granger was at fault in an accident which caused $900 damage to his own car and $1,200 damage to the other vehicle. (a) How much of this property damage will Granger's insurance company pay? (b) Suppose that a court suit results in a $12,000 award for personal injuries to the other driver. How much would the insurance company pay and how much is Granger's responsibility?

(a)

Property damage to the other car (under Granger's $5,000 property damage policy)	$1,200
Property damage to Granger's car ($900 − $250 deductible)	+ 650
Total property-damage settlement	$1,850

(b) Under Granger's $10,000 single and $20,000 total bodily injury policy, the maximum amount the insurance company will pay to any one victim is $10,000. Thus, Granger is personally liable for the remaining $2,000 of the $12,000 settlement (making his total obligation $2,250, including the $250 deductible).

SECTION 2 PROBLEMS

Compute the following motor-vehicle insurance problems using the tables in this section. Compute the total of liability and medical payment premiums for each of the following.

	Liability coverage	Medical payment	Territory	Driver classification	Total liability and medical pmt. premium
1. a.	15/30/10	$ 1,000	2	Female, 20, no driver's training, drives 6 miles to work	
b.	50/50/25	5,000	1	Female, 30, drives 12 miles to work	
c.	100/200/50	10,000	3	Unmarried male, 21, not principal operator, drives 5 miles to work	
2. a.	25/25/10	$ 1,000	1	Male, 50, drives 5 miles to work	
b.	50/100/50	2,500	2	Female, 16, no driver's training, drives for pleasure	
c.	300/300/100	10,000	3	Married male, 24, drives 15 miles to work	

Determine the total comprehensive/collision premiums for the following.

	Model	Age group	Territory	Deductible on collision	Driver classification	Total comprehensive/ collision premium
3. a.	A	3	1	$250	Female, 16, driver's training, pleasure driving	
b.	J	1	2	500	Married male, 20, no driver's training, drives 12 miles to work	
c.	L	2	2	500	Female, 65, drives 15 miles to work	
4. a.	N	2	1	$250	Female, 45, drives 9 miles to work	
b.	J	1	2	500	Unmarried male, 29, owner, drives less than 3 miles to work	
c.	A	3	3	500	Female, 20, driver's training, drives 10 miles to work	

5. Jim Chang, age 20 and unmarried, lives in Territory 2 and drives 6 miles each way to a part-time job. He has had driver's training and is not the principal operator. How much will he pay for liability coverage that includes $25,000 for single bodily injuries, $50,000 for total bodily injuries, $10,000 for property damages, and $1,000 for medical payment coverage?

6. Melissa Baylor, age 32, lives in Territory 1. Each day she drives 5 miles each way to the college where she teaches. Her liability insurance includes $25,000 for single bodily injury, $50,000 for total bodily injury, $25,000 for property damage, and $2,500 for medical payment. Determine her annual payment.

7. Becky Karolyi, age 27, lives in Territory 1 where she uses her car as a real estate agent. She drives a Model N, Age Group 1 car. Her bank requires that she carry comprehensive and $250-deductible collision insurance until her car loan is paid in full. What will be the annual cost of this insurance coverage?

8. John Poynor, unmarried and age 21, lives in Territory 3. He owns his car and drives a Model C, Age Group 4 car to work 6 miles each way. Determine his annual premium for comprehensive and $500-deductible collision coverage.

9. Bob Burton, age 48, lives in Territory 3. He drives 2½ miles each way daily to his computer analyst job in his Model B car that is 2 years old. His insurance coverage includes 100/100/50 liability and $10,000 medical payment. He also carries comprehensive and $500-deductible collision insurance. Determine his annual car insurance cost.

10. Nicole Sudduth, age 66, bought a new Model N car. She lives in Territory 2 and drives only for pleasure. Her auto insurance includes 25/50/5 liability coverage and $1,000 medical payment. She also has comprehensive and $250-deductible collision coverage. How much is her annual premium?

11. Erika Chavez, age 22, drives 11 miles each way to her position as a word processor at an insurance agency. She lives in Territory 2 and drives her Model K car that is 4 years old. Chavez's insurance includes 50/50/10 liability coverage, $2,500 medical pay, comprehensive, and $500-deductible collision protection. Find her annual cost.

12. Jabob Garber, age 24 and married, drives daily 4 miles each way to work in Territory 1. He drives a Model A car that is 1 year old. His insurance includes 100/100/50 liability, $10,000 medical pay, comprehensive, and $500-deductible collision. What is his annual premium?

13. Peter Conroy was liable for an accident in which two people were injured. His insurance included 50/100/10 liability protection. The court awarded one person $30,000 and the other $65,000 for personal bodily injuries. An award of $25,000 for damages to the other car was also part of the court settlement.
    a. How much of the court settlement will the insurance company pay?
    b. How much of the expense must Conroy pay personally?

14. Katie Rivers' car struck another car and injured two people inside. The other car received $15,000 damages. The court awarded $20,000 to one of the people and $40,000 to the other person for medical expenses. Miss Rivers carries 15/30/10 liability insurance.
    a. How much of this court settlement will the insurance company pay?
    b. How much is Miss Rivers' personal responsibility?

15. Janet Bruce's car was stolen while parked outside her home. The car had a market value of $20,000. Bruce carried comprehensive and $500-deductible collision insurance.
    a. How much indemnity will the insurance company pay?
    b. How much out-of-pocket expense will Bruce incur to replace the car, assuming that she can buy another one for $20,000?

16. George Messinger's car was vandalized in a parking garage, which caused $1,800 in damages. Messinger carried comprehensive and $500-deductible collision insurance.
    a. How much of the damages will his insurance cover?
    b. How much of the repair bill will Messinger pay?

17. A company van owned by Gault Service Corp. was responsible for an accident with a passenger car. The driver and an occupant of the car suffered bodily injuries of $48,000 and $60,000, respectively. Damage to the van totaled $7,000 and to the car $12,000. Gault Service Corp. carried 15/30/5 liability, comprehensive, and $500-deductible collision.

a. What indemnity will the insurance company pay?

b. How much of the total expense will Gault Service Corp. have to pay?

18. Sidwell Florist's delivery truck struck a car, causing $13,000 in damages to the car. The driver of the car and a passenger suffered internal injuries amounting to $10,000 and $75,000, respectively. Damages to the truck totaled $26,000. Sidwell carried 25/50/10 liability, comprehensive, and $250-deductible collision insurance.

a. What amount will Sidwell Florist's insurance company pay toward these costs?

b. How much will Sidwell Florist have to pay?

19. A truck driven by Tom Edmunds crashed into a car driven by Tim Gemini, killing the passenger of the car and injuring Mr. Gemini. After a court suit, the widow of the passenger was awarded $400,000 on the death of her husband, and $110,000 was awarded to Mr. Gemini. Mr. Edmunds suffered $3,000 of personal injuries. Property damages were $15,000 to the car and $2,000 to the truck. Mr. Edmunds carried 100/200/25 liability, comprehensive, $250-deductible collision, and medical payment of $1,000.

a. How much of the total cost will the insurance company pay?

b. How much will Mr. Edmunds have to pay?

20. A car driven by Ed Jerde crashed into a car driven by Susan Jennings, killing Miss Jennings and injuring her passenger. A court awarded $500,000 to the Jennings family and $50,000 to the injured passenger. Mr. Jerde suffered $25,000 of personal injuries. His car had $17,000 of damages. Property damages to the other car totaled $20,000. Mr. Jerde carried 200/300/100 liability, $10,000 medical payment, comprehensive, and $500-deductible collision insurance.

a. Determine the total amount the insurance company should pay.

b. For what amount is Mr. Jerde responsible?

LIFE INSURANCE

The basic purpose of life insurance is to provide compensation to survivors following the death of the insured. Whereas other types of insurance pay damages only up to the actual value of the insured property, life insurance companies make no attempt to assign a specific value to any life. A person (in good health and with normal expectancy of survival) may purchase any amount of insurance corresponding to the policyholder's income and standard of living; and, when death occurs, the full value of the life insurance coverage will be paid.

The responsibility that most heads of households feel for the financial security of their families induces them to purchase life insurance. Life insurance is owned by 78% of American households. The Life Insurance Marketing and Research Association reports that the average amount of life insurance is $143,100 per insured household in the United States (compared with only $111,600 per household when all households are considered). Some 71% of adult men have life insurance, whereas

65% of adult women are insured. The average (mean) amount of total life insurance on an adult man is $105,500, which is twice the average amount held by an insured adult woman ($52,700). Group life insurance is steadily increasing, having become a standard employee benefit. Group insurance amounts to 40.1% of the life insurance in force.

Many businesses purchase life insurance on key members of the firm whose death would cause a severe loss to the business. Business partners sometimes carry insurance on each other so that, in event of a death, the surviving partners can purchase the deceased partner's share of the business from the estate.

TERM

The **term** policy is so named because it is issued for a specific period of time. The policyholder is insured during this time, and, if the insured is still alive when the term expires, the coverage then ceases. Term insurance is considerably less expensive than the other types of policies. The principal reason for the low cost is that term policies do not build investment value for the policyholder. (Investment values are discussed in the section "Nonforfeiture Options.") Term insurance accounts for 22% of all new policies written today.

Term policies are commonly issued for terms of 5, 10, or 15 years. Many heads of families feel that their need for insurance is greatest while they are younger and have children to support and educate. Term policies can provide maximum insurance protection during these years at a minimum cost. Another typical use of term policies is the life insurance purchased for a plane or boat trip; the term in this case is the duration of the trip.

Term policies may include the provision that, at the end of the term, the policy may be renewed for another term or converted to another form of insurance (both at increased premium rates). A variation of the ordinary term policy is *decreasing term* insurance; under this policy, the face value is largest when issued and decreases each year until it reaches zero at the end of the term. A typical example of decreasing term insurance is mortgage insurance, which decreases as the home loan decreases.

WHOLE LIFE

The **whole life** (or **straight life**) policy provides protection throughout the insured's lifetime. Payment of the face amount is paid upon death of the insured regardless of when death occurs. The whole life policy is by far the life insurance coverage most often purchased in the United States. Approximately 58% of all policies purchased are whole life. In order to keep the original policy in force, premiums must be paid for as long as the insured lives. The premium cost depends upon the age of the policyholder when the insurance is first purchased, and this premium remains the same each year thereafter. In effect, the insurance cost is averaged over the policyholder's lifetime. An important feature of whole life insurance is that it accumulates a cash value, similar to a savings account.

LIMITED PAYMENT LIFE

The **limited payment life** policy is a variation of the whole life policy. With limited payment insurance, the policyholder makes payments for only a limited number of years. If the policyholder survives beyond this time, no further payments are required; however, the same amount of insurance coverage remains in force for the remainder of the policyholder's life. This coverage is very similar to whole life insurance, except that the payments are sufficiently larger so that the policyholder will pay, during the limited payment period, an amount approximately equivalent to the total that would be paid during an average lifetime for whole life insurance.

Limited payment policies are frequently sold with payment periods of 20 or 30 years. Such policies are called "20-payment life" or "30-payment life" policies. The limited payment policy is also commonly sold with a payment period so that the policy is "paid up" at age 65.

UNIVERSAL AND VARIABLE LIFE

The **universal life** policy is a flexible-premium, adjustable death benefit policy. Offered as an inducement to investors, universal life provides low-cost term insurance, with the balance of each premium going into a cash account that earns interest. A minimum rate is guaranteed (about 4% or 5%) but may go much higher depending on how successful the company's own investments are. The actual rate of growth will vary from year to year.

The universal life policy is highly flexible, allowing the policyholder to raise or lower the face amount as circumstances change. The premium payment may also vary, or, if the policyholder fails to make a payment, the premium is simply deducted from the cash account. (Regular payments would be necessary, however, to keep coverage in force for the full term and to build investment value.)

The **variable life** policy is similar to universal life, but was developed to offer investors more options, with the potential for higher rates of return. A set premium is paid, and the investment portion is distributed among any of several different options selected by the policyholder—typically, stock funds, bond funds, and money-market accounts. These funds are all subject to the interest-rate fluctuations of the national economic climate.

Certain minimums are guaranteed, but in general both the interest rate and the death benefit may rise or fall, depending on the performance of the policyholder's investment choices. Thus, with a variable life policy, the customer assumes more responsibility for the investment risk, whereas the return on a universal life policy depends on the success of the company's investments overall. Universal and variable life policies together account for about 20% of policies written today.

Since many people are uncomfortable with an uncertain death benefit, the insurance industry developed a policy intended to offer the best of both worlds. Called **variable universal life,** this policy provides a guaranteed death benefit while still allowing the policyholder to select from various types of funds for the cash investment. As with universal life, there is much flexibility regarding when premiums are paid and their amount, but if interest rates should fall, premiums must be paid throughout the term of this policy, to keep it in force.

Note. This policy may also be called "flexible-premium variable life."

There are arguments pro and con regarding the advisability of using insurance policies as an investment. The insurance industry therefore attempts to offer something for everyone—from pure insurance with no investment (term policies) to policies that are primarily investments (such as single-premium variable life policies that are fully paid upon purchase and are bought simply to build value that will provide monthly annuity payments after retirement).

PARTICIPATING AND NONPARTICIPATING

Insurance policies are further classified as participating or nonparticipating. **Participating** policies are so named because the policyholders participate in the profits earned on these policies. The premium rates charged for participating policies are usually somewhat higher, because the insurance companies charge more than they expect to need in order to pay claims and operating expenses. However, the excess funds at the end of the year are then refunded to policyholders in the form of *dividends.*

At the time a participating policy is purchased, the policyholder selects the form in which dividends will be paid. The usual choices available to the policyholder are that the dividends may be (1) paid in cash, (2) used to help pay the premiums, (3) used to purchase additional paid-up insurance (paid-up insurance is discussed in the section "Nonforfeiture Options"), or (4) left invested with the company to accumulate with interest. A fifth option sometimes available is that the dividends may be (5) used to purchase a 1-year term insurance policy.

Nonparticipating policies, on the other hand, generally offer lower premium rates than participating policies; however, no dividends are paid on nonparticipating policies. Thus, the actual annual cost of the life insurance is usually less for participating policyholders than for the owners of nonparticipating policies.

Most participating policies are sold by mutual companies. **Mutual life insurance companies** do not have stockholders, but are owned by the policyholders themselves. The companies are governed by a board of directors elected by the policyholders. Life insurance companies may also be *stock companies,* which are corporations operated for the purpose of earning profits and which are owned by stockholders. As is the usual case for corporations, the board of directors of a stock life insurance company is elected by the stockholders, and the profits earned by the company are paid to the stockholders. Thus, most stock companies issue nonparticipating policies, although a few do issue some participating policies.

PREMIUMS

The risk assumed by insurance companies for other types of policies (fire or automobile, for example) is the probability that a loss will occur times the expected size of the loss (it may be only a partial loss). With life insurance, however, the risk is not "whether" a loss will occur but "when" the loss will occur, for death will inevitably come to all policyholders, and the full value of each effective policy will always be paid after a death.

Premium rates for life insurance are computed by an *actuary,* a highly skilled person trained in mathematical probability as well as business administration. To help calculate premiums and benefits, the actuary uses a *mortality table;* this is an extensive table showing how many people die at each age and the death rate (per 100,000 people) for each age group. Table 6-6 shows an excerpt from a table published by the U.S. Department of Health and Human Services. (Collecting and tabulating these statistics requires a delay of two or more years between the date of the data and their publication.)

If a group of people, all of the same age, take out insurance during the same year, the mortality table enables the actuary to know how many claims to expect during each succeeding year. Premium rates must be set so that the total amount paid by all policyholders is sufficient to pay all these claims, as well as to finance business expenses.

Almost all life insurance sold today is issued on the *level premium system.* This means that the policyholder pays the same premium each year for life insurance protection. Thus, the policyholders pay more than the expected claims during the early years of their policies, and during the later years they pay less than the cost of the claims. The excess paid during the early years is invested by the insurance company, and interest earned on these investments pays part of the cost of the insurance. Thus, the total premium cost over many years is less than if the policyholders paid each year only the amount which the company expected to need.

A physical examination is often required before an individual life insurance policy will be issued. (Group policies, on the other hand, are generally issued without physical examinations.) Annual premiums on new policies are naturally less expensive for younger people, since their life expectancy is longer and the insurance company expects them to pay for more years than older persons would. (This emphasizes the advisability of setting up an adequate insurance program while one is still young. The high cost of insurance taken out at an older age makes new coverage too expensive for many older persons to afford.) Life insurance taken out at any given age is usually less expensive for women than for men, because women have a longer life expectancy.

Life insurance premiums on new policies are determined by the applicant's age to the nearest birthday. Thus, applicants who are $26\frac{1}{2}$ would pay premiums as if they were 27 at the time when the policy was issued.

Life insurance policies are written with face values in multiples of $1,000. The rate tables list the cost for a $1,000 policy, and this rate must be multiplied by the number of thousands of dollars in coverage that is being purchased. Table 6-7 gives typical annual premiums per $1,000 of face value for each type of life insurance policy, when taken out at various ages. (The policyholder would pay a slightly higher annual total if premiums were paid semiannually, quarterly, or monthly. Each semiannual premium is approximately 51.5% of the annual premium; quarterly premiums are approximately 26.3% of the annual; and each monthly premium would be approximately 8.9% of the annual.)

Table 6-6 Deaths and death rates by age, United States, 1996

[*Rates per* 100,000 *Population in Specified Group*]

Number

Age	All races: Both sexes	Male	Female
All ages	2,314,690	1,163,569	1,151,121
Under 1 year	28,487	15,965	12,522
1 - 4 years	5,948	3,349	2,599
5 - 9 years	3,780	2,200	1,580
10 - 14 years	4,550	2,803	1,747
15 - 19 years	14,663	10,682	3,981
20 - 24 years	17,780	13,631	4,149
25 - 29 years	20,730	14,825	5,905
30 - 34 years	30,417	21,245	9,172
35 - 39 years	42,499	28,806	13,693
40 - 44 years	53,534	35,498	18,036
45 - 49 years	67,032	43,056	23,976
50 - 54 years	77,297	47,815	29,482
55 - 59 years	96,726	58,974	37,752
60 - 64 years	136,999	82,196	54,803
65 - 69 years	200,045	116,660	83,385
70 - 74 years	273,849	152,513	121,336
75 - 79 years	321,223	167,377	153,846
80 - 84 years	342,067	157,882	184,185
85 years and over	576,541	187,719	388,822
Not stated	523	373	150

Rate

Age	All races: Both sexes	Male	Female
All ages[1]	872.5	896.4	849.7
Under 1 year[2]	755.7	828.0	680.0
1 - 4 years	38.3	42.2	34.3
5 - 9 years	19.4	22.1	16.7
10 - 14 years	24.0	28.8	18.9
15 - 19 years	78.6	111.0	44.0
20 - 24 years	101.3	151.5	48.5
25 - 29 years	109.1	155.4	62.4
30 - 34 years	142.4	199.4	85.7
35 - 39 years	188.2	255.9	121.0
40 - 44 years	257.2	344.3	171.7
45 - 49 years	363.6	475.2	255.7
50 - 54 years	554.8	705.6	411.9
55 - 59 years	851.3	1,081.1	639.1
60 - 64 years	1,370.1	1,744.8	1,036.3
65 - 69 years	2,022.4	2,588.3	1,548.6
70 - 74 years	3,119.8	3,995.1	2,446.1
75 - 79 years	4,674.0	5,934.2	3,796.8
80 - 84 years	7,505.6	9,477.2	6,369.8
85 years and over	15,327.2	17,547.7	14,444.7

[1]Figures for age not stated are included in "All ages" but are not distributed among age groups.
[2]Death rates under 1 year (based on population estimates) differ from infant mortality rates (based on live births).
Source: *1996 Monthly Vital Statistics Report,* Vol. 47, No. 9, National Center for Health Statistics, 1998.

TABLE 6-7 ANNUAL LIFE INSURANCE PREMIUMS

Per $1,000 of Face Value for Male Applicants[a]

AGE ISSUED (YEARS)	TERM 10-YEAR	WHOLE LIFE	VARIABLE UNIVERSAL	LIMITED PAYMENT 20-YEAR
18	$ 6.81	$14.77	$18.36	$24.69
20	6.88	15.46	19.81	25.59
22	6.95	16.12	20.28	26.53
24	7.05	16.82	21.33	27.53
25	7.10	17.22	22.17	28.06
26	7.18	17.67	23.43	28.63
28	7.35	18.64	24.79	29.84
30	7.59	19.73	26.05	31.12
35	8.68	23.99	30.87	35.80
40	10.64	28.26	35.55	40.34
45	14.52	33.79	41.24	46.01
50	22.18	40.77	48.11	53.24
55	32.93	51.38	57.62	64.77
60	—	59.32	—	70.86

[a]Because of women's longer life expectancy, premiums for women approximately equal those of men who are 5 years younger.

The premiums in Table 6-7 represent an approximate average between participating and nonparticipating rates. The premiums for variable universal policies can vary widely, depending on the policyholder's investment goal, but they probably average about the same as whole life policies. These rates are for example purposes only.

Since life expectancy is longer for women than for men, different premium rates are used. There would be a separate premium table for women, or an adjustment of 3 to 5 years can be made in the men's table. An adjustment or setback of 5 years in Table 6-7 will be used in this text for examples and problems involving women.

Example 1 Paul Kirkland is 30 and wishes to purchase $25,000 of life insurance. Determine the annual cost of (a) a whole life policy and (b) a 20-payment life policy.

In each case, the insurance coverage is to be $25,000. Thus, each premium rate from Table 6-7 must be multiplied by 25.

(a) *Whole Life*		(b) *20-Payment Life*
$ 19.73	← Premium per $1,000 →	$ 31.12
× 25		× 25
$493.25	← Premium on $25,000 →	$778.00

Kirkland would pay an annual premium of (a) $493.25 for a whole life policy or (b) $778 for similar coverage under a 20-payment life policy.

Example 2 Suppose that Paul Kirkland (Example 1) lived for 38 years after purchasing his life insurance. (a) How much would he pay altogether in premiums for each policy? (b) Which policy's premiums would cost more, and how much more?

(a)

Whole Life		*20-Payment Life*
$ 493.25	Annual premiums	$ 778.00
× 38	Number of premiums	× 20
$18,743.50	Total premium cost	$15,560.00

The total premium cost would be $18,743.50 for a $25,000 whole life policy and only $15,560 for an equivalent 20-payment life policy. Thus,

(b)

$18,743.50	Total premium cost of whole life policy
− 15,560.00	Total premium cost of 20-payment life policy
$ 3,183.50	Extra premium cost of whole life policy

The whole life policy would cost Kirkland $3,183.50 more if he lived to be 68.

Example 3 Donna Pitman, age 30, is considering the purchase of either a whole life policy or a variable universal policy with a face value of $50,000. Determine the annual cost of (a) a whole life policy and (b) a variable universal policy.

Using Table 6-7, the premium rates for Ms. Pitman will be "set back" 5 years from the male applicant rates, arriving at the age 25 rate.

(a) *Whole Life*		(b) *Variable Universal*
$ 17.22	← Premium per $1,000 →	$ 22.17
× 50		× 50
$860.00	← Premium on $50,000 →	$1,108.05

Actually, premiums can vary greatly from company to company on all policies, as do the nonforfeiture values, which will be discussed later. Besides the basic life insurance coverage, several types of additional protection may be purchased at a slight increase in premiums. Among the additional features most frequently purchased are the **waiver of premium benefit** (the company agrees to assume the cost of the policyholder's insurance in case of total and permanent disability) and the **double indemnity benefit** (the insurance company will pay twice the face value of the policy if death results from accidental causes rather than illness).

Since many different provisions are available on different policies, buyers trying to find the best value found themselves trying "to compare apples and oranges." To avoid this confusion, insurance companies are now required to disclose the **interest-adjusted net cost** of each policy—the total cost strictly for life insurance coverage, disregarding the investment values. This cost is also expressed as the **net cost index**—an annual cost equivalent to the total interest-adjusted net cost. Therefore, if a buyer is considering two similar insurance policies from equally reliable companies, the better choice may well be the policy with the lower interest-adjusted net cost.

NONFORFEITURE OPTIONS

The most obvious benefit from an insurance policy is the death benefit: The policyholder designates a person, called the **beneficiary,** to receive the face value of the policy following the death of the insured. However, there are a number of alternative benefits available to the policyholders themselves which should be taken into consideration when selecting a policy.

Whereas the premiums paid for other types of insurance policies purchase only protection, the premiums paid for life insurance policies (except term policies) also build an investment for the policyholder. This investment accumulates because the level premium system has caused excess funds to be paid into the company. Also, on the newer types of policies, the policyholder intentionally contributes extra funds for investment purposes.

Because of selling and bookkeeping costs, a policyholder is not usually considered to have accumulated any investment until 2 or 3 years after the policy was written. Any time thereafter, however, the policy possesses certain values, called **nonforfeiture values** (or **nonforfeiture options**) to which the policyholder is entitled. These values may often be claimed only if the insured stops paying premiums for some reason, or if the policyholder turns in the policy (and is thus no longer insured). The principal nonforfeiture options (cash value, paid-up insurance, and extended term insurance) are outlined briefly in the following paragraphs.

Included in each life insurance policy (excluding term policies) is a table showing the **cash value** of the policyholder's investment after each year. As discussed earlier, even the universal and variable life policies are guaranteed a minimum cash value, approximately equal to that of whole life policies with the same face value.

If the policyholder wishes to terminate the insurance coverage, (s)he may surrender the policy and receive this cash value. If the policyholder needs some money but prefers to maintain the insurance protection, (s)he may borrow up to the cash value of the policy. The loan must be repaid with interest at a moderate rate, and the full amount of insurance coverage will remain in force during this time.

If the insured wishes to stop paying premiums and surrender the policy, the cash value of the policy may be used to purchase a reduced level of **paid-up insurance.** This means that the company will issue a policy which has a smaller face value but which is completely paid for. That is, without paying any further premiums, the policyholder will have a reduced amount of insurance that will remain in effect for the remainder of his or her life. Life insurance policies (except term policies) also contain a table showing the amount of paid-up insurance that the cash value would purchase after each year. (The newer types of policies do not build paid-up insurance values.)

When a policyholder simply stops paying premiums without notifying the company of any intent and without selecting a nonforfeiture option, the company usually automatically applies the third nonforfeiture option: **extended term** insurance. Under this plan the policy remains in effect, at its full face value, for a limited period of time. Stated another way, the cash value of the policy is used to purchase a term policy with

TABLE 6-8 NONFORFEITURE OPTIONS[a] ON TYPICAL LIFE INSURANCE POLICIES
Issued at Age 25

YEARS IN FORCE	WHOLE LIFE				20-PAYMENT LIFE				VARIABLE UNIVERSAL
	CASH VALUE	PAID-UP INSURANCE	EXT. TERM		CASH VALUE	PAID-UP INSURANCE	EXT. TERM		
			YEARS	DAYS			YEARS	DAYS	CASH VALUE
3	$ 9	$ 19	1	190	$ 32	$ 94	10	84	$ 44
5	31	87	9	200	74	218	19	184	95
10	93	248	18	91	187	507	28	186	321
15	162	387	20	300	319	768	32	164	595
20	251	535	22	137	470	1,000	Life		1,030
40	576	827	—	—	701	—	—	—	2,979

[a]The "cash value" and "paid-up insurance" nonforfeiture values are per $1,000 of life insurance coverage. The time period for "extended term insurance" applies as shown to all policies, regardless of face value. The cash value for variable universal life represents a reasonable estimate of the increase in value rather than a guaranteed amount.

the same face value and for the maximum time period that the cash value will finance. Each policy also includes a table indicating after each year the length of time that an extended term policy would remain in force. This provision does not apply to term policies, since they do not build cash value. Also, if the policyholder of a universal or variable life policy fails to make payments, the company will simply deduct them from the cash account (rather than converting to extended term or paid-up insurance).

Note. Most insurance companies allow a *grace period* (usually 31 days following the date a premium is due) during which time the overdue premium may be paid without penalty. The policy remains in effect during this time.

The nonforfeiture options shown in Table 6-8 are typical provisions of life insurance policies. As stated previously, expanded tables are provided in individual policies. Example 4 illustrates the use of Table 6-8 in determining the three principal nonforfeiture options.

Example 4 Carl Best purchased a $50,000 20-payment life insurance policy at age 25. Determine the following values for his policy after it had been in force for 10 years: (a) the cash value, (b) the amount of paid-up insurance he could claim, and (c) the time period for which extended term insurance would remain in effect.

(a) As shown in Table 6-8, the cash value per $1,000 of face value after 10 years on a 20-payment life policy issued at age 25 is $187. Therefore, the cash value of a $50,000 policy is

$ 187	Cash value per $1,000
× 50	
$9,350	Cash value of a $50,000 policy

The cash value of Best's $50,000 policy would be $9,350 after the policy has been in effect for 10 years. Best could receive this amount by turning in his policy and forgoing his insurance coverage, or he could borrow this amount and pay it back with interest without losing his insurance protection.

(b) The amount of paid-up insurance available is

$ 507	Paid-up insurance per $1,000
× 50	
$25,350	Paid-up insurance on a $50,000 policy

If Best surrenders his policy, he could receive a policy of $25,350, which would remain in force until his death without any premiums being required.

(c) From Table 6-8, we see that if Best stops paying premiums, for whatever reason, extended term insurance would allow his full $50,000 coverage to remain in effect for 28 years and 186 days.

SETTLEMENT OPTIONS

When a life insurance policyholder dies, there are several ways in which the death benefits may be paid, called **settlement options.** According to the American Council of Life Insurance, life insurance companies paid out over $239 billion to beneficiaries in 1997. Whole life policies accounted for 56% of the payments. The beneficiary may receive the face value in one lump-sum payment, or the benefits may be left invested with the company in order to earn interest.

An alternative method of receiving benefits, which deserves consideration, is to receive the benefits in the form of an annuity. An **annuity** is a series of payments, equal in amount and paid at equal intervals of time. (We shall consider here only annuities paid monthly.)

There are several options available to a beneficiary or an *annuitant* (person receiving an annuity) concerning the type of annuity (or installment) to choose. These choices are described briefly here.

1. The beneficiary may choose a monthly installment of a *fixed amount.* In this case, the specified amount will be paid each month for as long as the insurance money (plus interest earned on it) lasts. For example, a beneficiary may decide to receive a payment of $150 per month; this will continue until the account is depleted. The amount of income is the primary consideration when choosing this option.
2. The beneficiary may prefer the security of a monthly payment for a *fixed number of years.* For example, the beneficiary may want to receive monthly payments for 15 years. The insurance agent will then determine how much the beneficiary may receive monthly in order for the funds to last the specified number of years.
3. A third choice is to take the benefits in the form of an *annuity for life.* That is, depending on the age and sex of the beneficiary, the insurance company will

agree to pay a monthly income to the beneficiary for as long as he or she lives. No further payments are made to anyone after the primary beneficiary dies.

4. The remaining possibility is to receive a *life annuity, guaranteed for a certain number of years.* For example, if the beneficiary chooses a life annuity guaranteed for 15 years, a monthly annuity will be paid for a minimum of 15 years or as long thereafter as the beneficiary lives. The monthly income from this annuity is somewhat lower than from the life annuity in plan 3, but a guaranteed annuity is selected much more often than an ordinary life annuity. The reason for this is that the unqualified life annuity is payable only to the beneficiary; when the beneficiary dies, even if only one annuity payment has been made, no further benefits will be paid. With the guaranteed annuity, however, installments are payable for as long as the primary beneficiary lives, or, if the primary beneficiary dies before a predetermined number of years, the company will continue the installments to a secondary beneficiary until the period ends. Thus, most people with dependents naturally choose the guaranteed annuity. Even persons without dependents often select the guaranteed annuity, feeling that they would rather have someone receive the payments if they should die, instead of risking that most of the face value of the policy might be lost.

It should be noted that these settlement options are also available to policyholders themselves, if they surrender the policy. Older people often feel that they no longer need the amount of life insurance coverage carried since they were younger and had children to support. Therefore, many retired persons convert some of their life insurance policies to annuities in order to supplement other retirement income. When whole life or limited payment life policies are converted to annuities, the monthly installment depends on the *cash value* of the policy rather than upon its face value.

Table 6-9 lists the monthly income per $1,000 of face value available under the various settlement options. Life insurance policies also include tables similar to Table 6-9.

Example 5 Martha Donaldson, age 60 and the beneficiary of a $30,000 life insurance policy, has decided to receive the money as monthly payments rather than one lump sum. (a) Find the monthly payment if she chooses to receive payments for 15 years. (b) If Mrs. Donaldson arranges a monthly income of approximately $255, how many years will the payments continue?

(a) Table 6-9 shows that for a $1,000 policy, the monthly installment for the fixed number of years (15) would be $6.89. Thus, Mrs. Donaldson could receive $206.70 monthly for 15 years, as follows:

$ 6.89	Monthly installment per $1,000
× 30	
$206.70	Monthly installment from $30,000 policy (for 15 years)

(b) Mrs. Donaldson will receive 30 times any amount shown in Table 6-9. Thus, 30 × $8.52 will provide approximately $255 per month. The table indicates that she could receive $255.60 monthly for 12 years before the money (and interest paid by the insurance company) has been depleted.

TABLE 6-9 SETTLEMENT OPTIONS

Monthly Installments per $1,000 of Face Value

OPTIONS 1 AND 2: FIXED AMOUNT OR FIXED NUMBER OF YEARS		OPTIONS 3 AND 4: INCOME FOR LIFE				
		AGE WHEN ANNUITY BEGINS				
YEARS	AMOUNT	MALE	FEMALE	LIFE ANNUITY	LIFE WITH 10 YEARS CERTAIN	LIFE WITH 20 YEARS CERTAIN
10	$9.60	40	45	$4.60	$4.56	$4.44
12	8.52	45	50	5.16	5.07	4.80
14	7.71	50	55	5.30	5.28	5.00
15	6.89	55	60	6.13	6.00	5.46
16	6.43	60	65	6.56	6.31	5.65
18	6.08	65	70	7.22	6.70	5.78
20	5.66					

Note. Notice that Mrs. Donaldson's age had no bearing on this problem, because she was not considering an annuity that would pay for the remainder of her life.

Example 6 Walter Carson became the beneficiary of a $25,000 life insurance policy at age 55. What would be the monthly income (a) from a life annuity and (b) from an annuity guaranteed for 20 years?

Carson's age affects both of these options because either annuity pays until his death. (Notice that an older person would receive more per month because payments are not expected to be made for as many years.) From Table 6-9 for a male, age 55, the two annuities would pay as follows:

(a) *Life Annuity*		(b) *Life, 20-Year Certain*
$ 6.13	← Monthly per $1,000 →	$ 5.46
× 25		× 25
$153.25	← Monthly for $25,000 policy →	$136.50

Option (a) indicates that Carson could receive $153.25 monthly for life. These payments would stop immediately upon his death, however, regardless of how few monthly installments might have been paid. Option (b) will pay $136.50 monthly for as long as Carson lives (even if this is more than 20 years). If he dies in less than 20 years, the payments will continue to his heirs until the guaranteed 20 years have passed.

Example 7 Mary Reynolds was widowed at age 60, inheriting a $40,000 life insurance policy. She chose to receive all the benefits over a 10-year period. (a) What was her monthly installment? (b) How much would she have received monthly by choosing a life annuity guaranteed for 10 years? (c) Suppose that Mrs. Reynolds actually lived another

13 years. From which option above would she have received more total income, and how much more?

(a)	*10-Year Annuity*		(b) *Life, 10-Year Certain*
	\$ 9.60	← Monthly per \$1,000 →	\$ 6.00
	× 40		× 40
	\$384.00	← Monthly for \$40,000 policy →	\$240.00

Option (a) paid Mrs. Reynolds \$384 per month for 10 years, although she lived longer. Had she chosen option (b), she would have received \$240.00 monthly for the entire 13 years she lived.

(c) The total payments received under each settlement would have been as follows:

10-Year Annuity		*Life Annuity Guaranteed 10 Years*
12	Months per year	12
× 10	Years	× 13
120	Payments	156
\$ 384	Monthly annuity	\$ 240
× 120	Payments	× 156
\$46,080	Total received	\$37,440

Mrs. Reynolds received \$46,080 − \$37,440 = \$8,640 more income by selecting the fixed 10-year annuity, despite the fact that the guaranteed life annuity would have paid for 3 years longer. Observe that no further payments would be made to her heirs, because the guaranteed period is past for both annuities.

SECTION 3 PROBLEMS

Unless other rates are given, use the tables in this chapter to compute the following life insurance problems. Assume that each age given is the "insurable age" (age to the nearest birthday) of the applicant. For female applicants, use a 5-year "setback" in age when obtaining rates from a table for males.

Using the information given and Table 6-7, compute the annual premium for each life insurance policy.

	APPLICANT, AGE	TYPE OF POLICY	FACE VALUE	ANNUAL PREMIUM
1.	a. Male, 20	10-year term	\$ 15,000	
	b. Male, 45	Whole life	50,000	
	c. Female, 25	20-payment life	20,000	
	d. Female, 30	Variable universal	100,000	

	Applicant, age	Type of policy	Face value	Annual premium
2.	a. Male, 18	10-year term	$10,000	
	b. Male, 35	Variable universal	75,000	
	c. Female, 40	Limited payment life	35,000	
	d. Female, 35	Whole life	50,000	

Determine the nonforfeiture values of the following policies, based on the rates in Table 6-8 (for policies issued at age 25).

	Years in force	Type of policy	Face value	Nonforfeiture options	Nonforfeiture value
3.	a. 20	Whole life	$100,000	Cash value	
	b. 5	20-payment life	50,000	Paid-up insurance	
	c. 10	Whole life	125,000	Extended term	
	d. 3	Variable universal	150,000	Cash value	
4.	a. 5	20-payment life	$ 60,000	Cash value	
	b. 15	Whole life	80,000	Paid-up insurance	
	c. 20	Whole life	25,000	Extended term	
	d. 40	Variable universal	140,000	Cash value	

Use Table 6-9 to find the monthly annuity (or total years of the annuity) that the beneficiary may select in settlement of the following policies.

	Beneficiary		Face value	Settlement options	Monthly annuity	
	Sex	Age			Years	Amount
5.	a. M	60	$ 55,000	Fixed number of years	16	
	b. F	50	75,000	Fixed amount per month		$425
	c. F	65	100,000	Life annuity	—	
	d. F	45	150,000	Guaranteed annuity	20	
6.	a. M	56	$100,000	Fixed number of years	15	
	b. M	50	45,000	Fixed amount per month		$290
	c. F	55	80,000	Life annuity	—	
	d. F	60	125,000	Guaranteed annuity	10	

7. Charles Simon purchased a 20-payment life policy with a face value of $85,000 when he was 25.

a. What was the annual premium for the policy?

b. If Simon purchased a whole life policy for the same face value, how much would the annual premium be?

Assume that Simon lives 30 years after purchasing the policy. What would his total premiums be for

c. The 20-payment life policy?

d. The whole life policy?

8. Zack Richards was 28 years old when he purchased his 20-payment life policy with a face value of $55,000.

a. What was the annual premium for the policy?

b. If Richards purchased a whole life policy for the same face value, how much would the annual premium be?

Assume that Richards lives 40 years after purchasing the policy. How much would he pay in total premiums for

c. The 20-payment life policy?

d. The whole life policy?

9. Christine Rubella purchased a $50,000, 10-year term policy at age 30.

a. What was the annual premium?

b. How much in total premiums will she pay over 10 years?

c. What would the equivalent whole life cost be during the same 10-year period?

d. How much would she save in premium cost by purchasing the term policy?

e. If Christine died the year after the 10-year term, how much life insurance would her beneficiary receive under each plan?

10. Jane Adams purchased a $20,000, 10-year term life policy at age 40.

a. What was the annual premium?

b. How much will she pay in total premiums?

c. If Adams had selected a whole life policy, how much would premiums have cost during the same 10 years?

d. How much would be saved in premiums by purchasing the 10-year term policy?

e. If Adams died in the following year after the term policy expired, how much life insurance would her beneficiary receive under each of the plans?

11. At age 30, James Godwin purchased a $75,000 whole life policy.

a. How much less would he pay annually on premiums if he had started this policy at age 20?

b. What total amount will he pay for the premiums to age 65?

c. What would the premiums have cost to age 65 if Godwin had started the policy at the age of 20?

d. What other factors should be considered?

12. When Mohammad Suzzman was 40, he bought a $100,000, variable universal policy.

a. What would the annual savings on premiums have been if Suzzman had taken out the policy at age 30?

b. How much will Mohammad's premiums cost to age 60?

c. What would the total premiums have been on a policy taken out at age 30, assuming that he lived to age 60?

d. What other factors should be considered?

13. Roland Burke, age 25, purchased a $100,000, 20-payment life policy.

a. What was his annual premium?

b. After 10 years, Burke surrendered his policy. What was the cash value of the policy at that time?

c. How much paid-up insurance for life would this policy provide after 10 years in force?

d. For how long would the same coverage remain in force if he discontinued paying premiums?

14. Sam Williamson, age 25, purchased a $200,000 whole life policy.

a. What was the annual premium?

b. After 10 years, Williamson decided to terminate the coverage. What was the cash value of the policy at this point?

c. How much paid-up insurance for life would this policy provide after 10 years?

d. How long would the same $200,000 coverage be continued under extended term?

15. At age 25, Tanya Dean purchased a $40,000, 20-payment life policy. It has been in force for 15 years. She is now faced with college tuition payments for her daughter. How much can she borrow on her $40,000 policy without surrendering her coverage?

16. Carol Siwek purchased a $65,000 whole life insurance policy at age 25. The policy has been in force for 20 years. How much can she borrow on her $65,000 policy without surrendering her coverage?

17. Marlow Jones purchased a $150,000 variable universal policy at age 25.

a. What is the cash value after 20 years?

b. What would the cash value be after 20 years on a whole life policy of the same amount?

18. Jack Shannon purchased a $100,000 variable universal life policy when he was 25.

a. How much cash value can he expect after 10 years?

b. What would the cash value be after 10 years on a 20-payment life policy of the same amount?

19. At age 52, Hilda Strothers became the beneficiary of an $80,000 life insurance policy.

a. If she decides to have the benefits paid over a 14-year period, what monthly annuity will she receive?

b. If she decides to receive a monthly annuity of about $770, for approximately how many years will she receive the benefits?

20. Faye Irvy, at age 55, became the beneficiary of her husband's $75,000 life insurance policy.

a. If she decides to have the benefits paid over a 20-year period, what annuity will Mrs. Irvy receive each month during that time?

b. If she wants to receive a monthly benefit of about $480, for approximately how many years will she receive the benefits?

21. Amilla Pashayev has become the beneficiary of a $125,000 life insurance policy at the age of 45.

a. Determine the monthly payment she will receive if she selects a life annuity.

b. If she selects a life annuity guaranteed for 20 years, how much will she receive each month?

22. Harry Paltrow inherited a $50,000 life insurance policy when he was 55.

a. Determine the monthly payment Paltrow will receive if he chooses a life annuity option.

b. If he chooses a life annuity guaranteed for 10 years, how much will he receive each month?

23. Sandra Tolson decided to convert the $50,000 cash value of her life insurance policy to a 10-year annuity when she reached age 55. She lived to age 68.

a. Determine the total amount she received from the 10-year annuity.

b. If she had chosen a life annuity with 10 years certain, what was the total amount that she would have received?

c. Did she gain or lose by her annuity choice, and how much? (Hint: An annuity is based on the cash value rather than the face value when the policyholder converts a policy.)

24. At age 50, Joe Cohen decided to convert the $45,000 cash value of his life insurance to an annuity. Mr. Cohen decided to pick a 20-year annuity. He lived to age 80.

a. Determine the total amount he received from the annuity.

b. If he had chosen a life annuity with 20 years certain, how much would he have received?

c. Did he gain or lose by his annuity choice, and how much? [Hint: Recall that an annuity is based on the cash value (rather than the face value) when the policyholder converts a policy.]

25. At age 50, Jennifer Holt became the beneficiary of a $90,000 life insurance policy. Ms. Holt chose a monthly annuity for life as her settlement option.

a. What was her monthly income from the annuity?

b. If Ms. Holt lived to age 67, what total amount did she receive?

c. If she had selected a life annuity guaranteed for 20 years, what monthly annuity would she have received?

d. Did she gain or lose by her choice of settlements, and how much?

26. Meg Denison, at age 60, became the beneficiary of a $70,000 life insurance policy. She selected as a settlement option a monthly annuity for life.

a. How much did she receive each month?

b. Mrs. Denison lived to be 76. What total amount did the annuity pay?

c. Suppose that Mrs. Denison had chosen a life annuity guaranteed for 20 years. What annuity would she have received each month?

d. Did she gain or lose by her choice of settlement options, and how much?

CHAPTER  GLOSSARY

Annuity. The payment of a set amount at a regular interval. (Often used in settlement of a life insurance claim, either for a fixed number of years or for life—sometimes for life with 10 or 20 years certain.)

Beneficiary. The person designated to receive the face value of a life insurance policy after death of the insured.

Bodily injury liability. The portion of standard motor vehicle insurance that pays for physical injuries to other persons when the insured is at fault in an accident.

Business interruption insurance. Insurance that provides income after a fire or natural disaster to meet continuing business expenses and net income until business operations resume.

Carrier. An insurance company (also called an "underwriter" or an "insurer").

Cash value. An investment value earned by all life insurance policies except term, which may be borrowed against or may be paid in cash to a policyholder who terminates the policy.

Coinsurance clause. A clause in many fire policies requiring coverage of a stipulated percent (often 80%) of value at the time of a fire in order to qualify for full reimbursement for damages.

Collision (and upset) insurance. Insurance that pays for damages to the policyholder's own vehicle (1) if the insured is at fault in an accident or (2) if the insured's vehicle is damaged when there is no other responsible driver whose liability insurance will pay.

Comprehensive insurance. Insurance that pays for damages of any kind to the policyholder's own vehicle, except those caused by collision or one-vehicle upset.

Disability insurance. Insurance that provides coverage for workers who become disabled and unable to work due to a physical or mental reason.

Double indemnity. An optional provision of many life insurance policies that pays twice the face value in case of accidental death.

Extended coverage. A supplement to standard fire policies that insures against related damages such as smoke and water damage.

Extended term. A nonforfeiture option earned by life insurance policies (except term or variable universal) whereby the same level of coverage may be maintained for a specified time after premiums are terminated.

Face value. The full (maximum) amount of insurance specified by an insurance policy.

Group insurance. Life and/or health/accident insurance available at a reduced cost to employees of a business or to members of a business/professional organization.

Indemnity. The payment made by an insurance company after an insured loss.

Insured. A person or business that purchases insurance coverage (also called a "policyholder"), or a person whose life is covered by a life insurance policy.

Insurer. An insurance company (also called a "carrier" or an "underwriter").

Interest-adjusted net cost. The method now used to determine the cost of life insurance protection only, not including the portion of premiums used to build investment value.

Interest-adjusted net cost index. The annual cost index of a life insurance policy: Total interest-adjusted net cost ÷ Years.

Liability insurance. The portion of standard motor vehicle insurance and general business insurance that provides financial protection when the person or business is held legally responsible for personal injuries and/or physical damages.

Limited payment life insurance. Insurance that provides life insurance and, within the specified payment period, builds paid-up insurance of the full face value of the policy.

Medical payment insurance. The optional portion of standard motor vehicle insurance that pays medical costs (supplementing other policies) for injuries to the policyholder's own self, family, and passengers, regardless of who is at fault in an accident.

Mutual life insurance company. A life insurance company owned by policyholders rather than stockholders; sells participating policies, which pay dividends to the policyholder.

Net cost index. An annual cost equivalent to the total interest-adjusted net cost. Also called the "interest-adjusted net cost index."

No-fault insurance. Vehicle insurance in which each motorist collects for bodily injuries from his or her insurance company after an accident, regardless of who is to blame.

Nonforfeiture value (or option). The investment value (cash value, paid-up insurance, or extended term) available when a life insurance policy is terminated (except term; variable/universal has cash value only).

Nonparticipating policy. A life insurance policy that pays no dividends and usually has lower premiums than participating policies; the type of policy usually sold by stock life insurance companies.

Paid-up insurance. A nonforfeiture value earned by life insurance policies (except term or variable/universal policies) whereby a specified level of insurance may be maintained after premiums are terminated.

Participating policy. A life insurance policy that pays dividends and usually has higher premiums than nonparticipating policies; the type of policy usually sold by mutual life insurance companies.

Policy. An insurance contract that specifies in detail the provisions and limitations of the coverage.

Policyholder. A person or business that purchases insurance (also called the "insured").

Premium. The payment made to purchase insurance coverage.

Property damage liability. The portion of standard motor vehicle insurance that pays for damages caused by the policyholder to another vehicle or other property.

Settlement option. Any of several types of annuities available to the beneficiary of a life insurance policy instead of a lump-sum payment.

Short-term policy. An insurance policy in force for less than 1 year (usually referring to a fire insurance policy).

Straight life insurance. (See "Whole life insurance.")

Term (of a policy). The time period for which a policy will remain in effect.

Term life insurance. Provides life insurance only and builds no nonforfeiture values for the policyholder; the least expensive type of life insurance policy.

Underwriter. An insurance company (also called the "carrier" or "insurer").

Uninsured motorists insurance. An optional portion of standard motor vehicle insurance that provides financial protection for the policyholder's own bodily injuries when hit by a driver without bodily injury liability insurance.

Universal life insurance. A policy providing term life coverage, with the remainder of each premium going into a cash account that earns interest at a rate reflecting the company's overall profit. Allows flexibility in paying premiums and changing policy provisions.

Variable life insurance. A policy providing term life coverage, with the remainder of each set premium going into stocks, bonds, money market accounts, or other funds the policyholder selects. The death benefit may fluctuate along with the value of the investment accounts.

Variable universal life insurance. Insurance that combines aspects of both variable life and universal life policies. Provides a set death benefit, invests cash into various financial funds, and allows flexibility in premium payments. (Also called flexible-premium variable life.)

Waiver of premium benefit. An optional provision of many life insurance policies whereby the company bears the cost of the premiums if the policyholder is totally and permanently disabled.

Whole (ordinary, or straight) life insurance. The most common type of policy, in which premiums are paid for as long as the insured lives in order to maintain coverage of full face value.

Worker's compensation. A form of business liability insurance that pays employees for sickness or accidental injury resulting directly from work responsibilities.

CHAPTER 7

CHECKBOOK AND CASH RECORDS

OBJECTIVES

Upon completion of Chapter 7, you will be able to:

1. Define and use correctly the terminology associated with each topic.
2. Complete check stub balances using the information given (Section 1: Problems 1, 2).
3. Prepare a bank reconciliation (Section 1: Example 1):
   a. Using given information (Problems 3–8)
   b. Using a check register and bank statement (Problems 9–12).
4. Complete the following cash forms (Section 2):
   a. Daily cash record (Example 1; Problems 1–4)
   b. Over-and-short summary (Example 2; Problems 5, 6).

Cash, checks, and electronic funds transfers are the means by which individuals and businesses pay for merchandise and services. Today, you can bank over the telephone, at an ATM, or via the Internet. Funds can be transferred, bills paid, accounts balanced, and financial information provided without your stepping into a bank or writing a check. Of course, each of these services generates a transaction fee. But checks are still the primary means for payment of financial obligations since they provide convenience, safety, and legal receipts. This chapter concentrates on checkbook records, cash records, and the reconciliation required for good internal control of the most liquid asset of an individual and a business: cash.

SECTION 1

CHECKBOOK RECORDS

A check, like a bank note, is a **negotiable instrument,** which means that it may be transferred from one party to another for its equivalent value; the second party then holds full title to the check. Virtually all checks, from both businesses and individuals, are written on preprinted forms obtained from the individual bank. Although checks may vary in appearance somewhat, the following requirements must be met: (1) It must be dated; (2) the bank (or **drawee**) must be positively identified (with address, if necessary); (3) the check must name the **payee** (the person, firm, "cash," or to whomever it is payable); (4) the amount must be shown; and (5) the check must be signed by the **drawer** or **maker** (the person writing the check).

Figure 7-1 illustrates the parts of a check. The amount of the check is written twice, once in figures and then in words. The numbers written in words are considered the legal amount of the check; that is, if there is a discrepancy between the fig-

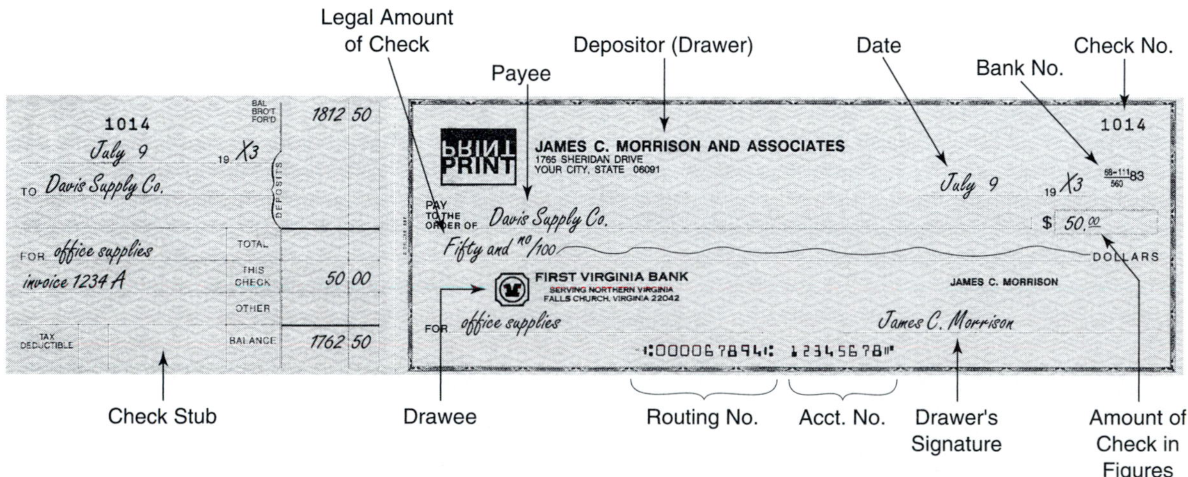

Courtesy of First Virginia Bank

Figure 7-1 Parts of Check

ures ($50) and the written words (Fifty and no/100 dollars), the written words are considered to be the correct and legal amount of the check.

The magnetic ink figures at the bottom of the check are used for identifying the check and for sorting the check electronically during the routing process from the payee's (Davis Supply Co.) bank through the Federal Reserve District Bank back to the drawer's (James C. Morrison and Associates) bank.

The check stub balance should always be kept current. All checks and deposits should be recorded immediately and the new balance computed. If a check is not used due to errors in writing the check or for other reasons, it should be voided. The word "void" should be written on the check stub and on the check itself to provide a record for that particular check. The check stub or **check register** also provides space to record pertinent information about the check: what the check was for, how much sales tax (or interest) was included, how much cash discount was given, and so forth.

The **voucher check** (Figure 7-2) is often used by businesses. The "voucher" attached to the check itself explains which invoice the check covers, itemizes any

ORIGINAL

PAYEE SHOULD DETACH TOP AND DEPOSIT VOUCHER AT ONCE

STATE BOARD OF EDUCATION

DEPARTMENT OF COMMUNITY COLLEGES

VOUCHER NO 1,308

PURCHASE ORDER NO	DATE OF INVOICE	BUDGET POS NO	PARTICULARS	CODE	GROSS AMOUNT	DEDUCTIONS	NET AMOUNT
819	Sept. 23	14	30 Lab Stools	16-D	$345.00	$ 7.90	$337.10
834	Sept. 26	21	60 #243A Desks	9-A	900.00	18.00	882.00
							$1,219.10

IBAA 35

DRAWN BY (NAME AND ADDRESS OF INSTITUTION)

Blue Ridge Community College
Asheville, North Carolina

STATE BOARD OF EDUCATION - DEPARTMENT OF COMMUNITY COLLEGES

PRESENT
TO STATE TREASURER—
STATE OF NORTH CAROLINA
RALEIGH, N. C.

66 1059
512

VOUCHER NO 1,308

PAYABLE AT PAR THROUGH THE FEDERAL RESERVE SYSTEM

DATE November 9 19

THIS CHECK VOID AFTER 60 DAYS FROM DATE

PAY Twelve hundred nineteen and 10/100 ------------------------ DOLLARS $ 1,219.10

TO THE ORDER OF
Whitman Technical Furniture, Inc.
Box 823
High Point, North Carolina

Specimen
Priscilla K. Stokes
AUTHORIZED SIGNATURE
William R. Morton
COUNTERSIGNED

G-185

⑆0512⑉1059⑆ 4⑉000⑉574⑈

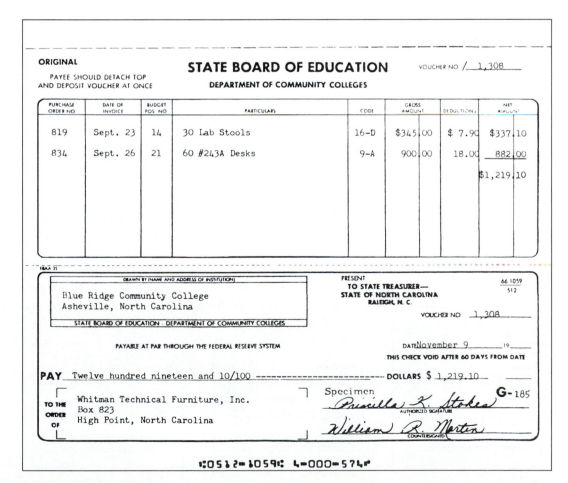

Figure 7-2 Voucher Check

DATE	CHECK NUMBER	ORDER OF	FOR	AMOUNT CHECK	AMOUNT DEPOSIT	BALANCE
20—						$ 919.81
July 16	362	The May Co.	Brass fixtures	$ 86.15		833.66
18	363	Warren's, Inc.	Office supplies	15.83		817.83
18		Deposit			$840.00	1,657.83
19	364	Payroll	Weekly payroll	789.25		868.58
21	365	Carson's, Inc.	Equipment rental	27.44		841.14

Figure 7-3 Check Register

deductions, or otherwise accounts for the amount shown on the check. The voucher is not negotiable and is torn off at the perforation before the check is cashed or deposited. The company issuing the check retains one or more carbon copies of the voucher for its records. In larger firms, these voucher checks are typically generated by a computer, which also updates the account for their bank balance at the same time.

Smaller firms using voucher checks ordinarily keep a **check register,** which is a concise record of checks and deposits. Figure 7-3 illustrates a check register.

DEPOSITS

Before a company can deposit or cash the checks it has received, the checks must be endorsed by the payee. The endorsement should be on the back lefthand side of the check in a designated, boxed area that is within $1\frac{1}{2}$ inches from the edge of the check. The space below this boxed area is used by the Federal Reserve.

There are three forms of **endorsement** in common use: the blank endorsement, the restrictive endorsement, and the special endorsement.

The **blank** endorsement is just the name of the person or the firm (and often the individual representing that firm) to which the check is made payable. This endorsement makes the check negotiable by anyone who has possession of it. That is, such a check could be cashed by anyone if it should become lost or stolen.

The other two endorsements are somewhat similar. The **restrictive** endorsement is usually applied to indicate that the check must be deposited in the firm's bank account. Many firms endorse all their checks in this manner as soon as they are received,

ENDORSE HERE
x James C. Morrison
DO NOT WRITE, STAMP OR SIGN BELOW THIS LINE
RESERVED FOR FINANCIAL INSTITUTION USE
BLANK

ENDORSE HERE
x For Deposit at
First Virginia Bank
Acct. # 12345678
DO NOT WRITE, STAMP OR SIGN BELOW THIS LINE
RESERVED FOR FINANCIAL INSTITUTION USE
RESTRICTIVE

ENDORSE HERE
x Pay to the order of
Union Fund
James C. Morrison
DO NOT WRITE, STAMP OR SIGN BELOW THIS LINE
RESERVED FOR FINANCIAL INSTITUTION USE
SPECIAL

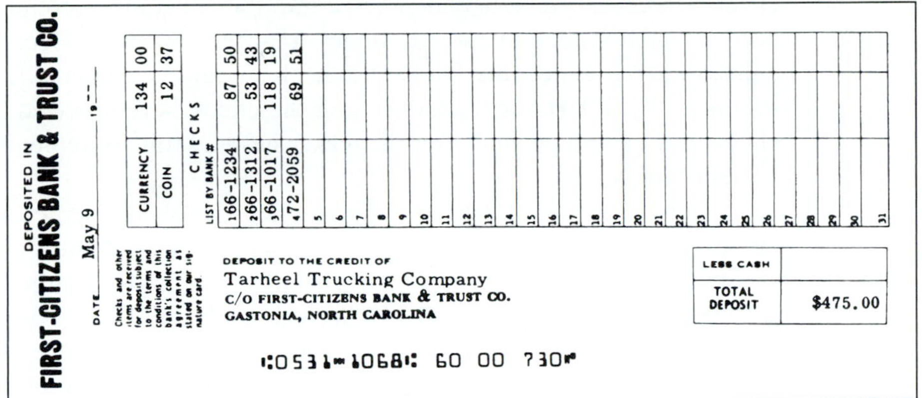

DEPOSITED IN
FIRST-CITIZENS BANK & TRUST CO.

DATE May 9 19__

Checks and other items are received for deposit subject to the terms and conditions of this bank's collection agreement as stated on our signature card

CURRENCY	134	00
COIN	12	37
CHECKS — LIST BY BANK #		
1 66-1234	87	50
2 66-1312	53	43
3 66-1017	118	19
4 72-2059	69	51
5		
6		
7		
8		
9		
10		
11		
12		
13		
14		
15		
16		
17		
18		
19		
20		
21		
22		
23		
24		
25		
26		
27		
28		
29		
30		
31		

DEPOSIT TO THE CREDIT OF
Tarheel Trucking Company
c/o FIRST-CITIZENS BANK & TRUST CO.
GASTONIA, NORTH CAROLINA

LESS CASH	
TOTAL DEPOSIT	$475.00

:0531-1068: 60 00 730

Figure 7-4 Deposit Slip

as a precautionary measure against theft. The **special** (or **full**) endorsement indicates that a check is to be passed on to another person, firm, or organization. The special endorsement's great advantage over the blank endorsement is that, should the check be lost or stolen, it is negotiable only by the party named in the special endorsement.

After being endorsed, the checks are listed individually on a bank deposit slip. Checks are often listed according to the name of the bank on which they were written, or by the name of the person or firm who wrote them; however, many banks prefer that checks be listed by *bank number.* Most printed checks contain the complete identification number of the bank on which the check is written. For example,

$$\frac{66\text{-}1059}{512}$$

The "512" represents the Federal Reserve District within which the bank is located; the "66" identifies the location (state, territory, or city) of the bank; and the "1059" indicates the particular bank in that location. Notice in Figure 7-2 that the bank number "0512 · · · 1059" is coded in magnetic ink at the bottom of the check. Some checks contain only the numbers indicating location and official bank number (for example, 66-1059). It is sufficient to use this hyphenated number when listing checks on the deposit slip. A completed deposit slip is shown in Figure 7-4.

BANK SERVICES AND STATEMENTS

Banks today offer numerous services. Charges are made for these services and deducted by the bank from the customer's (or drawer's) account. The customer many times does not know of these charges until the bank statement is received, and thus has not recorded them on the check stub or in the check register. Typical bank charges are **service charges** (SC) for checks written, **debit memos** (DM) (charges for printing new checks or other services), corrections in deposits, **returned checks** (RT), **overdrafts** (OD), and **insufficient funds** (ISF). A charge for a cash withdrawal from an **automatic teller machine (ATM)** is another example. A returned check is a check previously *deposited* into the checking account when the maker's account did not contain enough money to cover the check. Because the depositor's bank could not collect the funds, the check amount is deducted from the depositor's account—the RT (return) charge—and the check is returned to the depositor. The check may be redeposited after the maker has had time to replenish the funds in his or her account.

When the owner of the checking account (the maker) writes a check that would overdraw the account, the bank charges a fee that is deducted from the maker's account—an OD (overdraft) or ISF (insufficient funds) charge. Most banks offer **overdraft protection.** Some banks automatically transfer money from savings accounts to cover the insufficient funds. Other banks use the customer's credit card as backing. A third type of overdraft protection is the **line of credit.** A line of credit is the limit to the amount of credit that is given to a customer. In this case, a line of credit is assigned to each individual account. If the maker or drawer overdraws the account balance, the bank will honor checks up to this line of credit. The owner of the account must repay the bank for any amounts used under the provisions of overdraft protection, plus interest.

Electronic funds transfer is a computerized system for depositing and/or paying funds out of the customer's account. Many employees have their employer transfer their wages directly from the company's bank account into their personal accounts. Another variation of electronic banking is the use of point-of-sale terminals, which are located at the merchant's business and connected to a bank computer. When the customer makes a purchase using a **debit card,** money from the customer's bank account is automatically and instantly transferred to the merchant's account. Another variation of electronic banking is the use of personal computers that are linked by telephone to the bank computer. In response to instructions from the customer's personal computer, the bank automatically transfers money from the customer's account to the merchant's account. Wide acceptance of electronic funds transfer could effectively make us a cashless society.

If the customer forgets to record any of these deductions on the check stub or in the check register, the discrepancy will appear when the bank statement (Figure 7-5) and check register are reconciled.

In some instances, the bank statement contains a credit that will increase the checkbook balance—typically interest. Many financial institutions are presently offering **negotiable order of withdrawal (NOW) accounts,** a combination checking-savings account that earns interest on its average daily balance, provided the average meets or exceeds a specified minimum. The interest on this type of account is treated like a deposit, and the bookkeeper should add it to the checkbook balance. (Similar types of interest-earning checking accounts at a savings and loan association are sometimes called "money-market accounts.")

Frequently, some checks and possibly a deposit (particularly those near the end of the month) will not have been processed in time to be recorded and returned with the bank statement. Such checks are known as **outstanding checks.** The bookkeeper should compare the bank statement against the check stubs to determine which checks (and deposits) are outstanding. The bookkeeper must determine the total of the outstanding checks and deduct this amount from the bank statement balance to obtain the

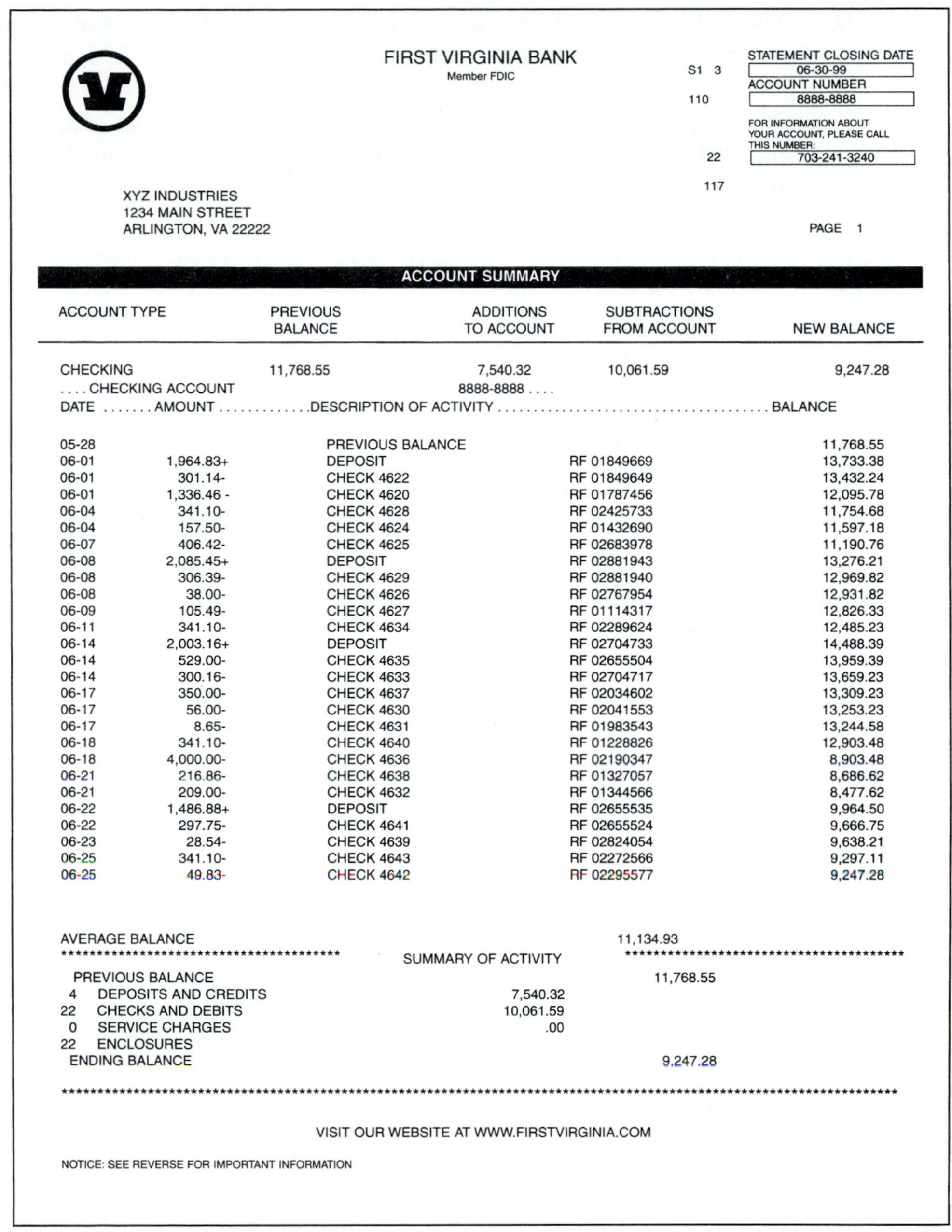

FIRST VIRGINIA BANK
Member FDIC

S1 3
110
22
117

STATEMENT CLOSING DATE 06-30-99
ACCOUNT NUMBER 8888-8888
FOR INFORMATION ABOUT YOUR ACCOUNT, PLEASE CALL THIS NUMBER: 703-241-3240

XYZ INDUSTRIES
1234 MAIN STREET
ARLINGTON, VA 22222

PAGE 1

ACCOUNT SUMMARY

ACCOUNT TYPE	PREVIOUS BALANCE	ADDITIONS TO ACCOUNT	SUBTRACTIONS FROM ACCOUNT	NEW BALANCE
CHECKING	11,768.55	7,540.32	10,061.59	9,247.28

. . . . CHECKING ACCOUNT 8888-8888 . . . .

DATE	AMOUNT	DESCRIPTION OF ACTIVITY		BALANCE
05-28		PREVIOUS BALANCE		11,768.55
06-01	1,964.83+	DEPOSIT	RF 01849669	13,733.38
06-01	301.14-	CHECK 4622	RF 01849649	13,432.24
06-01	1,336.46 -	CHECK 4620	RF 01787456	12,095.78
06-04	341.10-	CHECK 4628	RF 02425733	11,754.68
06-04	157.50-	CHECK 4624	RF 01432690	11,597.18
06-07	406.42-	CHECK 4625	RF 02683978	11,190.76
06-08	2,085.45+	DEPOSIT	RF 02881943	13,276.21
06-08	306.39-	CHECK 4629	RF 02881940	12,969.82
06-08	38.00-	CHECK 4626	RF 02767954	12,931.82
06-09	105.49-	CHECK 4627	RF 01114317	12,826.33
06-11	341.10-	CHECK 4634	RF 02289624	12,485.23
06-14	2,003.16+	DEPOSIT	RF 02704733	14,488.39
06-14	529.00-	CHECK 4635	RF 02655504	13,959.39
06-14	300.16-	CHECK 4633	RF 02704717	13,659.23
06-17	350.00-	CHECK 4637	RF 02034602	13,309.23
06-17	56.00-	CHECK 4630	RF 02041553	13,253.23
06-17	8.65-	CHECK 4631	RF 01983543	13,244.58
06-18	341.10-	CHECK 4640	RF 01228826	12,903.48
06-18	4,000.00-	CHECK 4636	RF 02190347	8,903.48
06-21	216.86-	CHECK 4638	RF 01327057	8,686.62
06-21	209.00-	CHECK 4632	RF 01344566	8,477.62
06-22	1,486.88+	DEPOSIT	RF 02655535	9,964.50
06-22	297.75-	CHECK 4641	RF 02655524	9,666.75
06-23	28.54-	CHECK 4639	RF 02824054	9,638.21
06-25	341.10-	CHECK 4643	RF 02272566	9,297.11
06-25	49.83-	CHECK 4642	RF 02295577	9,247.28

AVERAGE BALANCE 11,134.93

SUMMARY OF ACTIVITY

	PREVIOUS BALANCE		11,768.55
4	DEPOSITS AND CREDITS	7,540.32	
22	CHECKS AND DEBITS	10,061.59	
0	SERVICE CHARGES	.00	
22	ENCLOSURES		
	ENDING BALANCE		9,247.28

VISIT OUR WEBSITE AT WWW.FIRSTVIRGINIA.COM

NOTICE: SEE REVERSE FOR IMPORTANT INFORMATION

Courtesy of First Virginia Bank

Figure 7-5 Sample Bank Statement

adjusted bank balance. (A deposit not included on the bank statement must be added to the bank statement balance to determine the adjusted bank balance.)

After these adjustments have been made, the adjusted bank balance and the adjusted checkbook balance will be the same, provided that all transactions were recorded correctly and that there has been no mathematical mistake. This adjusted balance represents the true balance available to spend.

Example 1 A florist's checkbook shows a balance of $935.50, whereas the balance on the bank statement is $1,520.36. Outstanding checks amount to $707.90, and there is a deposit of $244.30 not recorded by the bank. The bank statement shows the following: a service charge of $2.50; interest on the NOW account of $4.76; an electronic funds deposit of $210; and a returned check for $100.00. The bookkeeper made a subtraction error of $9.00 in recording check number 125. Reconcile the bank statement.

BANK RECONCILIATION

Bank Statement Balance	$1,520.36	Checkbook Balance		$ 935.50
Add: Outstanding deposit	+ 244.30	Add: Interest	$ 4.76	
	$1,764.66	Electronic deposit	210.00	
		Recording error	9.00	+ 223.76
				$1,159.26
		Less: Service charge	$ 2.50	
		Returned check	100.00	− 102.50
Less: Outstanding checks	− 707.90			
Adjusted Balance	$1,056.76	Adjusted Balance		$1,056.76

SECTION 1 PROBLEMS

Complete the check stubs below with the information given, and update the account balance after each check is written.

1. Balance brought forward: $3,645.26

Check No. 360
Date: November 2, 20XX
Amount: $183.92
To: Gold Supplies
For: Office supplies, $176 plus sales tax, $7.92
Deposit: 0.

Check No. 361
Date: November 3, 20XX
Amount: $982.30
To: Baltimeier Sale
For: Computer, $940 plus sales tax, $42.30
Deposit: 0.

Check No. 362
Date: November 3, 20XX
Amount: $1,250.00
To: Garman, Inc.
For: Merchandise
Deposit: $374.46

Check No. 363
Date: November 4, 20XX
Amount: $64.00
To: U.S. Postal Service
For: Postage stamps
Deposit: 0.

Balance Brought Forward
20 ____
To ____
For ____
Amount
Check
Deposit
Balance

Balance Brought Forward
20 ____
To ____
For ____
Amount
Check
Deposit
Balance

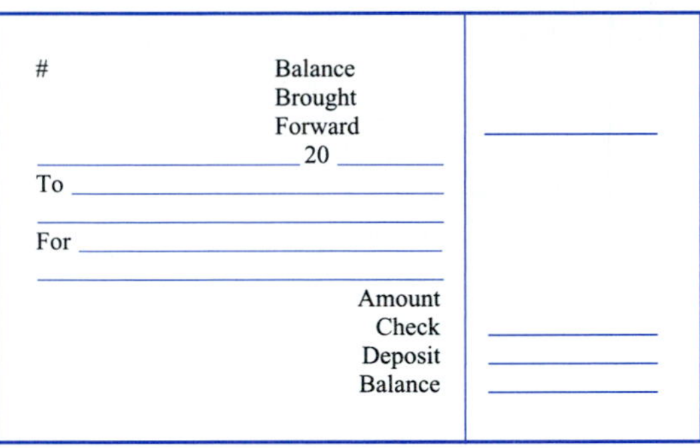
Balance Brought Forward
20 ____
To ____
For ____
Amount
Check
Deposit
Balance

Balance Brought Forward
20 ____
To ____
For ____
Amount
Check
Deposit
Balance

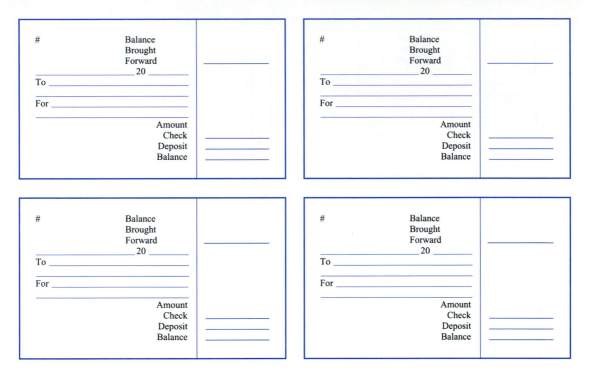

2. Balance brought forward: $5,246.48

Check No. 5622	
Date:	June 3, 20XX
Amount:	$3,100.00
To:	Hodges Manuf. Co.
For:	Merchandise
Deposit:	0.

Check No. 5623	
Date:	June 3, 20XX
Amount:	$550.00
To:	Scully Insurance Agency
For:	Quarterly liability insurance premium
Deposit:	$1,687.32

Check No. 5624	
Date:	June 4, 20XX
Amount:	$126.00
To:	Connelly Office Products
For:	Office supplies, $120.00 sales tax, $6.00
Deposit:	0.

Check No. 5625	
Date:	June 5, 20XX
Amount:	$75.42
To:	Charlie's Computer Services
For:	Computer repairs
Deposit:	0.

#	Balance Brought Forward	
20		
To		
For		
	Amount	
	Check	
	Deposit	
	Balance	

#	Balance Brought Forward	
20		
To		
For		
	Amount	
	Check	
	Deposit	
	Balance	

#	Balance Brought Forward	
20		
To		
For		
	Amount	
	Check	
	Deposit	
	Balance	

#	Balance Brought Forward	
20		
To		
For		
	Amount	
	Check	
	Deposit	
	Balance	

Prepare a bank reconciliation for each of the following.

3. The checkbook balance was $142.12, and the bank statement balance was $605.75. Outstanding checks were in the amounts of $86.20, $122.48, $250.39, and $12.56. A service charge of $8 had been deducted on the bank statement.

4. On a recent bank statement, the following checks were outstanding: $264.25, $188.60, $54.56, and $37.35. A deposit in transit for $600 was not recorded on the bank statement, but a service charge of $12 was listed. The checkbook balance was $1,539.90, and the balance on the bank statement was $1,472.66.

5. The bank statement balance was $1,814.41, and the checkbook balance was $1,077.97. A deposit of $334.82 had not been entered on the bank statement. A service charge of $10.45 and a check printing charge of $9.20 were listed on the statement. There was a returned check for $50. Interest on the NOW account was $5.15. Outstanding checks were for $176, $440.60, and $519.16.

6. The bank statement balance was $3,424.53, while the check register balance was $3,371.15. There were a returned check for $68 from a customer, a debit memo for printing checks of $15, and a service charge of $7.50 listed on the statement. Checks outstanding were $144.51, $76.00, and $400.62. A deposit of $477.25 was in transit and not listed on the statement.

7. The checkbook showed a balance of $1,348.29, whereas the bank statement had a balance of $845.32. An outstanding deposit totaled $800. Outstanding checks amounted to $477.75. The statement showed a charge of $2.00 to correct a deposit. A service charge was $6.55. A returned check amounted to $165.65. New checks cost $15 to print. The account earned interest of $8.48.

8. The bank statement showed a balance of $2,260.91, while the check register showed a balance of $1,313.55. The statement also showed a deposit by electronic transfer of $921.50 that was not recorded in the check register. Other items listed on the bank statement but not recorded in the check register were as follows: interest on the NOW account of $10.22; ATM withdrawals of $364 plus $4.50 ATM fees; and a debit memo of $20 for printing new checks. Outstanding checks amounted to $1,050.62. A deposit in transit of $646.48 was not included on the statement.

Four check registers and bank statements are given in Problems 9–12 on the following pages. Determine which checks and deposits have been processed, which are still outstanding, and what bank charges have been made. (Use the "√" column of the register to check off items that appear on the bank statement. Some have already been checked the preceding month.) Each check register shows the adjustment made last month to balance the register. Based on the information found, prepare a bank reconciliation for each problem, and determine the adjusted balance.

9.

CHECK REGISTER

CHECK NUMBER	DATE	CHECK ISSUED TO	AMOUNT OF CHECK	✓	AMOUNT OF DEPOSIT	BALANCE 2,648.21
220	2/14	McDonald Co.	85.26	√		2,562.95
221	2/15	Payroll	1,808.66			754.29
	2/15	DEPOSIT			2,004.60	2,758.89
222	2/18	Hardison Co.	264.34			2,494.55
223	2/19	VHG, Inc.	91.75			2,402.80
224	2/19	West Insurance Co.	376.28			2,026.52
	2/20	Adjustment for bank charges	3.27	√		2,023.25
225	2/20	Dean Gas Co.	559.42			1,463.83
226	2/21	Chavez Assoc.	130.90			1,332.93
227	2/21	HIH Services	228.48			1,104.45
228	2/21	Cain & Dole	240.30			864.15
229	2/22	Bradshaw Sales	54.62			809.53
	2/22	DEPOSIT			699.49	1,509.02
230	2/25	Powers & Co.	150.96			1,358.06
231	2/25	Smith Couriers	62.25			1,295.81
	2/26	DEPOSIT			1,359.00	2,654.81
232	2/27	Chairs, Inc.	282.44			2,372.37
233	2/27	Payroll	1,856.43			515.94
	2/27	DEPOSIT			836.24	1,352.18

BANK STATEMENT

CHECK NUMBER	CHECKS AND OTHER DEBITS	DEPOSITS	DATE	BALANCE
			2/14	2,559.68
		2,004.60	2/15	4,564.28
226	130.90		2/23	4,433.38
221	1,808.66	699.49	2/23	3,324.21
222	264.34		2/24	3,059.87
225	559.42		2/26	2,500.45
228	240.30		2/26	2,260.15
223	91.75	1,359.00	2/27	3,527.40
	400.00RT		2/27	3,127.40
227	228.48		2/28	2,898.92
230	150.96		2/28	2,747.96
231	62.25		2/29	2,685.71
	10.00SC		2/29	2,675.71

10.

CHECK REGISTER

CHECK NUMBER	DATE	CHECK ISSUED TO	AMOUNT OF CHECK	✓	AMOUNT OF DEPOSIT	BALANCE 4,950.60
420	3/14	Williams Interiors	62.44			4,888.16
421	3/15	Payroll	1,505.22			3,382.94
	3/15	DEPOSIT			2,500.00	5,882.94
422	3/15	Marshall Distributors, Inc.	127.29			5,755.65
423	3/18	Master's Insurance Co.	334.65			5,421.00
424	3/18	Lessin & Sons Plumbing Co.	100.00			5,321.00
	3/18	Adjustment for bank charges	10.00	√		5,311.00
425	3/19	Nuckolls Advertising Co.	295.00			5,016.00
426	3/20	ECG Enterprises	82.66			4,933.34
427	3/21	Howell Caterers	540.00			4,393.34
428	3/21	Shaw & Davis Co.	813.67			3,579.67
	3/22	DEPOSIT			1,070.61	4,650.28
429	3/25	Patel Transportation Co.	34.86			4,615.42
430	3/25	U.S. Postal Services	46.38			4,569.04
431	3/27	Klinko Printers	600.60			3,968.44
432	3/28	Billingsley Technologies	190.23			3,778.21
	3/29	DEPOSIT			1,667.50	5,445.71
433	3/29	Payroll	1,642.12			3,803.59

BANK STATEMENT

CHECK NUMBER	CHECKS AND OTHER DEBITS	DEPOSITS	DATE	BALANCE
			3/14	4,940.60
		2,500.00		7,440.60
	250.00RT			7,190.60
423	334.65		3/20	6,855.95
		1,070.61	3/23	7,926.56
425	295.00		3/23	7,631.56
420	62.44		3/23	7,569.12
421	1,505.22		3/24	6,063.90
428	813.67		3/25	5,250.23
424	100.00		3/27	5,150.23
422	127.29		3/28	5,022.94
426	82.66		3/28	4,940.28
	12.00SC		3/29	4,928.28

11. **CHECK REGISTER**

CHECK NUMBER	DATE	CHECK ISSUED TO	AMOUNT OF CHECK	✓	AMOUNT OF DEPOSIT	BALANCE 1,148.16
223	10/27	Allied Freight	18.53	√		1,129.63
224	10/28	IRS	57.49			1,072.14
225	10/28	Great Lakes Inc.	24.21			1,047.93
	10/30	DEPOSIT			1,476.00	2,523.93
226	11/3	Heavenly Beauty Supplies	35.89			2,488.04
227	11/4	Payroll	442.12			2,045.92
	11/5	Adjustment for bank charges	3.59	√		2,042.33
228	11/10	Midwest Realty	315.00			1,727.33
229	11/10	Barton Drug Supply	207.65			1,519.68
230	11/10	Commercial Sales Inc.	45.72			1,473.96
231	11/10	Protective Insurance Agency	78.33			1,395.63
232	11/10	Midwestern Bell Telephone	21.41			1,374.22
233	11/10	City of Chicago	86.64			1,287.58
234	11/10	Health Products Inc.	71.96			1,215.62
235	11/11	Payroll	407.78			807.84
	11/12	DEPOSIT			891.43	1,699.27
236	11/15	Great Lakes Lab.	169.35			1,529.92
237	11/17	Nationwide Sales Inc.	262.13			1,267.79
	11/18	DEPOSIT			455.52	1,723.31
238	11/18	Payroll	543.29			1,180.02
239	11/23	Allied Freight	4.34			1,175.68
	11/24	DEPOSIT			509.60	1,685.28
240	11/25	Payroll	428.80			1,256.48
241	11/27	Milford Laboratory	50.09			1,206.39
	11/28	DEPOSIT			234.03	1,440.42
242	11/29	Grange Drug Supply	361.15			1,079.27

BANK STATEMENT

CHECK NUMBER	CHECKS AND OTHER DEBITS	DEPOSITS	DATE	BALANCE
			10/31	1,101.83
225	24.21	1,476.00	11/5	2,577.83
227	442.12		11/8	2,135.71
	18.45RT		11/8	2,117.26
234	71.96		11/12	2,045.30
226	35.89		11/12	2,009.41
233	86.64		11/15	1,922.77
228	315.00		11/15	1,607.77
232	21.41		11/15	1,586.36
		891.43	11/16	2,477.79
235	407.78		11/17	2,070.01

CHECK NUMBER	CHECKS AND OTHER DEBITS	DEPOSITS	DATE	BALANCE
230	45.72		11/17	2,024.29
236	169.35	455.52	11/20	2,310.46
238	543.29		11/21	1,767.17
229	207.65		11/21	1,559.52
		509.60	11/26	2,069.12
240	428.80		11/28	1,640.32
224	57.49		11/28	1,582.83
241	50.09		11/30	1,532.74
	7.40SC		11/30	1,525.34

12.

CHECK REGISTER

CHECK NUMBER	DATE	CHECK ISSUED TO	AMOUNT OF CHECK	✓	AMOUNT OF DEPOSIT	BALANCE 5,445.88
258	12/5	Prime Freight Co.	32.46			5,413.42
	12/6	DEPOSIT			1,540.00	6,953.42
259	12/6	Treasurer, Arlington Co.	682.58			6,270.84
	12/7	Adjustment for bank charges	16.00	√		6,254.84
260	12/8	Payroll	2,660.74			3,594.10
	12/8	DEPOSIT			1,686.14	5,280.24
261	12/12	IRS	350.25			4,929.99
262	12/15	Payroll	2,542.25			2,387.74
263	12/15	Jefferson Service Co.	87.42			2,300.32
	12/16	DEPOSIT			1,800.40	4,100.72
264	12/19	Prime Freight Co.	44.47			4,056.25
265	12/20	Midtown Electric Co.	1,001.61			3,054.64
266	12/21	TRC Supplies Co.	223.30			2,831.34
267	12/22	Shelley Cable Co.	178.78			2,652.56
268	12/23	Payroll	2,530.46			122.10
	12/23	DEPOSIT			2,801.60	2,923.70
269	12/30	Payroll	2,535.23			388.47
270	12/30	HHH Inc.	90.64			297.83
271	12/30	Petty Cashier	74.20			223.63
	12/30	DEPOSIT			2,556.89	2,780.52

BANK STATEMENT

Check Number	Checks and Other Debits	Deposits	Date	Balance
				5,429.88
		1,540.00	12/6	6,969.88
	55.27RT		12/7	6,914.61
258	32.46		12/8	6,882.15
		1,686.14	12/8	8,568.29
260	2,660.74		12/12	5,907.55
259	682.58		12/12	5,224.97
262	2,542.25		12/16	2,682.72
261	350.25		12/17	2,332.47
		1,800.40	12/17	4,132.87
264	44.47		12/23	4,088.40
268	2,530.46		12/23	1,557.94
		2,801.60	12/23	4,359.54
266	223.30		12/26	4,136.24
267	178.78		12/27	3,957.46
269	2,535.23		12/30	1,422.23
	10.00SC		12/30	1,412.23

SECTION 2

CASH RECORDS

There is some variation, of course, but one good indication of business stability is the consistency of sales. These sales are recorded daily on the cash register tapes of a business and are given careful attention by officials of the company.

At the beginning of the business day, a stipulated amount is placed in each cash register drawer; this is the **change fund,** which must be available so that early customers may receive correct change. The fund ($20 to $200) will contain a certain number of the various denominations of money.

At the close of the business day, the clerk completes a **daily cash record.** (S)he counts each denomination of money separately, lists each check separately, and finds the total. The original change fund is then deducted from this total, and the remaining sum (the **net cash receipts**) should equal the day's total cash sales shown on the cash register tape.

Human error being what it is, however, it frequently occurs that the cash receipts and the cash register total are not equal. If the cash receipts are less than the total sales, the cash is **short.** If there is more cash than the total of sales, the cash is **over.**

A summary of these daily cash records, commonly called an **over-and-short summary,** may be kept by the week or by the month. Over a period of time, the totals of the amounts in the "cash over" and in the "cash short" tend nearly to cancel each other.

Example 1

THE PERFECT GIFT SHOP
Daily Cash Record

Date: February 8, 20X1
Register: 1-B
Clerk: Sharie Holmes

Pennies		$ 0.73
Nickels		1.85
Dimes		2.60
Quarters		6.75
Halves		2.50
Ones		21.00
Fives		50.00
Tens		40.00
Twenties		80.00
Other currency		
Checks (list separately):	$ 5.15	
	12.67	
	23.80	
	15.00	
	32.50	89.12
	Total	$294.55
	Less: Change fund	− 30.00
	Net cash receipts	$264.55
	Cash over (subtract)	
	Cash short (add)	+ 0.05
	Cash register total	$264.60

Example 2

THE PERFECT GIFT SHOP
Over-and-Short Summary

DATE	CASH REGISTER TOTALS	NET CASH RECEIPTS	CASH OVER	CASH SHORT
Feb. 8, 19X1	$ 552.59	$ 552.23		$0.46
9	615.87	616.12	$0.25	
10	581.69	580.69		1.00
11	643.81	643.08		0.73
12	577.16	578.25	1.09	
13	601.12	601.62	0.50	
Totals	$3,572.34	$3,571.99	$1.84	$2.19

Total cash receipts	$3,571.99
Total cash short (add)	+ 2.19
	$3,574.18
Total cash over (subtract)	− 1.84
Total cash register readings	$3,572.34

SECTION 2 PROBLEMS

1. Prepare a daily cash report (similar to that in Example 1) for clerk Tom Bruno on register 9 of Kitchen, Inc. using today's date. The number of each denomination of money in the cash drawer is as follows: pennies, 52; nickels, 68; dimes, 30; quarters, 74; halves, 2; ones, 34; fives, 12; tens, 15; twenties, 10; other currency, one $100. Checks were for $50.25, $36.36, $15.98, $48.00, and $67.57. The cash register total for the day is $689.48. There is a change fund of $100.
2. Prepare a daily cash report (similar to that in Example 1) for clerk Kate Calloway on register 4 of Prime Cut Meat Shop using today's date. The number of each denomination of money is as follows: pennies, 52; nickels, 65; dimes, 46; quarters, 50; halves, 3; ones, 71; fives, 18; tens, 6; twenties, 5; other currency, one $50. Checks were for $68.45, $81.80, $35.14, $108.00, $74.44, $56.85, $92.75, and $52.63. The cash register total for the day is $764.07. There is a $200 change fund.
3. Prepare a daily cash report for the cash drawer at Hair, Inc. using today's date. Several people collect money and make change from the drawer, so no one person is responsible for the register. The following denominations are in the drawer: pennies, 22; nickels, 74; dimes, 53; quarters, 15; halves, 1; ones, 36; fives, 18; tens, 12; twenties, 5; other currency, one $50. The checks in the drawer were $25, $30, $42, $30, and $35. The cash register total for the day was $496.22. There was a change fund of $75.
4. The following denominations are found in the cash drawer at the end of the day by Janet Yates, the clerk responsible for this cash drawer on register 2 at L & M Emporium: pennies, 60; nickels, 24; dimes, 48; quarters, 62; halves, 2; ones, 27; fives, 10; tens, 14; twenties, 15. Checks in the drawer were for $56.12, $89.95, $97.75, and $66.18. The change fund contained $75. The cash register tape reveals $774.48 for the day's cash sales. Prepare a daily cash record using the current date.
5. Compute a weekly over-and-short summary for USA Chips, Inc. Each day's cash register totals and net cash receipts are as follows:

USA CHIPS, INC.
Over-and-Short Summary

DATE	TOTAL SALES	NET CASH RECEIPTS	CASH OVER	CASH SHORT
June 3	$1,045.68	$1,045.70		
June 4	1,172.49	1,172.49		
June 5	1,099.46	1,100.00		
June 6	1,153.33	1,153.25		
June 7	1,155.21	1,155.16		
Totals				

Total cash receipts
Total cash short (add)

Total cash over (subtract)
Total cash register readings

6. Compute a weekly over-and-short summary for Sun and Surf Co. Each day's cash register totals and net cash receipts are as follows:

SUN AND SURF CO. Over-and-Short Summary				
DATE	TOTAL SALES	TOTAL NET CASH RECEIPTS	CASH OVER	CASH SHORT
May 15	$1,252.99	$1,253.44		
May 16	1,342.01	1,342.21		
May 17	1,304.68	1,303.68		
May 18	1,313.12	1,313.12		
May 19	1,359.75	1,359.88		
Totals				
		Total cash receipts		
		Total cash short (add)		
		Total cash over (subtract)		
		Total cash register readings		

CHAPTER 7 GLOSSARY

Automatic teller machine (ATM). A terminal used to perform simple banking transactions without the aid of a teller.

Blank endorsement. Name (signature) of person or firm to whom a check was payable and which transfers ownership of the check to anyone in possession of it.

Change fund. A set amount of money placed in a cash register at the start of each day, in order to provide change for early customers.

Check register. A record of checks and deposits made by the owner of a checking account.

Daily cash record. A form that compares the net cash receipts with the cash register's total cash sales, in order to determine whether the cash drawer is over or short.

Debit card. A card that decreases a person's bank account automatically when used to make purchases.

Debit memo. A fee that the bank charges for service to an account, such as the charge for printing checks.

Drawee. The bank from which funds of a check are drawn (paid).

Drawer (or maker). The person (or firm) who draws funds (pays) by check.

Electronic funds transfer. A computerized system for depositing funds and/or decreasing a bank account balance.

Endorsement. The inscription on the back of a check (or negotiable instrument) by which ownership is transferred to another party.

Insufficient funds. A situation in which a checking account does not contain enough money for the bank to honor a check written on the account.

Line of credit. The limit to the amount of credit allowed a customer.

Maker (or drawer). The writer of a check.

Negotiable instrument. A document (check or bank note) that may be transferred from one party to another for its equivalent value.

Net cash receipts. A clerk's total (of money and checks) remaining after the change fund has been deducted; equals the total cash sales, if cash is neither over nor short.

NOW (negotiable order of withdrawal) account. A combination checking-savings account that earns interest on its average daily balance, provided a certain minimum balance is maintained.

Outstanding check. A check given in payment but not yet deducted from the writer's bank account.

Over. Term meaning that net cash receipts exceeded the cash register total.

Over-and-short summary. A weekly or monthly list of the total daily cash records of a business.

Overdraft. Term meaning that a customer has written a check for which there are insufficient funds in the account.

Payee. The party named to receive the funds of a check.

Reconciliation. Adjustments to the checkbook and the bank statement whereby the owner finds the actual spendable balance and determines that both records agree.

Restrictive endorsement. An endorsement restricting the use of a check; usually indicates that the check must be deposited to the endorser's account.

Returned check. A check returned to the party who deposited it because the check writer's account did not contain sufficient funds.

Service charge. A fee paid to the bank by the owner of a bank account for checking account services.

Short. Term meaning that net cash receipts were less than the cash register total.

Special (or full) endorsement. An endorsement naming the next party to whom a check is to be transferred.

Voucher check. A check with a perforated attachment which explains or otherwise accounts for the amount shown on the check.

CHAPTER 8

WAGES AND PAYROLLS

OBJECTIVES

Upon completion of Chapter 8, you will be able to:

1. Define and use correctly the terminology associated with each topic.
2. Compute gross wages and payrolls based on:
   - a. Salary (Section 1: Example 1; Problems 1, 2)
   - b. Commission (including quota and override) (Section 2: Examples 1–4; Problems 3–20)
   - c. Hourly rates (Section 3: Example 1; Problems 1, 2)
   - d. Production (including incentives and dockings) (Section 4: Examples 1, 3, 4; Problems 1–6, 9, 10).
3. Complete the account sales and account purchase (Section 2: Examples 5, 6; Problems 21–24).
4. For overtime work, compute gross wages and payrolls using:
   - a. Hourly rates (at standard overtime or overtime excess) (Section 3: Examples 2, 3, 5; Problems 3–12)
   - b. Salary (Section 3: Example 4; Problems 7, 8)
   - c. Production (Section 4: Example 2; Problems 7, 8).
5. Complete net weekly payrolls, including standard deductions for (Section 5: Examples 1–3; Problems 1–12):
   - a. Social Security
   - b. Federal income tax
   - c. Other common deductions, as given.
6.
   - a. Complete the basic portion of the Employer's Quarterly Federal Tax Return (Section 6: Examples 1, 3; Problems 1–6, 13–16)
   - b. Determine employees' Social Security taxable wages and Medicare taxable wages (Example 2; Problems 7–12, 21, 22).
7.
   - a. Determine employees' taxable wages for unemployment (Section 6: Example 4; Problems 17–20)
   - b. Compute employers' state and federal unemployment taxes (Example 4: Problems 17–22).

In any business math course, many topics apply primarily to certain types of businesses. The payroll, however, is one aspect of business math that applies to every conceivable type of business, no matter how small or how large.

There is probably no other single factor so important to company–employee relations as the negotiation of wages and completion of the payroll. Wage negotiation within a company—whether by a union or on an individual basis—is not a responsibility of the payroll clerk, however. We shall thus be concerned only with computation of the payroll, which is also a very important responsibility. To find a mistake in their pay is very damaging to employees' morale. It is imperative, therefore, that everyone connected with the payroll exercise extreme care to ensure absolute accuracy.

There are four basic methods for determining **gross wages** (that is, wages before deductions): *salary, commission, hourly rate basis,* and *production basis.* We shall make a brief study of each.

SECTION 1

COMPUTING WAGES—SALARY

Employees on a salary receive the same wages each pay period, whether this be weekly, biweekly, semimonthly, or monthly. Most executives, office personnel, and many professional people receive salaries.

If a salaried person misses some work time because of illness or for personal reasons, (s)he will usually still receive the same salary on payday. However, if business conditions require a salaried person to work extra hours, the payment of additional compensation may vary according to the person's position in the firm. General office personnel who are subject to the Fair Labor Standards Act (a topic in Section 3) must receive extra compensation as though they were normally paid on an hourly rate basis. However, highly paid executive and administrative personnel, whose positions are at the supervisory and decision-making levels of the company, normally receive no additional compensation for extra work. Professional people (members of the medical professions, lawyers, accountants, educators, and so on) are similarly exempt from extra-pay provisions.

The employer will quote an annual salary to a prospective employee, but this amount normally is not paid annually or once a year. The annual salary will be paid weekly, biweekly, semimonthly, or monthly. The employee will receive approximately the same amount each pay period with the total of each period's wages equaling the annual salary.

Example 1 Jan Jones will receive an annual salary of $36,000. Determine her salary per pay period assuming that wages are paid (a) monthly, (b) semimonthly, (c) weekly, and (d) biweekly.

(a) There will be 12 pay periods if Ms. Jones receives her salary monthly.

$$\frac{\$36{,}000}{12} = \$3{,}000 \text{ monthly}$$

(b) There will be 24 pay periods if she is paid semimonthly.

$$\frac{\$36{,}000}{24} = \$1{,}500 \text{ semimonthly}$$

(c) If she receives her salary weekly, she will receive 52 payments.

$$\frac{\$36{,}000}{52} = \$692.31 \text{ weekly}$$

(d) If she is paid biweekly, she will be paid every 2 weeks or a total of 26 payments.

$$\frac{\$36{,}000}{26} = \$1{,}384.62 \text{ biweekly}$$

SECTION 2

COMPUTING WAGES—COMMISSION

Business people whose jobs consist of buying or selling merchandise are often compensated for their work by means of a commission. The true (or **straight**) **commission** is determined by finding some percent of the individual's net sales. Thus, we have another business formula based on the basic percent form "_____% of _____ = _____":

_____% of Net sales = Commission

or simply

$$\% \cdot S = C$$

Certain business people receive commissions without actually calling them commissions. For instance, a stockbroker's fee for buying stock is a certain percentage of the purchase price of that stock; however, this fee is called a **brokerage.** A lawyer's fee for handling the legal aspects of a real estate sale is generally some percentage of the value of the property; this constitutes a commission, although it probably will not be called one.

STRAIGHT COMMISSION

Example 1 (a) Edith Reilly receives an 8% commission on her net sales. Find her commission if her gross (original) sales were \$5,680 and sales returns and allowances were \$180.

Note. "Sales returns and allowances" are first deducted from gross sales, because the customer either returned the merchandise or was later allowed some reduction in sales price.

Gross sales	\$5,680	$\%S = C$
Less: Sales returns and allowances	− 180	$8\% = \$5{,}500 = C$
Net sales	\$5,500	$\$440 = C$

(b) A stockbroker received a \$240 commission on a stock sale. What was the value of the stock, if the broker charged 2%?

$$\%S = C$$

$$2\%S = \$240$$

$$\frac{\cancel{0.02}\,S}{\cancel{0.02}} = \frac{240.00}{0.02}$$

$$S = \$12{,}000$$

Travel, meals, entertainment, and similar expenses of salespeople often remain approximately the same regardless of their sales. Thus, once a salesperson has made sufficient sales to cover expenses, the company may be willing to pay a higher commission on additional sales. A commission rate that increases as sales increase is known as a **sliding-scale commission.**

Example 1 (cont.) (c) Determine George Butler's commission on his net sales of \$14,000. He receives 5% on his first \$5,000 sales, 6% on the next \$7,000, and 7.5% on all net sales over \$12,000.

$$\%S = C$$

0.05 ×	\$ 5,000 =	\$250	
0.06 ×	7,000 =	420	
0.075 ×	2,000 =	150	
Net sales	\$14,000	\$820	Commission

COMMISSION VARIATIONS

Salary. Some salespeople are guaranteed a minimum income in the form of a salary. They are often then paid commissions at a lower rate than is usually earned by those on straight commission.

Example 2

(a) Suppose that Edith Reilly [Example 1(a)] also receives a $100 salary. Her total gross wages would then be

Salary	$100
Commission	+ 440
Gross wages	$540

Drawing Account. A salesperson on commission may be allowed a **drawing account** from which funds may be withdrawn to defray business or personal expenses. This is an advance on earnings, and the drawings are later deducted from the salesperson's commission. Like a salary, a drawing account assures a minimum income. However, the earned commission must consistently either equal or exceed the drawings, or else the employee would be released from the company.

Example 2 (cont.)

(b) Suppose that Edith Reilly [Example 1(a)] is entitled to a drawing account of $300 and that she has withdrawn the $300 before pay day.

Commission	$440
Less: Drawings	− 300
Amount due	$140

Quota. Many persons on commission, particularly those who also receive a salary, are required to sell a certain amount (or **quota**) before the commission starts. This plan is designed to ensure the company a certain level of performance for the salary it pays.

Example 2 (cont.

(c) Bob Hendrix has net sales of $12,000 and must meet a $4,000 quota. His 5% commission is then computed as follows:

Total sales	$12,000	$\%S = C$
Less: Quota	− 4,000	$5\% \times \$8{,}000 = C$
Sales subject to commission	$ 8,000	$\$400 = C$

Override. Department heads in retail stores are often allowed a special commission on the net sales of all other clerks in their departments. This commission, called an **override,** is given to offset the time a department head must devote to nonselling responsibilities (such as pricing and stocking merchandise, taking inventory, or placing orders). An override is computed exactly like an ordinary commission, but at a lower rate than the commission rate for one person's sales. The override is paid in addition to the department head's own commission and/or salary.

Example 3 Dana Harvey is a department head in a retail store. From the following information, determine her week's wages:

Personal sales	\$ 4,734	Salary	\$ 325
Returns and allowances	54	Quota	\$1,800
Department sales	16,455	Commission rate	5%
Dept. returns and allowances	375	Override rate	$\frac{1}{2}$%

Personal sales	\$4,734	Department sales	\$16,455
Less: Sales returns	− 54	Less: Sales returns	− 375
	\$4,680	Sales subject to override	\$16,080
Less: Quota	− 1,800		
Sales subject to commission	\$2,880		

$$\%S = C \qquad \%S = OR$$
$$0.05 \times \$2{,}880 = C \qquad 0.005 \times \$16{,}080 = OR$$
$$\$144 = C \qquad \$80.40 = OR$$

Salary	\$325.00
Commission	144.00
Override	+ 80.40
Total wages	\$549.40

Example 4 The following are examples of commission payrolls. The second example includes the calculation of net sales, as well as quotas and salaries.

(a) **TOP DRAWER FASHIONS**
Payroll for Week Ending September 6, 20XX

NAME	NET SALES M	T	W	T	F	TOTAL	COMM. RATE	GROSS COMM.
Albright, K.	\$ 950	\$1,075	\$ 920	\$1,070	\$1,085	\$5,100	5%	\$ 255.00
Carver, D.	1,100	1,225	1,380	1,320	1,275	6,300	4	252.00
Foyt, B.	930	—	1,060	1,175	1,350	4,515	6	270.90
Gantt, M.	1,040	975	975	1,050	1,220	5,260	5	263.00
							Total	\$1,040.90

(b) **VALLEY SALES DISTRIBUTORS, INC.**
Payroll for Week Ending June 13, 20XX

NAME	SALES			QUOTA	COMM. SALES	COMM. RATE	GROSS COMM.	SALARY	GROSS WAGES
	GROSS	R&A	NET						
Barton, H.	$3,518	$18	$3,500	$1,200	$2,300	4%	$ 92.00	$ 275.00	$ 367.00
Dixon, R.	5,432	52	5,380	—	5,380	3	161.40	250.00	411.40
Hill, C.	4,776	86	4,690	2,000	2,690	4	107.60	280.00	387.60
Landers, J.	3,084	34	3,050	—	3,050	6	183.00	200.00	383.00
						Totals	$544.00	$1,005.00	$1,549.00

Note. On the bottom line, the total Gross Commissions plus total Salaries must equal the total Gross Wages ($544 + $1,005 = $1,549).

COMMISSION AGENTS

The entire operation of some agencies consists of buying and/or selling products for others. Agents employed for this purpose are known as **commission agents, commission merchants, brokers,** or **factors.** The distinguishing characteristic of such agents is that whereas they are authorized to act for another individual or firm, at no time do they become the legal owners of the property involved.

Many items—particularly foodstuffs—are produced in areas which are some distance from the market. The producer (or **consignor**) ships the goods "on consignment" to a commission agent (or **consignee**) who sells them at the best possible price. The price obtained by the agent is the **gross proceeds.** From these gross proceeds the agent deducts a commission, as well as any expenses incurred (such as storage, freight, insurance, and so forth). The remaining sum, called the **net proceeds,** is then sent to the producer. Accompanying the net proceeds is an **account sales,** which is a detailed account of the sale of the merchandise and of the expenses involved. (Gross proceeds − Total charges = Net proceeds.)

Example 5 The following example shows an account sales.

ACCOUNT SALES

HARCOURT & WYMAN

2519 South Shore Blvd.
Chicago, Illinois 60649

REC'D: January 20, 20XX NO: B27116
VIA: Air freight DATE: Feb. 1, 20XX

Aloha Floral Distributors
7401 Kamani Blvd.
Honolulu, Hawaii 96813

20--		Sales:		
Jan.	21	50 boxes—orchids @ $36	$1,800.00	
	22	25 boxes—bird of paradise @ $60	1,500.00	
	22	15 boxes—ti @ $40	600.00	
		Gross proceeds		$3,900.00
		Charges:		
		Commission, 6% of $3,900	$ 234.00	
		Refrigerator storage	24.75	
		Delivery charges	36.50	
				− 295.25
		Net proceeds		$3,604.75

Calculator Tip. Add the Sales and press [M+]; add the Charges and press [M−]; then pressing [MR] displays the Net proceeds.

Other commission agents act as buyers for their clients. When a store wishes to purchase an item not regularly carried in its inventory and not obtainable through regular wholesale distributors, the store may commission an agent to secure the items. Typical examples of such items might be art objects, specialized furniture, carpets, exclusive dry goods, exotic foods, and so on, which are specialties of a particular locality or country. Much international as well as domestic purchasing is accomplished in this manner.

In general, a commission agent is employed whenever it is inconvenient to send a company buyer or when a special knowledge of the product or the market conditions is required. The agent submits an itemized account of the purchase, called an **account purchase.** The account purchase includes the actual price paid by the agent (**prime cost**) plus the agent's commission on the prime cost, as well as any expenses incurred during the purchase. This total (or **gross cost**) is the actual cost of the merchandise to the company and the amount to be remitted to the commission agent. (Prime cost + Total charges = Gross cost.)

Example 6 The following example shows an account purchase.

ACCOUNT PURCHASE

IMPORT SPECIALTIES, INC.
32714 Santa Monica Blvd.
Los Angeles, California 90037

BOUGHT FOR THE ACCOUNT OF:

The Import House
7709 Granville Rd.
St. Paul, Minnesota 55101

YOUR ORDER: #5415 NO: G42081
DATE: April 9, 20XX DATE: May 1, 20XX

20XX		Purchases:		
April	18	15—Philippine wicker fan chairs @ $96	$1,440.00	
	20	12—Philippine wicker hooded swings @ $75	900.00	
	21	20—Wicker chests @ $25	500.00	
		Prime cost		$2,840.00
		Charges:		
		Shipping	$ 78.85	
		Insurance	15.95	
		Import duty	56.80	
		Commission, 5% of $2,840	142.00	
				+ 293.60
		Gross cost		$3,133.60

SECTIONS 1 AND 2 PROBLEMS

Determine the salary.

	Annual Salary	Monthly	Semimonthly	Weekly	Biweekly
1. a.	$48,000				
b.		$6,000			
c.			$1,000		
2. a.	$150,000				
b.		$2,800			
c.			$4,000		

Complete the following.

		SALARY	QUOTA	RATE	NET SALES	COMMISSION	GROSS WAGES
3.	a.	X	X	6%	$9,500		X
	b.	X	X	5½		$396	X
	c.	X	X		6,100	244	X
	d.	$200	X	5	4,400		
	e.	110	X	7			$530
	f.	X	$500	4½	9,900		X
	g.	175	200	8	5,700		
	h.	80	1,000	6			404
	i.		1,400	5	13,800		820
4.	a.	X	X	8%	$13,000		X
	b.	X	X	7		$5,600	X
	c.	X	X		25,000	1,875	X
	d.	$300	X	6	20,000		
	e.	400	X	9			$1,930
	f.	X	$500	10	18,000		X
	g.	500	1,000	8	40,000		
	h.	1,200	2,000	5½			2,850
	i.		600	9	37,000		3,476

5. Gail Johnson had sales totaling $7,500, with sales returns of $350. If she receives a 6% commission on net sales, what were her gross wages?

6. Donald Rodgers receives a 7% commission on his net sales. If his sales totaled $10,900 but customers returned $500 of merchandise, what were his gross wages for the week?

7. A real estate agent receives $10,500 as a commission on a house sale. If his commission rate is 7%, what was the amount of the house sale?

8. Antonio Cordova's commission was $10,800 for a house sale. If his commission rate was 8%, how much was the amount of the house sale?

9. Mike Perry receives a sliding-scale commission of 3½% on the first $2,000 of net sales, 5% on the next $4,000 of net sales, and 6% on all net sales over $6,000. He is allowed a drawing account of $350. Mr. Perry's net sales for the week were $13,600.

a. What was his total commission?

b. If Mr. Perry withdraws the $350 during the week, what amount is due him at the end of the week?

10. Martha Bedont earns a sliding-scale commission of 4% on the first $6,000 of net sales, 4½% on the next $5,000 of net sales, and 5% on all net sales over $11,000. She is allowed a monthly drawing account of $250. Ms. Bedont's net sales for the month were $16,000.

a. What was her total commission?

b. Ms. Bedont withdrew the $250 for expenses. What amount is still due her at the end of the month?

11. Lisa Wilson earned a gross wage of $300 which included a $120 salary and a 5% commission. How much were her net sales last week?

12. Lui Wen earned a gross wage of $810, which included a salary of $90 and a 6% commission. How much were her net sales?

13. Robin Alterman earns a weekly salary of $175 plus a 4% commission on net sales in excess of $200. What were her gross wages for a week in which she had sales of $3,000 and sales returns of $50?

14. Shannon Chastine earns a weekly salary of $150 plus a 6% commission on sales in excess of $300. Determine her gross wages for a week in which she had sales of $8,000 and sales returns of $100.

15. Rebecca Marshall manages the women's sportswear department of Woodwind Co. She receives a salary of $160; in addition, she receives a 6% commission on her net sales in excess of $500 and a 1½% override on the net sales of the sales associates in her department. Her sales totaled $2,200 last week. Other employee sales totaled $10,100 with sales returns of $50. Determine her weekly gross wages.

16. Edward Hamm, a manager of a men's clothing department, receives a salary of $150 each week, plus a 7% commission on his net sales in excess of $500 and a 1% override on the net sales of his associates. His sales last week were $5,000. The other employees had sales of $35,400 with sales returns of $400. What was his gross wage?

Calculate the commission payrolls in Problems 17–20.

17.

OZARK OFFICE SUPPLY
Payroll for Week Ending June 11, 20XX

	NET SALES								
NAME	M	T	W	T	F	S	TOTAL	COMMISSION RATE	GROSS COMM.
Dempsey, J.	$396	$375	$280	$195	$304	$400		8%	
Gold, K.	266	483	501	442	455	423		10	
Keller, E.	322	388	404	425	421	500		9	
Miller, M.	329	284	292	275	360	410		8	
Wright, C.	340	465	488	492	502	613		7	
								Total	

18.

WASHINGTON SPORTING GOODS, INC.
Payroll for Week Ending May 15, 20XX

	NET SALES								
NAME	M	T	W	T	F	S	TOTAL	COMMISSION RATE	GROSS COMM.
Dodd, D.	$ 925	$ 742	$ 957	$ 896	$ 850	$ 750		10%	
Hayes, B.	1,256	1,345	1,384	1,390	1,400	1,450		8	
Jenkins, P.	660	782	778	840	860	780		7	
Moody, A.	1,306	984	1,000	1,105	800	905		8	
Owens, K.	864	700	1,010	1,496	580	800		9	
								Total	

19.

ROZELLA COMPUTER CO.
Payroll for Week Ending October 4, 20XX

NAME	SALES			QUOTA	COMM. SALES	COMM. RATE	GROSS COMM.	SALARY	GROSS WAGES
	GROSS	R&A	NET						
Bellis, E.	$6,340	$40		—		8%		$ 80	
Chambers, D.	5,820	70		$ 800		6		—	
Duggan, C.	4,690	50		—		5		100	
Meyer, C.	6,500	—		1,000		6		120	
Quader, E.	6,332	82		500		8		—	
						Totals			

20.

COPPER ENTERPRISE, INC.
Payroll for Week Ending November 4, 20XX

NAME	SALES			QUOTA	COMM. SALES	COMM. RATE	GROSS COMM.	SALARY	GROSS WAGES
	GROSS	R&A	NET						
Barnes, D.	$ 7,500	$100		—		8%		$110	
Cordel, Y.	10,680	80		$1,000		7			
Leiter, L.	9,844	64		—		5		200	
Murphy, D.	12,000	—		1,200		9		150	
Ward, J.	8,956	26		630		8		—	
						Totals			

21. Prepare an account sales for the following transactions: Rivera Farms, Homestead, Florida, consigned a shipment to Thompson Produce, Nashville, Tennessee. The shipment contained 220 lbs. tomatoes sold on March 5 @ $0.65 per pound; 175 lbs. carrots sold on March 5 @ $0.52 per pound; 460 lbs. red potatoes sold on March 10 @ $0.48 per pound; 200 bu. lima beans sold on March 12 @ $6.30 per bushel. Freight charges totaled $98.40, and insurance charges totaled $100. The commission agents' charge was 6%.

22. Complete the following account sales.

ACCOUNT SALES

SIWEK & SIWEK, INC.
Commision Merchants
Clemson, South Carolina 29403

Sold for Account of:

Rankin Fruit Co.
Carlisle, PA 17014

20XX July	8	600 lbs. peaches @ $0.68/lb.		
	8	500 lbs. tomatoes @ $0.40/lb.		
	15	150 bu. white corn @ $4.80/bu.		
	22	175 bu. bush beans @ $6.40/bu.		
		Gross proceeds		
		Charges:		
		Freight	$72.00	
		Insurance	56.00	
		Commission, 5%		
		Net proceeds		

23. Prepare an account purchase for the following purchases: The China Closet in Jackson, Mississippi, commissioned Taylor Imports in Baltimore, Maryland, to purchase the following crystal pieces from Ireland: On October 1, 5 cake plates at $70 and 10 goblets at $36; on October 10, 3 7″ vases at $61; on October 18, 2 8″ × 10″ picture frames at $57 and 4 pairs of 6″ candlesticks at $45. Shipping charges were $96, and insurance totaled $55. Taylor Imports charges a 5% commission.

24. Complete the following account purchase.

ACCOUNT PURCHASE

McKenney & Sons
Fairfax, VA 22312

BOUGHT FOR THE ACCOUNT OF:

Hope Imports
Alexandria, VA 22201

20XX				
Feb.	15	25 ginger jar lamps @ \$76.00/ea.		
	23	16 Ming vases @ \$52.00/ea.		
	27	10 3′ × 5′ silkscreens @ \$48.00/ea.		
	27	6 5′ × 7′ Oriental carpets @ \$195.00/ea.		
		PRIME COST		
		Charges:		
		Shipping	\$98.00	
		Insurance	87.00	
		Commission, 4%		
		GROSS COST		

SECTION 3

COMPUTING WAGES—HOURLY RATE BASIS

The hourly rate basis for determining wages is the method under which wages are computed for the majority of the American working force. Under this plan, the employee receives a stipulated amount for each hour's work. One or more of several factors may determine the hourly rate: the individual's ability, training, and experience, as well as federal minimum-wage laws or union contracts. Once the hourly rate has been established, however, the employees' quality of performance or amount of production in no way influences their wages.

The **Fair Labor Standards Act** governs the employment of more than 80 million workers. Previously affecting primarily workers engaged in some form of interstate, international, or other large-scale commercial enterprises, the Act has now been extended to include a wide range of occupations. Among covered workers are nonprofessional employees of hospital and nursing homes, public or private schools, cleaners and laundries, transportation services, construction industries, domestic workers, and all levels of city, county, state, and federal governments.

As of September 1, 1997, the Act prescribed a 40-hour workweek and required a minimum wage for all covered employees of $5.15 per hour. The "training wage" allows employers to pay newly hired employees under the age of 20 a minimum of $4.25 per hour for the first 90 calendar days of employment. Many states have also established minimum wages for employees not subject to the federal regulations. The federal act applies equally to men and women. It should be emphasized, however, that certain employees are specifically excluded from the provisions of the Act—executive and administrative personnel in businesses, professional people (such as medical personnel, educators, lawyers, and so forth), as well as outside salesworkers, employees of very small retail or service establishments, and many seasonal and small-farm workers.

OVERTIME

In accordance with the federal law covering the 40-hour workweek, most employees' regular week consists of five 8-hour working days. Under the federal regulation, **overtime** must be paid to covered employees for work in excess of 40 hours. Many employers, even though not subject to the federal act, follow this policy voluntarily.

The standard overtime rate is $1\frac{1}{2}$ times the worker's regular hourly rate ("time-and-a-half"). "Double time" is often paid for work on Sundays or on holidays. It should be noted, however, that the Fair Labor Standards Act does *not* require extra pay for Saturdays, Sundays, or holidays (except for hours in excess of 40); neither does it require pay for time off during vacations or holidays. Whether premium rates are paid or time off is granted depends entirely on the employment (or union) agreement.

In some instances, overtime is paid when an employee works more than 8 hours on any one day. (Some union contracts contain this provision.) As a general practice, however, a worker is required to have worked 40 hours before receiving any overtime compensation; otherwise, the worker might be paid some overtime without having completed the prescribed 40-hour week. If a company nonetheless pays overtime on a daily basis, the payroll clerk must keep a careful record of each employee's hours to ensure that the same time is not counted twice; that is, hours which qualify as daily overtime must not be applied toward the regular 40-hour week, after which all work would be at overtime rates.

Union contracts often contain overtime provisions regarding any number of other situations, such as working two complete shifts in 1 day or working 6 or 7 consecutive days.

Large firms usually have timeclocks that record entering and leaving times on each employee's timecard. Total work hours are then calculated from this record. Just as employees are compensated for overtime work, they are likewise penalized for tardiness. In smaller firms, a supervisor may record the whole crew's hours; or, in some instances, employees report their own hours. Total hours worked are usually kept to the nearest quarter-hour each day.

Example 1 Compute the gross wages of each employee.

B & J PAINT CO.
Payroll for Week Ending May 9, 20XX

NAME	M	T	W	T	F	S	TOTAL HOURS	RATE PER HOUR	GROSS WAGES
Anderson, James	9	7	8	8	8	0	40	$7.50	$ 300.00
Bailey, Carl	8	8	10	0	10	4	40	8.75	350.00
Carson, W. J.	8	6	0	5	8	4	31	6.80	210.80
Davis, Willard	9	8	8	8	6	0	39	8.00	312.00
								Total	$1,172.80

Example 2 Compute the gross wages of each employee of Wilkins-Casey, Inc. Overtime is paid at $1\frac{1}{2}$ times the regular rate on hours in excess of 40 per week.

WILKENS-CASEY, INC.
Payroll for Week Ending December 12, 20XX

NAME	M	T	W	T	F	S	TOTAL HOURS	REG. HOURS	HOURLY RATE	OVERTIME HOURS	OVERTIME RATE	REG. WAGES	OVERTIME WAGES	TOTAL GROSS WAGES
Elliot, D.	8	8	10	9	10	4	49	40	$ 7.50	9	$11.25	$ 300.00	$101.25	$ 401.25
Farley, R.	10	8	9	7	7	4	45	40	10.00	5	15.00	400.00	75.00	475.00
Givens, T.	8	8	8	8	8	0	40	40	16.50	0	24.75	660.00	0	660.00
Harper, T.	5	8	6	7	8	4	38	38	14.00	0	21.00	532.00	0	532.00
Ivey, M.	9	8	6	8	9	4	44	40	15.00	4	22.50	600.00	90.00	690.00
											Totals	$2,492.00	$266.25	$2,758.25

Note. The Regular Wages total plus the Overtime Wages total must equal the total Gross Wages ($2,492 + $266.25 = $2,758.25).

Example 3 Compute David Pressley's gross wages based on the following: Overtime at time-and-a-half is paid daily on work in excess of 8 hours and weekly on work in excess of 40 regular hours; double time is paid for all Sunday work. His hours each day were as follows:

	M	T	W	T	F	S	S
Pressley, David	8	9	7	10	7	5	5

		M	T	W	T	F	S	S	TOTAL HOURS	RATE PER HOUR	BASE WAGES	TOTAL GROSS WAGES
Pressley, D.	RT[a]	8	8	7	8	7	2		40	$ 6.80	$272.00	
	1½		1		2		3		6	10.20	61.20	
	DT							5	5	13.60	68.00	$401.20

[a]RT indicates "Regular time"; 1½, "Time-and-a-half"; and DT, "Double time."

Calculator Tip. For this problem and also in the succeeding examples, place the regular and overtime components into memory and use [MR] to determine each person's total wages: 40 [×] 6.8 [M+] → 272; 6 [×] 10.2 [M+] → 61.2; 5 [×] 13.6 [M+] → 68; [MR] → 401.2.

OVERTIME FOR SALARIED PERSONNEL

The following example illustrates the computation of gross wages for salaried personnel who must be paid overtime wages. This procedure is required under the Fair Labor Standards Act for salaried personnel subject to overtime provisions. (Refer also to Section 1, "Computing Wages—Salary.")

Example 4 (a) A receptionist receives a $180 salary for her regular 40-hour week. What should her gross wages be for a week when she works 44 hours?

The clerk's regular wages are equivalent to $\frac{\$180}{40 \text{ hours}}$ or $4.50 per hour. She must be paid overtime at $1\frac{1}{2}$ times this rate, or $6.75 per hour. Her wages for a 44-hour week would thus be $207:

40 hours @ $4.50	= $180	Regular salary
4 hours @ $6.75	= 27	Overtime
44-hour week	= $207	Gross wages

(b) An administrative assistant receives a $217 weekly salary. If a regular workweek contains 35 hours, how much compensation will she receive when she works 43 hours?

$$\frac{\$217}{35 \text{ hours}} = \$6.20 \text{ per hour}$$

$$\text{Overtime rate} = 1\tfrac{1}{2} \times \$6.20 = \$9.30 \text{ per hour}$$

40 hours @ $6.20	= $248.00	
3 hours @ $9.30	= 27.90	
43-hour week	= $275.90	Gross wages

(c) A service manager's salary is $324, although his hours vary each week. What are his gross wages for a week when he works 45 hours?

$$\frac{\$324}{45 \text{ hours}} = \$7.20 \text{ per hour}$$

$$\text{Overtime rate} = 1\tfrac{1}{2} \times \$7.20 = \$10.80 \text{ per hour}$$

40 hours @ $ 7.20	= $288.00	
5 hours @ $10.80	= 54.00	
45-hour week	= $342.00	Gross wages

Observe that, for any employee whose hours vary each week as do this man's hours, the employee's regular rate per hour will also vary each week. That is, his regular rate is $7.20 per hour only during a week when he works 45 hours. If he works 54 hours the following week, then his regular rate that week will be

$$\frac{\$324}{54} = \$6.00 \text{ per hour on the 40 hours}$$

plus overtime for 14 hours at $9.00 per hour.

OVERTIME EXCESS

All the examples above show regular rates for 40 hours and an overtime rate of $1\frac{1}{2}$ times the regular rate. An alternative method will produce the same result: Calculate the total hours at the regular rate plus the overtime hours at one-half the regular rate. The additional compensation calculated by this method (overtime hours times one-half the regular rate) is called the **overtime excess.**

Example 5 A worker who earns \$10 per hour works 46 hours. Compare these earnings (a) by standard overtime and (b) by overtime excess.

(a)

40 hours @ \$10.00 =	\$400.00	Straight-time wages
6 hours @ \$15.00 =	90.00	Overtime wages
46 hours =	\$490.00	Total gross wages

(b) One-half the regular rate is $\frac{1}{2}$ of \$10.00 = \$5.00 per hour for the excess hours over 40:

46 hours @ \$10.00 =	\$460.00	Regular-rate wages
6 hours @ \$ 5.00 =	30.00	Overtime excess
46 hours =	\$490.00	Total gross wages

Some firms prefer the overtime excess method for their own cost accounting, since it emphasizes the extra expense (and lost profit) resulting from overtime work. (Notice above that if the extra 6 hours' work could have been delayed until the following week and all done at the regular rate, the company's cost would have been only \$460, rather than \$490.)

Use of the overtime excess method is becoming more widespread. Although either overtime method is permitted under the Fair Labor Standards Act, the Wage and Hour Division (which enforces the act) uses the overtime excess method. A major reason for using overtime excess is that workers' compensation insurance premiums are based on a worker's regular-rate wages (that is, the \$460 above).

SECTION 3 PROBLEMS

1. Complete the following gross payroll.

CRISTO CLEANING SERVICE
Payroll for Week Ending May 5, 20XX

NAME	M	T	W	T	F	S	TOTAL HOURS	RATE PER HOUR	GROSS WAGES
Becker, P.	8	8	7	7	6	0		\$6.50	
Doyle, D.	8	7	8	9	8	0		6.75	
Margo, D.	5	8	7	8	8	3		5.80	
Murray, A.	8	7	7	7	7	2		6.25	
Richards, B.	6	6	6	6	6	6		6.00	
Stark, R.	8	9	9	8	6	0		7.10	
Wu, Y.	9	8	8	7	5	0		7.00	
								Total	

2. Complete the following gross payroll.

THE HEARN CO.
Payroll for Week Ending June 2, 20XX

NAME	M	T	W	T	F	S	TOTAL HOURS	RATE PER HOUR	GROSS WAGES
Burke, A.	8	7	5	8	8	0		$ 9.50	
Carter, F.	6	8	8	8	8	2		11.25	
Keeney, K.	9	6	8	7	8	0		10.70	
Poole, R.	8	8	7	7	7	3		10.45	
Quinn, B.	5	7	6	9	9	3		12.00	
Schick, B.	9	8	7	8	8	0		9.40	
Weiman, T.	8	8	9	9	3	0		9.75	
								Total	

3. The employees of Saunders, Inc. are required to work a regular 40-hour week. Standard overtime is paid for any additional hours. Compute the following gross payroll.

SAUNDERS, INC.
Payroll for Week Ending December 6, 20XX

NAME	M	T	W	T	F	S	TOTAL HOURS	REG. HOURS	HOURLY RATE	OVERTIME HOURS	OVERTIME RATE	REG. WAGES	OVERTIME WAGES	TOTAL GROSS WAGES
Agnew	8	8	8	8	8	0			$17.50					
Barski	8	7	9	8	7	3			16.50					
Clancy	6	8	7	8	8	0			16.00					
Ibar	7	8	8	9	8	5			17.20					
Nozek	8	8	6	9	8	4			17.00					
											Totals			

4. The employees of Mahoney Electronics, Inc. are required to work a regular 40-hour week, after which time they receive standard overtime for any additional hours. Complete the payroll record below.

MAHONEY ELECTRONICS, INC.
Payroll for Week Ending June 14, 20XX

NAME	M	T	W	T	F	S	TOTAL HOURS	REG. HOURS	HOURLY RATE	OVERTIME HOURS	OVERTIME RATE	REG. WAGES	OVERTIME WAGES	TOTAL GROSS WAGES
Austin	9	8	8	7	8	0			$10.50					
Cote	8	9	9	8	8	0			11.50					
Jarvis	6	7	8	8	6	4			10.00					
Osuna	9	9	8	9	6	5			9.90					
Way	8	8	8	9	7	4			13.00					
											Totals			

5. Daily overtime (in excess of 8 hours) and weekly overtime (after 40 regular hours) are paid to the employees listed below. They also receive double time for Sunday work. From the total daily hours given, determine the number of hours in each category, and compute the gross wages.

	M	T	W	T	F	S	S
Coffey	6	8	7	9	7	0	4
Hanson	7	8	10	8	7	8	0
Yoder	9	5	8	9	8	4	3

NAME		M	T	W	T	F	S	S	TOTAL HOURS	RATE PER HOUR	BASE WAGES	TOTAL GROSS WAGES
Coffey	RT									$8.00		
	1½											
	DT											
Hanson	RT									8.50		
	1½											
	DT											
Yoder	RT									8.60		
	1½											
	DT											
											Total	

6. Daily overtime (in excess of 8 hours) and weekly overtime (after 40 regular hours) are paid to the employees at Pavlidis Office Systems, Inc. They also receive double time for work done on Sundays or national holidays. Monday of the week shown was a national holiday. Determine the number of hours in each category, and complete the gross payroll.

	M	T	W	T	F	S	S
Amsbry	6	8	9	7	8	0	0
Emory	8	8	10	8	9	8	0
Williams	0	10	8	8	8	5	5

NAME		M	T	W	T	F	S	S	TOTAL HOURS	RATE PER HOUR	BASE WAGES	TOTAL GROSS WAGES
Amsbry	RT 1½ DT									$10.00		
Emory	RT 1½ DT									9.00		
Williams	RT 1½ DT									15.00		
											Total	

Compute the gross weekly wages for the salaried employees shown who are all subject to the Fair Labor Standards Act.

POSITION	REGULAR HOURS PER WEEK	REGULAR SALARY	HOURS WORKED	GROSS WAGES
7. a. Sales manager	40	$336	45	
b. Office manager	38	304	44	
c. Plant manager	(Varies)	437	46	
8. a. Administrative assistant	40	$540	43	
b. Executive secretary	36	522	46	
c. Supervisor	(Varies)	506	44	

For each of the employees listed below, compute gross earnings (1) with overtime at time-and-a-half and (2) by using overtime excess.

	Employee	Hours	Rate per hour	Gross earnings
9. a.	G	43	$ 5.60	
b.	H	45	7.70	
c.	I	47	8.50	
10. a.	J	42	$10.00	
b.	K	46	15.40	
c.	L	48	13.50	

11. Use the overtime excess method in computing the following gross payroll.

KING APPLIANCES, INC.
Payroll for Week Ending February 15, 20XX

Name	M	T	W	T	F	S	Total hours	Rate per hour	Overtime hours	Overtime excess rate	Regular rate wages	Overtime excess wages	Total gross wages
Dinh	8	8	10	8	9	5		$5.50					
Horn	8	9	8	9	9	6		6.40					
Rook	7	8	10	10	8	4		7.30					
Soka	9	9	8	8	8	0		7.90					
Vern	8	9	9	6	10	8		7.60					
										Totals			

12. Use the overtime excess method in computing the following gross payroll.

LEARY VIDEO SALES AND SERVICE
Payroll for Week Ending December 3, 20XX

Name	M	T	W	T	F	S	Total hours	Rate per hour	Overtime hours	Overtime excess rate	Regular rate wages	Overtime excess wages	Total gross wages
Chung	8	8	9	9	8	0		$10.20					
Hampton	7	8	8	8	8	6		12.50					
Kramer	8	8	8	8	8	4		11.40					
Rhoades	9	8	7	8	8	8		10.50					
Sherman	8	8	5	8	6	11		12.00					
										Totals			

COMPUTING WAGES—PRODUCTION BASIS

In many factories, each employee's work consists of repeating the same task—either some step in the construction of a product or inspecting or packing finished products. The employee's value to the company can thus be measured by how often this task is completed. Therefore, many plants compensate employees at a specified rate per unit completed; this constitutes the **production** basis of computing wages. This type of wage computation is also known as the **piecework** or **piecerate** plan.

Naturally, some steps in the production of a product require more time than others. Thus, the rate per unit for that step would be more than the rate on another task which could be completed much more quickly. In order to keep the entire plant production moving at an even pace, more employees are hired to perform time-consuming tasks, and fewer workers are employed to perform easier tasks.

On the straight production plan, the employee receives the same rate for each piece; thus,

292 pieces at \$1.55 = \$452.60 wages

Example 1 The following example shows a payroll based on straight production.

TEXAS BOOT CO.
Payroll for Week Ending April 5, 20XX

NAME	NET PIECES PRODUCED						TOTAL NET PIECES	RATE PER PIECE	GROSS WAGES EARNED
	M	T	W	T	F	S			
Cruz, Ramon	78	89	90	81	80	72	490	\$1.25	\$ 612.50
Gillian, Art	39	46	39	37	34	19	214	2.50	535.00
Kearns, Don	125	138	140	131	129	65	728	1.00	728.00
Lotus, Kim	71	67	74	79	70	34	395	1.40	553.00
Steiner, Lois	76	82	81	77	80	40	436	1.75	763.00
								Total	\$3,191.50

INCENTIVES (PREMIUM RATES)

At first glance, the production plan would seem to be quite acceptable to employees, since compensation is determined entirely by individual performance. Such has not been the case, however. Most companies that use the piecework plan have established quotas or minimum production standards for each job. As an extra incentive to production, workers receive a higher rate per unit on all units they complete in excess of this minimum. It is these quotas which historically have caused such dissatisfaction with the production plan. In the past, many companies set "production" (quotas) at such high levels that only a few of the most skillful and experienced employees could attain them. Furthermore, employees complained that when they finally did master the task well enough to "make production," the company would reassign them to a new, unfamiliar task.

As a result of the general dissatisfaction over piecework plans, many union contracts now contain provisions that members must be paid on the hourly rate basis. Those plants which do still use the production basis have tended to reestablish quotas that are more in line with the production that average employees can attain.

OVERTIME

Piecework employees who are subject to the Fair Labor Standards Act must likewise be paid time-and-a-half for work in excess of 40 hours per week. This overtime may be computed similarly to that of salaried workers (by finding the hourly rate to which their regular-time earnings are equivalent and paying $1\frac{1}{2}$ this hourly rate for overtime hours). However, the overtime wages of a production worker may be computed at $1\frac{1}{2}$ times the regular rate per piece, provided that certain standards are met to ensure that workers are indeed being paid the required minimum wage for regular-time hours, as well as time-and-a-half for overtime hours. This method is preferred because it is simpler to compute and will therefore be used for our problems. We shall not, however, attempt to compute overtime where premium production is also involved.

Example 2 In the following production payroll, all work performed on Saturday is overtime and is paid at $1\frac{1}{2}$ times the regular piece rate.

PRECISION INSTRUMENTS, INC.
Payroll for Week Ending August 30, 20XX

	NET PIECES PRODUCED						REG. TIME	RATE PER	OVERTIME	OVERTIME	REG.	OVERTIME	TOTAL GROSS
NAME	M	T	W	T	F	S	PROD.	PIECE	PROD.	RATE	WAGES	WAGES	WAGES
Ball	214	220	218	216	219	112	1,087	$0.25	112	$0.375	$ 271.75	$ 42.00	$ 313.75
Grange	94	90	97	99	101	51	481	0.50	51	0.75	240.50	38.25	278.75
Mills	68	70	72	69	71	30	350	0.65	30	0.975	227.50	29.25	256.75
Trent	151	155	153	150	152	75	761	0.36	75	0.54	273.96	40.50	314.46
										Totals	$1,013.71	$150.00	$1,163.71

Calculator Tip. Use the same technique as for Example 2 of Section 3. That is, find (a) the total production for each worker and (b) each person's overtime rate. Then compute (c) the Regular Wages column and (d) the Overtime Wages column, and find the total for the two columns. Finally, (e) total each person's wages with [M+], and use [MR] to verify that the Total Gross Wages column shows the same total you obtained after step (d).

DOCKINGS (SPOILAGE CHARGEBACKS)

Although companies are willing to pay premium rates as incentives to high production, they also wish to discourage haste resulting in carelessness and mistakes. An employee who makes an uncorrectable mistake has not only destroyed the raw materials involved but has also wasted the work done previously by other employees. Thus, an unreasonably high amount of spoilage would be quite expensive to a company.

To discourage carelessness, many plants require employees to forfeit a certain amount for each piece they spoil. These penalty **dockings** (or **chargebacks**) are generally at a lower rate than the employee ordinarily receives for each task, because a small amount of spoilage is to be expected. The worker is actually penalized *twice* for the spoilage: First, the number of spoiled units is subtracted from the worker's gross units to obtain the net units earning at the base (or quota) rate; later, the spoiled units are multiplied by the chargeback rate, and this penalty is subtracted from earnings. In general,

$$\text{Gross wages} = \text{Base production} + \text{Premium} - \text{Chargeback}$$

"Base production" is computed on the number of pieces (gross production − spoilage) that apply toward meeting quota. "Premium" is computed on any excess pieces (those still above quota). "Chargeback" is computed on the spoiled pieces. (See Example 3 for an illustration.)

Example 3 Cynthia Simmons sews zippers into dresses for a garment manufacturer. Her base rate is $1.25 per dress, and a premium rate of $1.52 per dress is paid after she exceeds the weekly production quota of 240 zippers. She is docked 75¢ for each zipper that does not pass inspection. Compute her gross earnings for the week:

	M	T	W	T	F	TOTAL
Gross production	54	51	55	53	52	265
Less: Spoilage	2	0	0	3	1	6
Net production	52	51	55	50	51	259

Cynthia will receive the premium rate on 259 − 240 = 19 zippers. Her Gross wages = Base production + Premium − Chargeback is as follows:

	TOTAL NET PRODUCTION	×	RATE	=	GROSS WAGES
Base production	240	@	$1.25		$300.00
Premium production	19	@	1.52		+ 28.88
					$328.88[a]
Spoilage chargeback	6	@	0.75		− 4.50
					$324.38

[a]Wage computation would end at this point for any production tasks not subject to chargeback.

Example 4 Complete the following production payroll, which includes incentive and penalty chargeback rates:

HANOVER GAME CO.
Payroll for Week Ending October 17, 20XX

NAME		DAILY PRODUCTION M	T	W	T	F	TOTAL PRODUCTION	NET QUOTA	PRODUCTION	RATE (CENTS)	BASE WAGES	TOTAL GROSS WAGES
Eaton, C. C.	GP[a]	159	162	152	155	158	786		Base: 750	45¢	$337.50	
	S	7	3	0	2	2	14		Premium: 22	56	12.32	$344.92
	NP	152	159	152	153	156	772	750	Chargeback: 14	35	4.90	
Gray, N. S.	GP	126	128	131	125	129	639		Base: 632	47¢	$297.04	
	S	2	1	3	0	1	7		Premium: —	59	—	294.38
	NP	124	127	128	125	128	632	650	Chargeback: 7	38	2.66	
Spann, B. R.	GP	143	141	145	142	144	715		Base: 700	46¢	$322.00	
	S	2	0	0	1	2	5		Premium: 10	58	5.80	325.95
	NP	141	141	145	141	142	710	700	Chargeback: 5	37	1.85	
											Total	$965.25

[a]GP indicates "gross production"; S, "spoilage"; and NP, "net production."

SECTION 4 PROBLEMS

1. Tom Stone operates a machine that bolts two pieces of an automobile engine together. His base rate is 52¢ per bolt, with a premium of 70¢ per bolt in excess of the first 500 bolts per week. Determine his gross wages for the following week:

	M	T	W	T	F	TOTAL
Stone, Tom	128	140	155	158	167	

2. Melissa Taylor is an inspector in a textile mill, for which she is paid 95¢ per dozen pairs of socks inspected. She receives a premium rate of $1.10 after she exceeds her base production of 300 dozen per week. Determine her wages for the following week:

	M	T	W	T	F	TOTAL
Taylor, Melissa	85 dozen	92 dozen	100 dozen	98 dozen	105 dozen	

3. Alice Pollen, working in a paper plant, operates a machine that attaches glue to assembled pads of paper. Her base rate is 15¢ per pad on the first 1,000 pads of net production, and a premium rate of 24¢ per pad is paid for net production over 1,000 pads. She is docked 8¢ for each pad that does not pass inspection. Calculate her wages for the following week:

	M	T	W	T	F	TOTAL
Gross production	325	344	346	350	342	
Less: Spoilage	5	6	2	0	1	
Net production						

4. Dave Berry assembles parts in microcomputers. His base rate is $2.00 per assembly on the first 100 assemblies. He receives an incentive of $2.60 per assembly for his net production in excess of 100 assemblies. He is docked 50¢ for each assembly that does not pass inspection. Calculate his wages for the week when he had the following production:

	M	T	W	T	F	TOTAL
Gross production	26	34	28	34	29	
Less: Spoilage	2	0	0	3	1	
Net production						

5. Complete the following payroll based on straight production:

PACIFIC TOOL CO.
Payroll for Week Ending April 5, 20XX

	NET PIECES PRODUCED					TOTAL	RATE	GROSS
NAME	M	T	W	T	F	NET PIECES	PER PIECE	WAGES EARNED
Abbott, A.	64	78	76	84	90		$1.00	
Breck, J.	71	77	78	79	86		1.03	
Dunn, M.	56	58	62	65	67		1.01	
Jones, B.	80	89	88	85	97		1.04	
Simon, C.	75	75	82	86	89		1.02	
Woods, V.	84	85	89	86	91		1.05	
Yoe, T.	88	85	88	90	92		1.06	
							Total	

6. Complete the following production payroll:

RICHMOND FABRICS
Payroll for Week Ending April 16, 20XX

NAME	NET PIECES PRODUCED					TOTAL NET PIECES	RATE PER PIECE	GROSS WAGES EARNED
	M	T	W	T	F			
Adams, K.	90	88	100	101	102		$0.75	
Green, N.	95	100	105	108	108		0.80	
Hart, M.	115	112	114	114	104		0.90	
Lester, A.	109	110	120	125	121		0.92	
Potter, D.	112	126	120	121	120		0.80	
Randolph, T.	116	122	106	110	110		0.96	
Wilkins, E.	109	101	104	105	106		0.87	
							Total	

7. Complete the following production payroll. Overtime will be paid for Saturday work at 1½ times the standard rate per piece.

STROZIER MANUFACTURING CO.
Payroll for Week Ending January 18, 20XX

NAME	NET PIECES PRODUCED						REG. TIME PROD.	RATE PER PIECE	OVERTIME PROD.	OVERTIME RATE	REG. WAGES	OVERTIME WAGES	TOTAL GROSS WAGES
	M	T	W	T	F	S							
Chang	56	58	60	62	66	50		$0.60					
Evans	60	61	64	67	70	45		0.68					
Hall	72	75	75	79	78	70		0.70					
Long	65	70	76	71	75	63		0.74					
Thomas	61	71	73	76	74	68		0.68					
										Totals			

8. Complete the following production payroll. Overtime will be paid for Saturday work at 1½ times the standard rate per piece.

JAMESTOWN CABLE CO.
Payroll for Week Ending July 27, 20XX

NAME	NET PIECES PRODUCED						REG. TIME PROD.	RATE PER PIECE	OVERTIME PROD.	OVERTIME RATE	REG. WAGES	OVERTIME WAGES	TOTAL GROSS WAGES
	M	T	W	T	F	S							
Baylor	80	82	81	88	85	46		$0.68					
Downes	92	101	102	99	104	66		0.72					
Kelly	89	91	73	84	89	45		0.94					
Harvey	80	94	95	90	81	36		0.88					
Scully	77	68	78	80	84	61		0.92					
										Totals			

9. Find the gross wages for the following payroll:

SCHNELL ATHLETIC EQUIPMENT
Payroll for Week Ending August 12, 20XX

NAME		DAILY PRODUCTION					TOTAL PRODUCTION	NET QUOTA	PRODUCTION	RATE	BASE WAGES	TOTAL GROSS WAGES
		M	T	W	T	F						
Allen	GP	39	48	50	51	46		200	Base:	$0.88		
	S	1	2	0	1	2			Premium:	0.95		
	NP								Chargeback:	0.75		
Dill	GP	36	41	49	54	50		150	Base:	0.92		
	S	0	2	2	0	0			Premium:	0.99		
	NP								Chargeback:	0.60		
Edwards	GP	54	58	64	60	63		250	Base:	0.70		
	S	3	1	1	0	0			Premium:	0.80		
	NP								Chargeback:	0.50		
Jenks	GP	48	49	53	59	66		300	Base:	0.85		
	S	1	0	2	1	0			Premium:	0.93		
	NP								Chargeback:	0.40		
Strong	GP	65	68	73	75	79		275	Base:	0.86		
	S	3	1	1	2	1			Premium:	0.96		
	NP								Chargeback:	0.45		
											Total	

10. Find the gross wages for the following payroll:

SIMON TEXTILES, INC.
Payroll for Week Ending May 6, 20XX

NAME		DAILY PRODUCTION M	T	W	T	F	TOTAL PRODUCTION	NET QUOTA	PRODUCTION	RATE	BASE WAGES	TOTAL GROSS WAGES
Ballard	NP	100	94	95	90	86		300	Base:	$0.88		
	S	1	0	2	2	2			Premium:	0.94		
	NP								Chargeback:	0.30		
Galano	NP	65	72	85	92	90		250	Base:	0.95		
	S	0	1	2	1	1			Premium:	0.98		
	NP								Chargeback:	0.20		
Maddox	GP	101	106	110	112	100		350	Base:	0.92		
	S	0	0	2	4	4			Premium:	0.96		
	NP								Chargeback:	0.40		
Pierson	GP	75	70	63	68	76		350	Base:	0.98		
	S	1	0	3	2	0			Premium:	1.00		
	NP								Chargeback:	0.44		
Stewart	GP	99	97	100	101	99		300	Base:	0.89		
	S	0	1	2	1	1			Premium:	0.95		
	NP								Chargeback:	0.30		
											Total	

SECTION 5

NET WEEKLY PAYROLLS

Employers are required by law to make certain deductions from the gross earnings of their employees. The amount remaining after deductions have been withheld is known as **net wages.** The same methods are used to determine net wages from gross wages regardless of which payroll plan has been used to compute gross wages.

Two compulsory federal deductions are made in all regular payrolls: FICA and personal income tax. In many states, the employer is also required to withhold state income tax. Certain other deductions, either compulsory or voluntary, may also be made. The following paragraphs discuss important aspects of the various deductions.

FEDERAL INSURANCE CONTRIBUTIONS ACT (FICA)

The original **Federal Insurance Contributions Act** was one of the emergency measures passed during the Great Depression of the 1930s; it has been amended and extended many times and now draws payments from over 148 million workers. **FICA** includes both **Social Security** and **Medicare.** Social Security is not a tax in the normal sense; it could be described more accurately as a compulsory savings or insurance program. Social Security funds are used to pay monthly benefits to retired or disabled workers and to pay benefits to the surviving spouse and minor children of a deceased worker. Social Security accounted for 40% of total income of senior citizens and kept 1.4 million children out of poverty in 1996. Currently, 48 million people receive payments from Social Security.

Under a law enacted in 1999, the Social Security Administration must provide an annual statement estimating the retirement, disability, and survivor benefits that each working American is eligible to receive. The mailing of these statements is triggered by the employee's birth month. Information provided should be checked against pay stubs, W-2 forms, and tax returns to be sure that the records at Social Security are correct. If an error is found, it should be reported to the Social Security Administration.

The Social Security program has been the subject of much concern in recent years, because paying benefits to retirees and other qualified recipients has required spending all the funds that current workers pay in—and the number of recipients continues to increase each year. Although the program was originally intended only to supplement individual savings and retirement plans, for many older persons, their FICA benefits have in reality become their sole source of income. Moreover, as the number of older people in the population increases, a smaller proportion of younger workers will be paying into the Social Security fund. Thus, the prospect of a bankrupt Social Security system has serious national impact. So far, the situation has been handled primarily by frequently raising both the percentage rate and the base amount of wages on which Social Security is computed. Other major changes can also be expected in the near future to ensure that the Social Security program survives.

The Medicare program was enacted by Congress in 1965. Medicare is a two-part program of health insurance available to persons 65 or older who are also eligible for Social Security benefits or Railroad Retirement benefits. There are exceptions for certain disabled workers and for disabled widows and widowers under age 65. Part A pays for hospital, hospice, and skilled nursing facility services. Part B is optional and pays for most doctor bills, medical equipment, and certain outpatient services. There is a monthly premium for Part B.

Prior to 1991, there was one tax for both Social Security and Medicare levied on earnings up to a maximum taxable wage. A 1990 tax law provided for two separate tax rates and separate maximum taxable wages, one for Social Security and one for Medicare. The maximum taxable wages for Medicare was dropped in 1994.

This text uses the 2000 withholding rate for Social Security, 6.2% on the first $76,200, and for Medicare, 1.45% on all gross earnings. All FICA deductions must be matched by an equal contribution from the employer (making the total rate 15.3%). Self-employed persons paid 12.4% on the first $76,200 for Social Security, and 2.9%

on gross earnings for Medicare. The rates and base income figures for FICA withholding are considered annually by Congress.

The Internal Revenue Service provides separate Social Security and Medicare tax tables that payroll clerks can use. This text will use the percentage method: tax rates of 6.2% for Social Security and 1.45% for Medicare.

Today, most payroll systems are computerized, but an understanding of the calculations is essential for employer and employee. Careful attention must be given to total gross earnings for any employee reaching the $76,200 maximum taxable wages for Social Security. Computerized payroll systems are programmed to cease Social Security deductions when the maximum taxable wages have been reached. Gross wages after $76,200 are still subject to the Medicare tax of 1.45%.

Example 1 Ewald Straiton earns a gross salary of $720 each week. What deductions must be made for Social Security and Medicare each week?

$$0.062 \times \$720 = \$44.64 \text{ Social Security deduction*}$$

$$0.0145 \times \$720 = \$10.44 \text{ Medicare deduction}$$

Although the $76,200 base for Social Security means that most workers make contributions from their entire year's earnings, many executives and some highly paid workers have earnings above $76,200 that are not subject to Social Security.

FEDERAL INCOME TAX

Employers are required to make deductions from each worker's wages for federal **income tax,** and most businesses use federal income tax **withholding** tables to determine these deductions. Beginning in 1984, the wage intervals have been adjusted annually for inflation—so that a worker's earnings could increase with inflation without the person having to pay tax at a higher percentage rate. This text uses the income tax withholding tables effective in 1999.

Married and single persons are subject to different income tax rates, the rates being higher for single persons. The income tax withholding tables (Table 11, *Tables Booklet*) therefore contain separate listings for married and single persons. The tax to be withheld is dependent upon three factors: (1) the worker's gross earnings, (2) whether the worker is married or single, and (3) the number of allowances claimed by the worker.

Immediately after a new employee is hired, the payroll clerk should ask the employee to complete **Form W-4,** the **Employee's Withholding Allowance Certificate** (Figure 8-1). This form indicates the employee's marital status, Social Security number, and total number of withholding allowances claimed. One allowance may be taken for each member of the family. This total number may include a

*Mr. Straiton will earn $37,440 annually ($720 × 52 weeks). This amount is under the maximum Social Security taxable wages for the year, $76,200. If he earns this annual amount, his total Social Security tax will be $2,321.28 ($37,440 × 6.2%).

Cut here and give Form W-4 to your employer. Keep the top part for your records.

Form **W-4** — Department of the Treasury, Internal Revenue Service

Employee's Withholding Allowance Certificate

► For Privacy Act and Paperwork Reduction Act Notice, see page 2.

OMB No. 1545-0010 — **2000**

1 Type or print your first name and middle initial: Sarah B.	Last name: Bingham	2 Your social security number: 987 65 4321
Home address (number and street or rural route): 4120 Frosty Court	3 ☑ Single ☐ Married ☐ Married, but withhold at higher Single rate. **Note:** *If married, but legally separated, or spouse is a nonresident alien, check the Single box.*	
City or town, state, and ZIP code: Birimingham, AL 35209	4 If your last name differs from that on your social security card, check here. **You must call 1-800-772-1213 for a new card** . . . ► ☐	

5 Total number of allowances you are claiming (from line **H** above **OR** from the applicable worksheet on page 2) — 5: 1

6 Additional amount, if any, you want withheld from each paycheck . . . . — 6: $

7 I claim exemption from withholding for 2000, and I certify that I meet **BOTH** of the following conditions for exemption:
- Last year I had a right to a refund of **ALL** Federal income tax withheld because I had **NO** tax liability **AND**
- This year I expect a refund of **ALL** Federal income tax withheld because I expect to have **NO** tax liability.

If you meet both conditions, write "EXEMPT" here . . . . ► 7

Under penalties of perjury, I certify that I am entitled to the number of withholding allowances claimed on this certificate, or I am entitled to claim exempt status.

Employee's signature (Form is not valid unless you sign it) ► Sarah B. Bingham — Date ► August 1, 2000

8 Employer's name and address (Employer: Complete lines 8 and 10 only if sending to the IRS.)	9 Office code (optional)	10 Employer identification number
MacArthur Electronics, Inc; 2001 Second Avenue Birmingham, AL 35200		11 222222

Cat. No. 10220Q

Figure 8-1 Employee's Withholding Allowance Certificate, Form W-4

grandparent or other persons who live with the family and/or qualify as dependents.* These allowances are the same ones that are claimed on the personal income tax return, Form 1040.

Care should be taken to ensure the accuracy of this form, as it controls the amount of federal and in many cases state income tax withheld from the employee's gross pay. Civil and criminal fines can be imposed for the claiming of too many withholding allowances, thereby reducing the amount of income taxes withheld.

Generally, the number of withholding allowances permitted is governed by the number of taxpayers and qualifying dependents to be claimed on the employee's income tax return. The employee can claim fewer withholding allowances than (s)he is entitled to if (s)he wishes to increase the amount of withholding. Although the withholding laws are complex, some general guidelines are given:

1. Husband and wife who will file a joint tax return may split the withholding allowances as they choose, but they should be careful not to duplicate the allowances. Married persons who will file separate income tax returns should realize that only the amounts withheld from their wages can be credited on their income tax return, and that the rules for claiming taxpayer and dependency exemptions on their income tax return are less flexible.
2. Changes in the claimed withholding allowances can be made as new situations occur.

*To qualify, the persons (1) must either be specified relatives or reside with the taxpayer, (2) must have received more than one-half their support from the taxpayer, (3) must have earned less than a certain gross income of their own (except youths under 19 or qualified students), (4) must not have filed a joint return with a spouse, and (5) must be a citizen of the United States or of certain other specified places.

3. In addition to the withholding allowances for the taxpayer and qualifying dependents, additional allowances may be claimed in certain situations to avoid overwithholding of taxes:
   a. A single taxpayer or a married taxpayer with a nonworking spouse may claim an additional allowance.
   b. Additional allowances are permitted when the employee has large itemized deductions, alimony payments, business losses, capital losses, and certain other deductions.
4. Some employees may be eligible for exemption from income tax withholding by filing a special form, W-4E.

Ultimately the employee is responsible for the withholding allowances claimed. The claiming of too many allowances can result in fines and penalties for the underpayment of taxes. Conversely, the claiming of too few can result in the overwithholding of taxes and the interest-free use of the employee's money by the government.

Example 2 Sharie Chiles and Sue Watson both earn salaries of $350 per week. Each claims only herself as a dependent; however, Chiles is married and Watson is single. How much income tax will be withheld from each salary?

Since Sharie Chiles is married and claims only herself, the allowance indicated on her payroll sheet is "M-1." Using the income tax withholding table for "Married Persons" paid on a "Weekly Payroll Period," we find the gross wage interval containing wages that are "at least $350 but less than $360." The columns at the right have headings indicating the number of allowances claimed. For Chiles, we use the column which shows "1 allowance claimed" and see that the tax to be withheld is $27. If Chiles's salary does not change, she will have income tax of $27 withheld each week during the entire year.

Sue Watson's payroll sheet indicates that her allowance is "S-1." Using the income tax withholding table for "Single Persons" paid on a "Weekly" basis, we find the same gross wage interval of "at least $350 but less than $360." In the column for "1 allowance claimed," we see that income tax of $38 will be deducted from Watson's salary each week. Thus, Watson, because she is single, pays $38 − $27, or $11 more income tax each week than Chiles pays.

State Deductions

In many states, state income tax withholding is also a compulsory deduction. Withholding tables are usually available for the state tax, which is typically at a considerably lower rate than the federal income tax.

The Social Security Act (FICA) contains provisions for an **unemployment insurance tax.** The funds from this tax provide compensation to able-bodied workers who are willing to work but who have been "laid off" by their employers because there is no work available. Under the federal act, this unemployment tax is paid by the *employer only,* with most of the funds being collected by the individual states. (That is, no deduction is made from the worker's pay for federal unemployment tax.) In a few states, however, the state has levied its own unemployment tax in addition to the federal tax. In these states, a compulsory unemployment tax may be deducted from the worker's wages.

OTHER DEDUCTIONS

There are any number of other deductions that may be withheld from the worker's wages. One of the most common is group insurance. Many firms sponsor a group policy whereby employees may purchase health, accident, hospitalization, and sometimes even life insurance at rates substantially lower than they could obtain individually.

Union dues or dues to professional organizations may be deducted. Employees often contribute to a company or state retirement plan which will supplement their Social Security retirement benefits. Contributions to charitable causes (such as the United Way Fund) may be withheld. Some businesses participate in a payroll savings plan whereby deductions are made to purchase U.S. Savings Bonds or to buy shares of the corporation's stock (often at a reduced price). Numerous other examples could be cited.

Example 3 Below is a net payroll of Everett Paint Co. In addition to the standard deductions, 1.5% of each employee's gross wages is withheld for a company retirement plan.

EVERETT PAINT CO.
Payroll for Week Ending June 3, 20XX

			DEDUCTIONS						
EMPLOYEE	ALLOWANCE	GROSS WAGES	SOCIAL SECURITY	MEDICARE	FEDERAL INCOME TAX	OTHER (RET.)	OTHER	TOTAL DEDUCTIONS	NET WAGES DUE
Camden, S.	M-3	$ 345	$ 21.39	$ 5.00	$ 9.00	$ 5.18		$ 40.57	$ 304.43
Duncan, E.	M-2	345	21.39	5.00	17.00	5.18		48.57	296.43
Hager, W.	S-2	345	21.39	5.00	28.00	5.18		59.57	285.43
Ingles, C.	M-1	278	17.24	4.03	15.00	4.17		40.44	237.56
Mott, C.	M-3	367	22.75	5.32	12.00	5.51		45.58	321.42
Reid, M.	S-1	290	17.98	4.21	29.00	4.35		55.54	234.46
Trent, B.	M-4	380	23.56	5.51	7.00	5.70		41.77	338.23
	Totals	$2,350	$145.70	$34.07	$117.00	$35.27		$332.04	$2,017.96

Note. The totals of the individual deductions, when added together, must equal the total in the "Total Deductions" column ($145.70 + $34.07 + $117.00 + $35.27 = $332.04). The Total Gross Wages minus the Total Deductions must give the Total Net Wages Due ($2,350 − $332.04 = $2,017.96).

Hint. A net payroll can be completed more efficiently if all deductions of the same type are done at one time. That is, calculate Social Security deductions for all employees; then determine Medicare tax deductions for all employees; and so forth.

SECTION 5 PROBLEMS

1. Compute each person's net wages due. Deduct Social Security, Medicare, and federal income tax. All employees also pay $10.00 for their own group insurance plus $4.00 for each additional dependent. (The number of "Additional Dependents" for all employees is shown in parentheses beside their names.)

LESSIN PRINTING CO.
Payroll for Week Ending February 3, 20XX

EMPLOYEE (ADDL. DEPS.)	ALLOW-ANCES	GROSS WAGES	DEDUCTIONS					NET WAGES DUE
			SOCIAL SECURITY	MEDI-CARE	FED. INC. TAX	OTHER (INS.)	TOTAL DED.	
Beasley (0)	M-2	$545						
Kirtley (1)	M-2	860						
Mason (0)	S-1	772						
Musser (2)	M-3	751						
Smith (3)	M-4	687						
	Totals							

2. Complete the following salary payroll. Withhold Social Security, Medicare, and federal income tax. All workers pay $8.00 for their own health insurance plus $4 for each additional dependent. (The number of "Additional Dependents" for all employees is shown in parentheses beside their names.)

ATKINS GARDEN SHOP
Payroll for Week Ending April 26, 20XX

EMPLOYEE (ADDL. DEPS.)	ALLOW-ANCES	GROSS WAGES	DEDUCTIONS					NET WAGES DUE
			SOCIAL SECURITY	MEDI-CARE	FED. INC. TAX	OTHER (INS.)	TOTAL DED.	
Donley (0)	M-2	$735						
Moran (2)	M-3	766						
Pepper (0)	S-1	643						
Robb (1)	S-2	695						
Ticer (3)	M-4	728						
	Totals							

3. Complete the weekly payroll for Graystone Hardware Co. The employees are paid on an hourly basis, and deductions are withheld for Social Security, Medicare, federal income tax, and union dues. These dues are deducted according to the employee's hourly rate as follows:

RATE PER HOUR	DUES	RATE PER HOUR	DUES
\$12.00–\$12.99	\$7.50	\$14.00–\$14.99	\$8.50
13.00–13.99	8.00	15.00–15.99	9.00

GRAYSTONE HARDWARE CO.
Payroll for Week Ending June 14, 20XX

EMPL. NO.	ALLOW-ANCES	TOTAL HOURS	RATE PER HOUR	GROSS WAGES	DEDUCTIONS: SOCIAL SECURITY	MEDI-CARE	FED. INC. TAX	OTHER (UNION DUES)	TOTAL DEDS.	NET WAGES DUE
44	S-1	40	\$13.50							
45	M-1	39	14.50							
46	M-2	40	14.00							
47	S-2	40	15.75							
48	S-1	38	15.00							
			Totals							

4. Compute the net payroll for Rameriz Computer Services, whose employees are paid on an hourly basis. Deductions are made for Social Security, Medicare, federal income tax, and union dues. Their union dues are deducted according to the chart in Problem 3.

RAMERIZ COMPUTER SERVICES
Payroll for Week Ending March 22, 20XX

EMPL. NO.	ALLOW-ANCES	TOTAL HOURS	RATE PER HOUR	GROSS WAGES	DEDUCTIONS: SOCIAL SECURITY	MEDI-CARE	FED. INC. TAX	OTHER (UNION DUES)	TOTAL DEDS.	NET WAGES DUE
205	M-1	40	\$14.50							
206	M-2	40	12.80							
207	S-1	35	15.00							
208	S-2	40	14.75							
209	M-3	39	14.00							
			Totals							

5. The following commission payroll has standard deductions for Social Security, Medicare, and federal income tax. The sales representatives have each pledged 10% of their gross wages for the week as contributions to the United Way campaign. Determine the net payroll for the week.

SWARTS SALES CORP.
Payroll for Week Ending February 17, 20XX

EMPL. NO.	ALLOW-ANCES	NET SALES	COMM. RATE	GROSS WAGES	DEDUCTIONS: SOCIAL SECURITY	MEDI-CARE	FED. INC. TAX	OTHER (UNITED WAY)	TOTAL DEDS.	NET WAGES DUE
G-4	S-0	$7,600	8.5%							
G-5	M-2	7,500	7							
G-6	M-1	7,800	8							
G-7	M-3	8,000	8.5							
G-8	S-1	7,400	7.5							
			Totals							

6. Employees at Erry Music Co. are paid on a commission basis. Standard deductions are made. Each employee has pledged 5% of gross wages for the week as contributions to a company charity fund. Determine the net payroll for the week.

ERRY MUSIC CO.
Payroll for Week Ending April 5, 20XX

EMPL. NO.	ALLOW-ANCES	NET SALES	COMM. RATE	GROSS WAGES	DEDUCTIONS: SOCIAL SECURITY	MEDI-CARE	FED. INC. TAX	OTHER (CHARITY FUND)	TOTAL DEDS.	NET WAGES DUE
A-10	S-0	$5,600	12%							
A-11	M-1	6,000	11							
A-12	M-2	5,600	10.5							
A-13	S-1	6,200	10							
A-14	M-3	5,400	11.5							
			Totals							

7. The production method is used by King Associates, Inc. to pay its employees. Deductions are made for Social Security, Medicare, and federal income tax, plus health insurance at the rate of $10 for each employee and $5 for each additional dependent. (The total number of additional dependents is listed in parentheses beside each "Employee Number.")

KING ASSOCIATES, INC.
Payroll for Week Ending June 11, 20XX

EMPL. NO.	ALLOWANCES	NET PRODUCTION	RATE PER PIECE	GROSS WAGES	DEDUCTIONS					NET WAGES DUE
					SOCIAL SECURITY	MEDICARE	FED. INC. TAX	OTHER (INS.)	TOTAL DEDS.	
54 (2)	S-2	6,200	$0.10							
55 (1)	M-2	5,800	0.12							
56 (2)	M-3	4,400	0.15							
57 (0)	S-1	5,000	0.13							
58 (1)	S-1	4,000	0.16							
			Totals							

8. Palguta Sales and Services, Inc. pays its employees on the production plan. Deductions are made for Social Security, Medicare, and federal income tax, plus health insurance at the rate of $12 for each employee and $7 for each additional dependent. (The number of additional dependents is listed in parentheses beside each employee number.)

PALGUTA SALES AND SERVICES, INC.
Payroll for Week Ending September 21, 20XX

EMPL. NO.	ALLOWANCES	NET PRODUCTION	RATE PER PIECE	GROSS WAGES	DEDUCTIONS					NET WAGES DUE
					SOCIAL SECURITY	MEDICARE	FED. INC. TAX	OTHER (INS.)	TOTAL DEDS.	
22 (2)	S-2	6,500	$0.11							
23 (3)	M-2	5,400	0.13							
24 (2)	M-3	3,800	0.15							
25 (0)	S-0	5,000	0.12							
26 (1)	S-1	4,500	0.14							
			Totals							

9. McVeigh Industries pays standard overtime for work over 40 hours per week. Group life insurance of $10 is withheld for each employee. Determine the company's net payroll for the week.

MCVEIGH INDUSTRIES
Payroll for Week Ending October 8, 20XX

						DEDUCTIONS					
EMPL.	ALLOW-ANCES	HOURS WORKED	REG. WAGES	OVER-TIME WAGES	TOTAL GROSS WAGES	SOCIAL SECURITY	MEDI-CARE	FED. INC. TAX	OTHER (INS.)	TOTAL DEDS.	NET WAGES DUE
E	M-3	41	$700								
F	M-2	45	600								
G	S-0	48	580								
H	S-1	43	660								
		Totals									

10. L. D. Pape Tutorial Services, Inc. pays standard overtime for work over 40 hours per week. Group insurance of $10 is withheld for each employee. Determine the company's net payroll for the week.

L. D. PAPE TUTORIAL SERVICES, INC.
Payroll for Week Ending February 22, 20XX

						DEDUCTIONS					
EMPL.	ALLOW-ANCES	HOURS WORKED	REG. WAGES	OVER-TIME WAGES	TOTAL GROSS WAGES	SOCIAL SECURITY	MEDI-CARE	FED. INC. TAX	OTHER (INS.)	TOTAL DEDS.	NET WAGES DUE
A	S-2	42	$580								
B	M-2	44	640								
C	S-0	41	600								
D	M-3	43	540								
		Totals									

11. Complete the net payroll for Dawes Equipment, Inc. Overtime is paid at 1½ times the regular piecerate. In addition to standard deductions, deduct 5% for the company retirement plan.

DAWES EQUIPMENT, INC.
Payroll for Week Ending January 5, 20XX

								DEDUCTIONS					
EMPL.	ALLOW-ANCES	NET REG. PRODUC-TION	OVER-TIME PRODUC-TION	REG. PIECE-RATE	REG. WAGES	OVER-TIME WAGES	TOTAL GROSS WAGES	SOCIAL SECURITY	MEDI-CARE	FED. INC. TAX	OTHER (INS.)	TOTAL DEDS.	NET WAGES DUE
#20	M-2	890	75	$0.60									
21	M-0	850	50	0.64									
22	S-2	825	35	0.72									
23	M-1	800	45	0.88									
			Totals										

12. J. V. Gray Accounting Software, Inc. pays its employees on a production basis. Standard overtime is paid at 1½ times the regular piecerate. Deduct 4% for a company retirement plan.

J. V. GRAY ACCOUNTING SOFTWARE, INC.
Payroll for Week Ending November 15, 20XX

								DEDUCTIONS					
EMPL.	ALLOW-ANCES	NET REG. PRODUC-TION	OVER-TIME PRODUC-TION	REG. PIECE-RATE	REG. WAGES	OVER-TIME WAGES	TOTAL GROSS WAGES	SOCIAL SECURITY	MEDI-CARE	FED. INC. TAX	OTHER	TOTAL DEDS.	NET WAGES DUE
#1	S-2	1,000	100	$0.50									
2	M-0	1,100	70	0.56									
3	M-2	850	30	0.60									
4	M-3	900	40	0.70									
			Totals										

SECTION 6

QUARTERLY EARNINGS REPORT AND RETURNS

EMPLOYEE'S QUARTERLY EARNINGS REPORT

Employers must keep an individual payroll record showing each employee's earnings, deductions, and other pertinent information. Figure 8-2 presents an **employee's**

UNLIMITED HORIZONS
Quarterly Earnings Report
For the Period From Jan 1, 1995 to Mar 31, 1995

Employee ID Employee SS No	Date Reference	Amount	Gross Fed_Income FUTA_ER	Soc_Sec State_Incom SUI_ER	Medicare Soc_Sec_ER	Health Ins Medicare_ER
Beginning Balance for						
JOHN HORTON						
HORTON	1/2/95	343.61	414.00	-25.67	-6.00	-9.50
JOHN HORTON	1264		-29.22		-25.67	-6.00
198-78-2498			-3.31	-10.35		
HORTON	1/9/95	385.37	468.00	-29.02	-6.79	-9.50
JOHN HORTON	1400		-37.32		-29.02	-6.79
198-78-2498			-3.74	-11.70		
HORTON	1/16/95	354.05	427.50	-26.51	-6.20	-9.50
JOHN HORTON	1425		-31.24		-26.51	-6.20
198-78-2498			-3.42	-10.69		
HORTON	1/23/95	301.84	360.00	-22.32	-5.22	-9.50
JOHN HORTON	1456		-21.12		-22.32	-5.22
198-78-2498			-2.88	-9.00		
HORTON	1/30/95	395.83	481.50	-29.85	-6.98	-9.50
JOHN HORTON	1489		-39.34		-29.85	-6.98
198-78-2498			-3.85	-12.04		
HORTON	2/6/95	298.37	355.50	-22.04	-5.15	-9.50
JOHN HORTON	1565		-20.44		-22.04	-5.15
198-78-2498			-2.84	-8.89		
HORTON	2/13/95	322.73	387.00	-23.99	-5.61	-9.50
JOHN HORTON	1595		-25.17		-23.99	-5.61
198-78-2498			-3.10	-9.68		
HORTON	2/20/95	298.37	355.50	-22.04	-5.15	-9.50
JOHN HORTON	1645		-20.44		-22.04	-5.15
198-78-2498			-2.84	-8.89		
HORTON	2/27/95	374.94	454.50	-28.18	-6.59	-9.50
JOHN HORTON	1685		-35.29		-28.18	-6.59
198-78-2498			-3.64	-11.36		
HORTON	3/6/95	333.17	400.50	-24.83	-5.81	-9.50
JOHN HORTON	1789		-27.19		-24.83	-5.81
198-78-2498			-3.20	-10.01		
HORTON	3/13/95	280.95	333.00	-20.65	-4.83	-9.50
JOHN HORTON	1815		-17.07		-20.65	-4.83
198-78-2498			-2.66	-8.32		
HORTON	3/20/95	427.15	522.00	-32.36	-7.57	-9.50
JOHN HORTON	1864		-45.42		-32.36	-7.57
198-78-2498			-4.18	-13.05		
HORTON	3/27/95	364.50	441.00	-27.34	-6.39	-9.50
JOHN HORTON	1987		-33.27		-27.34	-6.39
198-78-2498			-3.53	-11.03		
		4,480.88	5,400.00	-334.80	-78.29	-123.50
Total 1/1/95 thru 3/31/95			-382.53		-334.80	-78.29
			-43.19	-135.01		
Total for		4,480.88	5,400.00	-334.80	-78.29	-123.50
JOHN HORTON			-382.53		-334.80	-78.29
			-43.19	-135.01		

Figure 8-2 Employee's Quarterly Earnings Report

quarterly earnings report. The report shows quarterly and weekly, biweekly, semimonthly, or monthly totals, which are used in computing the state and federal quarterly payroll reports that every company must file. The format of the report will vary depending on the software used if the payroll is computerized. Some explanation is needed for this particular report. In the heading, "Reference" refers to the check number issued. "Gross" is the gross weekly wages, and "Fed_Income" is the federal income tax deduction. "FUTA_ER" is the employer's federal unemployment tax liability. "Soc_Sec" refers to the employee's Social Security deduction, whereas "Soc_Sec_ER" is the employer's contribution of the Social Security tax. "SUI_ER" is the employer's liability for state unemployment insurance. "Medicare_ER" is the employer's share of the Medicare tax; the employee's deduction is shown under "Medicare."

In addition, the quarterly earnings report will show the cutoff points for Social Security and unemployment taxes when cumulative earnings hit those maximum amounts. It also contains the quarterly and yearly totals of earnings and deductions required for personal income tax purposes, which are distributed to the employee and the Internal Revenue Service on **Form W-2,** the **Wage and Tax Statement** (Figure 8-3).

1 Control number	22222	For Official Use Only ▶ OMB No. 1545-0008	
2 Employer's name, address, and ZIP code Maxwell Restaurant 456 Peachtree Dr. Atlanta, GA 30306		6 Statutory employee ☐ Deceased ☐ Pension plan ☐ Legal rep. ☐	942 emp. ☐ Subtotal ☐ Deferred compensation ☐ Void ☐
		7 Allocated tips	8 Advance EIC payment
		9 Federal income tax withheld 1,860.00	10 Wages, tips, other compensation 16,000.00
3 Employer's identification number 12-3456789	4 Employer's state I.D. number	11 Social security tax withheld 992.00	12 Social security wages 10,000.00
5 Employee's social security number 111-22-3333		13 Social security tips 6,000.00	14 Medicare wages and tips 16,000.00
19a Employee's name (first, middle, last)		15 Medicare tax withheld 232.00	16 Nonqualified plans
William O. Little 128 Green Street Atlanta, GA 30305 19b Employee's address and ZIP code		17 See Instrs. for Form W-2	18 Other
20	21	22 Dependent care benefits	23 Benefits included in Box 10

24 State income tax	25 State wages, tips, etc.	26 Name of state	27 Local income tax	28 Local wages, tips, etc.	29 Name of locality
372.00	16,000.00	GA			

Copy A For Social Security Administration — Department of the Treasury—Internal Revenue Service

Figure 8-3 Wage and Tax Statement, Form W-2

QUARTERLY PAYROLL RETURNS

There are several payroll reports which must be completed at the end of each quarter. The exact reports vary somewhat from state to state, but there are two types which must be completed in every state: the **Employer's Quarterly Federal Tax Return (Form 941)** and the state **unemployment returns** required under the Federal Insurance Contributions Act.

Employer's Quarterly Federal Tax Return. This form reports the amount of Social Security and Medicare paid (by both the employer and the employee) and the income tax that has been withheld by the company. Figure 8-4 shows a complete Form 941, **Employer's Quarterly Federal Tax Return.**

Notice that lines 2–5 of Form 941 report the total (gross) wages and tips earned during the quarter and the income tax that has been deducted on them. Taxable Social Security wages (line 6a), taxable Social Security tips (line 6c), and taxable Medicare wages and tips (line 7a) are listed separately. The amounts on lines 6a and 6c are multiplied times 12.4% to obtain the total Social Security obligation for both the employees and the employer. Likewise, the amount on line 7a is multiplied times 2.9% to obtain the total Medicare obligation. For convenience, the Internal Revenue Service allows taxpayers to round amounts to whole dollars. Line 8 contains the total Social Security and Medicare taxes. If there are no adjustments to Social Security or Medicare on line 9, then line 10 shows the same amount as line 8. The quarter's total taxes, shown on line 11, is the sum of the total income tax (line 5) plus total Social Security and Medicare (line 10).

Line 17 divides the quarter into its three months. The employer must indicate, for each month, the total tax liability (income tax plus FICA) associated with each payroll. The 3-month total on line 17d must equal the net taxes computed on line 11 above. The amount previously deposited (as subsequently explained) is entered on line 14. This total should normally equal the total taxes for the quarter (shown on line 13), assuming that no adjustments were required. Any taxes still due are computed on line 15 and may be submitted with the quarterly return, if less than $500.

Whenever the Social Security and Medicare (of both the employee and the employer) plus income tax withheld (plus any backup withholding) during one month exceeds $500, the employer is required to deposit these funds in an authorized commercial or Federal Reserve bank on a monthly basis, rather than waiting until the quarter ends. (Similarly, the taxes must be deposited when the cumulative total for two months exceeds $500.) Most firms with only a few employees will have at least $500 in taxes for any one month and will thus have already deposited all taxes before the 941 Quarterly Return is completed. Large firms whose total taxes exceed $3,000 in less than one month must deposit their taxes immediately after the $3,000 level is reached. Electronic deposits are required to be submitted three days after the distribution of the payroll for these large companies.

Most employee earnings reports include (after the end of each month and each quarter) a total of the worker's cumulative earnings and deductions for the year (see Figure 8-2). Most of the figures required for the quarterly return can thus be obtained

Form **941** (Rev. January 1999) Department of the Treasury Internal Revenue Service

Employer's Quarterly Federal Tax Return

▶ See separate instructions for information on completing this return.

Please type or print.

Enter state code for state in which deposits were made ONLY if different from state in address to the right ▶ (see page 2 of instructions).

Name (as distinguished from trade name): **James V. Gray**

Date quarter ended: **Sept., 2000**

Trade name, if any: **JVG Diner**

Employer identification number: **12-7654321**

Address (number and street): **346 Braddock Road**

City, state, and ZIP code: **Springfield, VA 22300**

OMB No. 1545-0029

T / FF / FD / FP / I / T

If address is different from prior return, check here ▶ ☐ — IRS Use

If you do not have to file returns in the future, check here ▶ ☐ and enter date final wages paid ▶

If you are a seasonal employer, see **Seasonal employers** on page 1 of the instructions and check here ▶ ☐

Line	Description				Line	Amount
1	Number of employees in the pay period that includes March 12th ▶	1				
2	Total wages and tips, plus other compensation				2	23,200
3	Total income tax withheld from wages, tips, and sick pay				3	3,528
4	Adjustment of withheld income tax for preceding quarters of calendar year				4	
5	Adjusted total of income tax withheld (line 3 as adjusted by line 4—see instructions)				5	3,528
6	Taxable social security wages	6a	22,000	× 12.4% (.124) =	6b	2,728
	Taxable social security tips	6c	1,200	× 12.4% (.124) =	6d	149
7	Taxable Medicare wages and tips	7a	23,200	× 2.9% (.029) =	7b	673
8	Total social security and Medicare taxes (add lines 6b, 6d, and 7b). Check here if wages are not subject to social security and/or Medicare tax ▶ ☐				8	3,550
9	Adjustment of social security and Medicare taxes (see instructions for required explanation) Sick Pay $ ____ ± Fractions of Cents $ ____ ± Other $ ____ =				9	
10	Adjusted total of social security and Medicare taxes (line 8 as adjusted by line 9—see instructions)				10	3,550
11	**Total taxes** (add lines 5 and 10)				11	7,078
12	Advance earned income credit (EIC) payments made to employees				12	
13	Net taxes (subtract line 12 from line 11). **If $1,000 or more, this must equal line 17, column (d) below (or line D of Schedule B (Form 941))**				13	7,078
14	Total deposits for quarter, including overpayment applied from a prior quarter				14	7,078
15	**Balance due** (subtract line 14 from line 13). See instructions				15	0

16 **Overpayment.** If line 14 is more than line 13, enter excess here ▶ $ ________ and check if to be: ☐ Applied to next return **OR** ☐ Refunded.

- **All filers:** If line 13 is less than $1,000, you need not complete line 17 or Schedule B (Form 941).
- **Semiweekly schedule depositors:** Complete Schedule B (Form 941) and check here ▶ ☐
- **Monthly schedule depositors:** Complete line 17, columns (a) through (d), and check here ▶ ☐

17 **Monthly Summary of Federal Tax Liability.** Do not complete if you were a semiweekly schedule depositor.

(a) First month liability	(b) Second month liability	(c) Third month liability	(d) Total liability for quarter
2,000	2,363	2,715	7,078

Sign Here Under penalties of perjury, I declare that I have examined this return, including accompanying schedules and statements, and to the best of my knowledge and belief, it is true, correct, and complete.

Signature ▶ J V Gray, Jr. Print Your Name and Title ▶ JAMES V. GRAY PRESIDENT Date ▶ 10/31/00

For Privacy Act and Paperwork Reduction Act Notice, see back of form. Cat. No. 17001Z Form **941** (Rev. 1-99)

Figure 8-4 Employer's Quarterly Federal Tax Return, Form 941

simply by computing company totals from these earnings reports—the total gross wages (and tips, if applicable) for line 2, and the total income tax withheld for line 3. During the early quarters, all wages earned are FICA taxable; thus, line 6a will be the same as the company's gross wages.

Example 1 Complete lines 2–15 of the Employer's Quarterly Federal Tax Return for Taylor Appliance Sales, based on the following information: Total wages (all FICA taxable) for the quarter total $12,000, and income tax of $1,872 was withheld. Verify that all taxes have been paid, if monthly deposits of $1,295, $1,218, and $1,195 have been made. (No tips are earned by these employees.)

The completed upper portion (lines 2–15) of Form 941 as shown here indicates that the total taxes due are $3,708. The total monthly deposits for the quarter are $1,295 + $1,218, + $1,195 = $3,708. Thus, all taxes have been deposited before Form 941 is filed.

2	Total wages and tips, plus other compensation				2	12,000
3	Total income tax withheld from wages, tips, and sick pay				3	1,872
4	Adjustment of withheld income tax for preceding quarters of calendar year				4	
5	Adjusted total of income tax withheld (line 3 as adjusted by line 4—see instructions)				5	1,872
6	Taxable social security wages	6a	12,000	× 12.4% (.124) =	6b	1,488
	Taxable social security tips	6c		× 12.4% (.124) =	6d	
7	Taxable Medicare wages and tips	7a	12,000	× 2.9% (.029) =	7b	348
8	Total social security and Medicare taxes (add lines 6b, 6d, and 7b). Check here if wages are not subject to social security and/or Medicare tax ▶ ☐				8	1,836
9	Adjustment of social security and Medicare taxes (see instructions for required explanation) Sick Pay $ _____ ± Fractions of Cents $ _____ ± Other $ _____ =				9	
10	Adjusted total of social security and Medicare taxes (line 8 as adjusted by line 9—see instructions)				10	1,836
11	**Total taxes** (add lines 5 and 10)				11	3,708
12	Advance earned income credit (EIC) payments made to employees				12	
13	Net taxes (subtract line 12 from line 11). **If $1,000 or more, this must equal line 17, column (d) below (or line D of Schedule B (Form 941))**				13	3,708
14	Total deposits for quarter, including overpayment applied from a prior quarter				14	3,708
15	**Balance due** (subtract line 14 from line 13). See instructions				15	0

Particular attention must be paid, however, to the "taxable Social Security wages" (line 6a of Form 941). Recall that Social Security applies only to the first $76,200 earned; thus, many executives and some highly paid workers will earn wages during the latter quarter(s) that are exempt from Social Security deductions. When that occurs, the "taxable Social Security wages" on line 6a will be less than the "total wages and tips" on line 2. All employees' earnings will continue to have Medicare taxes withheld. (All earnings, of course, are subject to income tax deductions.)

For instance, if an executive had earned \$65,700 during the first three quarters, then his Social Security wages during the fourth quarter would be computed as follows:

Maximum Social Security wages	−	Previous total wages	=	Taxable Social Security wages for quarter
\$76,200	−	\$65,700	=	\$10,500

Thus, only \$10,500 of the executive's fourth quarter wages would be included in the firm's taxable Social Security wages on line 6a.

Assume that the executive earns \$15,000 during the fourth quarter. Then his Medicare wages for the fourth quarter, \$15,000, would be included on line 7a.

Example 2 Chris Dean, a sales representative for computer software, earned quarterly wages of \$19,000, \$19,200, \$19,500, and \$19,600. How much of her 4th quarter wages were taxable for Social Security?

The quarterly totals of Dean's earnings were

\$19,000 + \$19,200 + \$19,500 = \$57,700 End of 3rd quarter

\$19,000 + \$19,200 + \$19,500 + \$19,600 = \$77,300 End of 4th quarter

Since Dean passed the \$76,200 base amount for Social Security during the 4th quarter, her taxable Social Security earnings were

\$76,200 − \$57,700 = \$18,500 Social Security taxable wages for 4th quarter

Therefore, Social Security tax was deducted on \$18,500 of Dean's \$19,600 4th quarter earnings, and \$18,500 is the amount included in the company's Social Security taxable wages on line 6a. However, \$19,600 is included in line 7a for Medicare taxable wages, and income tax was also deducted on her entire \$19,600, which is included on line 2 for "Total wages subject to (income tax) withholding."

Example 3 Complete lines 2–15 of the employer's quarterly tax return for the small computer firm of Example 2. Assume that the *other* employees of the company earned gross wages (all Social Security taxable) of \$28,200. (See Example 2 for Chris Dean's wages.) Income tax (for everyone) of \$6,200 was withheld. Also determine whether all required taxes have been paid if deposits were made in the amounts of \$4,416, \$4,502, and \$4,459.

Line 2 of Form 941 contains the "Total (gross) wages subject to (income tax) withholding" for the other employees and for Dean (\$28,200 + \$19,600 = \$47,800). Line 6a computes 12.4% of taxable Social Security wages of \$28,200 + \$18,500 = \$46,700. Line 7a computes 2.9% of taxable Medicare wages of \$28,200 + \$19,600 = \$47,800. Since the previous deposits total \$4,416 + \$4,502 + \$4,459 = \$13,377, the entire amount due for this quarter has already been paid.

Line	Description				Line	Amount
2	Total wages and tips, plus other compensation				2	47,800
3	Total income tax withheld from wages, tips, and sick pay				3	6,200
4	Adjustment of withheld income tax for preceding quarters of calendar year				4	
5	Adjusted total of income tax withheld (line 3 as adjusted by line 4—see instructions)				5	6,200
6	Taxable social security wages	6a	46,700	× 12.4% (.124) =	6b	5,791
	Taxable social security tips	6c		× 12.4% (.124) =	6d	
7	Taxable Medicare wages and tips	7a	47,800	× 2.9% (.029) =	7b	1,386
8	Total social security and Medicare taxes (add lines 6b, 6d, and 7b). Check here if wages are not subject to social security and/or Medicare tax ▶ ☐				8	7,177
9	Adjustment of social security and Medicare taxes (see instructions for required explanation) Sick Pay $ ______ ± Fractions of Cents $ ______ ± Other $ ______ =				9	
10	Adjusted total of social security and Medicare taxes (line 8 as adjusted by line 9—see instructions)				10	7,177
11	**Total taxes** (add lines 5 and 10)				11	13,377
12	Advance earned income credit (EIC) payments made to employees				12	
13	Net taxes (subtract line 12 from line 11). **If $1,000 or more, this must equal line 17, column (d) below (or line D of Schedule B (Form 941))**				13	13,377
14	Total deposits for quarter, including overpayment applied from a prior quarter				14	13,377
15	**Balance due** (subtract line 14 from line 13). See instructions				15	0

Unemployment Tax Returns. As mentioned earlier, the employer's **Federal Unemployment Tax Act (FUTA)** was passed as part of the FICA Act. The funds collected under this tax are used to provide limited compensation, for a limited period of time, to workers who are unemployed because there is no work available to them. The federal tax is levied only against employers, with most of the funds being collected by the states; the federal unemployment tax is *not* deducted from workers' earnings. (Only a few states have levied additional unemployment taxes of their own, which may be deducted from the workers' gross earnings.) The portion of unemployment taxes paid to the state must be reported quarterly; the federal return, Form 940, is submitted each January for the preceding calendar year, although deposits may be required quarterly. If the tax liability for any quarter is over $100, deposits must be made to an authorized depository. Since unemployment forms are not uniform from state to state, however, we shall omit any forms and study only the computation of these taxes.

Employers are subject to federal unemployment tax if they employ one or more persons during any 20 weeks of the year or if they paid $1,500 in salaries during any quarter. Employers of certain farmworkers and household workers must also pay. The unemployment tax is similar to the Social Security tax in that it applies only to wages up to a specified maximum—currently the first $7,000 earned by each employee. The FUTA rate for 2000 was as follows:

5.4%	Paid to the state
0.8%	Paid to the federal government
6.2%	Total federal unemployment tax

Employers who have a good history of stable employment may have their state rates substantially reduced, while still paying only 0.8% to the federal government, so that their actual rate is well below 6.2%. Conversely, employers with a history of high unemployment may be required to pay substantially more than 5.4% for the state portion of their unemployment insurance. However, 5.4% is the maximum that could be claimed against the 6.2% federal unemployment rate; that is, all employers must pay at least 0.8% to the federal government. Employers in some states must pay more than 0.8% (that is, they cannot claim the full 5.4% to the state) because their state has borrowed from the federal fund and has kept an outstanding balance for at least two years. The following example illustrates the computation of the federal unemployment tax (FUTA).

Example 4 Angela Klein is a salaried part-time employee who earned $2,500 each quarter.

(a) Determine the amount of Ms. Klein's wages that are subject to unemployment tax each quarter.

The cumulative totals of her earnings after each quarter are:

$1 \times \$2{,}500 = \$\ 2{,}500$ 1st quarter

$2 \times \$2{,}500 = \$\ 5{,}000$ 2nd quarter

$3 \times \$2{,}500 = \$\ 7{,}500$ 3rd quarter

$4 \times \$2{,}500 = \$10{,}000$ 4th quarter

She had earned $5,000 at the end of the 2nd quarter; thus, her wages passed the $7,000 cutoff point for unemployment tax during the 3rd quarter. In general, for the quarter when the maximum is reached,

$$\text{Maximum unemployment wages} - \text{Previous total wages} = \text{Taxable wages for quarter}$$

$$\$7{,}000 - \$5{,}000 = \$2{,}000$$

Thus, Ms. Klein's employer must pay unemployment tax on $2,000 of the wages she earns during the 3rd quarter. No tax will be due on her 4th quarter earnings.

(b) Assume that the employer in Part (a) has a reduced state unemployment tax rate of 4.0%. Compute the company's state unemployment taxes for each quarter, based on the following wages:

	TAXABLE WAGES FOR UNEMPLOYMENT			
	1ST QUARTER	2ND QUARTER	3RD QUARTER	4TH QUARTER
Other employees	$6,000	$5,300	$3,800	0
Ms. Klein	2,500	2,500	2,000	0
Totals	$8,500	$7,800	$5,800	0

$\$8{,}500 \times 4.0\% = \$340$ 1st quarter

$7{,}800 \times 4.0\% = 312$ 2nd quarter

$5{,}800 \times 4.0\% = 232$ 3rd quarter

No unemployment tax will be owed to the state during the fourth quarter.

(c) Find the federal unemployment tax due each quarter, assuming the standard federal rate.

$\$8{,}500 \times 0.8\% = \$68.00$ 1st quarter

$7{,}800 \times 0.8\% = \$62.40$ 2nd quarter

$5{,}800 \times 0.8\% = \$46.40$ 3rd quarter

No federal unemployment tax will be due in the 4th quarter.

Note. The company is not required to make a deposit until the second quarter, when total federal unemployment tax exceeds $100.

SECTION 6 PROBLEMS

Complete lines 2–15 of the Employer's Quarterly Federal Tax Return (Form 941) for Problems 1–6, based on the information given. Assume that only taxable Social Security wages are earned, except where noted. Determine whether the full tax obligation has already been deposited.

	TOTAL WAGES AND TIPS	TAXABLE SOCIAL SECURITY WAGES	TAXABLE TIPS	INCOME TAX WITHHELD	MONTHLY DEPOSITS OCT.	NOV.	DEC.
1.	$15,000	$15,000	—	$2,250	$1,505	$1,615	$1,425
2.	40,000	40,000		5,900	4,520	4,300	3,200
3.	50,000	44,000	$6,000	7,500	4,050	5,500	5,600
4.	41,000	38,000	3,000	5,400	3,607	3,866	4,200
5.	29,500	26,000	1,500	4,200	2,596	2,800	3,070
6.	57,000	50,000	5,000	9,000	6,073	6,000	5,400

Quarterly earnings for three executives are listed in each of the following problems. Assuming that $76,200 is the maximum Social Security taxable wages, determine each person's taxable earnings for each quarter.

	Employee	Gross Wages Earned 1st Quarter	2nd Quarter	3rd Quarter	4th Quarter
7.	G	$19,500	$19,000	$19,000	$18,200
	H	21,000	25,000	29,000	20,000
	I	26,000	26,000	26,000	26,000
8.	J	$ 7,000	$15,500	$16,000	$16,000
	K	18,900	18,400	19,600	19,600
	L	31,000	30,000	31,000	31,000

2	Total wages and tips, plus other compensation	2	
3	Total income tax withheld from wages, tips, and sick pay	3	
4	Adjustment of withheld income tax for preceding quarters of calendar year	4	
5	Adjusted total of income tax withheld (line 3 as adjusted by line 4—see instructions) . . .	5	
6	Taxable social security wages . . . . . 6a ___ × 12.4% (.124) =	6b	
	Taxable social security tips . . . . . . 6c ___ × 12.4% (.124) =	6d	
7	Taxable Medicare wages and tips . . . 7a ___ × 2.9% (.029) =	7b	
8	Total social security and Medicare taxes (add lines 6b, 6d, and 7b). Check here if wages are not subject to social security and/or Medicare tax . . . . ▶ ☐	8	
9	Adjustment of social security and Medicare taxes (see instructions for required explanation) Sick Pay $ ___ ± Fractions of Cents $ ___ ± Other $ ___ =	9	
10	Adjusted total of social security and Medicare taxes (line 8 as adjusted by line 9—see instructions)	10	
11	**Total taxes** (add lines 5 and 10)	11	
12	Advance earned income credit (EIC) payments made to employees	12	
13	Net taxes (subtract line 12 from line 11). **If $1,000 or more, this must equal line 17, column (d) below (or line D of Schedule B (Form 941))**	13	
14	Total deposits for quarter, including overpayment applied from a prior quarter	14	
15	**Balance due** (subtract line 14 from line 13). See instructions	15	

The cumulative earnings for the first 3 quarters are listed below. Determine each person's earnings during the 4th quarter that will be subject to Social Security tax deductions and to Medicare deductions.

	Employee	Cumulative Earnings	4th Quarter Earnings	Social Security Earnings	Medicare Earnings
9.	#1	$ 66,000	$18,000		
	2	55,000	29,000		
	3	80,000	28,000		
10.	#4	$ 68,000	$20,000		
	5	64,000	15,500		
	6	100,000	25,000		

Listed in each problem below are all the employees of a small business, along with their total gross earnings and income tax deductions for each quarter.

(a) Determine each employee's Social Security taxable wages for each quarter.
(b) Find the firm's total Social Security taxable wages for each quarter.

		1ST QUARTER			2ND QUARTER			3RD QUARTER			4TH QUARTER		
	EMPL.	GROSS WAGES	INC. TAX	SOC. SEC. WAGES	GROSS WAGES	INC. TAX	SOC. SEC. WAGES	GROSS WAGES	INC. TAX	SOC. SEC. WAGES	GROSS WAGES	INC. TAX	SOC. SEC. WAGES
11.	A	$ 6,800	$ 980		$ 7,400	$1,036		$ 7,500	$1,050		$ 7,900	$1,106	
	B	17,600	1,960		20,800	2,072		18,000	2,100		19,300	2,242	
	C	18,800	2,492		18,600	2,604		20,700	2,818		20,500	2,775	
	D	29,000	4,060		29,500	3,990		28,900	4,046		28,800	4,032	
	Totals												
12.	E	$ 6,200	$ 940		$12,800	$1,470		$12,800	$1,470		$13,000	$1,506	
	F	15,500	1,900		16,400	2,200		16,800	2,350		16,900	2,358	
	G	20,000	2,520		20,000	2,520		20,000	2,520		20,000	2,520	
	H	30,000	3,030		35,000	3,200		35,000	3,200		36,000	3,240	
	Totals												

Compute lines 2–15 of the 941 Employer's Quarterly Federal Tax Return for Problems 13–16, given the monthly deposits of withheld income tax and FICA tax (Social Security and Medicare).

2	Total wages and tips, plus other compensation	2	
3	Total income tax withheld from wages, tips, and sick pay	3	
4	Adjustment of withheld income tax for preceding quarters of calendar year	4	
5	Adjusted total of income tax withheld (line 3 as adjusted by line 4—see instructions) . . .	5	
6	Taxable social security wages . . . . . . 6a ________ × 12.4% (.124) =	6b	
	Taxable social security tips . . . . . . 6c ________ × 12.4% (.124) =	6d	
7	Taxable Medicare wages and tips . . . 7a ________ × 2.9% (.029) =	7b	
8	Total social security and Medicare taxes (add lines 6b, 6d, and 7b). Check here if wages are not subject to social security and/or Medicare tax . . . . ▶ ☐	8	
9	Adjustment of social security and Medicare taxes (see instructions for required explanation) Sick Pay $ ________ ± Fractions of Cents $ ________ ± Other $ ________ =	9	
10	Adjusted total of social security and Medicare taxes (line 8 as adjusted by line 9—see instructions)	10	
11	**Total taxes** (add lines 5 and 10)	11	
12	Advance earned income credit (EIC) payments made to employees	12	
13	Net taxes (subtract line 12 from line 11). **If $1,000 or more, this must equal line 17, column (d) below (or line D of Schedule B (Form 941))**	13	
14	Total deposits for quarter, including overpayment applied from a prior quarter	14	
15	**Balance due** (subtract line 14 from line 13). See instructions	15	

	REFER TO PROBLEM #	QUARTER TO COMPUTE	MONTHLY DEPOSITS 1ST MO.	2ND MO.	3RD MO.
13.	11	3rd	$6,226	$6,390	$7,500
14.	12	3rd	6,040	6,543	6,949
15.	11	4th	6,590	6,611	4,790
16.	12	4th	7,427	6,930	3,474

17. Refer to the employees in Problem 7.

a. Determine the taxable wages for unemployment for each worker for each quarter.

b. Find the total taxable wages for unemployment of the company for each quarter, assuming that these are the only employees.

c. How much unemployment tax will the company pay to its state each quarter? What are the annual taxes due to the federal government? Assume that the firm has a state rate of 3.6%. (The federal tax rate is never reduced.)

18. Refer to the employees in Problem 8.

a. Determine the taxable wages for unemployment of each worker for each quarter.

b. Determine the total taxable wages for unemployment of the company for each quarter, assuming that these are the only employees.

c. How much unemployment tax will the company pay to its state each quarter? What are the annual taxes due to the federal government? Assume that the firm has a state rate of 3.6%. (The federal tax rate is never reduced.)

19. Refer to the employees in Problem 11.

a. Determine the taxable wages for unemployment of each worker for each quarter.

b. Determine the total taxable wages for unemployment of the company for each quarter, assuming that these are the only employees.

c. How much unemployment tax will the company pay to its state each quarter? What are the annual taxes due to the federal government? Assume that the firm has a reduced state rate of 3.6%. (The federal tax rate is never reduced.)

20. Refer to the employees in Problem 12.

a. Determine the taxable wages for unemployment of each worker for each quarter.

b. Determine the total taxable wages for unemployment of the company for each quarter, assuming that these are the only employees.

c. How much unemployment tax will the company pay to its state each quarter? What are the annual taxes due to the federal government? Assume that the firm has a reduced state rate of 2.9%. (The federal tax rate is never reduced.)

21. The chief executive officer of a corporation earned $500,000 for the year. Determine

a. His total Social Security tax deduction for the year

b. His total Medicare tax deduction for the year

c. The amount of federal unemployment tax the corporation must pay

d. The amount of state unemployment tax the corporation must pay, assuming a state rate of 5.4%

22. A professional athlete has gross earnings of $1,000,000 for the year. Determine
    a. Her total Social Security tax deduction for the year
    b. Her total Medicare tax deduction for the year
    c. The amount of federal unemployment tax her employer must pay for the year
    d. The amount of state unemployment tax her employer must pay for the year, assuming a state rate of 5.4%

CHAPTER  GLOSSARY

Account purchase. Itemized statement (Prime cost + Charges = Gross cost) for merchandise bought by a commission agent.

Account sales. Itemized statement (Gross proceeds − Charges = Net proceeds) for the merchandise sold by a commission agent.

Broker. (See "Commission agent.")

Brokerage. (See "Commission.")

Chargeback (or docking). A deduction made from base production or base earnings, because of production work that did not pass inspection.

Commission. Wages computed as a percent of the value of items sold or purchased; this is also called "straight commission" or "brokerage."

Commission agent. One who buys or sells merchandise for another party without becoming legal owner of the property; also called a "commission merchant," "broker," or "factor."

Consignee. Agent to whom a shipment is sent in order to be sold.

Consignor. Party who sends a shipment to an agent who will sell it.

Docking. (See "Chargeback.")

Drawing account. Fund from which a salesperson may receive advance wages to defray business or personal expenses.

Employee's quarterly earnings report. The summary of an employee's gross wages, itemized deductions, and net pay for each pay period during a quarter.

Employee's Withholding Allowance Certificate (Form W-4). Federal form on which an employee lists marital status and computes the number of qualified allowances for income tax withholding.

Employer's Quarterly Federal Tax Return (Form 941). Quarterly form on which a company reports wages and tips earned (taxable and total) and the amount of Social Security, Medicare, and income tax that has been withheld (and matched, for Social Security and Medicare).

Factor. (See "Commission agent.")

Fair Labor Standards Act. Federal law requiring the payment of a minimum hourly wage and overtime (at "time-and-a-half" for hours worked in excess of 40 per week) to all employees covered.

Federal Insurance Contributions Act (FICA). Deduction of a specified percent made on a base amount of each worker's yearly wages (and matched by the employer) to fund benefits for Social Security and Medicare. (See also "Social Security" and "Medicare.")

Federal Unemployment Tax Act (FUTA) Federal tax (paid partially to the state) to provide some compensation to workers who are unemployed because there is no work available; paid by the employer only.

Gross cost. Total cost, after commission and other charges are added, of merchandise purchased by a commission agent.

Gross proceeds. Initial sales value, before commission and other charges are deducted, of merchandise sold by a commission agent.

Gross wages. Wages earned by an employee before FICA, income tax, or other deductions.

Income tax (federal). Tax withheld from each employee's pay, as determined by the person's gross earnings, marital status, and number of allowances.

Medicare. Percentage deduction based on an employee's gross earnings (and matched by the employer) to fund health insurance for persons age

65 or older and eligible for Social Security or Railroad Retirement benefits.

Net proceeds. Remaining sales value, after commission and other charges are deducted, of merchandise sold by a commission agent.

Net wages. The actual wages (after all deductions) paid to a worker.

Override. A small commission on the net sales of all other salespersons, paid to reimburse a supervisor for nonselling duties.

Overtime. (See "Fair Labor Standards Act.")

Overtime excess. The extra amount (at one-half the regular rate) that must be paid for work done during overtime hours.

Piecework or piecerate. (See "Production method.")

Premium rates. Higher-than-regular rates, paid for overtime work or work that exceeds a quota.

Prime cost. Initial cost, before commission and other charges are added, of merchandise purchased by a commission agent.

Production method. Wage computation determined by the number of times a task is completed; also called the "piecework" or "piecerate" method.

Quarterly return. (See "Employer's Quarterly Federal Tax Return.")

Quota. Minimum sales or production level required in many jobs (usually in order to qualify for pay at a higher rate).

Sliding-scale commission. A commission rate that increases, once sales exceed the specified level(s), on all sales above the set level(s).

Social Security. Percentage deduction based on an employee's gross earnings up to a maximum amount (and matched by the employer) to fund benefits for retired or disabled workers, and for surviving spouses and children.

Straight commission. (See "Commission.")

Unemployment insurance tax. (See "Federal Unemployment Tax Act [FUTA].")

Wage and Tax Statement (Form W-2). The form completed by the employer to record the employee's yearly totals of earnings and deductions. The W-2 form is distributed to the employee and to the Internal Revenue Service; it must also be included with the employee's personal income tax return.

Withholding. Deductions made by employers from the gross earnings of their employees—including required FICA and income tax deductions. (See also "Employee's Withholding Allowance Certificate.")

CHAPTER 9

DEPRECIATION AND OVERHEAD

OBJECTIVES

Upon completion of Chapter 9, you will be able to:

1. Define and use correctly the terminology associated with each topic.
2. Compute depreciation or cost recovery (including schedules) by the following four methods (Section 1):
   a. Straight-line (Example 1; Problems 1, 2, 3–8, 23, 24)
   b. Declining-balance (Example 2; Problems 1, 2, 9–14, 23, 24)
   c. Units-of-production (Example 3; Problems 21, 22)
   d. Modified accelerated cost recovery/MACRS (Examples 1, 2; Problems 1–8).
3. Prorate depreciation for partial years by the following methods (Section 2):
   a. Straight-line (Example 1; Problems 1–4)
   b. Declining-balance (Example 2; Problems 1, 2, 5, 6).
4. Compute overhead as a ratio of (Section 3):
   a. Floor space per department (Example 1; Problems 1, 2, 7, 8)
   b. Net sales (Example 2; Problems 3, 4, 9, 10)
   c. Employees per department (Example 3; Problems 5, 6, 11, 12).

When expenses are calculated for a business, two of the most important items are **depreciation** and **overhead.** To determine these expenses, you will have to learn certain mathematical techniques. The procedures you will need to know are explained in the following sections.

SECTION 1

DEPRECIATION

The **plant assets** of a business are its buildings, machinery, equipment, land, and similar tangible properties that will be used for more than one year. All plant assets except land should be depreciated. In business, **depreciation** is the systematic allocation of the asset's cost over its period of economic utility (its **useful life**). The allocation of cost is necessary because the asset loses its economic utility because of such things as technological obsolescence, functional inadequacy, and/or physical deterioration. Its usefulness is *consumed* in the process of conducting business.

Because a plant asset will be used for several years, the Internal Revenue Service does not allow a business to list the entire business cost of the asset as an expense during the year in which it was obtained. However, a business is permitted to recover the cost of a plant asset over a prescribed period of years. Thus, the depreciation or cost recovery claimed each period does not normally represent money actually spent during the present year; rather, it represents this year's share of a larger expenditure that occurred during some previous year. This annual calculation is considered an operating expense (just as are salaries, rent, insurance, and so forth) and is deducted from business profits when determining the year's taxable income. Because depreciation or cost recovery is a tax-deductible item on business income tax returns, the Internal Revenue Service carefully regulates the conditions under which it may be computed.

The following sections present the traditional or conventional methods of depreciation: the **straight-line method,** the **declining-balance method,** and the **units-of-production method.** Then the *Modified Accelerated Cost Recovery System* is discussed. The similarities and differences in methods are explained in the following sections. Generally, the conventional depreciation methods are used for internal or financial accounting purposes, while the Modified Accelerated Cost Recovery System is required for income tax accounting.

STRAIGHT-LINE METHOD

The **straight-line method** is the simplest and most commonly used method for computing depreciation. By this method, the total allowable depreciation is divided evenly among all the years. That is, the same amount of depreciation is charged each full year.

Use of the straight-line method requires determining a depreciable base—the total amount to be depreciated. When using this method, the business first estimates the asset's **residual value,** which is the expected value of the asset at the end of its useful life. The residual value is often called the asset's **trade-in value** (the expected amount to be received if the asset is traded), **resale value** (the expected amount to be received if the asset is sold), or **scrap value** or **salvage value** (the expected amount if the asset is junked). The asset's residual value is subtracted from the asset's cost to determine the **depreciable base.** The depreciable base is the maximum total depreciation expense that can be claimed on the asset over its useful life.

Example 1 A machine that manufactures keys is purchased for $5,100. It is expected to last for 5 years and have a trade-in value of $900. Prepare a depreciation schedule by the straight-line method.

First, the depreciable base must be computed and divided by the years of use to determine each year's depreciation charge:

Original cost		$5,100
Trade-in value	−	900
Depreciable base		$4,200

$$\frac{\text{Depreciable base}}{\text{Years}} = \frac{\$4{,}200}{5}$$

$$= \$840 \text{ Annual depreciation}$$

Another way to solve Example 1 is to multiply the depreciable base by the straight-line rate. The **straight-line rate** is the reciprocal of the number of years of useful life ($\frac{1}{5}$ in this example). So, instead of dividing the depreciable base by the number of years, you can multiply the straight-line rate times the depreciable base, as follows:

$$\frac{1}{5} \text{ of } \$4{,}200 = \$840 \quad \text{Annual depreciation}$$

A **cost-recovery schedule** or **depreciation schedule** is usually kept for every asset class acquired each year. This is a record showing the amount of cost recovered each year, the total amount recovered to date, and the amount as yet unrecovered. This remaining portion of the cost may also be referred to as the **book value.** Book value is used in calculating the value of a company's capital assets for the year.

The cost-recovery schedule is completed as follows:

1. A year "0" is marked down to indicate when the property was new, and the original cost is entered under "book value." Since no depreciation (cost recovery) has yet occurred when the property is new, no other entries are made on line 0.
2. Each year the "annual depreciation" is entered.

3. The "accumulated depreciation" is found each year by adding the current year's depreciation amount to the previous year's accumulated depreciation.
4. The "book value" is determined by subtracting the current year's (annual) depreciation from the previous year's book value.

Example 1 (cont.) As computed above, depreciation of $840 would be claimed each year for 5 years by the traditional straight-line method. A depreciation schedule for the foregoing depreciation is completed below.

DEPRECIATION SCHEDULE—STRAIGHT-LINE METHOD

YEAR	BOOK VALUE (END OF YEAR)	ANNUAL DEPRECIATION	ACCUMULATED DEPRECIATION
0	$5,100	—	—
1	4,260	$840	$ 840
2	3,420	840	1,680
3	2,580	840	2,520
4	1,740	840	3,360
5	900	840	4,200

↙ Residual value ↙ Depreciable base

If the straight-line depreciation schedule is done correctly, a visual check for its accuracy can be done very quickly. The last year's book value will equal the residual value, and the last year's total accumulated depreciation will equal the depreciable base.

QUICK PRACTICE—STRAIGHT-LINE METHOD

Compute the annual depreciation by the straight-line method using the data given. Construct depreciation schedules for the (c) parts only.

	COST	RESIDUAL VALUE	USEFUL LIFE (YEARS)
1. a.	$ 7,500	$ 300	6
b.	5,800	400	9
c.	2,900	700	4
2. a.	$20,000	$2,000	10
b.	12,000	800	8
c.	6,000	400	7

DECLINING-BALANCE METHOD

Another method frequently used for computing depreciation is the **declining-balance method** (sometimes called the *constant percent method*). This method also offers the advantage of greatest annual depreciation during the early years, progressively declining as time passes. The method is often referred to as "double declining-balance" because the rate most often used is twice the straight-line rate. For example, the straight-line rate of an asset using 6 years is 1/6 each year. The typical declining-balance rate (percent) for the same 6 years is thus

$$2 \times \frac{1}{6} = \frac{2}{6} = \frac{1}{3} \quad \text{or} \quad 33\frac{1}{3}\% \text{ annually}$$

Stated differently, the declining-balance rate is usually equal to twice the reciprocal of the number of years of useful life. (This is normally the maximum rate allowed; a lower rate may be used for internal accounting.) Because this method uses a factor of 2, it is also often called the "200% declining-balance" method.

When the declining-balance method is used, depreciation is computed using the same rate (percent) each year. For the *first year,* this *constant rate* is multiplied by the full *cost* (not the depreciable base) to determine the first year's depreciation. This amount is then subtracted from the asset's cost to yield the book value of the asset at the end of the first year. For each succeeding year, the book value at the end of the previous year is multiplied by the constant rate. This method gets its name from the procedure of applying a constant rate to a decreasing book value (declining balance) each year.

If there is a residual value, caution must be used to ensure that the asset is not depreciated to a value below its residual value. This often requires adjusting the annual depreciation amount at some point so that the ending book value is equal to the residual value.

Example 2

(a) A new vending machine that cost $4,800 has an expected residual value of $600 and a life of 5 years. Prepare a depreciation schedule by the declining-balance method.

Because the machine will be used for 5 years, it can be depreciated at a rate of

$$2 \times \frac{1}{5} = \frac{2}{5} = 40\% \text{ annually}$$

Each succeeding book value will be multiplied by 40% (or 0.4) to find the next annual depreciation. The following depreciation schedule shows dollar amounts rounded to whole dollars for this example. For practical purposes, most accountants round dollar amounts to whole dollars because depreciation and cost recovery are only estimates. We will follow this practice in the examples and problems in this chapter.

DEPRECIATION SCHEDULE—DECLINING-BALANCE METHOD

YEAR	BOOK VALUE (END OF YEAR)	ANNUAL DEPRECIATION	ACCUMULATED DEPRECIATION
0	$4,800	—	—
1	2,880	$1,920	$1,920
2	1,728	1,152	3,072
3	1,037	691	3,763
4	622	415	4,178
5	600	22	4,200

[a]This is an adjusted final depreciation value, not a declining-balance calculation.

Notice that an adjustment was made to the annual depreciation in the fifth year so that the end-of-year book value would equal the residual value, $600. This adjustment may sometimes be necessary before the final year. As the estimated residual value increases in amount, the probability of the adjustment to annual depreciation before the final year increases. For example, assume that the residual value is $800. The book value at the end of the third year is $1,037, so the fourth-year annual depreciation will be $215 (the difference between the $1,037 and $800), and the book value at the end of that year will be $800. No further depreciation would be taken. This is illustrated in the following example.

Example 2 (cont.)

(b) Cost = $4,800
Residual value = $800
Life = 5 years

DEPRECIATION SCHEDULE—DECLINING-BALANCE METHOD

YEAR	BOOK VALUE (END OF YEAR)	ANNUAL DEPRECIATION	ACCUMULATED DEPRECIATION
0	$4,800	—	—
1	2,880	$1,920	$1,920
2	1,728	1,152	3,072
3	1,037	691	3,763
4	800	237	4,000
5	—	—	—

A characteristic of declining-balance depreciation is that the book value never reaches zero in the natural course of events. It must be forced to do so by the accountant. Otherwise, the process can go on for many years before it even approaches zero.

If the same business assumed a residual value of zero, then the adjustment would be made in the fifth year, as shown in Example 2(c).

Example 2 (cont.)

(c) Cost = \$4,800
Residual value = 0
Life = 5 years

YEAR	BOOK VALUE (END OF YEAR)	ANNUAL DEPRECIATION	ACCUMULATED DEPRECIATION
0	\$4,800	—	—
1	2,880	\$1,920	\$1,920
2	1,728	1,152	3,072
3	1,037	691	3,763
4	622	415	4,178
5	—	622	4,800

From a commercial standpoint, most machinery actually undergoes greater depreciation during its early years and less depreciation during later years. The declining-balance method of depreciation is thus more "realistic" than the straight-line method, even for internal accounting purposes. However, when listing the net worth of assets on its balance sheet, a business may want the assets to be shown at the greatest possible value. In that case, the straight-line method would be preferred, because that method would result in a slower loss of value.

QUICK PRACTICE—DECLINING-BALANCE METHOD

Compute the annual depreciation and construct a depreciation schedule by the declining-balance method, using the data given for each part.

	COST	RESIDUAL VALUE	USEFUL LIFE (YEARS)
1. a.	\$1,900	\$ 0	5
b.	800	0	4
c.	1,800	150	6
2. a.	\$4,200	\$ 0	5
b.	5,000	0	8
c.	1,500	50	3

UNITS-OF-PRODUCTION METHOD

Another conventional depreciation method, the **units-of-production method,** directly ties the amount of depreciation expense claimed for the period to the actual use or output of the asset. The more the asset is used or the more the asset produces, the higher the depreciation expense claimed. If the asset is not used, or if it does not produce any units, no depreciation expense is claimed. When using this method, the first step is to calculate the fixed rate of depreciation per unit (per barrel, per hour, per mile, and so on). This is found by dividing the asset's depreciable base (cost − residual value) by the total expected output that the asset is expected to produce or to be used during its service life. The amount of depreciation expense to be claimed each year is then found by multiplying the fixed rate per unit times the actual number of units (hours, miles, and so on) for that year. A depreciation schedule would then be completed as for other methods.

Example 3 Use the units-of-production method to determine the annual depreciation for the key machine in Example 1. (A \$5,100 machine was expected to be worth \$900 after 5 years.) The number of units (keys) produced annually is given below. The machine is expected to produce 560,000 keys during its lifetime.

The depreciable value (\$5,100 − \$900 = \$4,200) is divided by the expected 560,000 units to obtain the per-unit depreciation:

$$\frac{\text{Depreciable value}}{\text{Total units}} = \frac{\$4{,}200}{560{,}000} = \$0.0075 \text{ per-unit depreciation}$$

YEAR	ANNUAL UNITS	×	UNIT RATE	=	ANNUAL DEPRECIATION
1	120,000	×	\$0.0075	=	\$ 900
2	160,000	×	0.0075	=	1,200
3	40,000	×	0.0075	=	300
4	80,000	×	0.0075	=	600
5	160,000	×	0.0075	=	1,200
Totals	560,000				\$4,200

Although the units-of-production method is convenient mathematically, it has not been recommended by accountants. It is difficult to estimate in advance how many units a machine will produce, and thus the depreciated value may be too high or too low. If the machine becomes obsolete and is taken out of service, it would then appear not to be depreciating—an extremely unrealistic situation. By contrast, the units-of-production method can cause more depreciation in the latter years (as shown above), which reverses the pattern of actual economic depreciation.

QUICK PRACTICE—UNITS-OF-PRODUCTION METHOD

Compute the annual depreciation by the units-of-production method using the data given.

	COST	RESIDUAL VALUE	USEFUL LIFE (UNITS)	ANNUAL PRODUCTION
1. a.	$ 7,500	$ 300	90,000	18,000; 20,000; 25,000; 19,000; 8,000
b.	8,100	600	30,000	10,000; 8,000; 7,000; 5,000
c.	2,600	200	200,000	35,000; 44,000; 40,000; 33,000; 28,000; 20,000
2. a.	$17,000	$1,000	400,000	62,000; 80,000; 78,000; 75,000; 64,000; 41,000
b.	32,000	2,000	500,000	100,000; 125,000; 105,000; 90,000; 80,000
c.	3,000	1,200	60,000	10,000; 25,000; 16,000; 9,000

COST RECOVERY WITH ACRS METHOD

The Economic Recovery Tax Act of 1981 made enormous changes in both the concept of depreciation and the method of calculation. Under this act, "depreciation" was replaced by the concept of **cost recovery:** Assets were placed into designated classes, and each year a business was entitled to recover a certain percentage of the asset's original cost. Eventually, the entire cost would be claimed as an operating expense. Cost recovery thus has little connection with the actual value that an asset may have or the actual length of its useful life. Also, no distinction is made between new or used property.

The **accelerated cost recovery system (ACRS)** greatly simplified previous methods of tax depreciation. All plant assets were placed in cost-recovery classes of either 3, 5, 10, 15, 18, or 19 years. All property of the same class acquired during one year was grouped together, and the ACRS tables stipulated the percent to be multiplied times the total cost of all assets assigned to that class during each year of the recovery period. The ACRS method remains in effect for assets in the 15- and the 19-year classes which were purchased and placed in service during the years 1985 and 1986.

MODIFIED ACRS METHOD (MACRS)

The Tax Reform Act of 1986 modified the ACRS method considerably for assets placed into service in 1987 or after. Additional cost-recovery classes were created, and percentage calculations were replaced by other prescribed methods of cost recovery. As before, there is no distinction between the purchase of new versus used assets for cost-recovery purposes. A primary difference in the methods is that most assets were shifted from their previous class to the next longer class; thus, more years are now required for a business to recover its investment cost.

TABLE 9-1 CLASSES OF PROPERTY USED FOR MACRS COST RECOVERY

CLASS	PROPERTY
3-year class	Tractors, racehorses that are 2 years old, or any other horse that is over 12 years old
5-year class	Automobiles, taxis, trucks, office equipment, computers, office machines, research assets
7-year class	Office furniture, fixtures, and single-purpose agricultural and horticultural structures, and any property not designated by law to be in any other class
10-year class	Vessels, barges, tugs, and other water transportation equipment
15-year class	Land improvements such as roads, fences, bridges, shrubbery
20-year class	Farm buildings and municipal sewers
27.5-year class	Residential real property
31.5-year class	Nonresidential real property placed in service before May 13, 1993
39-year class	Nonresidential property placed in service after May 12, 1993

Table 9-1 shows examples of property in the different classes. The 3-, 5-, 7-, and 10-year classes are depreciated using a 200% declining-balance method, switching to the straight-line method about halfway through the recovery period. The 15- and 20-year classes use a 150% declining-balance method, with a later switch to straight-line. The 27.5-, 31.5-, and 39-year classes use straight-line depreciation.

An important characteristic of cost recovery is that assets are depreciated to a zero residual value, thereby making the total cost recovery over the asset's life equal to 100% of its cost. By contrast, an asset's book value, if depreciated by the declining-balance method, will never reach zero in the natural course of events. It must be forced to do so by the accountant.

An asset's cost recovery is often computed using a table similar to Table 12 in the Tables Booklet and Table 9-2, which contain multiplicative factors for each year during the cost-recovery period. The smaller first-year factors for the 3-year through 20-year classes assume that the various assets were acquired throughout the year and thus average a half-year of service. This practice is known as the *half-year convention.* Assets in the 27.5-, 31.5-, and 39-year classes are assumed to have been purchased mid-month. Thus, these real properties use a *mid-month convention.* The IRS publishes 12 different factors or tables for the first years in these classes. Only the factors for 27.5-, 31.5-, and 39-year class assets purchased in January are shown in Table 9-2.

If 40% or more of all assets are acquired during the last quarter of the business year, the firm must use a different method that depreciates them according to the quarter in which they were acquired. Over all the years in each class, 100% of the purchase cost is recovered.

Note. Cost recovery or depreciation may never be claimed on land and ordinarily may not be claimed by an individual on personal property. However, some plant assets, particularly automobiles, are frequently used for both business and pleasure. Careful records must be kept of the amount of business use, and a corresponding portion of standard cost recovery may then be claimed by the owner (unless IRS-approved reimbursement has been paid for the business use).

TABLE 9-2 MACRS COST RECOVERY FACTORS

For Property Placed into Service in 1987 and Thereafter

RECOVERY YEAR(S)	3-YR. CLASS	5-YR. CLASS	7-YR. CLASS	10-YR. CLASS
1	0.333333	0.200000	0.142857	0.100000
2	0.444444	0.320000	0.244898	0.180000
3	0.148148	0.192000	0.174927	0.144000
4	0.074074	0.115200	0.124948	0.115200
5	—	0.115200	0.089249	0.092160
6	—	0.057600	0.089249	0.073728
7	—	—	0.089249	0.065536
8	—	—	0.044624	0.065536
9	—	—	—	0.065536
10	—	—	—	0.065536
11	—	—	—	0.032768

RECOVERY YEAR(S)	15-YR. CLASS	20-YR. CLASS	27.5-YR. CLASS[a]	31.5-YR. CLASS[a]	39-YR. CLASS[a]
1	0.050000	0.037500	0.034848	0.030423	0.024573
2	0.095000	0.072188	0.036364	0.031746	0.025641
3	0.085500	0.066773	0.036364	0.031746	0.025641
4	0.076950	0.061765	0.036364	0.031746	0.025641
5	0.069255	0.057133	0.036364	0.031746	0.025641
6	0.062330	0.052848	0.036364	0.031746	0.025641
7	0.059049	0.048884	0.036364	0.031746	0.025641
8	0.059049	0.045218	0.036364	0.031746	0.025641
9–15	0.059049	0.044615	0.036364	0.031746	0.025641
16	0.029525	0.044615	0.036364	0.031746	0.025641
17–20	—	0.044615	0.036364	0.031746	0.026541
21	—	0.022308	0.036364	0.031746	0.025641
22–27	—	—	0.036364	0.031746	0.025641
28	—	—	0.019697	0.031746	0.025641
29–31	—	—	—	0.031746	0.025641
32	—	—	—	0.017196	0.025641
33–39	—	—	—	—	0.025641
40	—	—	—	—	0.001068

[a]Assume that the asset was purchased in January.

Example 4 New processing equipment (7-year class) cost Nu-way Cleaners $14,000 during 1999. Determine the amount of cost recovery the cleaners can claim each year.

Using the MACRS table, the years of use appear in the left-hand column, and the corresponding factor for each year is found on the same line in the column entitled "7-yr. class." Each year's cost recovery amount is then found by multiplication.

YEAR	COST	×	FACTOR	=	COST RECOVERY
1	$14,000	×	0.142857	=	$ 2,000.00
2	14,000	×	0.244898	=	3,428.57
3	14,000	×	0.174927	=	2,448.98
4	14,000	×	0.124948	=	1,749.27
5	14,000	×	0.089249	=	1,249.49
6	14,000	×	0.089249	=	1,249.49
7	14,000	×	0.089249	=	1,249.49
8	14,000	×	0.044624	=	624.74
				Total	$14,000.03

Thus, the total $14,000 cost is recovered over 7 years, although the equipment will probably be used for much longer. The MACRS method is called "accelerated" because owners can recover their investment cost in fewer years than the asset's expected useful life.

Example 5 Nu-way Cleaners wishes to prepare a cost-recovery schedule for its 7-year property shown in Example 4.

We follow the practice of whole-dollar accounting in the examples and problems that follow.

COST-RECOVERY SCHEDULE—MACRS METHOD

YEAR	BOOK VALUE (END OF YEAR)	ANNUAL RECOVERY	ACCUMULATED COST RECOVERY
0	$14,000	—	—
1	12,000	$2,000	$ 2,000
2	8,571	3,429	5,429
3	6,122	2,449	7,878
4	4,373	1,749	9,627
5	3,124	1,249	10,876
6	1,875	1,249	12,125
7	626	1,249	13,374
8	0	626	14,000

Note. Observe that the final book value should be zero and that the final accumulated cost recovery should equal the cost of the property. That is the reason the annual cost recovery in year 8 is $626 instead of the $625 as shown in Example 4. An adjustment was made in that year so that the annual cost recovery equals the book value at the beginning of that year. The $1 difference is a rounding error. Also observe that a mistake at any point in a recovery schedule will cause the succeeding entries to be incorrect, so you should also use the following checks:

(a) For any year, the sum of all the annual amounts to that point should equal that year's accumulated cost recovery. Example:

$$\text{Year 3: } \$2{,}000 + \$3{,}429 + \$2{,}449 = \$7{,}878$$

(b) During any year, the book value plus the accumulated cost recovery must equal the original cost. Example:

$$\text{Year 4: } \$4{,}373 + \$9{,}627 = \$14{,}000$$

Calculator techniques . . . FOR EXAMPLE 5

For the MACRS method, you should calculate the entire Annual Recovery column first, using the factors as shown in Example 4. Then complete the Accumulated Cost Recovery column by adding the annual amounts. As you press [+] preceding the third value (and for every year thereafter), notice that the current subtotal briefly appears; this provides the corresponding year's accumulated value, which you can fill in before proceeding:

2,000 [+] 3,429 [+] [→ 5,429]
2,449 [+] [→ 7,878]
1,749 [+] [→ 9,627] and so on.

Similarly, the Book Value column can be completed by successively subtracting the annual recovery amounts from the cost. The current subtotal appears as you press [−], prior to the second and all subsequent annual recovery amounts:

14,000 [−] 2,000 [−] [→ 12,000]
3,429 [−] [→ 8,571]
2,449 [−] [→ 6,122] and so on.

Also check that the book value plus the accumulated cost recovery for any year equals the original cost, as illustrated in Part (b) of the Note preceding this Calculator Technique. The calculator procedures described here for an MACRS cost-recovery schedule also apply for depreciation schedules under the straight-line method, described next.

The MACRS method of cost recovery was established by Congress to provide incentives to businesses to make capital investments. With 100% of asset cost being recoverable—and in less time than previously allowed—the extra depreciation claimed each year means that companies have higher tax deductions and thus more after-tax funds. This was seen as a way to stimulate widespread sales and production and thereby to improve the country's economic status.

Because depreciation accounting has been a standard business procedure for so long, calculations under the MACRS system of cost recovery continue to be referred to as "depreciation" in common practice. It must be emphasized that in most cases the MACRS method applies *only* for tax accounting purposes, and a business may use any of the single, traditional depreciation methods for its own internal accounting. The straight-line method is acceptable for tax purposes in many cases as an alternative to the MACRS method, and it is used entirely for real estate (in a variation which assumes that the usage began at mid-month).

QUICK PRACTICE—MACRS METHOD

Compute the annual cost recovery for each of the following using the MACRS method and the data given. Construct a cost recovery schedule for (c) parts only.

	COST	CLASS/YEARS
1. a.	$3,000	3
b.	5,800	7
c.	4,500	5
2. a.	$9,000	7
b.	7,200	5
c.	2,800	3

REVIEW

Stated simply, every depreciation method involves multiplying a depreciation *rate times a base value* to find an annual amount of depreciation. The methods differ in how to determine the depreciation rate (most use the same rate every year; one doesn't) and in how to determine the base value to which the rate will be applied (here again, some methods use the same base value every year; others don't). In review, let's look at the various depreciation formulas and see how the methods differ. The following review assumes full years.

Straight-Line Method

Under the straight-line method, a constant rate is multiplied times the depreciable base each year. The depreciable base is the difference between the asset's cost and its residual value. The annual depreciation will be the same amount each year.

$$\frac{\text{Cost} - \text{Residual value}}{\text{Years}} = \text{Annual depreciation}$$

Note from the preceding formula that dividing by years is the same as multiplying the depreciable value by $\frac{1}{\text{Years}}$.

Declining-Balance Method

The declining-balance method also uses a constant rate each year—typically twice the straight-line rate. This rate is multiplied times the full cost the first year (not times the depreciable base). For each succeeding year, the constant rate is multiplied times the preceding year's book value.

1.	2 × (straight-line rate)	= Rate
2.	Rate × cost	= First year's depreciation
3.	Cost − depreciation (from previous step)	= Book value
4.	Rate × book value	= Subsequent year's depreciation
5.	Book value − depreciation	= New book value
6.	Continue steps 4 and 5.	

Units-of-Production Method

The units-of-production method is very similar to the straight-line method. First determine the depreciable base, and then divide by the total number of units to be produced. This yields a rate per unit. The rate per unit is multiplied times each year's production to give the annual depreciation each year.

1. $\dfrac{\text{Cost} - \text{Residual value}}{\text{Total number of units}} = \text{Rate}$
2. Rate × each year's production = Annual depreciation

Note. Although many of the problems in this section describe a single asset, keep in mind as you work these problems that the MACRS method requires all assets of the same class that are purchased during one year to be computed as a group.

SPREADSHEETS AND DEPRECIATION

For Problems 3–8 and 15–20, directions are provided for spreadsheet preparation. These problems are found on the data disk that accompanies this textbook. You may wish to look at the two demonstration problems before attempting the problems.

Directions for Demo 1 (Straight-line Method)

- Before starting the spreadsheet problems in this chapter, you should review the general instructions for Excel® in Chapter 4, pages 83–85.
- Start Excel. Insert the data disk in *Drive A.*
- Click the *File* button on the Menu Bar. Click *Open.*
- At the *Open* Dialog Box, change the Look-in Box to *A*.
- Click the *Chapter 9* folder. Click the *Open* button.
- Click *Demo 1*. Click the *Open* button.

- Begin each formula with an equal (=) symbol. You can copy formulas to adjacent cells, which will save time. If a dollar ($) symbol precedes a row number or a column letter in a formula, the value represented by that row number or column letter becomes a constant value. This is referred to as **absolute referencing.** We want the values in Cells B1, B2, and B3 to be constants (values that do not change when the formula is copied for straight-line depreciation).

TO FIND . . .	CLICK IN . . .	KEY IN . . .	✓
Book value, end of year 0	B7	6000	
First year's depreciation	C8	=($B$1-$B$2)/$B$3. Press *Enter.*	
Remaining years' annual depreciation (years 2 through 7)	C8	Click the fill handle in the lower right corner of the cell. Drag down through Cell C14. This copies the formula from C8.	
Book value, end of first year	B8	=B7-C8. *Enter.*	
Remaining years' book values	B8	Click the fill handle in the lower right corner of the cell. Drag down through Cell B14. This copies the formula from B8.	
First year's accumulated depreciation	D8	=D7+C8. *Enter.*	
Remaining years' accumulated depreciation	D8	Click the fill handle in the lower right corner of the cell. Drag down through the D cells. This copies the formula from D8.	

Directions for Demo 2 (MACRS)

TO FIND . . .	CLICK IN . . .	KEY IN . . .	✓
Book value, end of year 0	B3	850	
Factors	Beginning in C4	Factors from the MACRS table for the 3-year class (Table 9-2 on page 239).	
First year's annual cost recovery	D4	=C4*$B$3. *Enter.*	
Remaining years' annual cost recovery (years 2 through 4)	D4	Click the fill handle in the lower right corner of the cell and drag down to D7. This copies the formula from D4.	
Book value, end of year 1	B4	=B3-D4. *Enter.*	
Remaining years' book value	B4	Click the fill handle in the lower right corner of the cell and drag down through B7. This copies the formula in B4.	
First year's accumulated cost recovery	E4	=E3+D4. *Enter.*	
Remaining years' accumulated cost recovery	E4	Click the fill handle in the lower right corner of the cell and drag down through E7. This copies the formula in E4.	

For the straight-line problems, Problems 3–8, use the instructions for Demo 1. Change the amount for the "Book value, end of year 0" to the amount given in each problem. For the MACRS problems, Problems 15–18, use the instructions for Demo 2. Change the amount for the "Book value, end of year 0" to the amount given in each problem. For Problems 19 and 20, use the following instructions:

FOR PROBLEM 19: TO FIND . . .	CLICK IN . . .	KEY IN . . .	✓
Book value, end of year 0 for the Warehouse	B3	600000	
Factors	Beginning in C4	Factors from the MACRS table (Table 9-2 on page 239) for the 27.5-year class.	
First year's cost recovery	D4	=$B$3*C4. *Enter.*	
Remaining years' annual cost recovery	D4	Click the fill handle in the lower right corner of D4 and drag down through D6. This copies the formula.	
Remaining years' book value	B4	=B3-D4. *Enter.* Click the fill handle of B4 and drag down through B6. This copies the formula.	
First year's accumulated cost recovery	E4	=E3+D4. *Enter.*	
Remaining years' accumulated cost recovery	E4	Click the fill handle of E4 and drag down through E6. Excel stores numbers with full precision (decimal places) even though you rounded to whole dollars. The final figure in Cell E6 (64,546) is precisely 64,545.60.	
Book value for year 0 for the office equipment	B11	25000	
Factors	Beginning in C12	Factors from the MACRS table (Table 9-2 on page 239) for the 5-year class.	
First year's cost recovery	D12	=$B$11*C12. *Enter.*	
Remaining years' cost recovery	D12	Click the fill handle of D12 and drag down through D14.	
Remaining years' book value	B12	=B11-D12. *Enter.* Click the fill handle of B12 and drag down through B14.	
First year's accumulated cost recovery	E12	=E11+D12. *Enter.*	
Remaining years' accumulated cost recovery	E12	Click fill handle of E12 and drag down through E14.	
Year 1 Warehouse annual cost recovery	B19	=D4. *Enter.*	
Years 2 and 3 Warehouse annual cost recovery	B20 and B21	=D5 and =D6, respectively.	
Year 1 Office Equipment annual cost recovery	C19	=D12	
Years 2 and 3 Office Equipment annual cost recovery	C20 and C21	=D13 and =D14, respectively.	
Total of first year's cost recovery	D19	=B19+C19. Note that the two columns do not total correctly. Excel stores and saves the numbers with full precision even though you may have rounded to whole dollars. Accumulated totals may appear incorrectly due to rounding.	
Remaining totals	D19	Click the fill handle of D19 and drag down through D21.	

FOR PROBLEM 20: TO FIND . . .	CLICK IN . . .	KEY IN . . .	✓
Book value, end of year 0 for the Building	B3	2000000	
Factors	Beginning in C4	Factors from the MACRS table (Table 9-2 on page 239) for the 39-year class.	
First year's cost recovery	D4	=$B$3*C4. *Enter.*	
Remaining years' annual cost recovery	D4	Click the fill handle in the lower right corner of D4 and drag down through D6. This copies the formula.	
Remaining years' book value	B4	=B3-D4. *Enter.* Click the fill handle of B4 and drag down through B6. This copies the formula.	
First year's accumulated cost recovery	E4	=E3+D4. *Enter.*	
Remaining years' accumulated cost recovery	E4	Click the fill handle of E4 and drag down through E6.	
Book value for year 0 for the Office Machines	B11	4500	
Factors	Beginning in C12	Factors from the MACRS table (Table 9-2 on page 239) for the 5-year class.	
First year's cost recovery	D12	=$B$11*C12. *Enter.*	
Remaining years' cost recovery	D12	Click the fill handle of D12 and drag down through D14.	
Remaining years' book value	B12	=B11-D12. *Enter.* Click the fill handle of B12 and drag down through B14.	
First year's accumulated cost recovery	E12	=E11+D12. *Enter.*	
Remaining years' accumulated cost recovery	E12	Click fill handle of E12 and drag down through E14.	
Year 1 Building annual cost recovery	B19	=D4. *Enter.*	
Years 2 and 3 Building annual cost recovery	B20 and B21, respectively	=D5 and =D6, respectively.	
Year 1 Office Machine annual cost recovery	C19	=D12	
Years 2 and 3 Office Machines annual cost recovery	C20 and C21, respectively	=D13 and =D14, respectively.	
Total of first year's cost recovery	D19	=B19+C19. *Enter.*	
Remaining totals	D19	Click the fill handle of D19 and drag down through D21.	

SECTION 1 PROBLEMS

Compute the annual cost recovery (depreciation) and complete a schedule for each of the following, using the data and method indicated. Round your answers to whole dollars.

	Cost	Residual Value	Class/ Years	Cost Recovery or Depreciation Method
1. a.	$ 6,000	$400	8	Straight-line
b.	4,000	200	5	Declining-balance
c.	1,400	0	3	MACRS
d.	3,600	0	7	MACRS
2. a.	$17,000	$500	10	Straight-line
b.	8,000	600	6	Declining-balance
c.	5,500	0	3	MACRS
d.	7,400	0	5	MACRS

3. DDG purchased office equipment costing $6,800. The useful life will be 7 years. The residual value is $500. Construct a depreciation schedule using the straight-line method.

4. Carter Career Services purchased equipment costing $4,450 with a useful life of 5 years and a residual value of $400. Construct a depreciation schedule using the straight-line method.

5. State Supply Store purchased equipment for $1,200. The salvage value was estimated to be $300, and the useful life was 6 years. Determine the annual depreciation by the straight-line method.

6. Determine the annual depreciation by the straight-line method for factory equipment for the Coulter Realty Co. costing $18,000, with a salvage value of $200 and a useful life of 8 years.

7. The Mason-Dixon Restaurant paid $38,000 for new kitchen equipment. The equipment will be worth $6,000 after 8 years of use. What is the annual depreciation expense by the straight-line method?

8. The Entree Shoe Co. paid $43,000 for new factory equipment. The equipment will be worth $1,000 after 12 years. What is the annual depreciation by the straight-line method?

9. Using the declining-balance method, construct a depreciation schedule for office furniture costing $12,000. After the estimated 6-year life, the furniture should have a residual value of $800.

10. An accounting firm purchased a new conference table and chairs for $15,000. For financial reporting, the firm has chosen the declining-balance method for depreciation. The life of the table and chairs is estimated to be 10 years, and the resale value at that time will be $2,000. Construct a depreciation table.

11. Factory equipment that cost $6,000 will be obsolete after 4 years of use. The scrap value should be zero. Determine the annual depreciation by the declining-balance method, and construct a depreciation schedule.

12. Modern Plumbing and Heating purchased tools costing $700. The tools will be obsolete in 5 years, and the scrap value will be zero. Determine the annual depreciation by the declining-balance method, and construct a depreciation schedule.

13. The Aerials Gymnastics Center spent $800 on a new balance beam. The useful life of the beam is 8 years, at which time the Center wants to sell it for $100. Construct a depreciation schedule using the declining-balance method.

14. Colarulli Market purchased a freezer for $6,300. The estimated useful life will be 7 years, at which time the market wants to sell it for $500. Construct a depreciation schedule using the declining-balance method.

15. HHI Industries purchased $6,200 of equipment in 1999. Prepare a cost-recovery schedule by the MACRS method for this 5-year class equipment.

16. Norcio Scuba Diving purchased $10,000 of scuba diving equipment. Prepare a cost-recovery schedule using the MACRS method for this 7-year class equipment.

17. Afshar Investments purchased an office building in January for $800,000. Using the 31.5-year class, determine the cost recovery for only the first 3 years by the MACRS method.

18. In January, the McNally Co. purchased an office building for $2,500,000. Using the 27.5-year class and the MACRS method, determine the cost-recovery for the first 3 years only.

19. Charleston Manufacturing Co. purchased a warehouse for $600,000 in the 27.5-year class and office equipment for $25,000 in the 5-year class under MACRS. Determine the cost recovery for tax purposes each year during the first 3 years. Compute each class separately, and sum the amount for each year.

20. In January, 2000, CAM Roofing Co. purchased a building for $2,000,000 and office machines totaling $4,500. Under MACRS, the real property is classified in the 39-year class, and the office machines are classified in the 5-year class. How much cost recovery will be included during the first 3 years on its tax returns? Compute each class separately, and sum the amount for each year.

Use the units-of-production method to compute the unit rate and the annual depreciation for each machine below. Do not round the cost per unit to the nearest cent. In each case, assume that the machine is expected to produce 300,000 units during its useful life.

	Cost	Scrap Value	Useful Life (Years)	Annual Production
21. a.	$ 3,000	$ 300	4	75,000; 82,000; 76,000; 67,000
b.	5,700	300	4	70,000; 80,000; 88,000; 62,000
c.	6,800	500	6	56,000; 58,000; 60,000; 55,000; 40,000; 31,000
22. a.	$ 8,100	$ 600	4	62,000; 79,000; 85,000; 74,000
b.	9,700	100	5	55,000; 76,000; 84,000; 78,000; 7,000
c.	15,000	3,000	6	50,000; 53,000; 56,000; 52,000; 44,000; 45,000

23. A computer was purchased by the Nu Thang Co. for $4,200. The business assumes that it will have a useful life of 5 years and will have a trade-in value of $200. Compare the straight-line and declining-balance methods of depreciation with the MACRS cost recovery for the first 3 years only.

24. The Prather Laundromat purchased a delivery truck for $19,000. The business estimates a useful life of 5 years, at which time the trade-in value should be $3,000. Compare the straight-line method and declining-balance method of depreciation with the MACRS cost recovery for the first 3 years only.

PARTIAL-YEAR DEPRECIATION

The discussion so far has described plant assets depreciated for full years. Not all assets, however, are purchased and placed into service on January 1 nor sold or discarded on December 31. How does a business prorate the depreciation for a partial year of use? With the MACRS method, this is not a concern because its table uses the *half-year convention* or *mid-month convention.* The half-year convention automatically allows six months of depreciation during the first year, regardless of when the asset is actually placed into service. (Remember the 40% exception mentioned on page 238.) Under MACRS, a plant asset acquired April 1 would have the same cost recovery in the first year as if it had been acquired October 1.

Under the traditional depreciation methods (straight-line and declining-balance), the annual depreciation should be prorated for partial years of service, normally by using the mid-month convention. That is, if an asset is placed into service on or before the 15th of the month, it is depreciated as if it were placed into service on the first of the month. Conversely, if the asset is placed into service after the 15th of the month, depreciation does not start until the first day of the next month.

Example 1 A machine costing $6,800 is to be depreciated by the straight-line method. It was purchased on April 1. The residual value is to be $800, and the life is estimated to be 5 years.

$$\text{April 1 to December 31} = 9 \text{ months} = \frac{9}{12} \text{ year}$$

$$\frac{\$6{,}800 - \$800}{5} = \frac{\$6{,}000}{5} = \$1{,}200 \quad \text{Annual depreciation}$$

$$\frac{9}{12}(\$1{,}200) = \$900 \qquad \text{First year's depreciation}$$

The first year's depreciation will be $900, and $1,200 will be the depreciation for each of the remaining full years. However, the business is entitled to 5 full years of depreciation, so the depreciation schedule will cover 6 calendar years as shown subsequently. The first year's depreciation is for 9 months, the depreciation for the 2nd through 5th years represent 12 months each, and the 6th year's depreciation represents 3 months.

YEAR	BOOK VALUE (END OF YEAR)	ANNUAL DEPRECIATION	ACCUMULATED DEPRECIATION
0	$6,800	—	—
1	5,900	$ 900	$ 900
2	4,700	1,200	2,100
3	3,500	1,200	3,300
4	2,300	1,200	4,500
5	1,100	1,200	5,700
6	800	300	6,000

Example 2 Assume that the business in Example 1 decides to depreciate the asset by the declining-balance method. How much annual depreciation will be taken each year?

$$\text{Rate} = 2\left(\frac{1}{5}\right) = \frac{2}{5} = 40\%$$

In the following yearly calculations, the depreciable value for each successive year is obtained by first computing a new book value in the subsequent depreciation schedule.

Year	
1	$\frac{9}{12} \times 0.40 \times \$6{,}800 = \$2{,}040$
2	$0.40 \times \$4{,}760 = 1{,}904$
3	$0.40 \times \$2{,}856 = 1{,}142$
4	$0.40 \times \$1{,}714 = 686$
5	$0.40 \times \$1{,}028 = 411$ (adjusted to \$228)

YEAR	BOOK VALUE (END OF YEAR)	ANNUAL DEPRECIATION	ACCUMULATED DEPRECIATION
0	$6,800	—	—
1	4,760	$2,040	$2,040
2	2,856	1,904	3,944
3	1,714	1,142	5,086
4	1,028	686	5,772
5	800	228	6,000

Proration does not apply to the units-of-production method because that method is based on actual usage regardless of the time period that the asset is held.

SECTION 2 PROBLEMS

Compute the annual depreciation and complete a schedule for each of the following assets, using the data and method indicated. Use the mid-month convention where applicable to prorate depreciation according to the date of purchase. Assume that the asset was placed in service on the purchase date.

	COST	RESIDUAL VALUE	USEFUL LIFE (YEARS)	PURCHASE DATE	METHOD
1. a.	$ 9,000	$ 0	5	10/21	MACRS
b.	4,900	400	5	9/10	Straight-line
c.	2,600	0	4	4/4	Declining-balance
2. a.	$100,000	$ 0	5	8/18	MACRS
b.	6,000	600	4	9/13	Straight-line
c.	1,800	0	3	2/20	Declining-balance

3. Construct a depreciation schedule using the straight-line method on equipment costing $4,600 if the trade-in value is estimated to be $400, the useful life will be 7 years, and the equipment was purchased on November 1.
4. On June 5, Sonny's Auction paid $1,860 for a display case. The business plans to use the case for 5 years. It estimates that the residual value will be $60. Construct a depreciation schedule using the straight-line method.
5. On August 5, a business purchased a printer costing $900. The estimated life is 3 years, and the residual value is expected to be $50. Using the declining-balance method, construct a depreciation schedule.
6. Construct a depreciation schedule using the declining-balance method for office furniture costing a business $27,000 if the trade-in value is estimated to be $2,000, the useful life is 6 years, and the furniture was purchased on May 19.

OVERHEAD

In addition to the cost of materials or merchandise, other expenses are incurred in the operation of any business. Examples of such *operating expenses* are salaries, rent, utilities, office supplies, taxes, depreciation, insurance, and so on. These additional expenses are known collectively as **overhead.**

Overhead contributes indirectly to the actual total cost of the merchandise being manufactured or sold by the concern. In order to determine the efficiency of various departments, to determine the total cost of manufacturing various items, or to determine whether certain items are being sold profitably, a portion of the total plant overhead is assigned to each department. From the many methods of distributing overhead, we shall consider three: according to total floor space, according to total net sales, and according to the number of employees in each department.

Example 1 The mean monthly overhead of $96,000 at Johnson Manufacturing Corp. is distributed *according to the total floor space* of each department. The floor space of each department is as follows:

DEPARTMENT	FLOOR SPACE (SQ. FT)
A—Receiving and raw materials	4,000
B—Assembling	9,000
C—Inspecting and shipping	8,000
D—Administration	+ 3,000
Total	24,000

Each department therefore has the following ratio of floor space and is assigned the following amount of overhead expense:

DEPARTMENT	RATIO OF FLOOR SPACE	OVERHEAD CHARGE	
A	$\frac{4,000}{24,000} = \frac{1}{6}$	$\frac{1}{6} \times \$96,000 =$	\$16,000
B	$\frac{9,000}{24,000} = \frac{3}{8}$	$\frac{3}{8} \times \$96,000 =$	36,000
C	$\frac{8,000}{24,000} = \frac{1}{3}$	$\frac{1}{3} \times \$96,000 =$	32,000
D	$\frac{3,000}{24,000} = \frac{1}{8}$	$\frac{1}{8} \times \$96,000 =$	+ 12,000
		Total overhead =	\$96,000

Example 2 Avery Hardware wishes to distribute its \$20,000 July overhead *according to the total net sales* of each department. These sales and the overhead distribution are as follows:

DEPARTMENT	NET SALES	RATIO OF SALES	OVERHEAD CHARGE	
Lumber/glass	\$24,000	$\frac{\$24,000}{\$80,000} = \frac{3}{10}$	$\frac{3}{10} \times \$20,000 =$	\$ 6,000
Tools/electrical	16,000	$\frac{\$16,000}{\$80,000} = \frac{1}{5}$	$\frac{1}{5} \times 20,000 =$	4,000
Building supplies	20,000	$\frac{\$20,000}{\$80,000} = \frac{1}{4}$	$\frac{1}{4} \times 20,000 =$	5,000
Housewares	8,000	$\frac{\$8,000}{\$80,000} = \frac{1}{10}$	$\frac{1}{10} \times 20,000 =$	2,000
Sports equipment	\$12,000	$\frac{\$12,000}{\$80,000} = \frac{3}{20}$	$\frac{3}{20} \times 20,000 =$	+ 3,000
Total net sales =	\$80,000		Total overhead =	\$20,000

Example 3 The \$15,000 monthly office overhead at White Plumbing & Heating is charged to the various divisions *according to the number of employees* in the office who work directly with each division. This results in the following overhead charges:

DEPARTMENT	NUMBER OF EMPLOYEES	RATIO OF EMPLOYEES	OVERHEAD CHARGE	
Plumbing	3	$\frac{3}{10}$	$\frac{3}{10} \times \$15{,}000 =$	$ 4,500
Heating	1	$\frac{1}{10}$	$\frac{1}{10} \times \$15{,}000 =$	1,500
Air conditioning	4	$\frac{4}{10} = \frac{2}{5}$	$\frac{2}{5} \times \$15{,}000 =$	6,000
Appliances	+ 2	$\frac{2}{10} = \frac{1}{5}$	$\frac{1}{5} \times \$15{,}000 =$	+ 3,000
Total employees = 10			Total overhead =	$15,000

SECTION 3 PROBLEMS

In Problems 1–6, distribute overhead according to floor space, net sales, or number of employees, depending on the information given.

	DEPARTMENT	FLOOR SPACE (SQ. FT.)	TOTAL OVERHEAD
1. a.	#1	800	$13,500
	2	1,000	
	3	1,200	
	4	1,500	
b.	#16	500	32,000
	17	300	
	18	400	
	19	800	

	DEPARTMENT	FLOOR SPACE (SQ. FT.)	TOTAL OVERHEAD
2. a.	M	$2,000	$48,000
	N	1,000	
	O	1,600	
	P	1,400	
b.	Q	900	15,000
	R	800	
	S	300	
	T	500	

	Department	Net Sales	Total Overhead
3. a.	W	$30,000	$56,000
	X	15,000	
	Y	14,000	
	Z	11,000	
b.	AA	20,000	36,000
	BB	22,000	
	CC	14,000	
	DD	16,000	

	Department	Net Sales	Total Overhead
4. a.	O	$45,000	$84,000
	V	24,000	
	E	36,000	
	R	15,000	
b.	H	62,000	96,000
	E	68,000	
	A	64,000	
	D	46,000	

	Department	Number of Employees	Total Overhead
5. a.	#22	40	$18,000
	33	10	
	44	30	
	55	20	
b.	10	2	30,000
	11	4	
	12	3	
	13	6	

	Department	Number of Employees	Total Overhead
6. a.	#66	12	$54,000
	77	15	
	88	13	
	99	20	
b.	14	8	19,000
	15	4	
	16	5	
	17	3	

7. A department store allocated operating expenses of $60,000 according to the total floor space occupied by each department. Find the overhead for each department.

Department	Floor Space (sq. ft.)	Overhead
Accounting	200	
General office	300	
Marketing	400	
Service/repairs	600	

8. Billy's Drug Store allocates its operating expenses of $14,400 according to the total floor space occupied by each department. Find the overhead for each department.

DEPARTMENT	FLOOR SPACE (SQ. FT.)	OVERHEAD
Pharmacy	400	
Home health equipment	1,000	
Cosmetics and toiletries	800	
Greeting cards and magazines	200	

9. William's Book Store distributes its overhead according to each department's ratio of total net sales. Allocate operating expenses of $40,000.

DEPARTMENT	NET SALES	OVERHEAD
Biographies	$60,000	
Adventure/mysteries	44,000	
Computers/software	42,000	
Romance	38,000	
Sports	16,000	

10. Brantley Department Store distributes its overhead according to each department's ratio of total net sales. If the September operating expenses were $20,000, what was each department's share of the overhead?

DEPARTMENT	NET SALES	OVERHEAD
Men's clothing	$10,000	
Women's clothing	12,000	
Shoes	8,000	
Small appliances	20,000	

11. The overhead expenses at George's Hardware are apportioned among its departments according to the number of employees in each department. If the total overhead during February was $90,000, determine how much should be charged to each department.

DEPARTMENT	NUMBER OF EMPLOYEES	OVERHEAD
Lawn/garden	7	
Appliances	8	
Automotive	5	
Building supplies	6	
Paint	4	

12. Allocate the overhead expenses of $50,000 among the branch offices of Pickford Associates according to the number of employees.

BRANCH	NUMBER OF EMPLOYEES	OVERHEAD
A	5	
B	9	
C	8	
D	6	
E	12	

CHAPTER 9 GLOSSARY

Absolute referencing. A technique used in a spreadsheet that keeps a cell reference constant when a formula is copied to another cell.

ACRS and MACRS. "Accelerated cost recovery system" and "modified accelerated cost recovery system" for determining the amount of asset cost to recover as annual tax deductions. As substantially revised by the Tax Reform Act of 1986, the MACRS method prescribes recovery categories (five for personal property and two for real estate) that enable 100% of the purchase cost to be deducted over the entire recovery period. This method employs a combination of the declining-balance method and the straight-line method of depreciation to accelerate cost recovery for tax purposes. (See also "Cost recovery.")

Book value. The calculated value of a plant asset after depreciation or cost recovery has been deducted.

Cost recovery. Reduction in value of a plant asset, based on set annual portions of the original cost; a tax-deductible business expense. The amount of cost recovery should not be confused with the decrease in value that an asset undergoes from the standpoint of its actual market value.

Cost recovery (or depreciation) schedule. A record showing annual cost recovery (or depreciation), current book value, and accumulated recovery (or depreciation) of a plant asset.

Declining-balance method. The depreciation method in which a constant percent is annually multiplied times book value (starting with cost). Formerly called "double (or 200%) declining-balance." The double declining-balance method employs twice the straight-line rate.

Depreciable base. The total amount of depreciation that is computed over the life of an asset. Depreciable base = Total cost of asset − Residual value.

Depreciation. The systematic allocation of the asset's cost over its period of economic utility (its usefulness).

Overhead. Business expenses other than the cost of materials or merchandise. (Examples: Salaries, rent, utilities, supplies, advertising.)

Plant asset. A business possession that will be used for more than 1 year. (Examples: Buildings, vehicles, machinery, equipment.)

Resale value. The expected value to be received if the asset is sold.

Residual value. The expected value of the asset at the end of its useful life.

Salvage (scrap) value. The value of the materials in a plant asset at the end of its useful life.

Straight-line method. The depreciation method in which the depreciable value of an asset is evenly divided among the years. Legal for tax purposes as an alternative to the MACRS method.

Straight-line rate. The depreciation rate that is the reciprocal of the number of years of useful life. (Example: For an asset used 5 years, the annual straight-line rate is $\frac{1}{5}$.)

Trade-in value. The expected residual value to be received if the asset is traded.

Units-of-production method. The depreciation method in which depreciable value is divided by total expected production to obtain a per-unit depreciation rate, which is then multiplied times the number of units produced each year, in order to determine annual depreciation.

Useful life. The number of years that a plant asset is efficiently used and depreciated in a business. (Useful life differs from the recovery period used in the MACRS method.)

CHAPTER 10

FINANCIAL STATEMENTS AND RATIOS

OBJECTIVES

Upon completion of Chapter 10, you will be able to:

1. Define and use correctly the terminology associated with each topic.
2. On an income statement with the individual amounts given (Section 1: Examples 1, 2; Problems 1, 2):
   a. Compute the net profit.
   b. Find the percent of net sales that each item represents.
3. On a balance sheet with individual amounts given, compute (Section 2: Example 1; Problems 3, 4):
   a. Total assets
   b. Total liabilities
   c. Net worth
   d. Percent each item represents.
4. On comparative income statements and comparative balance sheets, compute (Section 1: Example 3; Problems 5, 6; Section 2: Example 2; Problems 7, 8):
   a. Subtotals and totals
   b. Increase or decrease between years (amount and percent)
   c. Percent of net sales for each year (on income statements)
   d. Percent of total assets for each year (on balance sheets).
5. Select from income statements and balance sheets the amounts required to compute common business ratios (Section 2: Example 3; Problems 9–12).
6. Compute (both at cost and at retail) (Section 3):
   a. Average inventory (Example 1; Problems 1–4, 7–12)
   b. Inventory turnover (Examples 2, 3; Problems 5–12).
7. **a.** Evaluate inventory by the methods of (Section 4: Example 1; Problems 1–8):
   (1) Weighted average, (2) FIFO, and (3) LIFO.
   b. On an income statement, compare gross profit (amount and percent) under each inventory method.

The accounting records of every public company are audited at least once a year to determine the financial condition of the business: what the volume of business was, how much profit (or loss) was made, how much these things have changed in the past year, how much the business is actually worth, and so on. As a means of presenting this important information, two financial statements that are normally prepared and analyzed are the **income statement** and the **balance sheet.**

SECTION 1

INCOME STATEMENT

The **income statement** shows a business's sales, expenses, and profit (or loss) during a certain period of time. (The time might be a month, a quarter, 6 months, or a year.) The basic calculations of an income statement are these:

	Net sales
−	Cost of goods sold
	Gross profit
−	Operating expenses
	Net profit (or net loss)

Notice that **gross profit** (also known as **margin** or **markup**) is the amount that would remain after the merchandise has been paid for. Out of this gross profit must be paid all of the **operating expenses** (**overhead,** such as salaries, rent, utilities, supplies, insurance, and so forth). The amount that remains is the clear or spendable profit—the **net income** or **net profit.**

In order to effectively analyze expenses or to compare an income statement with preceding ones, it is customary to convert the dollar amounts to percents of the total net sales. These percents are computed as follows:

$$\_\_\_\_\% \text{ of Net sales} = \text{Each amount}$$

$$\_\_\_\_\% = \frac{\text{Each amount}}{\text{Net sales}}$$

Example 1 During April, Williams Pharmacy had net sales of $60,000. The drugs and merchandise cost $45,000 wholesale, and operating expenses totaled $12,000. How much net profit was made during April, and what was the percent of net sales for each item?

WILLIAMS PHARMACY
Income Statement for Month Ending April 30, 20XX

Net sales	\$60,000	100%
Cost of goods sold	− 45,000	− 75%
Gross profit	\$15,000	25%
Operating expenses	− 12,000	− 20%
Net profit	\$ 3,000	5%

The percents above were obtained as follows:

$$\_\_\_\% = \frac{\text{Each amount}}{\text{Net sales}}$$

(a) $$\frac{\text{Cost of goods}}{\text{Net sales}} = \frac{\$45,000}{\$60,000} = 75\%$$

(b) $$\frac{\text{Gross profit}}{\text{Net sales}} = \frac{\$15,000}{\$60,000} = 25\%$$

(c) $$\frac{\text{Operating expenses}}{\text{Net sales}} = \frac{\$12,000}{\$60,000} = 20\%$$

(d) $$\frac{\text{Net profit}}{\text{Net sales}} = \frac{\$3,000}{\$60,000} = 5\%$$

Thus, we find that 75% of the income from sales was used to pay for the drugs and merchandise itself, leaving a gross profit (\$15,000) of 25% of sales. Another 20% of the income from sales was used to pay the \$12,000 operating expenses. This leaves a clear, net profit of \$3,000, which is 5% of the sales income.

Note. Observe that the cost of goods, the overhead, and the net profit together account for all the income from sales. Also, the overhead (operating expenses) plus the net profit comprise the margin (gross profit). In equation form,

$$\text{Cost} + \text{Overhead} + \text{Net profit} = \text{Sales}$$

$$\$45,000 + \underbrace{\$12,000 + \$3,000} = \$60,000$$

$$\text{Cost} + \quad \text{Margin} \quad = \text{Sales}$$

$$\$45,000 + \quad \$15,000 \quad = \$60,000$$

This fundamental relationship is the basis for the equation $C + M = S$, which is applied by merchants to determine the selling price of their merchandise. (This is the subject of Chapter 13.)

These relationships, which are true for the amounts of money shown on the income statement, are also true of the percents for each item:

$$\text{Cost} + \text{Overhead} + \text{Net profit} = \text{Sales}$$

$$75\% + \underbrace{20\% + 5\%}\ = 100\%$$

$$\text{Cost} + \text{Margin} = \text{Sales}$$

$$75\% + 25\% = 100\%$$

The foregoing income statement was a highly simplified example. In actual practice, the basic topics used to calculate profit (net sales, cost of goods sold, and operating expenses) are broken down in detail, showing the contributing factors of each. The next example presents a more complete income statement.

Of the various topic breakdowns, only cost of goods sold needs particular mention. To the value of the inventory on hand at the beginning of the report period is added the total of all merchandise bought during the period (less returned merchandise and plus freight charges paid separately). This gives the total value of all merchandise that the company had available for sale during the period. Since some of this merchandise is not sold, however, the value of the inventory on hand at the end of the period must be deducted, in order to obtain the value of the merchandise that was actually sold.

Notice that, just as the individual amounts are added or subtracted to obtain subtotals, so may their corresponding percents be added or subtracted to obtain the percents of the subtotals. This is possible because each amount is represented as a percent of the same number (net sales).

The percents indicated in the right-hand column are the ones of particular interest to management and potential investors: (1) What percent of sales was the cost of goods? (2) What percent of sales was the gross profit (margin)? (3) What percent of sales was the operating expense (overhead)? But most important by far to the owners and possible investors is the final percent: (4) What percent of sales was the net profit? This is usually the first question asked by anyone considering investing money in a business.

Example 2

RANIER ELECTRONICS, INC.
Income Statement for Year Ending December 31, 20XX

Income from sales:					
Total sales		$123,000		102.5%	
Less: Sales returns and allowances		− 3,000		− 2.5	
Net sales			$120,000		100.0%
Cost of goods sold:					
Inventory, January 1		$ 31,300			
Purchases	$71,600				
Less: Returns and allowances	− 1,300				
Net purchases	$70,300				
Add: Freight in	+ 800				
Net purchase cost		+ 71,100			
Goods available for sale		$102,400			
Inventory, December 31		− 27,400			
Cost of goods sold			− 75,000		− 62.5
Gross profit on sales			$ 45,000		37.5%
Operating expenses:					
Office salaries		$ 24,000		20.%	
Rent and utilities		4,800		4.0	
Office supplies		1,200		1.0	
Insurance		720		0.6	
Advertising		1,800		1.5	
Depreciation		3,000		2.5	
Miscellaneous		+ 480		+ 0.4	
Total operating expenses			− 36,000		− 30.0
Net income from operations			$ 9,000		7.5%

Calculator Tip. Store the net sales, $120,000, into memory with [M+]. Then each financial item [÷] [MR] [%] → its corresponding percent. For instance, 9,000 [÷] [MR] [%] → 7.5. Recall that the [%] key includes the "=" process as it converts the result to its equivalent percent.

The percents derived in Examples 1 and 2 represent **vertical analysis.** That is, each separate item is represented as a percent of the total during a *single* time period. In order to analyze business progress and potential more effectively, however, recent figures should be compared with corresponding figures from the previous period (or periods). When comparisons are made between corresponding items during a *series* of two or more time periods, this is known as **horizontal analysis.**

The following example illustrates horizontal analysis. On this comparative income statement, both the dollar increase (or decrease) and the percent of change are shown. Figures for the earlier year are used as the basis for finding the percent of change, using the formula "_____% of original = change?"

Example 3

GOLDSTEIN'S INC.

Comparative Income Statement for Fiscal Years Ending June 30, 20X2 and 20X1

			INCREASE OR (DECREASE)		PERCENT OF NET SALES	
	20X2	20X1	AMOUNT	PERCENT	20X2	20X1
Income:						
Net sales	$275,000	$250,000	$25,000	10.0%	100.0%	100.0%
Cost of goods sold:						
Inventory, July 1	$ 60,000	$ 70,000	($10,000)	(14.3%)	21.8%	28.0%
Purchases	205,000	165,000	40,000	24.2	74.5	66.0
Goods available for sale	$265,000	$235,000	$30,000	12.8%	96.4%	94.0%
Inventory, June 30	75,000	60,000	15,000	25.0	27.3	24.0
Cost of goods sold	190,000	175,000	15,000	8.6	69.1	70.0
Gross margin	$ 85,000	$ 75,000	$10,000	13.3%	30.9%	30.0%
Expenses:						
Administration	$ 50,000	$ 45,000	$5,000	11.1%	18.2%	18.0%
Occupancy	7,500	5,000	2,500	50.0	2.7	2.0
Sales	16,000	10,000	6,000	60.0	5.8	4.0
Miscellaneous	1,500	2,500	(1,000)	(40.0)	0.5	1.0
Total expenses	75,000	62,500	12,500	20.0	27.3	25.0
Net income (before taxes)	$ 10,000	$ 12,500	($2,500)	(20.0%)	3.6%	5.0%

The following example (for net sales) illustrates how the preceding percents of change were computed. Net sales increased from $250,000 to $275,000, a "change" of $25,000. The "original" is the first year, 20X1. Thus,

_____% of Original = Change

_____% of $250,000 = $25,000

$$\frac{\cancel{250{,}000}\,x}{\cancel{250{,}000}} = \frac{25{,}000}{250{,}000}$$

$$x = 10\%$$

Notice that, whereas the "amounts" of increase or decrease may be added or subtracted to obtain the amounts of the subtotals, the "percents" of increase or decrease *cannot* be added or subtracted to obtain the percents of change in the subtotals. This is because each percent of change is based on a different number (the 20X1 amount of that particular item).

Calculator techniques . . . FOR EXAMPLE 3

After the 20X2 and 20X1 columns are totaled, you work *horizontally* to calculate the Amount and Percent of increase or decrease. To illustrate, this computation determines Inventory (July 1) and Purchases. Recall that parentheses indicate you should use a displayed number without reentering it. Although the [%] includes "=," you must round each resulting percent to the nearest tenth.

60,000 [−] 70,000 [=] ⟶ −10,000
(−10,000) [÷] 70,000 [%] ⟶ −14.285714;
205,000 [−] 165,000 [=] ⟶ 40,000
(40,000) [÷] 165,000 [%] ⟶ 24.242424; and so on.

Later, the two columns for Percent of Net Sales are calculated *vertically*, using the year's Total Sales from memory. For 20X2, this technique finds the percent of net sales represented by Inventory and Purchases:

275,000 [M+]
60,000 [÷] [MR] [%] ⟶ 21.818181;
205,000 [÷] [MR] [%] ⟶ 74.545454; and so on.

The 20X1 column is computed similarly, with net sales of $250,000 in memory. This technique also applies for comparative balance sheets, presented in the next section.

SECTION 2

BALANCE SHEET

The **balance sheet** shows the financial position of a business at a specific point in time. It is similar to a "snapshot" of the business. It freezes the action and shows how much is owned, how much is owed, and the book value of the owners' investment on a particular day, usually at the end of its **fiscal** or financial period. On the balance sheet are listed all of a business's **assets** (resources owned and money owed to the business) and all of its **liabilities** (debts that the business owes to someone else). The balance remaining after the liabilities are subtracted from the assets is the **net equity** of the owner(s). Thus, the assets must balance the liabilities and the owners' equity, or

$$\text{Assets} = \text{Liabilities} + \text{Net worth}$$

The general heading of assets on a balance sheet is ordinarily broken down into current and plant categories, as follows:

CURRENT ASSETS	PLANT ASSETS
Cash	Buildings
Accounts receivable	Furnishings
Notes receivable	Machinery
Office supplies	Equipment
Merchandise inventory	Land

Current assets are the resources that are owned by the business that will be converted to cash or consumed within one operating cycle or one year of the balance sheet date. Cash includes the cash on hand and cash in checking and savings accounts in the bank. **Accounts receivable** are accounts owed to the business by customers who have bought merchandise or services "on credit." Notes receivable indicate money owed by customers who have signed a written promise to pay by a certain date.

Plant assets are the long-term, tangible assets that are used repeatedly by the business, such as equipment. Equipment and other plant assets have a long life and can be used for several years. Because these assets are used for more than one year, their cost is spread over the useful life of the assets; that is, the plant assets are depreciated and are shown on the balance sheet at their current book value. The only exception is for land, which is not depreciated.

The liabilities of a business are separated according to current liabilities and long-term liabilities, as indicated:

CURRENT LIABILITIES	LONG-TERM LIABILITIES
Accounts payable	Mortgages
Notes payable	Long-term notes payable
Interest payable	Bonds
Taxes payable	

Current liabilities are debts that will be paid within one operating cycle or one year of the balance sheet date. They include **accounts payable,** which are accounts owed to others for goods or services bought on credit by the business. The notes payable, interest, and taxes are also items that are due within a short time.

Long-term liabilities are the debts that do not mature or come due until more than one year after the balance sheet date. A good example is a 25-year mortgage on a building. The long-term notes payable are those promissory notes that will be due after one year from the balance sheet date.

Other common terms used instead of net worth are **net ownership, net investment, proprietorship** (for an individually owned business), **owner's equity,** and **stockholders' equity** (for a corporation). Stockholders' equity is composed of two parts: **Contributed capital** represents the investments of owners in the business, and **retained earnings** represents the part of net profits that is kept for use in the business. Cash dividends are paid from a corporation's retained earnings.

The balance sheet in the following example again illustrates vertical analysis. That is, each asset is represented as a percent of the total assets; and each liability, as well as stockholders' equity (net worth), is presented as a percent of the total liabilities plus stockholders' equity (which is the same as the total assets).

Example 1

COLLIER-WHITWORTH CORP.
Balance Sheet, December 31, 20X1

Assets				Percent	
Current assets:					
Cash on hand		$ 500		0.2%	
Cash in bank		9,000		3.6	
Accounts receivable		42,000		16.8	
Inventory		23,500		9.4	
Total current assets			$ 75,000		30.0%
Plant assets:					
Plant site		$25,000		10.0%	
Building	$150,000				
Less: Depreciation	− 55,000				
		95,000		38.0	
Equipment	$75,000				
Less: Depreciation	− 20,000				
		55,000		22.0	
Total plant assets			175,000		70.0
Total assets			$250,000		100.0%
Liabilities and Stockholders' Equity					
Current liabilities:					
Accounts payable		$37,500		15.0%	
Notes payable		10,000		4.0	
Total current liabilities			$ 47,500		19.0%
Long-term liabilities:					
Mortgage			102,500		41.0
Total liabilities			$150,000		60.0%
Stockholders' equity			100,000		40.0
Total liabilities and stockholders' equity			$250,000		100.0%

The following illustrations show how the preceding percents were computed. The percents are found by dividing each amount by total assets:

(a) Total current assets are what percent of total assets?

$$\_\_\_\_\% \text{ of Total assets} = \text{Current assets}$$

$$\_\_\_\_\% \text{ of } \$250{,}000 = \$75{,}000$$

$$250{,}000x = 75{,}000$$

$$x = \frac{75{,}000}{250{,}000}$$

$$x = 30\%$$

(b) The plant site is what percent of total assets?

$$\_\_\_\_\% \text{ of Total assets} = \text{Plant site}$$

$$\_\_\_\_\% \text{ of } \$250{,}000 = \$25{,}000$$

$$250{,}000x = 25{,}000$$

$$x = \frac{25{,}000}{250{,}000}$$

$$x = 10\%$$

Just as horizontal analysis was helpful in studying the income statement, so also is it useful in determining the significance of a balance sheet. The next example presents such a balance sheet.

Example 2

THE DANVILLE CORP.
Comparative Balance Sheet, January 31, 20X2 and 20X1

			INCREASE OR (DECREASE)		PERCENT OF TOTAL ASSETS	
	20X2	20X1	AMOUNT	PERCENT	20X2	20X1
Assets						
Cash	$ 11,000	$ 14,000	($ 3,000)	(21.4%)	4.6%	7.0%
Accounts receivable	32,000	28,000	4,000	14.3	13.3	14.0
Inventory	57,000	38,000	19,000	50.0	23.8	19.0
Current assets	$100,000	$ 80,000	$20,000	25.0%	41.7%	40.0%
Plant assets (net)	140,000	120,000	20,000	16.7	58.3	60.0
Total assets	$240,000	$200,000	$40,000	20.0%	100.0%	100.0%
Liabilities and Stockholders' Equity						
Current liabilities	$ 45,000	$ 36,000	$ 9,000	25.0%	18.7%	18.0%
Long-term liabilities	90,000	74,000	16,000	21.6	37.5	37.0
Total liabilities	$135,000	$110,000	$25,000	22.7%	56.2%	55.0%
Common stock	$ 75,000	$ 65,000	$10,000	15.4%	31.3%	32.5%
Retained earnings	30,000	25,000	5,000	20.0	12.5	12.5
Total stockholders' equity	105,000	90,000	15,000	16.7	43.8	45.0
Total liabilities and stockholders' equity	$240,000	$200,000	$40,000	20.0%	100.0%	100.0%

In the preceding balance sheet, a horizontal analysis was done to find the "increase or decrease" amounts and percents. After the amount of increase or decrease is determined, the "percent of change" formula is used to find the percent of increase or decrease.

(a) What is the change in cash from 20X1 to 20X2?

$14,000	20X1
− 11,000	20X2
($ 3,000)	Decrease in cash

(b) What is the percent of change from 20X1 to 20X2?

_____% of Original year = Change

_____% of $14,000 = ($3,000)

$$14{,}000x = (3{,}000)$$

$$x = \frac{(3{,}000)}{14{,}000}$$

$$x = (21.4\%)$$

A vertical analysis is done for each year separately to find the Percent of Total Assets. The percents in the columns are found by dividing each item in a particular year by the total assets for that same year.

(c) Cash in 20X2 is what percent of total assets?

______% of Total assets = Cash

______% of \$240,000 = \$11,000

$$240{,}000x = 11{,}000$$

$$x = \frac{11{,}000}{240{,}000}$$

$$x = 4.6\%$$

(d) Cash in 20X1 is what percent of total assets?

______% of Total assets = Cash

______% of \$200,000 = \$14,000

$$200{,}000x = 14{,}000$$

$$x = \frac{14{,}000}{200{,}000}$$

$$x = 7\%$$

BUSINESS RATIOS

In addition to the analysis possible from single or comparative financial statements, a number of other business ratios are commonly used to provide a more thorough indication of a business's financial condition. These ratios are computed using figures readily available from the income statement and the balance sheet. The next example contains an explanation and illustration of some of the more common ratios.

Example 3 The six ratios given below are based on the balance sheet of The Danville Corp. in Example 2. Its income statement for 20X2 was as follows:

THE DANVILLE CORP.

Condensed Income Statement for Fiscal Year Ending January 31, 20X2

Net sales	$300,000	100.0%
Cost of goods sold	− 195,000	− 65.0
Gross margin	$105,000	35.0%
Operating expenses	− 90,000	− 30.0
Net income (before taxes)	$ 15,000	5.0%
Income taxes	− 5,000	− 1.7
Net income (after taxes)	$ 10,000	3.3%

Working Capital Ratio

The **working capital ratio** is often called the **current ratio,** because it is the ratio of current assets to current liabilities. This ratio indicates the ability of the business to meet its current obligations. The Danville Corp.'s 20X2 current ratio is computed as follows:

$$\frac{\text{Current assets}}{\text{Current liabilities}} = \frac{\$100{,}000}{\$45{,}000}$$

$$= \frac{2.2}{1} \quad \text{or} \quad 2.2:1$$

The Danville Corp.'s current ratio is 2.2:1. A current ratio of 2:1 is generally considered the minimum acceptable ratio; bankers would probably hesitate to lend money to a firm whose working capital ratio was lower than 2:1.

Acid-Test Ratio

The current ratio included all current assets. In order to provide some measure of a firm's ability to obtain funds quickly, the **acid-test** or **quick ratio** is computed. This is a ratio of the quick assets to the current liabilities. **Quick assets** (sometimes called **liquid assets**) are those which can be easily and quickly converted to cash, for example, accounts receivable, notes receivable, and marketable securities, as well as cash itself. (The inventory and supplies are not included in the acid-test ratio, because it often proves difficult to convert both of these assets to cash quickly.)

$$\frac{\text{Quick assets}}{\text{Current liabilities}} = \frac{\$43{,}000}{\$45{,}000}$$

$$= \frac{0.96}{1} \quad \text{or} \quad 1.0:1$$

As an indication of a firm's ability to meet its obligations, the acid-test ratio should be at least 1 : 1. The Danville Corp. barely meets this minimum requirement in 20X2. This may be only a temporary drop, however (from 1.2 : 1 the previous year). Observe that during 20X2, Danville greatly increased its inventory, with a corresponding increase in current liabilities and a drop in cash. If the increased inventory results in higher net sales and an improved cash position, the acid-test ratio may also soon improve.

Net Income (after Taxes) to Average Net Worth

Since businesses are organized to make profits, of primary concern to the owners is the return on their investment. The **ratio of net income (after taxes) to average net worth** indicates the rate of return on the average owners' equity. Using the 20X1 and 20X2 figures for total stockholders' equity of The Danville Corp., we obtain

$$\text{Average net worth} = \frac{\$90{,}000 + \$105{,}000}{2}$$

$$= \frac{\$195{,}000}{2}$$

$$= \$97{,}500$$

$$\frac{\text{Net income (after taxes)}}{\text{Average net worth}} = \frac{\$10{,}000}{\$97{,}500}$$

$$= 10.3\%$$

Net Sales to Average Total Assets

The **ratio of net sales to average total assets** gives some indication of whether a business is using its assets to best advantage—that is, of whether its sales volume is in line with what the company's capital warrants. This ratio (sometimes called the **total capital turnover**) would indicate inefficiency if it is too low.

Note. The "total assets" used here should exclude any assets held by the company which do not contribute to the business operation.

$$\text{Average total assets} = \frac{\$200{,}000 + \$240{,}000}{2}$$

$$= \frac{\$440{,}000}{2}$$

$$= \$220{,}000$$

$$\frac{\text{Net sales}}{\text{Average total assets}} = \frac{\$300{,}000}{\$220{,}000}$$

$$= \frac{1.4}{1} \quad \text{or} \quad 1.4{:}1$$

The ratio of net sales to average total assets at The Danville Corp. is 1.4:1; or the turnover of total capital was 1.4 times during 20X2.

Accounts Receivable Turnover

Accounts receivable turnover is the ratio of net sales to average accounts receivable. (Notes receivable, if they are held, should also be included with the accounts receivable.) This is a measure of how fast a business converts its accounts receivable into cash.

$$\text{Average accounts receivable} = \frac{\$28{,}000 + \$32{,}000}{2}$$

$$= \frac{\$60{,}000}{2}$$

$$= \$30{,}000$$

$$\frac{\text{Net sales}}{\text{Average accounts receivable}} = \frac{\$300{,}000}{\$30{,}000}$$

$$= 10 \text{ times}$$

Accounts receivable are not actually an asset to a business until they are collected. Prompt collection of the accounts receivable makes funds available to the business to meet its own obligations and to take advantage of cash discounts and special offers. Furthermore, the older an account becomes, the less the likelihood that it will be collected. Thus, a minimum of the firm's assets should be tied up in receivables, and they should turn over frequently.

Average Age of Accounts Receivable

This ratio, the **average age of accounts receivable,** in conjunction with the preceding one, indicates whether a business is keeping its accounts receivable up to date. The average age in days is found by dividing 365 days by the turnover in accounts receivable.

$$\frac{365}{\text{Accounts receivable turnover}} = \frac{365}{10}$$

$$= 36.5 \text{ days}$$

The average age of the accounts receivable is 36.5 days. If The Danville Corp. has an allowable credit period of 30 days, this average age of 36.5 days indicates that it is not keeping up with collections.

There are quite a few other ratios of interest to owners and management. One of these, the *equity ratio,* is already indicated on the balance sheet; the equity ratio is the ratio of owners' equity to total assets (43.8%). The *ratio of current liabilities to net worth* and the *ratio of total liabilities to net worth* are ratios based on balance sheet figures. Other common ratios relating to net sales are the *ratio of net sales to net worth,* the *total asset turnover* (the ratio of net sales to average total assets), and the **ratio of net sales to net working capital:**

$$\frac{\text{Net sales}}{\text{Current assets} - \text{Current liabilities}}$$

Also, the **inventory turnover** (the ratio of net sales to average inventory) is discussed in Section 3.

Additional ratios involving net income are the *ratio of net income to total assets* and the **rate (percent) of return per share of common stock:**

$$\frac{\text{Net income after taxes} - \text{Preferred dividends}}{\text{Average common stockholders' equity}}$$

Of interest also is the **book value of the stock** when there is only one type of stock; this is computed as follows:

$$\frac{\text{Owners' equity}}{\text{Number of shares of stock}}$$

Any number of other ratios could be computed if it were thought they would be useful.

Although the financial statements and the ratios that can be computed from them are of great value in evaluating the overall efficiency and financial condition of a business, such reports are not foolproof. Similar information from other businesses of the same type, as well as the general economic condition of the country as a whole, must also be taken into consideration as part of any thorough analysis by management or by a potential investor. On this basis, reports which originally seemed acceptable might appear less acceptable or might indicate that the business did unusually well in comparison with competing firms.

SPREADSHEETS AND FINANCIAL STATEMENTS

The problems for Sections 1 and 2 may be worked out manually or by a computer using Microsoft's Excel software. The disk that accompanies this text contains the spreadsheets for the following problems. Refer to the general instructions for spreadsheets in Chapter 4, pages 83–85.

If you work the problems via a computer, open the **folder** entitled *Chapter 10* on the data disk. A folder on a disk is similar to a folder in a file drawer—it is a place where related documents can be stored together. Each problem in Chapter 10 will be found in the *Chapter 10* folder. Once the appropriate file or problem is open, you can move from one cell to another with the *Tab* key, or you can use the mouse and click inside the cell. To enter a formula in a cell, first enter the equal sign (=). For addition, use the plus sign (+), and for subtraction, use the hyphen (-). For multiplication, use the asterisk (*), and for division, use the slash (/). *Note:* It does not matter if you use uppercase or lowercase letters when entering formulas.

Demo 1 Directions

Before working on the Sections 1 and 2 Problems, let's complete the Demo 1 Problem, an income statement, by completing the following steps:

- Start Excel.
- Click on *File* from the Menu Bar.
- Click on *Open* from the File Menu.
- In the *Open File* window, change the *Look-in* Box to *Drive A* (Floppy Disk 3.5″).
- The *Chapter 10* folder will be displayed in the *Name* Box, along with other folders on the disk. Click the *Chapter 10* folder, and click the *Open* button.
- All the documents or files will be shown. Click the *Demo 1* file. The spreadsheet for the Smithfield Supply Co. income statement appears on your screen. The next table will walk you through each step for completing the spreadsheet. Put a check mark in each row as you complete it.

TO FIND . . .	CLICK IN CELL . . .	KEY IN . . .	✓
Net sales	C5	=B3-B4	
What % of Net Sales is Sales	D3	=B3/C5	
What % of Net Sales is Sales Discount	D4	=B4/C5	
What % of Net Sales is Net Sales	E5	100%	
Cost of Goods Available for Sale	B9	=B7+B8	
Cost of Goods Sold	C11	=B9-B10	
What % of Net Sales is Cost of Goods Sold	E11	=C11/C5	
Gross Profit	C12	=C5-C11	
What % of Net Sales is Gross Profit	E12	=C12/C5	
Total Expenses	C21	=SUM(B14:B20)	
What % of Net Sales is Total Expenses	E21	=C21/C5	

TO FIND . . .	CLICK IN CELL . . .	KEY IN . . .	✓
What % of Net Sales is each expense	Formulas similar to that in E21. For each expense, enter its B# divided by C5 (for example, in D14: =B14/C5).		
Net Income From Operations	C22	=C12-C21	
What % of Net Sales is Net Income from Operations	E22	=C22/C5	
What % of Net Sales is Income Taxes	E23	=C23/C5	
Net Income After Taxes	C24	=C22-C23	
What % of Net Sales is Net Income After Taxes	E24	=C24/C5	

Note. You may have to format the cells. The cells containing dollar amounts should not include decimal places (cents). The cells containing percents should have percents rounded to tenths (one decimal place), as follows:

- If the cells for dollar amounts contain decimals, click on the *Decrease Decimal* button on the Tool Bar until the decimals are removed.

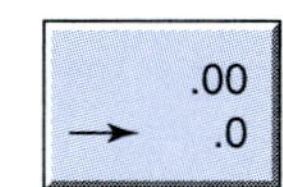

- If the percent cells contain the decimal equivalents, click or select each cell that contains a decimal; then click the *Percent* button (%) on the Tool Bar. While each cell is still selected, click the *Increase Decimal* button once.

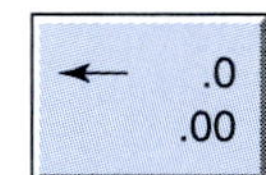

Even if you round percents to one decimal place for display purposes, the Excel spreadsheet stores each percent with full precision. Otherwise, the rounded percents might not add to a correct, known total. For example, suppose that current assets equaled 33.45% and plant assets equaled 66.55%. The total assets would then equal 100.0%. However, if the two percents had been rounded to 33.5% and 66.6%, then the total assets would be 100.1% if Excel were not maintaining full precision internally.

The Net Income After Taxes for Demo 1 should be $95,000, and its corresponding Percent of Net Sales should be 11.9%. As you work through these demonstration problems, fewer directions will be given so that you will start thinking of which formulas to use.

Demo 2 Directions

Follow the steps (bullets) given in Demo 1. At the last bullet, click on *Demo 2 Problem* instead of *Demo 1 Problem.* Remember, lowercase for SUM and column letters is acceptable in formulas, but you must start writing a formula with the equal sign (=). Demo 2 is a balance sheet problem.

TO FIND . . .	CLICK IN CELL . . .	KEY IN . . .	✓
Total Current Assets	D8	=SUM(C4:C7)	
Net Equipment	C12	=B10-B11	
Net Building	C15	=B13-B14	
Total Plant Assets	D17	=SUM(C12:C16)	
Total Assets	D18	=SUM(D8:D17)	
Total Current Liabilities	D24	=SUM(C22:C23)	
Total Liabilities	D27	=SUM(D24:D26)	
Total Equity	D31	=SUM(C29:C30)	
Total Liabilities and Equity	D32	=SUM(D27:D31)	
What % of Total Assets is Cash	E4	=C4/D18	
What % of Total Assets is Accounts Receivable	E5	=C5/D18	
What % of Total Assets is Supplies	E6	=C6/D18	
What % of Total Assets is Inventory	E7	=C7/D18	
What % of Total Assets is Total Current Assets	F8	=D8/D18	
The remaining percents	Corresponding cells	Formulas similar to those in E7 or F8	

Note. You may have to format the dollar cells by clicking on the *Comma* button and the *Decrease Decimal* button on the Tool Bar. Click on the $ button on the Tool Bar to add a dollar sign to the cells containing: Total Current Assets, Total Assets, Total Current Liabilities, Total Liabilities, and Total Liabilities and Equity. The percents should be formatted with the *Percent* button and the *Increase Decimal* button (or *Decrease Decimal* button) to one decimal place. The Total Assets should be $321,500, and the Total Equity is 68.3% of the Total Assets.

Demo 3 Directions

Start Excel and open *Demo 3* in the *Chapter 10* folder on *Drive A*. Follow the steps in Demo 1, except click the *Demo 3* file instead of the *Demo 1* file, as shown in the last bullet.

TO FIND . . .	CLICK IN CELL . . .	KEY IN . . .	✓
20X2 Goods Available for Sale	B9	=SUM(B7:B8)	
20X2 Cost of Goods Sold	B11	=B9-B10	
20X2 Gross Profit	B12	=B5-B11	
20X2 Total Expenses	B19	=SUM(B14:B18)	
20X2 Net Profit	B20	=B12-B19	

Now, determine the formulas for the Year 20X1 (found in Column C). There is a shortcut that you might like to try. You will use the fill handle to copy a formula to an adjacent cell. This is how it is done: Click B9. This encloses the cell with a heavy border. At the bottom right of the cell is a tiny square (fill handle). Point the mouse to this square until it turns into a cross-hair symbol (+). Click and drag the fill handle to C9; then release the mouse button. If you were successful, $133,000 will appear in C9. Continue to find the remaining 20X1 figures with this click-and-drag procedure, or enter formulas in each cell.

TO FIND . . .	CLICK IN CELL . . .	KEY IN . . .	✓
Amount of Increase (Decrease) for Net Sales	D5	=B5-C5	
The remainder of Column D	Use the copy procedure: Click in Cell D5; point the mouse to the fill handle in the right corner; click and drag the formula to D20. Release the mouse button.		
Percent of Increase (Decrease) for Net Sales	E5	=D5/C5	
The remaining %s in Column E	Corresponding Column E cells	Formulas similar to the one in E5, using D cells divided by C cells, respectively. You cannot copy this formula.	
What percent of 20X2 Net Sales is Net Sales	F5	100	
What percent of 20X2 Net Sales is January 1 Inventory	F7	=B7/B5	
The remaining %s in Column F	Corresponding Column F cells	Formulas similar to the one in F7, using B cells divided by B5 as the divisor.	
What percent of 20X1 Net Sales is Net Sales	G5	100	
What percent of 20X1 Net Sales is the January 1 Inventory	G7	=C7/C5, or click in F7 and use the fill handle to click/drag the formula to G7.	
The remaining %s in Column G	Repeat the above procedure for each adjacent cell in Column G.		

Net profit for 20X2 should be $10,000, and for 20X1, $12,600. The amount of increase for Cost of Goods Sold should be $4,000, and the percent of increase should be 2.4%. The Total Expense as a percent of Net Sales should be 43.4% for 20X2 and 45.5% for 20X1. Check the format of the cells. The percents should show one decimal place.

Demo 4 Directions

Follow the steps (bullets) given in Demo 1. At the last bullet, click on *Demo 4 Problem*. Remember, lowercase for SUM and column letters is acceptable in formulas, but you must start writing a formula with the equal sign (=).

TO FIND . . .	CLICK IN CELL . . .	KEY IN . . .	✓
20X2 Total Current Assets	B10	=SUM(B6:B9)	
20X1 Total Current Assets	B10 and use the fill handle to click/drag the formula to C10. Release the mouse button.		
20X2 Total Assets	B12	=SUM(B10:B11)	
20X1 Total Assets	B12 and copy the formula by clicking/dragging the fill handle to C12. Release the mouse button.		
20X2 Total Liabilities	B16	=SUM(B14:B15)	
20X1 Total Liabilities	B16 and copy the formula by clicking/dragging the fill handle to C16. Release the mouse button.		
20X2 Total Equity	B21	=SUM(B18:B20)	
20X1 Total Equity	Copy B21 formula to C21.		
20X2 Total Liabilities and Equity	B22	=B16+B21	
20X1 Total Liabilities and Equity	Copy B22 formula to C22.		
Amount of Increase (Decrease) for Cash	D6	=B6-C6	
Amount of Increase (Decrease) for remaining accounts	Copy the formula in D6 down through D22.		
Percent of Increase (Decrease) for Cash	E6	=D6/C6	
Percent of Increase (Decrease) for remaining accounts	Corresponding E cells	Formulas similar to the one in E6, using the adjacent D cells divided by the adjacent C cells.	
What % of 20X2 Total Assets is Cash	F6	=B6/B12	
Remaining %s for 20X2	Corresponding F cells	Formulas similar to the one in F6, using the adjacent B cells divided by B12.	
Remaining %s for 20X1	Copy the formulas from each cell in Column F to the adjacent cells in Column G.		

Remember, you may have to format the dollar answers so that they show no decimal places and format the percents to show one decimal place.

SECTIONS 1 AND 2 PROBLEMS

Complete the following financial statements, computing percents to the nearest tenth. Compute percents only at the points indicated with tinted boxes in Problems 1–4. In Problems 5–8, compute values on all rows.

1.

DAVIE & ASSOCIATES
Income Statement for Year Ending December 31, 20X2

Income from sales:				
Sales	$520,000		%	
Sales discount	20,000			
Net sales		$		%
Cost of goods sold:				
Inventory, January 1	$ 82,000			
Net purchases	150,000			
Goods available for sale	$			
Inventory, December 31	83,000			
Cost of goods sold				%
Gross profit		$		%
Operating expenses:				
Salaries	$200,000		%	
Depreciation	32,000			
Utilities	18,000			
Maintenance	16,500			
Advertising	10,000			
Insurance	8,500			
Office supplies	8,000			
Miscellaneous	7,000			
Total expenses				
Net income from operations		$		%
Income tax		14,000		
Net income after taxes		$		%

2.

PRENTICE SPORTING GOODS
Income Statement for Year Ending December 31, 20X2

Income from sales:				
Sales	$760,000		%	
Sales discount	40,000			
Net sales		$		%
Cost of goods sold:				
Inventory, January 1	$ 55,000			
Net purchases	300,000			
Goods available for sale	$			
Inventory, December 31	48,000			
Cost of goods sold				%
Gross profit		$		%
Operating expenses:				
Salaries	$196,000		%	
Rent	80,000			
Depreciation	18,000			
Insurance	12,000			
Utilities	10,000			
Office supplies	6,000			
Advertising	2,500			
Miscellaneous	1,500			
Total expenses				
Net income from operations		$		%
Income taxes		26,000		
Net income after taxes		$		%

3.

MADISON & CO.
Balance Sheet, December 20X1

Assets					
Current assets:					
Cash		$ 27,000		%	
Accounts receivable		32,000			
Notes receivable		20,000			
Inventory		45,000			
Total current assets			$		%
Plant assets:					
Building	$300,000				
Less: Accum. depreciation	30,000				
Net building		$		%	
Trucks	38,000				
Less: Accum. depreciation	18,000				
Net trucks					
Land		160,000			
Total plant assets					
Total assets			$		%
Liabilities and Owner's Equity					
Current liabilities:					
Accounts payable		$ 34,000		%	
Notes payable		46,000			
Total current liabilities			$		%
Long-term liabilities:					
Mortgage			229,000		
Total liabilities			$		%
Owner's equity			265,000		
Total liabilities and owner's equity			$		%

4.

TUCKER, CONNOR, AND RIVA CO.
Balance Sheet, December 31, 20X1

Assets					
Current assets:					
Cash		$ 36,000			
Accounts receivable		45,000			
Notes receivable		10,000			
Inventory		83,000			
Total current assets			$		%
Plant assets:					
Building	$450,000				
Less: Accum. depreciation	25,000				
Net building		$		%	
Land		200,000			
Total plant assets			$		
Total assets			$		%
Liabilities and Owner's Equity					
Current liabilities:					
Accounts payable		62,000		%	
Notes payable		20,000			
Total current liabilities			$		%
Long-term liabilities:					
Mortgage			350,000		
Total liabilities			$		%
Owner's equity			367,000		
Total liabilities and owner's equity			$		%

5.

NOOR FLOOR CO.
Comparative Income Statement for Years Ending Dec. 31, 20X2 and 20X1

			INCREASE OR (DECREASE)		PERCENT OF NET SALES	
	20X2	20X1	AMOUNT	PERCENT	20X2	20X1
Income:						
Net sales	$325,000	$300,000				
Cost of goods sold:						
Inventory, January 1	$ 90,000	$ 82,000				
Purchases	135,000	150,000				
Goods available for sale	$	$				
Inventory, December 31	81,000	90,000				
Cost of goods sold						
Gross profit	$	$				
Expenses:						
Salaries	$ 90,000	$ 60,000				
Rent	44,000	42,500				
Advertising	6,000	6,000				
Depreciation	5,000	5,500				
Utilities	3,000	2,800				
Miscellaneous	5,000	2,200				
Total expenses						
Net income	$	$				

6. **ELLETT HEATING AND AIR CONDITIONING CO.**
Comparative Income Statement for Years Ending Dec. 31, 20X2 and 20X1

	20X2	20X1	Increase or (Decrease)		Percent of Net Sales	
			Amount	Percent	20X2	20X1
Income:						
Net sales	$512,000	$500,000				
Cost of goods sold:						
Inventory, January 1	$100,000	$ 95,000				
Purchases	200,000	225,000				
Goods available for sale	$	$				
Inventory, December 31	96,000	100,000				
Cost of goods sold						
Gross profit	$	$				
Expenses:						
Salaries	$170,000	$149,000				
Office supplies	25,000	22,000				
Shop supplies	15,000	17,500				
Depreciation	10,000	11,500				
Utilities	8,000	6,300				
Advertising	2,000	1,000				
Miscellaneous	1,000	800				
Total expenses						
Net income	$	$				

7.

DAVIE & ASSOCIATES, INC.
Comparative Balance Sheet, December 31, 20X2 and 20X1

			INCREASE OR (DECREASE)		PERCENT OF TOTAL ASSETS	
	20X2	20X1	AMOUNT	PERCENT	20X2	20X1
Assets Current assets:						
Cash	$ 27,000	$ 33,000				
Accounts receivable	52,000	48,000				
Inventory	83,000	82,000				
Total current assets	$	$				
Total plant assets	300,000	322,000				
Total assets	$	$				
Liabilities and Equity						
Current liabilities	$ 79,000	$ 85,000				
Long-term liabilities	150,000	172,000				
Total liabilities	$	$				
Stockholders' equity:						
Preferred stock	$ 45,000	$ 45,000				
Common stock	80,000	78,000				
Retained earnings	108,000	105,000				
Total equity						
Total liabilities and equity	$	$				

8. **PRENTICE SPORTING GOODS**
Comparative Balance Sheet, December 31, 20X2 and 20X1

	20X2	20X1	INCREASE OR (DECREASE)		PERCENT OF TOTAL ASSETS	
			AMOUNT	PERCENT	20X2	20X1
Assets						
Current assets: Cash	$ 15,000	$ 12,000				
Accounts receivable	62,000	50,000				
Inventory	48,000	55,000				
Total current assets	$	$				
Total plant assets	98,000	90,000				
Total assets	$	$				
Liabilities and Owners' Equity						
Current liabilities	57,000	50,000				
Long-term liabilities	10,000	8,000				
Total liabilities	$	$				
Stockholders' equity:						
Preferred stock	25,000	25,000				
Common stock	70,000	68,000				
Retained earnings	61,000	56,000				
Total equity						
Total liabilities and equity	$	$				

9. Compute the following ratios or percents, based on the balance sheet for Madison & Co. in Problem 3. For Parts (c) and (d), assume that Madison & Co. had net sales of $300,000 and had issued only common stock of 5,000 shares. Round all answers to tenths.

a. Working capital ratio

b. Acid-test ratio

c. Ratio of net sales to net working capital

d. Book value of stock

10. Compute the following ratios or percents, based on the balance sheet for Tucker, Connor, and Riva Co. in Problem 4. For Parts (c) and (d), assume that the business had net sales of $425,000 and had issued only common stock of 8,000 shares.

a. Working capital ratio

b. Acid-test ratio

c. Ratio of net sales to net working capital

d. Book value of stock

11. Calculate the ratios or percents requested below, which pertain to the 20X2 business year of Davie & Associates, Inc. Use the income statement in Problem 1 and the balance sheet in Problem 7. Round all answers to tenths.

a. Working capital ratio

b. Acid-test ratio

c. Rate (percent) of net income (after taxes) to average net worth

d. Ratio of net sales to average total assets

e. Accounts receivable turnover

f. Average age of accounts receivable

12. Compute the ratios or percents requested below, concerning the fiscal year ending December 31, 20X2 for Prentice Sporting Goods. The figures needed to compute these ratios are found in Problems 2 and 8. Round all answers to tenths.

a. Working capital ratio

b. Acid-test ratio

c. Rate (percent) of net income (after taxes) to average net worth

d. Ratio of net sales to average total assets

e. Accounts receivable turnover

f. Average age of accounts receivable

INVENTORY TURNOVER

One indication of a business's stability and efficiency is the number of times per year that its inventory is sold—its **inventory** (or **stock**) **turnover.** This varies greatly with different types of businesses. For instance, a florist might maintain a relatively small inventory but sell the complete stock every few days. A furniture dealer, on the other hand, might stock a rather large inventory and have a turnover only every few months.

A business should compare the current period's inventory turnover with previous years' figures as well as with the inventory turnover of other similar businesses. The Village Bookstore in Example 2 will compare its turnover of 3.55 times with that of other bookstores. It will also compare the current year's turnover to its past years' inventory turnovers. If an inventory turnover is below average, the business may have too much inventory on hand. Thus, too much money will be tied to its inventory instead of being used elsewhere within the business.

Regardless of what might be considered a desirable turnover for a business, the methods used to calculate this turnover are similar. One business statistic that is used in this calculation is the average value of the store's inventory. Inventory may be evaluated at either its cost or its retail value, although it is probably more often valued at retail.

Average inventory for a period is found by adding the various inventories and dividing by the number of times the inventory was taken.

Example 1 Inventory at retail value at the Village Bookstore was $15,950 on January 1, $14,780 on June 30, and $17,270 on December 31. Determine the average inventory.

$$\begin{array}{r} \$15{,}950 \\ 14{,}780 \\ \underline{17{,}270} \\ 3\overline{)\$48{,}000} \end{array} = \$16{,}000 \qquad \text{Average inventory (at retail) for the year}$$

Stock turnover is found as follows:

$$\text{Inventory turnover (at retail)} = \frac{\text{Net sales}}{\text{Average inventory (at retail)}}$$

$$\text{Inventory turnover (at cost)} = \frac{\text{Cost of goods sold}}{\text{Average inventory (at cost)}}$$

Turnover is more frequently computed at retail than at cost. Many modern stores mark their goods only with retail prices, which makes it almost impossible to take inventory at cost. Another reason is that turnover at cost does not take into consideration the markdown on merchandise that was sold at a reduced price or the loss on goods that were misappropriated by shoplifters—a constant problem in many retail businesses. Turnover at retail and at cost will be the same only if all merchandise is sold at its regular price; ordinarily, the turnover at retail is slightly less than at cost.

Example 2 The bookstore in Example 1 had net sales during the year of \$56,800. What was the stock turnover for the year at retail?

$$\text{Inventory turnover} = \frac{\text{Net sales}}{\text{Average inventory (at retail)}}$$

$$= \frac{\$56{,}800}{\$16{,}000}$$

$$= 3.55 \text{ times} \qquad \text{(at retail)}$$

Example 3 The income statement of the store above shows that its \$43,200 "cost of goods sold" is 75% of net sales. Find the bookstore's inventory turnover at cost.

The average inventory at retail (sales) value was \$16,000. In general, the store's goods cost 75% of sales value. Thus, the average inventory cost is computed as follows:

$$\text{Cost} = 75\% \text{ of sales}$$

$$C = 0.75 \times \$16{,}000$$

$$C = \$12{,}000$$

The average inventory at cost was \$12,000. Thus, inventory turnover at cost is

$$\text{Inventory turnover} = \frac{\text{Cost of goods}}{\text{Average inventory (at cost)}}$$

$$= \frac{\$43{,}200}{\$12{,}000}$$

$$= 3.6 \text{ times} \quad \text{(at cost)}$$

Note. When inventory is taken at retail, the firm must then calculate the cost of its inventory for the financial statements using historical percentages (similar to the illustration above).

SECTION 3 PROBLEMS

1. Determine the average inventory at retail for a convenience store. The inventory counts at four different times during the period were: \$75,000, \$63,000, \$72,000, and \$84,000.
2. Inventories were taken at retail four times during the year at an automotive parts store as follows: \$120,000, \$104,000, \$116,000, and \$124,000. What was the average inventory at retail?

3. The first quarterly inventory of the year taken at cost was $74,000. The other inventories for the year were taken at retail and valued at $100,000, $104,000, and $102,000.
   a. What was the first inventory at retail if the store adds 40% of cost to obtain its selling price?
   b. What was the average inventory at retail?
4. The first inventory of the year taken at cost was $180,000. The other three inventories were taken at retail valued at $236,000, $245,000, and $200,000.
   a. What was the first inventory at retail if the store adds 30% of cost to obtain its selling price?
   b. What was the average inventory at retail?
5. Net sales for the store in Problem 1 were $279,300, and the cost of goods sold was $176,400. What was the inventory turnover at retail?
6. Net sales for the store in Problem 2 were $430,000, and the cost of goods sold was $340,000. What was the inventory turnover at retail?
7. Refer to the convenience store in Problems 1 and 5. The value of the store's inventory at cost is usually 60% of the value of their inventory at retail. Using this information,
   a. Determine their average inventory at cost.
   b. Calculate their inventory turnover at cost.
8. An analysis of the pricing structure used by the auto parts store in Problems 2 and 6 shows that the value of its inventory at cost is typically 75% of the value of its inventory at retail. Using this information,
   a. Determine the average inventory at cost.
   b. Calculate the stock turnover at cost.
9. A business took the following inventories at retail: $116,000, $124,000, $125,000, and $126,000. Their income statement showed net sales to be $392,800, and the cost of goods sold was $250,410. What was their inventory turnover at retail?
10. A shop took the following inventories at retail: $68,000, $72,000, $70,000, and $64,000. Its income statement reveals net sales to be $239,750 and cost of goods sold to be $145,000. What was its inventory turnover at retail?
11. Refer to the business in Problem 9. If their inventory value at cost is typically equal to 60% of their inventory value at retail,
    a. What is their average inventory at cost?
    b. What is their inventory turnover at cost?
12. Refer to the business in Problem 10. If its inventory value at cost is typically equal to 70% of its inventory at retail,
    a. What is the average inventory at cost?
    b. What is the inventory turnover at cost?

SECTION 4

INVENTORY VALUATION

We have used inventory values without previously considering the process by which the inventory was evaluated. Some firms keep a **perpetual inventory,** an inventory system that is updated constantly with each purchase and sale. The firm knows at all times the exact content and value of its inventory. A method of keeping up with the

Figure 10-1
Used with permission of Hunt-Wesson, Inc.

number of units on hand is the **uniform product code (UPC)**, an electronic bar code found on most products that identifies the product and its price. This code is "read" by electronic scanners at the checkout station. The cashier passes the code over the scanner that reads the code. The computer keeps track of the inventory by subtracting each sold item from the number of units on hand. There are advantages of using the bar codes to both the customer (faster checkouts) and to the business (automatic inventory control).

Other firms, however, keep a **periodic inventory.** Under this system, the exact content and value of the inventory are updated at the end of the period or periodically. With both methods, it is still necessary for a **physical inventory** to be taken during the fiscal period. The physical inventory is a count of the units of each product on hand. This number is then compared to the book inventory.

There are four common methods by which a periodic inventory may be taken: **specific identification; weighted average; first-in, first-out;** and **last-in, first-out.**

If each item is cost-coded, the periodic inventory may be taken by the **specific identification** method, where the inventory value is simply the total of the individual items. Evaluating items is not always an obvious process, however; inventories often contain many indistinguishable items that have been purchased at different times and for different amounts (such as valves in a plumbing supply house). If the items are small, it may be impractical to cost-code each item; thus, there would be some question about what value to attach to each. (This is a primary reason for the popularity of taking inventory at retail value—the current selling price of merchandise is always known.) One solution is to use the **weighted average method** of inventory valuation, where an average cost is computed like the weighted means we studied in Chapter 4.

The **first-in, first out (FIFO) method** is probably most commonly used. This method assumes that those items that were bought first were also sold first and that the items still in the inventory are those that were purchased most recently. This method conforms to actual selling practice in most businesses. The number of items on hand are valued according to the invoice prices most recently paid for the corresponding number of items.

The **last-in, first-out (LIFO) method** assumes that the items most recently bought are those items that have been sold and that the items still in stock are the oldest items. Although this method is contrary to most actual selling practice, there is certain justification from an accounting standpoint: On the income statement, the LIFO method uses "current" costs (which are usually higher) for the cost of goods sold and thus produces a smaller gross profit, which many accountants argue is more realistic in an inflationary economy. Accepted accounting procedure requires that income statements specify the method(s) used to evaluate the inventories listed.

Example 1 Assume that a new valve came on the market this year and that the following purchases were made during the year: 15 at \$5, 9 at \$6, and 12 at \$8. Twenty valves were sold (leaving 16 in inventory) for a total net sales of \$300. Evaluate the ending inventory by the following methods: (a) weighted average, (b) FIFO, and (c) LIFO. (d) Compare the effect of each inventory method on the income statement.

(a) **Weighted Average Method**

15 @ \$5 =	\$ 75	
9 @ \$6 =	54	
+12 @ \$8 = +	96	
36 items purchased	\$225	Total cost

$$\frac{\$225}{36} = \$6.25 \qquad \text{Average cost}$$

16 items @ \$6.25 = \$100	Inventory value at cost by weighted average method

(b) **FIFO Method**

The 16 items remaining are assumed to be the last 12 plus 4 from the previous order:

12 @ \$8 =	\$ 96	
+ 4 @ \$6 = +	24	
16	\$120	Inventory value at cost by FIFO

(c) **LIFO Method**

The 16 items remaining are assumed to be the first 15 plus 1 from the second order:

15 @ $5 =	$75	
+ 1 @ $6 = +	6	
16	$81	Inventory value at cost by LIFO

(d) **Income Statements**

	WEIGHTED AVERAGE		FIFO		LIFO	
Net sales		$300		$300		$300
Cost of goods sold:						
Beginning inventory	$ 0		$ 0		$ 0	
Purchases	+ 225		+ 225		+ 225	
Goods available	$225		$225		$225	
Ending inventory	− 100		− 120		− 81	
Cost of goods		− 125		− 105		− 144
Gross profit		$175		$195		$156
Gross profit as a % of net sales		58.3%		65%		52%

SECTION 4 PROBLEMS

Evaluate the following inventories by (1) weighted average, (2) FIFO, and (3) LIFO. If necessary, round the weighted average cost per unit to the nearest cent.

	PURCHASES	ENDING INVENTORY	PURCHASES	ENDING INVENTORY
1.	a. 10 @ $5.10	15 units	b. 25 @ $7.40	40 units
	30 @ 5.20		45 @ 7.50	
	20 @ 5.25		30 @ 7.65	
2.	a. 50 @ $8.60	35 units	b. 18 @ $25.00	28 units
	40 @ 8.65		22 @ 25.10	
	30 @ 9.00		26 @ 25.05	

3. An electronics store purchased television sets from a manufacturer for the following prices during the year: 31 @ $116; 32 @ $118; 40 @ $120; 44 @ $121; 28 @ $125; and 25 @ $126. At the end of the year, there were 10 sets left in inventory. Evaluate the inventory cost by the following methods:

a. Weighted average

b. FIFO

c. LIFO

4. A retail department store purchased coffee makers from a manufacturer at various times during the year at the following prices: 17 @ $17.30; 16 @ $17.40; 15 @ $18.45; and 24 @ $18.50. Evaluate the inventory cost by the following methods, assuming that 20 coffee makers are left in the inventory at the end of the accounting period.
   a. Weighted average
   b. FIFO
   c. LIFO
5. On June 30, the end of the semiannual accounting period at Raylon Co., there were 60 units in inventory. The beginning inventory and the purchases for the Raylon Co. were as follows:

January 1 inventory	40 @ $35
March 3 purchases	80 @ $34
May 5 purchases	60 @ $32
June 10 purchases	70 @ $30

   Determine the cost of the ending inventory by the following methods:
   a. Weighted average
   b. FIFO
   c. LIFO
6. On April 1, the end of the quarter, there were 26 units in the inventory at Mabel's Emporium. The beginning inventory and purchases for the store were as follows:

January 1 inventory	25 @ $52
February 15 purchases	22 @ $52.20
March 9 purchases	13 @ $52.30

   Determine the cost of the ending inventory by the following methods:
   a. Weighted average
   b. FIFO
   c. LIFO
7. Net sales for the electronics store in Problem 3 were $40,000. Complete a partial income statement that shows gross profit (dollars and percents) for each of the previous inventory methods.
8. Net sales for the coffee makers in Problem 4 were $5,000. Complete a partial income statement that shows gross profit (dollars and percent) for each of the preceding inventory methods.

CHAPTER 10 GLOSSARY

Accounts receivable (or payable). Money owed to a business (or owed by a business) for merchandise or services bought on credit.

Accounts receivable turnover. The ratio of net sales to average accounts receivable.

Acid-test (or quick) ratio. The ratio of quick assets to current liabilities; the minimum acceptable ratio is 1 : 1.

Assets. Resources owned by a business.

Average age of accounts receivable. In days: 365 divided by the accounts receivable turnover.

Balance sheet. A financial statement indicating how much a business is worth on a specific date: Assets = Liabilities + Net worth.

Book value of stock. The owners' equity divided by the number of shares of stock.

Contributed capital. The portion of stockholders' equity derived from the owners' investment in the business.

Current assets. The resources that are owned by a business and that will be converted to cash or consumed within one operating cycle or one year of the balance sheet date.

Current liabilities. The debts of a business that will be paid within one operating cycle or one year of the balance sheet date.

First-in, first-out (FIFO) method. The inventory costing method which assumes that the first items purchased were also sold first and that present inventory items are the ones most recently purchased.

Gross profit (margin or markup). The amount remaining when cost of merchandise is subtracted from sales income: $S - C = M$.

Horizontal analysis. Financial statement analysis in which each item in one year is compared (amount and/or percent) with that same item in the previous year.

Income statement. A financial statement showing a business's sales, expenses, and profit (or loss) during a specified time period: Sales − Cost of goods − Expenses = Net profit.

Inventory turnover. The number of times per year that the average inventory (stock) is sold:

$$\begin{pmatrix}\text{Turnover}\\ \text{at cost}\end{pmatrix} = \frac{\text{Cost of goods sold}}{\text{Average inventory (at cost)}}$$

$$\begin{pmatrix}\text{Turnover}\\ \text{at retail}\end{pmatrix} = \frac{\text{Net sales}}{\text{Average inventory (at retail)}}$$

Last-in, first-out (LIFO) method. The inventory costing method which assumes that the items most recently bought are the ones that have been sold and that present inventory contains the oldest items.

Liabilities. Money that a business owes to others.

Long-term liabilities. Debts of a business that do not mature or come due until more than one year after the balance sheet date.

Margin or markup. (See "Gross profit.")

Net income (or net profit). The amount remaining after both cost of goods and operating expenses are deducted from sales income.

Net income (after taxes) to average net worth. A ratio that indicates the owners' return on investment.

Net sales to average total assets. A ratio indicating whether sales volume is in line with the capital assets of a business.

Net worth (net ownership or net investment). The owners' share of a firm's assets; the amount remaining when liabilities are subtracted from assets.

Operating expenses (or overhead). Business expenses other than the cost of materials or merchandise. (Examples: Salaries, occupancy expenses, advertising, office supplies, etc.)

Owners' equity. (See "Net worth.")

Periodic inventory. An inventory taken at a regular interval by a physical count of items.

Perpetual inventory. An inventory that is continuously updated by recording the purchase and subsequent sale of each item.

Plant assets. The long-term, tangible assets that are used repeatedly by a business in the production of income.

Quick (or liquid) assets. Assets that can easily and quickly be converted to cash; all current assets except inventory and supplies.

Rate of return per share of common stock. Net income after taxes minus the preferred dividend, divided by the average common stockholders' equity.

Ratio of net sales to net working capital. Net sales divided by (the total current assets minus the total current liabilities).

Retained earnings. That part of net profit that is kept for use in the business (rather than being distributed to the owners).

Specific identification. The inventory costing method in which each item is cost-coded and the total inventory is the sum of these costs.

Stockholders' equity. Net worth owned by stockholders of a corporation.

Total capital turnover. (See "Net sales to average total assets.")

Uniform Product Code (UPC). The electronic bar code on a product that identifies the product and its price.

Vertical analysis. Financial statement analysis (pertaining to one year only) in which each item is compared to one significant figure on the statement (that is, as a percent of net sales on the income statement or as a percent of total assets on the balance sheet).

Weighted average method. The inventory costing method whereby each item is valued at the weighted average (or mean) of all costs paid during the time interval.

Working capital (or current) ratio. A ratio of current assets to current liabilities; the minimum acceptable ratio is 2 : 1.

CHAPTER 11

DISTRIBUTION OF PROFIT AND LOSS

OBJECTIVES

Upon completion of Chapter 11, you will be able to:

1. Define and use correctly the terminology associated with each topic.
2. Compute the dividend to be paid per share of common and preferred stock, given the total dividend declared and information about the number and types of stock (Section 1: Examples 1–5; Problems 1–22).
3. Compute the earnings per share on common stock, given the net income and information about the number and types of stock (Section 1: Example 5; Problems 3, 4, 21, 22).
4. Distribute profits (or losses) to the members of a partnership, given the total profit (or loss) and information about each partner's investment. Distributions will be according to (Section 2):
   a. Agreed ratios or no agreement (Examples 1–3, 5; Problems 1–12, 15–22)
   b. Average investments per month (Example 4; Problems 13–16)
   c. Salaries (Examples 5, 6; Problems 1, 2, 7, 8, 11, 12, 17–22)
   d. Interest on investments (Example 6; Problems 1, 2, 15–22)
   e. Combinations of these methods (Examples 5, 6; Problems 1, 2, 7, 8, 11, 12, 15–22).

Within our system of free enterprise, businesses are operated for the ultimate purpose of making a profit. After the amount of net profit (or loss) of a business has been determined, it is necessary to apportion that amount among the owners. Methods of distribution depend largely upon the type of business. In general, there are three categories of private business organization: **sole proprietorship, partnership,** and **corporation.**

SOLE PROPRIETORSHIP

Sole proprietorship is the simplest form of business. A single person makes the initial investment and bears the entire responsibility for the business. If a profit is earned, the entire profit belongs to the individual owner. By the same token, the sole proprietor must assume full responsibility for any loss incurred. (One principal disadvantage of this form of business is that the owner's liability extends even to personal possessions not used in business; for instance, in many states the proprietor's home might have to be sold to discharge business debts.)

Sole proprietorships are usually small businesses, since they are limited to the resources of one person. There often is no formal organization or operating procedure; an individual with sufficient skill or financial resources simply goes into business by making services available to potential clients.

Since all profit (or loss) resulting from a sole proprietorship belongs to the individual owner, there is no mathematical distribution required. Hence, the problems in this section will be limited to those concerning partnerships and corporations.

PARTNERSHIPS

A **partnership** (or general partnership) is an organization in which two or more persons engage in business as co-owners. A partnership may provide the advantages of increased investments and more diversified skills, as well as more manpower available to operate the business. (For example, individuals A, B, and C might form a partnership because A had the funds to start the business, B had the skill and experience required to manufacture their product, and C was an expert in sales and marketing. Thus, by pooling their resources, both financial and personal, the partners can form a profitable business; however, none of the partners could have established a profitable business independently.) Partnerships often include a **silent partner**—one who takes no active part in the business operations, whose contribution to the partnership is entirely financial, and whose very existence as a partner may not be known outside the business.

A partnership may be entered into simply by mutual consent. To avoid misunderstanding and disagreement, however, all partners should sign a legal agreement (called *articles of partnership*) setting forth the responsibilities, restrictions, procedures, method of distribution of profits and losses, and so forth, of the partnership. (For instance, the abrupt withdrawal of a partner, if not prohibited by written agreement, might place the entire partnership in bankruptcy.)

Taking into consideration the contribution made by each partner, a partnership is permitted to distribute profits or losses in any way that is acceptable to the partners involved. In the absence of any formal written agreement, the law would require that profits be shared equally among the partners. Despite whatever agreement the partners may have among themselves, however, in most states the partners are responsible "jointly and separately" for the liabilities of the partnership. This means that if the personal resources of the partners together (jointly), when calculated according to their terms of agreement, are not sufficient to meet the partnership's liabilities, then a single partner with more personal wealth will have to bear the liabilities of the partnership individually (separately). Also, a business commitment made by any partner, even if it is not within that person's agreed area of responsibility, is legally binding upon the entire partnership.

A partnership is automatically dissolved whenever there is any change in the partners involved, such as when a partner withdraws from the business, sells to another person, or dies, or when an additional person enters the partnership. However, a new partnership may be created immediately to reflect the change, without any lapse in business operations.

Almost all states also allow a form of partnership called a **limited partnership,** which requires a legal document to establish. Certain members, known as *limited partners,* contribute only capital funds and take no active part in managing the endeavor. One or more *general partners* assume full responsibility for managing the business, as well as for its financial obligations. The distinguishing characteristic of a limited partnership is that the financial responsibility of the limited partners is restricted to the amount of capital they have invested. A common example of limited partnerships is in real estate projects, where numerous limited partners make investments and a few general partners develop, build, or sell the property. Our study, however, will concentrate on the general partnership, where all members function equally as co-owners.

CORPORATIONS

The **corporation** is the most complex form of business. It is considered to be an artificial entity, created according to state laws. Its capital is derived from the sale of **stock certificates,** which are certificates of ownership in the business. A person's ownership in a corporation is measured according to the number of shares of stock that (s)he owns. Figure 11-1 illustrates a stock certificate.

The corporation offers a distinct advantage over both the sole proprietorship and the partnership in that its liability extends only to the amount invested in the business by the stockholders. That is, the personal property of the owners (**stockholders**) may not be confiscated to meet business liabilities. (It should be noted that even small businesses may be incorporated in order to provide this protection.) The corporation also differs from the sole proprietorship and the partnership in that the firm must pay income taxes on its net operating income.

When a corporation is formed, it receives a *charter* from the state, authorizing the operation of business, defining the privileges and restrictions of the firm, and

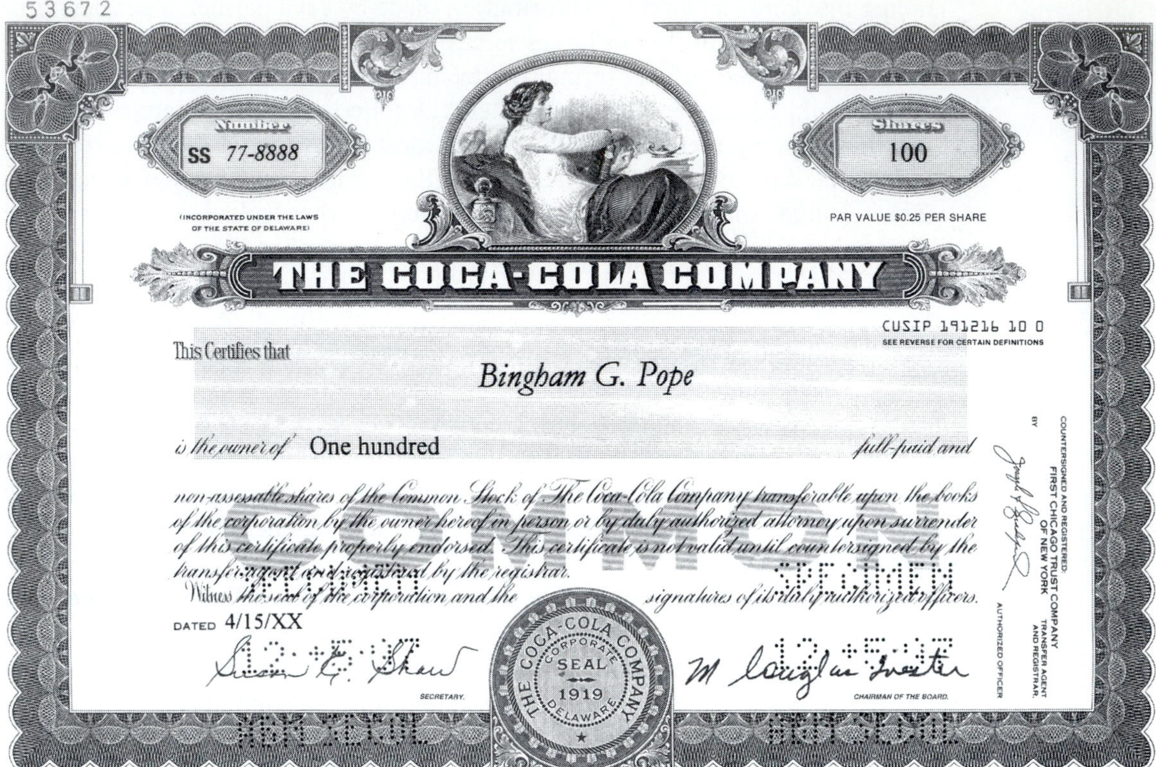

Courtesy of Coca-Cola Company

Figure 11-1 Specimen Stock Certificate

specifying the number of shares and type of stock that may be issued. The stockholders elect a **board of directors,** which selects the *administrative officers* of the corporation. Although it is ultimately responsible to the stockholders, the board actually makes decisions regarding policy and the direction that the business will take, and the officers are responsible for carrying out the board's decisions. The board of directors also makes the decision to distribute profits to the stockholders; the income declared on each share of stock is called a **dividend.**

When a corporation issues stock, the certificates may be assigned some value, called the **par** or **face value.** The par value may be set at any amount, although it is typically some round number. Stock is sold at par value or above par when a corporation is first organized. After the firm has been in operation for some time, the *market value* (the amount for which stockholders would be willing to sell their stock) will increase or decrease, depending on the prosperity of the business, and the par value becomes insignificant. Thus, stock is often issued without any face value, in which case it is known as **non-par-value** or **no-par-value stock.**

There are two main categories of stock: **common stock** and **preferred stock.** Since there are fewer methods for distributing profits to stockholders than to partners, we shall consider first the distribution of corporate profits.

SECTION 1

DISTRIBUTION OF PROFITS IN A CORPORATION

COMMON STOCK

Common stock carries no guaranteed benefits in case a profit is made or the corporation is dissolved. It does entitle its owner to one vote in a stockholders meeting for each share (s)he owns. Many corporations issue only common stock. In this case, distributing the profits is simply a matter of dividing the profit to be distributed by the **stock outstanding** (the number of shares of stock that have been issued).

Example 1 The board of directors of the Wymore Corp. authorized $258,000 of the net profit to be distributed among its stockholders. If the corporation has issued 100,000 shares of common stock, what is the dividend per share?

$$\frac{\text{Total dividend}}{\text{Number of shares}} = \frac{\$258{,}000}{100{,}000} = \$2.58 \text{ per share}$$

A dividend of $2.58 per share of common stock will be declared.

PREFERRED STOCK

The charters of many corporations also authorize the issuance of **preferred stock.** This stock has certain advantages over common stock; most notably, preferred stock carries the provision that dividends of a specified percent of par value (or of a specified amount per share, on non-par-value stock) must be paid to preferred stockholders before any dividend is paid to common stockholders. For example, a share of 6%, $100 par-value stock would be expected to pay a $6 dividend each year (usually in quarterly payments of $1.50 each).

Also, in the event that the corporation is dissolved, the corporation's assets may first have to be applied toward refunding the par value of preferred stock, before any compensation is returned to common stockholders. Holders of preferred stock usually do not have a vote in the stockholders meeting. Also, most preferred stock is **nonparticipating,** which means it cannot earn dividends above the specified rate. (However, some preferred stock is **participating,** which permits dividends above the stated percent when sufficient profits exist. Participating stock is frequently restricted so that additional earnings above the specified rate are permitted only after the common stockholders have received a dividend equal to the special preferred dividend.)

Even when a net profit exists, a board of directors may not think it advisable to declare a dividend. Thus, the owners of a corporation may receive no income on their investment during some years. **Cumulative preferred stock** provides an exception to this situation, however. As the name implies, unpaid dividends accumulate; that is, dividends due from previous years when no dividend was declared, as well as dividends from the current year, must first be paid to owners of cumulative preferred stock before any dividends are paid to other stockholders.

As an alternative to paying cash dividends, boards of directors may declare *stock dividends,* where each stockholder receives an additional share for each specified number of shares already owned. The stockholders thus obtain an increased investment but do not have to pay income taxes as they would on cash dividends. When a stock dividend is declared, the dividend for many stockholders would include a fractional share. Thus, the board assigns a value to each whole share (often slightly less than market value) and either pays the stockholder this value for any fractional share or else allows the stockholder to purchase the remaining portion of another whole share.

Example 2 Artcraft Sales, Inc., has 1,000 shares of 5%, \$100 par-value preferred stock outstanding and 10,000 shares of common stock. The board declared a \$50,000 dividend. What dividends will be paid on (a) the preferred stock and (b) the common stock?

(a) Dividends on preferred stock are always computed first, as follows:

$$\text{Total preferred dividend} = \underbrace{\% \times \text{Par value}} \times \begin{pmatrix}\text{Number of}\\ \text{preferred shares}\end{pmatrix}$$

$$= \begin{pmatrix}\text{Dividend}\\ \text{per share}\end{pmatrix} \times \begin{pmatrix}\text{Number of}\\ \text{preferred shares}\end{pmatrix}$$

$5\% \times \$100$ par value $= \$5$ — Dividend per share of preferred stock

$\$5$ per share $\times$ 1,000 shares $= \$5{,}000$ — Required to pay dividends on all preferred stock

(b)

\$50,000	Total dividend
− 5,000	Dividend on all preferred stock
\$45,000	Dividend allotted to common stock

$$\frac{\text{Total common stock dividend}}{\text{Number of common shares}} = \frac{\$45{,}000}{10{,}000} = \$4.50 \quad \text{Dividend per share of common stock}$$

Hence, the preferred stockholders receive their full dividend of \$5 per share, and a dividend of \$4.50 per share will be paid to common stockholders.

Example 3 Because of low sales during the following year of operations, Artcraft Sales, Inc., declared a dividend of only \$4,500. If the corporation still has 1,000 shares of 5%, \$100 par-value preferred and 10,000 shares of common outstanding, what dividend per share will be paid?

Because \$4,500 is not even enough to pay the required \$5,000 preferred dividend (5% × \$100 × 1,000 shares), the \$4,500 dividend will be divided by the 1,000 shares of preferred stock.

$$\frac{\$4{,}500}{1{,}000} = \$4.50 \quad \text{Per share for preferred}$$

0 Per share for common

Example 4 No dividend was paid last year by McGill, Inc. A dividend of \$325,000 has now been declared for the current year. Stockholders own 100,000 shares of common stock and 25,000 shares of 6%, \$50 par-value cumulative preferred stock. How much dividend will be paid on each share of stock?

Dividends for 2 years must first be paid on the cumulative preferred stock.

6% × \$50 par value =	\$3	Dividend per year on each share of cumulative preferred stock
	× 2	Years
	\$6	Dividend due on each share of cumulative preferred stock

\$6 per share × 25,000 shares = \$150,000 Payable to preferred stockholders

\$325,000	Total dividend
− 150,000	Dividend on preferred stock
\$175,000	Total dividend on common stock

$$\frac{\$175{,}000}{100{,}000 \text{ shares}} = \$1.75 \quad \text{Dividend per share of common stock}$$

Thus, a 2-year dividend of \$6 per share will be paid to owners of the cumulative preferred stock, and \$1.75 per share will go to the common stockholders.

EARNINGS PER SHARE

As a means of evaluating a firm's past and potential future performance, investors want to know the **earnings per share** of common stock. This figure represents the amount of net income earned by common shareholders during the year. The number is usually different from the cash dividend per share for common stock. If the board of directors declares a cash dividend for less than the total net income, the earnings per share and dividends per share will be different. Also, if the number of common shares outstanding changes during the year, the two numbers will be different. When the number of shares changes, a weighted average calculation is required to determine the average number of shares. An in-depth discussion of earnings per share is beyond the scope of this text. We will limit the consideration of earnings per share to simple capital structures and situations in which the common shares outstanding do not change. The calculation for the earnings per share is

$$\frac{\text{Net income} - \text{Preferred dividend}}{\text{Common shares outstanding}}$$

Example 5 The Powers Co. had a net income of \$1,000,000 last year. The board of directors declared a cash dividend of only \$600,000, retaining the remainder for future plant expansion. There were 50,000 shares of common stock outstanding and 75,000 shares of 8%, \$50 par-value preferred stock.

(a) What was the dividend per share?
(b) What was the earnings per share on common stock?

(a) $8\% \times \$50 = \$4$ Dividend on each preferred share

$\$4 \times 75{,}000 \text{ shares} = \$300{,}000$ Total preferred dividend

\$600,000	Total dividend
− 300,000	Total preferred dividend
\$300,000	Allocation for common dividend

$\frac{\$300{,}000}{50{,}000} = \$6$ Dividend on each common share

(b) $\frac{\text{Net income} - \text{Preferred dividend}}{\text{Common shares outstanding}} = \text{Earnings per share}$

$\frac{\$1{,}000{,}000 - \$300{,}000}{50{,}000} = \$14$ Earnings per share

MUTUAL FUNDS

Stocks, bonds, and other financial investments that are issued by corporations and government bodies and are bought and sold in securities markets are called **securities.** Instead of investing in one security, an investor may wish to diversify with ownership in many different securities by buying shares of a mutual fund. A **mutual fund** is a pooling of money from many investors into a portfolio investment that owns a wide selection of stock and often bonds or other income-producing securities. Many mutual funds specialize in a particular kind of investment, such as income-producing stocks or stocks of companies believed to have outstanding growth potential.

Today, the most popular security investment in the United States is the mutual fund. The popularity of mutual funds is due to their professional management and the reduction of the individual investor's risk, since the investment is spread among many different securities. The investor earns dividends on stocks and interest on bonds that are owned through the mutual fund. Since the investor owns only fractions of numerous stocks and bonds (instead of personally owning whole securities), the investor's return on each security in the fund is a fractional part of its total return.

SECTION 1 PROBLEMS

Determine the dividend that should be paid on each share of stock.

	NUMBER OF SHARES	TYPE OF STOCK	TOTAL DIVIDEND DECLARED
1. a.	45,000	Common	$ 63,000
b.	50,000	Common	100,000
	10,000	6%, $100 par-value preferred	
c.	20,000	Common	220,000
	40,000	5%, $50 par-value preferred, cumulative for 2 years	
2. a.	65,000	Common	$ 91,000
b.	40,000	Common	174,000
	30,000	5%, $100 par-value preferred	
c.	60,000	Common	150,000
	20,000	6%, $50 par-value preferred, cumulative for 2 years	

Determine the dividend per share on each class of stock and the earnings per share on common stock.

	NUMBER OF SHARES	TYPE OF STOCK	NET INCOME	TOTAL DIVIDEND DECLARED
3. a.	30,000	Common	$900,000	$450,000
	60,000	7%, $100 par-value preferred		
b.	10,000	Common	200,000	150,000
	20,000	8%, $50 par-value preferred		
4. a.	50,000	Common	$600,000	$450,000
	70,000	5%, $100 par-value preferred		
b.	25,000	Common	800,000	400,000
	90,000	7%, $50 par-value preferred		

5. The board of directors of Pape Industries, Inc. declared a $136,000 cash dividend. The firm has 80,000 shares of common stock outstanding. What will the dividend per share be?

6. Harkness Equipment, Inc. has 50,000 shares of common stock outstanding. The board of directors has voted a $137,500 cash dividend. What is the dividend per share?

7. Smith Transport, Inc. declared a cash dividend of $175,000. If there are 10,000 shares of common stock outstanding and 20,000 shares of 7%, $100 par-value preferred stock outstanding, what dividend per share will be paid to both classes of stockholders?

8. The board of directors of Yates Systems, Inc. declared a cash dividend of $135,000. If there are 5,000 shares of common stock and 15,000 shares of 8%, $100 par-value preferred stock outstanding, what is the dividend per share paid to both classes of stockholders?

9. Gofreed Products Corporation declared a cash dividend of $10,000 at the end of its first year of operations. The corporation has 500 shares of common stock outstanding and 1,500 shares of 8%, $50 par-value preferred stock outstanding. Compute the dividend on each share of stock.

10. The board of directors of Marbury Networks Corp. declared a cash dividend of $12,000. The corporation has 1,000 shares of common stock and 3,000 shares of 7%, $50 par-value preferred stock outstanding. What dividend per share is paid to each class of stockholders?

11. The Magic Mountain Coal Co. has 100,000 shares of common stock outstanding and 65,000 shares of 6%, $100 par-value preferred stock outstanding. Compute the dividend per share on a declared dividend of $325,000.

12. This year, a cash dividend of $150,000 has been declared by the board of directors of Messinger Insurance Co., Inc. If there are 20,000 shares of common stock and 30,000 shares of 6%, $100 par-value preferred stock outstanding, what dividend per share will be paid to each class of stockholders?

13. Stewart Art Distributors, Inc. did not declare or pay a cash dividend for the preceding year. This year, however, the board declared a cash dividend of $77,000. Determine the dividend per share on 5,000 shares of common stock and 4,000 shares of 7%, $50 par-value cumulative preferred stock.

14. Daniels Electronics did not declare or pay a cash dividend last year. This year, however, the board of directors declared a cash dividend of $425,000. Determine the dividend per share on 10,000 shares of common stock and 25,000 shares of 8%, $100 par-value cumulative preferred stock.

15. Everett Heating and Air Conditioning, Inc. did not pay a cash dividend for the preceding two years. The board has declared a dividend of $40,000 this year. What dividend per share will be paid on 8,000 shares of common stock and 3,000 shares of 8%, $50 par-value cumulative preferred stock?

16. For the previous two years, the board of directors of the Furman Group declared no dividends. This year, a $140,000 cash dividend was declared. What dividend per share will be paid on 5,000 shares of common stock and 6,000 shares of 7%, $100 par-value cumulative preferred stock outstanding?
17. Determine the dividend per share this year for the Midas Touch, Inc. Last year, no dividend was paid, but a $130,000 dividend was declared this year. There are 6,000 shares of common stock and 10,000 shares of 5%, $100 par-value cumulative preferred stock outstanding.
18. Determine the dividend per share this year for a corporation that has a declared dividend of $50,000. Last year, no dividend was paid. There are 15,000 shares of common stock and 4,000 shares of 5%, $50 par-value cumulative preferred stock outstanding.
19. The Prentice Group, Inc. declared a cash dividend of $45,000 this year. The board did not pay any dividend last year. Compute the dividend per share for 6,000 shares of common and 10,000 shares of 6%, $50 par-value noncumulative preferred stock.
20. Tangier Real Estate, Inc. declared a cash dividend of $65,000 this year. Last year, no dividend was declared or paid. There are 4,000 shares of common stock and 16,000 shares of 8%, $50 par-value noncumulative preferred stock outstanding. What is the dividend per share this year?
21. Food Crown, Inc. earned $150,000 in net income last year. Its board declared a cash dividend of $90,000. The supermarket had 10,000 shares of common stock outstanding and 15,000 shares of 5%, $50 par-value preferred stock outstanding. Calculate
    a. Dividends per share paid to the stockholders
    b. Earnings per share on the common stock
22. Scattolini Enterprises, Inc. had $78,000 in net income, but the board of directors declared a cash dividend of $64,000. There are 7,000 shares of common and 8,000 shares of 9%, $50 par-value preferred stock outstanding. Determine
    a. The dividend per share paid to each class of stock
    b. The earnings per share on the common stock

SECTION 2
DISTRIBUTION OF PROFITS IN A PARTNERSHIP

There are numerous ways in which the profits or losses of a partnership may be divided. We shall consider some of the typical methods. Recall that profits and losses are shared equally among partners unless another specific agreement is made.

Example 1 **Equal Distribution**

The partnership of Wynn, Bright & Phillips was organized without a formal agreement regarding the distribution of profits. A net profit of $39,000 was earned during the business year now ending. Their individual investments included $14,000 by Wynn, $8,000 by Bright, and $11,000 by Phillips. How much should be allocated to each partner?

$$\frac{\$39{,}000}{3} = \$13{,}000 \text{ each}$$

In the absence of a formal agreement, each partner would receive an equal distribution of $13,000.

Example 2 **Agreed Ratio**

After considering both the capital investment and the time devoted to the business by each partner in Example 1, partners Wynn (W), Bright (B), and Phillips (P) agreed to share profits (or losses) in the ratio of 4:3:6. Divide their $39,000 profit.

The sum of $4 + 3 + 6 = 13$ indicates that the profit will be divided into 13 shares: W will receive 4 of the 13 shares, B will receive 3 shares, and P will receive 6 shares. Thus, W receives $\frac{4}{13}$ of the profit, B gets $\frac{3}{13}$, and P earns $\frac{6}{13}$.

$$\begin{aligned} &\text{W:} && \tfrac{4}{13}(\$39{,}000) = \$12{,}000 \\ &\text{B:} && \tfrac{3}{13}(\$39{,}000) = \quad 9{,}000 \\ &\text{P:} && \tfrac{6}{13}(\$39{,}000) = \underline{\ 18{,}000\ } \\ &\text{Total profit} && \qquad\qquad\quad = \$39{,}000 \end{aligned}$$

Partner W will receive $12,000, B will receive $9,000, and P gets $18,000. Notice that the sum of all partners' shares equals the total profit of the partnership.

Note. Agreed distributions are also often expressed as percents, such as a 30%/20%/50% distribution. The given percents must total 100% of the profit, and finding each partner's share is simply a matter of calculating that given percent of the total profit (or loss).

Example 3 **Original Investment**

Partners A and B opened a business with a $10,000 investment by A and an $8,000 investment by B. Partner B later invested another $6,000. Their partnership agreement stipulated that profits or losses would be divided according to the ratio of their original investments. What portion of a $7,200 profit would be allocated to each partner?

$$\$10{,}000 + \$8{,}000 = \$18{,}000 \qquad \text{Total original investment}$$

$$\text{A had } \frac{\$10{,}000}{\$18{,}000} \text{ or } \frac{5}{9} \text{ of the original investment}$$

$$\text{B had } \frac{\$8{,}000}{\$18{,}000} \text{ or } \frac{4}{9} \text{ of the original investment}$$

Thus, A will receive $\frac{5}{9}$ of the profit and B will receive the other $\frac{4}{9}$:

$$\begin{aligned} &\text{A:} && \frac{5}{9}(\$7{,}200) = \$4{,}000 \\ &\text{B:} && \frac{4}{9}(\$7{,}200) = \underline{\ 3{,}200\ } \\ & && \text{Total profit} = \$7{,}200 \end{aligned}$$

A's share is $4,000 and B's share is $3,200.

Example 3 (cont.)

Investment at Beginning of Year

A variation of the foregoing "original-investment" distribution is to divide profits according to each partner's investment at the beginning of the year. Suppose that partners A and B (above) use this method during their second year of operations. Partner A still has $10,000 invested, while B's investment is now $8,000 + $6,000 = $14,000, giving a total investment by both partners of $24,000.

Therefore, following the second year of operations,

$$\text{A would receive } \frac{\$10{,}000}{\$24{,}000} \quad \text{or} \quad \frac{5}{12} \text{ of the profit}$$

$$\text{B would receive } \frac{\$14{,}000}{\$24{,}000} \quad \text{or} \quad \frac{7}{12} \text{ of the profit}$$

Example 4

Average Investment per Month

When partners C and D began operations on January 1, C had $4,000 invested, which did not change during the year. D had $5,800 invested on January 1; on May 1, he withdrew $200; and on August 1, he reinvested $800. If profits are to be shared according to each partner's average investment per month, how much of a $12,500 profit should be allocated to each?

Partner C's average investment is $4,000, since his investment did not change during the year. D's average investment is computed as follows:

DATE	CHANGE	AMOUNT OF INVESTMENT		MONTHS INVESTED		
January 1	—	$5,800	×	4	=	$23,200
May 1	−$200	5,600	×	3	=	16,800
August 1	+$800	6,400	×	+ 5	=	+ 32,000
				12		$72,000

$$\frac{\$72{,}000}{12} = \$6{,}000 \qquad \text{D's average investment per month}$$

$ 4,000	C's average investment
+ 6,000	D's average investment
$10,000	Total of average investments

Thus, C receives $\frac{\$4{,}000}{\$10{,}000}$ or $\frac{2}{5}$ of the profit. D's investment receives $\frac{\$6{,}000}{\$10{,}000}$ or $\frac{3}{5}$ of the profit.

$$\text{C:} \quad \tfrac{2}{5}(\$12{,}500) = \$\ 5{,}000$$
$$\text{D:} \quad \tfrac{3}{5}(\$12{,}500) = \underline{\ \ 7{,}500}$$
$$\text{Total profit} = \$12{,}500$$

C earns $5,000 of the total profit, and D receives $7,500.

Example 5 Salary and Agreed Ratio

Partners X, Y, and Z agreed to pay X a salary of $10,000, to pay Y a salary of $12,000, and to divide the remaining profit or loss in the ratio of 4:3:5. If the partnership earned a $40,000 net profit, how much will each partner receive?

The salaries should be deducted first, and the remaining profit is then divided into $4 + 3 + 5 = 12$ shares:

Salaries: $10,000 + $12,000 = $22,000

$40,000	Total net profit
− 22,000	Salaries
$18,000	To be divided in the ratio 4:3:5

X receives $\frac{4}{12}$ or $\frac{1}{3}$ of the $18,000: $\frac{1}{3}(\$18{,}000) = \$\ 6{,}000$

Y gets $\frac{3}{12}$ or $\frac{1}{4}$ of the $18,000: $\frac{1}{4}(\$18{,}000) = 4{,}500$

Z has earned $\frac{5}{12}$ of the $18,000: $\frac{5}{12}(\$18{,}000) = +\ 7{,}500$

$18,000

The total net profit is distributed as follows:

	X	Y	Z	CHECK	
Salary	$10,000	$12,000	—	$16,000	X's share
Ratio	6,000	4,500	$7,500	16,500	Y's share
Total	$16,000	$16,500	$7,500	7,500	Z's share
				$40,000	Total net profit

Thus, X's share of the net profit is $16,000, Y earns $16,500, and Z receives $7,500.

Example 6 Interest on Investment, Salary, and Agreed Ratio

Partners R and S began operations with an investment of $15,000 by R (a silent partner) and $5,500 by S. They agreed to pay 8% interest on each partner's investment, to pay S a $10,000 salary, and to divide any remaining profit or loss equally. (a) Distribute a profit of $18,640. (b) Distribute a $10,840 net profit.

(a) The various parts of the distribution are computed in the same order in which the agreement is stated. The interest earned by each partner is thus:

	R	S
Investment	$15,000	$5,500
Interest rate	× 8%	× 8%
Interest	$ 1,200	$ 440

The interest is deducted from the net profit, and then the salary is paid:

$1,200		$18,640	Total profit
+ 440		− 1,640	Interest
$1,640	Total interest	$17,000	
		− 10,000	Salary
		$ 7,000	To be shared equally

$$\frac{\$7{,}000}{2} = \$3{,}500 \quad \text{Additional profit due each partner}$$

Each partner thus receives the following portion of the net profit:

	R	S	CHECK	
Interest	$1,200	$ 440	$ 4,700	R's share
Salary	—	10,000	13,940	S's share
Ratio	3,500	3,500	$18,640	Total net profit
Total	$4,700	$13,940		

R receives $4,700 of the net profit, and S earns $13,940.

(b) The same interest and salary would be due each partner as computed in Part (a). The remaining calculations would then be

$10,840	Total profit
− 1,640	Interest
$ 9,200	
− 10,000	Salary
− $ 800	Shortage to be shared equally

The net profit of the partnership is $800 less than the interest and salary due. According to the terms of their agreement, this shortage will be shared equally by the partners. That is, $800 ÷ 2 or $400 of the amount due each partner will not be paid. Each partner's share is thus computed as follows:

	R	S	CHECK	
Interest	$1,200	$ 440	$ 800	R's share
Salary	—	10,000	10,040	S's share
	$1,200	$10,440	$10,840	Total profit
Ratio (shortage)	− 400	− 400		
Total	$ 800	$10,040		

R receives $800 of the net profit, and S receives $10,040.

SECTION 2 PROBLEMS

Divide the net profit among the partners according to the conditions given.

	PARTNER	INVESTMENT	METHOD OF DISTRIBUTION	NET PROFIT
1.	a. K	$12,000	Ratio of 1:3	$60,000
	L	10,000		
	b. A	5,000	Ratio of their investments	21,600
	B	6,000		
	C	7,000		
	c. S	30,000	6% interest on investment,	16,000
	T	35,000	and remainder in the ratio	
	U	15,000	of 2:2:1	
	d. V	30,000	5% interest on investment,	43,700
	W	28,000	$15,000 salary to V, and	
	X	24,000	remainder in ratio of their	
			investments	
2.	a. G	$50,000	Ratio of 2:1	$90,000
	H	50,000		
	b. I	18,000	Ratio of their investments	72,000
	J	16,000		
	K	14,000		
	c. L	40,000	7% interest on investment,	20,000
	M	30,000	and remainder in the ratio	
	N	10,000	of 3:2:1	
	d. P	10,000	6% interest on investment,	29,000
	Q	15,000	$8,000 salary to R, and	
	R	17,000	remainder in ratio of their	
			investments	

3. Catlin and Mary entered into a partnership without an agreement stipulating how profits or losses would be distributed. Catlin invested her life savings of $50,000, and Mary invested $10,000. Catlin spent 50 hours and Mary spent 40 hours a week in the business. How much of their first year's profit of $50,000 should each partner receive?

4. Graham and Brent entered into a partnership without an agreement stipulating how profits and losses would be allocated. Graham invested $6,000 in the business, while Brent invested $12,000. Graham spent 20 hours a week and Brent spent 40 hours a week in the business. How much of their first year's profit of $18,000 would be distributed to each partner?

5. Partners Q, R, and S agreed to share profits or losses in the ratio of 2:3:1.

a. What is each partner's share of a $42,000 profit?

b. How much of an $18,000 loss is each partner's share?

6. Partners L, M, and N have agreed to share profits and losses in the ratio of 3:2:2.

a. How much of a $35,000 profit would be allocated to each partner?

b. How much of a $5,600 loss would be allocated to each partner?

7. A partnership was formed by Bob, Chris, and Doug. The partnership agreement stipulated salaries of $10,000 to each partner, and any additional profits or losses would be divided 25%, 30%, and 45%, respectively. Find each person's share of a $50,000 profit.

8. Partners Lynn, Monica, and Nancy agreed to pay salaries of $12,000 to each partner; any additional profit or loss would be divided 25%/35%/40%, respectively. How much of a $40,000 profit would be distributed to each partner?

9. Julie and Greta entered into a partnership for interior design. Julie invested $16,000, and Greta invested $24,000 initially. Later during the first year, Julie invested another $4,000. The partners agreed to distribute profits and losses according to the ratio of their initial investments. How much of a $75,000 profit would each partner receive?

10. Matt and Bob entered into a partnership. Matt invested $20,000, and Bob invested $25,000. Later in the same year, Matt invested another $10,000. The partners had agreed to distribute profits and losses according to the ratio of their initial investments. How much of a $27,000 profit would be allocated to each partner?

11. During the second year of operations of Julie and Greta's interior design business (Problem 9), their distribution agreement was changed as follows: Each person would draw a salary of $6,000, and the remaining profit or loss would be divided according to the ratio of their investments at the beginning of the current year. What is each person's share of a $64,800 profit?

12. During the second year of operations, the partnership of Matt and Bob (Problem 10) changed their agreement: Matt and Bob would draw salaries of $19,000 and $22,000 respectively. Any remaining profit or loss would be divided according to their investment at the beginning of the current year. How much of a $74,000 profit would be allocated to each partner?

13. Partners Steve and Tom distribute profits and losses from their construction business in the ratio of their average investment per month. Steve invested $26,000 and did not change his investment during the year. Tom invested $20,000 on January 1 but withdrew $4,000 on June 1 and reinvested $1,000 on September 1.

a. What was Tom's average investment?

b. How much of a $77,000 profit would each partner receive?

14. George and James distribute profits and losses from their partnership in the ratio of their average investment per month. George invested $50,000 and did not change his investment during the year. James invested $40,000 on January 1 but withdrew $5,000 on May 1 and reinvested $20,000 on November 1.

a. What was James's average investment?

b. How much of a $270,000 profit would be distributed to each partner?

15. Mark and Pete became partners on January 1. Mark invested $22,000 and did not change his investment during the year. Pete invested $25,000 on January 1 but withdrew $3,000 on March 1 and reinvested $6,000 on July 1. The partners have agreed to distribute profits by paying 6% interest on their average investments, with the remaining profits or losses to be divided equally.

a. What was Pete's average investment?

b. If the profit was $30,000, how much should each partner receive?

16. Amin and Kalid have agreed to distribute profits and losses as follows: 8% interest will be paid on their average investment, and the remaining profit or loss will be divided equally. Amin's investment of $26,000 on January 1 was unchanged. Kalid invested $18,000 on January 1. He invested $3,000 more on June 1, and he withdrew $1,000 on October 1.

a. What was Kalid's average investment per month?

b. How much of a $40,000 profit will be distributed to each?

17. Margaret and Katherine formed a partnership by Margaret investing $16,000 and Katherine investing $30,000. They have agreed to pay 6% interest on investments and a salary of $5,000 to Margaret. The remaining profits or losses are to be shared equally.

a. Determine each partner's share of a $14,000 profit.

b. Determine each partner's share of a $7,000 profit.

18. Nelly and Jane formed a partnership with Nelly investing $15,000 and Jane investing $18,000. Their partnership agreement was to pay 7% interest on investments, pay a salary of $10,000 to Jane, and share the remaining profit or loss equally.

a. Distribute a net profit of $24,000.

b. Distribute a net profit of $12,000.

19. Partners Tim, Chrisy, and Barbara have agreed to allocate profits and losses in the following way: 7% interest on investments of $10,000, $9,000, and $8,000, respectively; salaries of $4,000 each; and any remainder to be divided in the ratio of their investments.

a. What is each partner's share of a $27,390 profit?

b. What is each partner's share of a $13,350 profit?

20. Hal, Jeff, and Steve have agreed to allocate profits and losses in the following way: 6% interest on their respective investments of $50,000, $40,000, and $30,000; salaries of $12,000 to each; and any remainder to be divided in the ratio of their investments.

a. What is each partner's share of a $48,000 net profit?

b. What is each partner's share of a $39,000 net profit?

21. Partners Richard, Sarah, and Ted have agreed to allocate profits and losses in the following way: 5% interest on investments of $20,000, $40,000, and $30,000, respectively; salaries of $5,000 to be paid to Sarah and Ted; and any remainder in the ratio of 1:3:2.

a. Distribute a profit of $16,000.

b. Distribute a profit of $13,300.

22. Connor, Dave, and Edwin have agreed to allocate profits and losses in the following way: 6% interest on their respective investments of $22,000, $28,000, and $30,000; salaries of $10,000 to each; and any remainder to be divided in the ratio of 1:1:2.

a. Distribute a net profit of $45,000.

b. Distribute a net profit of $32,000.

CHAPTER 11 GLOSSARY

Board of directors. Officials elected by stockholders to set policy and decide the direction of a corporation, including determining whether/when dividends will be paid.

Common stock. A share of corporation ownership that includes no guarantee about the size of the dividend.

Corporation. A business established as an artificial entity according to state law and owned by stockholders, whose liability for the business is limited to the amount invested in the stock.

Cumulative preferred stock. Preferred stock which guarantees that dividends unpaid in any year(s) will accumulate until paid in full the next time(s) a dividend is declared.

Dividend. The return on owner's investment (cash or additional stock) paid per share of stock in a corporation.

Earnings per share. The amount of net income earned per share of common stock during the year: Net income minus preferred dividends divided by the number of common shares outstanding.

Limited partnership. A partnership in which the liability of the limited partners, who take no active part in the business, is restricted to the amount they invested. The business is managed by one or more general partners, who are also responsible for any financial obligations that exceed the amount invested.

Nonparticipating stock. Preferred stock that stipulates that dividends will not exceed the specified rate.

Non-par-value stock. Stock issued without a par (face) value printed on it.

Participating stock. Preferred stock that may earn dividends above the specified rate.

Partnership. A business for which two or more individuals legally are jointly and separately liable, although responsibilities and profits may be divided in any way that is mutually agreeable.

Par (or face) value. A value assigned to and printed on a share of stock when a corporation is organized (or when it issues new stock at a later time).

Preferred stock. Stock for which the dividend is preset as a certain percent of the par value, and on which dividends must be paid before dividends are paid on common stock.

Proprietorship. A business having a single (sole) owner who is also personally liable for the business's debts.

Silent partner. A member of a partnership who invests financially but takes no active part in the business.

Stock certificate. A certificate of ownership in a corporation; classified as either common or preferred stock.

Stockholder. A person who owns stock in (and thus owns a portion of) a corporation.

Stock outstanding. The number of shares of stock that a corporation has sold.

CHAPTER 12

COMMERCIAL DISCOUNTS

OBJECTIVES

Upon completion of Chapter 12, you will be able to:

1. Define and use correctly the terminology associated with each topic.
2. When one or more trade discount percents are given (Section 1):
   a. Find the net cost rate factor (Examples 2, 5; Problems 1, 2, 7, 8).
   b. Use the net cost rate factor to compute:
      (1) Net cost (Examples 2, 3; Problems 1–8, 19–24)
      (2) The single equivalent discount percent (Examples 4–6; Problems 1, 2, 7–10, 15–18).
3. Using the trade discount formula "% Paid × List = Net" and given two of the following amounts, find the unknown third (Section 1):
   a. Net cost (Examples 1–3; Problems 1–8, 17–24)
   b. List price (Example 5; Problems 1, 2, 11–14)
   c. Discount percent (Examples 6, 7, 8; Problems 15–24).
4. Use sales terms given on an invoice to determine (Section 2: Examples 1–2; Problems 3–14):
   a. Cash discount
   b. Net payment due
   c. The last day on which a cash discount may be claimed and the applicable discount rate (Example 3; Problems 1, 2).
5. Use sales terms and a variation of the formula "% Paid × List = Net" to find (Section 2: Examples 4, 5; Problems 5, 6, 13–16):
   a. The credit granted toward an account when a partial payment is made
   b. The partial payment that must be made to obtain a given credit toward an account.

A business that purchases merchandise that is then sold directly to the **consumer** (user) is known as a **retail** business or store. The retailer may have purchased this merchandise either from the manufacturer or from a **wholesaler** (a business that buys goods for distribution only to other businesses and does not sell to the general public). The purchasing, pricing, and sale of merchandise in a retail store require a certain kind of mathematics, which is the subject of this unit.

Competition for the customer has led to widespread use of discounts, in both wholesale and retail sales. The following sections examine two of these commercial discounts—**trade discounts** and **cash discounts.**

SECTION 1

TRADE DISCOUNTS

Manufacturers and wholesalers usually issue catalogs to be used by retailers. These catalogs are often quite bulky, containing photographs, diagrams, and/or descriptions of all items sold by the company. As would be expected, the prices of items shown in the catalogs are subject to change periodically. To reprint an entire catalog each time there is a price change would be unduly expensive; thus, several methods have been developed whereby a business can keep the customer informed of current prices without having to reprint the entire catalog.

Some catalogs are issued in looseleaf binders, and new pages are printed to replace only those pages where a price change has occurred. Other catalogs contain no prices at all; instead, a separate price list, which is revised periodically, accompanies the catalog.

We shall be concerned in this course with a third type of catalog. The prices shown in this catalog are the **list prices** or **suggested retail prices,** usually the prices that the consumer pays for the goods. The retailer buys these goods at a reduction of a certain percentage off the catalog price. This reduction is known as a **trade discount.** The actual cost to the retailer, after the trade discount has been subtracted, is the retailer's **net cost** or **net price** of the item. The percents of discount that are to be allowed (which may vary from item to item in the catalog) may be printed on a discount sheet that supplements the catalog, or they may be quoted directly to a company placing a special order.

The seller will issue to the buyer an **invoice** (see Figure 12-1), which is a bill giving a detailed list of goods shipped or services rendered, with an accounting of all costs. The invoice lists not only a description of the goods, unit prices, and the total amount of the goods, but also sales terms, trade discounts, freight charges, telephone expenses, or any other charges associated with the transaction.

As you can see in Figure 12-1, the quantity is multiplied by the list price to give the extension (the total price of each item). The trade discount of 20% is applied to the total list price of $264. After this discount of $52.80 is subtracted, the net price is $211.20. The freight charge of $37.80 is added as the last step to give the total amount due on the invoice of $249.

		EAST STREET HARDWARE SUPPLY P.O. Box 239 Richmond, VA 23222		Invoice # 2414
TO:	Handy Hardware Co. 1241 Post Drive Alexandria, VA 22301			
DATE:	Sept. 9,20xx			
TERMS:	Net 30			
Your Order #534	Received 11/6/xx	Shipped 11/8/xx	Freight F.O.B.Richmond	
Quantity	Cat. #	Description	List	Extension
5	S301	Lounge Chair	$18.50 ea.	$ 92.50
6	S302	Folding Chair	12.75 ea.	76.50
10	D20	Aluminum Grill	9.50 ea.	95.00
		SubTotal		$264.00
		Less 20% discount		52.80
		Net		$211.20
		Freight		37.80
		Total Due		$249.00

Figure 12-1 Invoice from Wholesale Supplier

Freight charges or shipping charges may be expressed as **F.O.B.** (freight on board or free on board) **shipping point** or **F.O.B. destination.** The term *F.O.B. shipping point* means that the buyer takes responsibility for the merchandise at the vendor's or seller's location. That is, the freight charges will be paid by the buyer. By contrast, if the term *F.O.B. destination* appears on the invoice, the seller will retain title to the goods until they are delivered to the buyer's warehouse or store; thus, the seller is paying the transportation charges. In Figure 12-1, the invoice indicates that the shipping charges are F.O.B. shipping point (Richmond). As a courtesy to Handy Hardware Co.—the buyer—the vendor is prepaying the freight charge. It should be noted that this charge has been added to the net cost of the merchandise as a last step. The trade discount applies *only* to the cost of goods or merchandise and not to other charges.

Example 1 A tape recorder/radio is shown at a catalog list price of $60. The discount sheet indicates that a 25% trade discount is to be allowed.

(a) What is the net price to the retail store?

List price	$60
Less trade discount (25% × $60)	− 15
Net price	$45

Observe that when a 25% reduction is allowed, this means that the retailer still pays the other 75% of the list price. Indeed, the percent of discount subtracted from 100% (the whole list price) will always indicate what percent of the list price is still being paid for the goods. For example,

10% discount means $100\% - 10\% = 90\%$ paid

35% discount means $100\% - 35\% = 65\%$ paid

18% discount means $100\% - 18\% = 82\%$ paid

The **complement** of any given percent (or of its decimal or fractional equivalent) is some number such that their sum equals a whole, or 100%. Each "percent paid" above is thus the complement of the corresponding discount percent. By using the complement of the given discount, the previous problem can be solved by a shorter method, applying the formula % Paid × List = Net.

Hint. Keep in mind that percents must be converted to their decimal or fractional equivalent before being used. Many problems can be worked more conveniently when percents are changed to fractions.

Example 1 (cont.)

(b) A 25% discount on the tape recorder/radio above means that $100\% - 25\% = 75\%$ of the list price is still being paid. (The complement of 25% is 75%.)

Using Decimals	*Using Fractions*
% Paid × List = Net	% Paid × List = Net
$75\% \times \$60 = \text{Net}$	$75\% \times \$60 = \text{Net}$
$0.75 \times 60 = \text{Net}$	$\frac{3}{4} \times 60 = \text{Net}$
$\$45 = \text{Net}$	$\$45 = \text{Net}$

(c) A merchant purchased 15 calculators at a list price of $7 each. The trade discount allowed on this purchase is 35%. There is a freight charge of $10 on the invoice. What is the amount due on this invoice?

The merchandise itself cost $15 \times \$7 = \$105$. Thus,

$$\% \text{ Paid} \times \text{List} = \text{Net}$$

$$0.65 \times \$105 = \text{Net}$$

$$\$68.25 = \text{Net}$$

Net price	=	$68.25
Freight	=	+ 10.00
Amount due	=	$78.25

Two or more successive trade discounts are often quoted on the same goods; these multiple discounts are known as **series** or **chain discounts.** If discounts of 25%, 20%, and 10% (often written 25/20/10) are offered, the net price may be found by calculating each discount in succession. (However, the order in which the discounts are taken does not affect the final result.) *The discounts in a chain are never added together.* That is, a chain discount of 25%, 20%, and 10% is *not* the same as a single discount of 55%.

Example 2 An upholstered chair is advertised at $200 less 20%, 10%, and 12.5%. Find the net cost to the furniture store.

(a)

List	$200	
Less 20%	− 40	(20% × $200 = $40)
	$160	
Less 10%	− 16	(10% × $160 = $16)
	$144	
Less 12.5%	− 18	(12.5% × $144 = $18)
Net cost	$126	

This net cost may be computed much more quickly using the method % Paid × List = Net.

(b) 20%, 10%, and 12.5% discounts have complements of 80%, 90%, and 87.5% paid:

$$\%\ \text{Pd} \times L = N$$

$$(0.8)(0.9)(0.875)(\$200) = N$$

$$(0.63)200 = N$$

$$\$126 = N$$

The number that is the product of all the "percents paid" is very significant and is called the **net cost rate factor.** As demonstrated above, the net cost rate factor is a number that can be multiplied times the list price to determine the net cost. All items subject to the same discounts have the same net cost rate factor. Thus, instead of reworking the entire problem for each item, the buyer determines the net cost rate factor that applies and then uses it to find the net price of all the items.

Example 3 A coffee table subject to the same discount rates given in Example 2 lists for \$80. What is the net cost?

From Example 2, the net cost rate factor $= 0.8 \times 0.9 \times 0.875 = 0.63$. Then

$$0.63 \times \$80 = \$50.40 \qquad \text{Net cost}$$

For purposes of cost comparison, it is often desirable to know what single discount rate is equivalent (**single equivalent discount**) to a series of discounts. This is also an application of the complement, found by subtracting the percent paid (the net cost rate factor) from 100%, as follows.

Example 4 What single discount rate is equivalent to the series 20/10/12.5 offered in Example 2?

The net cost rate factor, 0.63, means that 63% of the list price was paid. Finding the complement indicates that

$$100\% - 63\% = 37\% \qquad \text{Single equivalent discount rate}$$

Observe that series discounts of 20%, 10%, and 12.5% are *not* the same as a single discount of 42.5%. If a single discount of 42.5% had been advertised on the chair in Example 2, the net cost would *not* have been \$126. Rather, the percent paid would have been 57.5%, and the net cost would have been

$$\% \text{ Pd} \times L = N$$

$$0.575(\$200) = N$$

$$\$115 = N$$

Example 5 Following trade discounts of 20% and 15%, a furnace cost a dealer \$374. (a) What is the list price? (b) What is the net cost rate factor? (c) What single discount percent is equivalent to discounts of 20% and 15%?

(a)

$$\% \text{ Pd} \times L = N$$

$$(0.8)(0.85)L = \$374$$

$$0.68L = 374$$

$$L = \frac{374}{0.68}$$

$$L = \$550$$

(b) The net cost rate factor = 0.68.

(c) The single discount equal to discounts of 20% and 15% is

$$100\% - 68\% = 32\%$$

Example 6 The suggested retail price of a television set is $400. If the net cost is $320, what single discount rate was offered?

The solution can be found either by the % Paid formula or by the % of Change formula:

% Paid Formula

$$\% \text{ Pd} \times L = N$$

$$x \cdot 400 = 320$$

$$x = \frac{320}{400}$$

$$x = \frac{4}{5}$$

$$x = 80\%$$

But x = % Paid. Thus, a 20% discount was given.

or

% of Change Formula

The price changed by $400 − $320 = $80:

___% of Original = Change?

$$x \cdot 400 = 80$$

$$x = \frac{80}{400}$$

$$x = \frac{1}{5}$$

$$x = 20\% \text{ discount}$$

Often the retailer is offered similar merchandise from various manufacturers with different trade discounts. All other factors being the same, the retailer will want to buy from the manufacturer with the largest trade discount or lowest price.

Example 7 Fong Denim Manufacturing Co. offers merchandise with trade discounts of 40% and 10%. REC Inc. offers similar merchandise with discounts of 20%, 10%, and 15%. Which company offers the better discount?

The net cost rate factor for each company is as follows:

Fong Denim Manuf. Co.	*REC Inc.*
(0.60)(0.90) = 0.54	(0.80)(0.90)(0.85) = 0.612

Thus, the single equivalent discount for each is

100% − 54% = 48%	100% − 61.2% = 38.8%

Fong Denim Manufacturing Co. offers a higher discount and a lower cost to the retailer.

Similarly, a merchant may want to know what additional discount must be offered to meet its competitor's price.

Example 8 Canvas bags are priced at \$50 less 20% at Vincent Bag Co., while the Shelley Co. lists the bags at \$40 less 25%. What additional discount percent will Vincent Bag Co. have to offer to match its competitor's price?

First, find the current net cost at each business:

Vincent Co.	*Shelley Co.*
% Pd × List = Net	% Pd × List = List
0.80 × \$50 = \$40	0.75 × \$40 = \$30

Next, find by what percent Vincent must reduce its price to meet its competitor's price.

Vincent's current price is \$40, and the competitor's price is \$30. So the amount of reduction, or change in price, is \$40 − \$30 = \$10. Thus, using the x% of Change formula,

$$x\% \text{ of Original} = \text{Change}$$

$$x\% \text{ of } \$40 = \$10$$

$$x = \frac{\$10}{\$40}$$

$$x = 25\% \qquad \text{Additional discount \%}$$

Or, using the % Pd formula,

$$x\% \text{ Pd} \times L = N$$

$$x\% \text{ Pd} \times \$40 = \$30$$

$$x = \frac{\$30}{\$40}$$

$$x = 75\% \qquad \text{\% still paid}$$

The buyer will still pay 75% of the current price. So the additional discount percent is 100% − 75% = 25%.

SECTION 1 PROBLEMS

Complete the following.

		TRADE DISCOUNT	% PAID (OR NET COST RATE FACTOR)	LIST PRICE	NET COST	SINGLE EQUIVALENT DISCOUNT PERCENT
1.	a.	20%		$ 72		X
	b.	25%		240		X
	c.	40%			$102	X
	d.		66⅔%		72	X
	e.	20%, 25%		290		
	f.	22 2/9%, 10%		150		
	g.	37½%, 30%			70	
	h.	20/40/30			168	
2.	a.	15%		$240		X
	b.	45%		30		X
	c.	12½%			$ 7	X
	d.		90%		36	X
	e.	20%, 15%		50		
	f.	25%, 10%		280		
	g.	33⅓%, 40%			144	
	h.	20/25/10			324	

3. Compute the net cost of the following items:
   a. Stationary bike listed at $109 less 45%
   b. Treadmill listed for $500 less 35%
   c. Roller blades listed for $40 less 37½%
   d. Aerobic slide listed for $18 less 11 1/9%

4. Determine the net cost for each of the following items:
   a. Fax machine listed at $190 less 20%
   b. Laser jet printer listed at $750 less 10%
   c. Scanner listed at $110 less 30%
   d. Electronic labeler listed at $80 less 15%

5. Determine the net cost and the total due for the following merchandise on a recent invoice. All items are subject to a 25% discount. Freight charges are $27.00.

20 boxes of formatted diskettes at $7 per box
36 packages of double-pocket portfolios at $3 per package
5 dozen ballpoint pens at $6 per dozen
10½ dozen dry chalkboard markers at $8 per dozen
12 boxes of letter-size file folders at $2 per box

6. Complete the following invoice, determining the total list price, the trade discount, the net price, and the amount due on the invoice.

Gulledge Computer Co.
345 Pender Road
Davidson, NC 29036

TO: Bell Castle, Inc.
1818—First Avenue
Charlotte, NC 28201

INVOICE NO.: 4367

DATE: September 5, 20xx

TERMS: Net 30

PURCHASE ORDER NO.	RECEIVED	SHIPPED	VIA
8109	9/2/xx	9/3/xx	Truck

QUANTITY	CAT NO.	DESCRIPTION	LIST	EXTENSION
5 cases	X020	Multi-purpose, 20-lb., 8½″ × 11″ copy paper	$20/case	
2 dozen	C036	36X CD ROM Drive	70 ea.	
6	H049	6.8 GB IDE Hard Drive	150 ea.	
10 boxes	D035	3.5″ HD Diskettes, IBM Formatted	29/box	
8	M056	56K Internal Modem	45 ea.	
			Total List	
			Less 30%	
			Net	
			Freight	$12.50
			Total Due	

7. Determine (1) the net cost rate factor, (2) the net cost, and (3) the single equivalent discount rate for each of the following items:

a. A food processor listed for $45 less 20% and 20%
b. A tea kettle listed for $20 less 15% and 25%
c. A toaster listed for $40 less 10%, 25½%, and 33⅓%
d. A pasta maker listed for $200 less 25%, 20%, and 15%

8. Determine (1) the net cost rate factor, (2) the net cost, and (3) the single equivalent discount rate for each of the following items:
   a. A fleece V-neck pullover for \$20 less 10% and 15%
   b. Wrinkle-resistant pants for \$26 less 20% and $12\frac{1}{2}$%
   c. A silk tie for \$35 less 20%, 30%, and 5%
   d. A hooded sweatshirt for \$30 less 25%, $33\frac{1}{3}$%, and 10%
9. Cotrone Co. advertises discounts of 25/20/10. Hodgkins Distributors offers 25% and 40%. Which company offers the higher discount? (*Hint:* Compare single equivalent discount rates.)
10. Home Decor offers discounts of 35% and 10%, while Carpet City offers 20%, 20%, and 5% discounts on carpeting. Which company offers the better discount? (*Hint:* Compare single equivalent discount rates.)
11. The retailer's net cost of a package of acrylic yarn was \$2.04 after discounts of 20% and 15%. What was the list price?
12. After trade discounts of 25% and 10%, the retailer's net price of a lipstick is \$5.40. What was the list price?
13. After trade discounts of 35% and 10%, the net cost for a woman's swimsuit was \$29.25. What had been the catalog list price of the swimsuit?
14. The net price for replacement gas caps is \$6.30 after trade discounts of $33\frac{1}{3}$% and 10%. What was the catalog list price?
15. What single discount percent has been offered, if a surge protector power outlet that lists for \$10 is sold for \$6?
16. What single discount percent has been allowed if a VCR listing for \$330 is sold for \$297?
17. The Rodgers Co. currently sells a dinnerware set for \$60 less 20%. Their competitor's price is \$42 for the same set.
   a. What additional discount percent must Rodgers offer to match their competitor's price?
   b. What single discount percent would produce the same net price?
18. A winter jacket sells for \$180 less 25% at Murphy Bros. The Deupree Clothing Co. lists the same jacket for \$108.
   a. What additional discount percent must Murphy Bros. offer to match Deupree's price?
   b. What single discount percent would produce the same net price?
19. Crabtree Pet Supplies Co. has pet beds for \$30 less $16\frac{2}{3}$%. Hannah's Pet Mart offers the same beds for \$35 less 35%.
   a. What net price does each company now offer?
   b. What additional discount percent must Crabtree offer so that its net price will be the same as Hannah's?
20. A breadmaker lists for \$70 less 40% at Jerde's Kitchen Bazaar. Melrose Inc. sells the same machine for \$60 less 20%.
   a. What net price does each store offer now?
   b. What additional discount will Melrose have to allow to meet Jerde's price?
21. A leather chair and ottoman list for \$480 less $33\frac{1}{3}$% at Shenandoah Furniture Co., while Dickens Co. sells the same set for \$400 less 25%.
   a. Find the net price of the set at each business.
   b. What additional discount must Shenandoah offer to match Dickens's net price?
22. Plymouth Distributors has priced a 35-mm camera for \$150 less 20%. Thomas Wholesalers has the same camera for \$180 less 40%.
   a. What is the net price for the camera at each business?
   b. What additional discount must Plymouth offer to match Thomas's price?

23. Thelma's Accessories carries a wall mirror for \$200 less 12½%. The same mirror is listed for \$280 less 50% at Johnson Department Store.
   a. What is the current net price of the mirror at each business?
   b. What further discount percent must be offered by the higher-priced company in order to meet the competitor's net price?

24. A freezer sells for \$180 less 20% and 12½% at Martin Co., while it sells for \$240 less 37½% at Jergan's Outlet.
   a. What is the net price of the freezer at each company?
   b. What further discount percent must be offered by the higher-priced company in order to meet its competitor's net price?

CASH DISCOUNTS

Merchants who sell goods on credit often offer the buyer a reduction on the amount due as an inducement to prompt payment. This reduction is known as a **cash discount.** The cash discount is a certain percentage of the price of the goods, but it may be deducted only if the account is paid within a stipulated time period. The invoice or the monthly statement contains **sales terms** which indicate the cash discount rate and allowable time period. There are several variations of the ordinary method, as described below.

ORDINARY DATING

By the most common method, **ordinary dating,** the discount period begins with the date on the invoice. A typical example would be "Terms: 2/10, n/30." These sales terms are read "two ten, net thirty." This means that a 2% discount may be taken if payment is made within 10 days from the invoice date; from the eleventh to thirtieth days, the net amount is due; after 30 days, the account is overdue and may be subject to interest charges. (If given sales terms do not specify the day by which the net amount must be paid, which is typical in several of the following methods, it is commonly assumed that the net amount is due within 30 days after the discount period began.) Traditionally, payment is considered to have been made within the discount period if the check is postmarked by the last day of the discount period. For instance, if an invoice is dated October 5 with sales terms of 2/10, n/30, the check must be mailed in an envelope postmarked no later than October 15. Some businesses, however, expressly state on the invoice that payment must be *received* by the end of the discount period in order to take advantage of the cash discount.

Contrary to the way it sounds, "cash discount" does not generally refer to a discount allowed only for the payment of cash on the day of purchase. On the other hand, cash discounts may be taken only on the net cost of goods, not on freight or other charges.

END OF MONTH, OR PROXIMO

An invoice containing terms such as "1/10, EOM" means that the 1% discount may be taken if paid within 10 days following the **end of the month,** that is, if paid by the tenth of the coming month. Terms of **"1/10, PROX"** mean essentially the same thing, since **proximo** is a Latin word indicating the next month after the present. The EOM method is often used on merchandise purchased near the end of the month, although the practice is not limited only to end-of-month purchases. Some monthly statements (bills covering the whole month's purchases) often contain these sales terms. It has become common practice for merchants purchasing merchandise after the twenty-fifth of the month to add a month to the cash discount period. In other words, for merchandise purchased on February 26 with sales terms of 1/10, EOM, the cash discount period ends on April 10. This text will assume the more traditional approach of payment by the tenth of the coming month (in this instance, by March 10).

Example 1 Determine the amount due on the following statement of account:

STATEMENT

NAPIER FURNITURE DISTRIBUTORS
3608 Longworth Blvd.
Los Angeles, California 90069

Phone:
213-767-8445

Golden West Furniture Co.
Box 4018
Pasadena, California 90472

DATE: November 25, 20XX
TERMS: 3/10, E.O.M.

Date	Invoice No.	Charges	Credits	Balance
11/1				$545.50
11/8	5391	$239.20		784.70
11/12			$545.50	239.20
11/16	5437	137.35		376.55
11/20	5562	123.45		500.00
11/25	Less 20% and 10% trade discount		140.00	360.00
11/25	Total Ppd. Fgt.*	25.80		385.80
				↑ PAY LAST AMOUNT IN THIS COLUMN

*"Ppd. Fgt." indicates "prepaid freight."

The charges shown for merchandise are $239.20 + $137.35 + $123.45, or $500 list. This is reduced by the trade discounts allowed. If the bill is paid by December 10, the 3% cash discount may be taken on the $360 net cost of the merchandise. (If no trade discount were given, the cash discount would be computed on the full $500 cost of the merchandise.) Recall that the cash discount does not apply to the freight.

Net cost of goods	$360.00	Due on goods	$349.20
Less discount (3% × $360)	− 10.80	Freight	+ 25.80
Payment due on goods	$349.20	Total amount due	$375.00

A check for $375 will be mailed with a notation or a voucher indicating that a $10.80 cash discount has been taken. The next statement will show a credit for the full $385.80. If the statement is paid December 11 or after, a check for the entire $385.80 must be sent.

Example 2 Three invoices from L. R. Hinson & Co. will be paid on May 15. Sales terms on these invoices are 4/15, 2/30, n/60. The invoices are dated April 8 for $930, April 23 for $1,125, and May 9 for $2,975. What total amount must be paid to retire the entire obligation?

		Amount Due
No discount may be taken on the April 8 invoice, since it is over 30 days old:		$ 930.00
The April 23 invoice is less than 30 days old, so a 2% discount may be taken:		
	$1,125.00	
(2% × $1,125)	− 22.50	
		1,102.50
The third invoice is less than 15 days old, so the full 4% discount may be taken:	$2,975.00	
(4% × $2,975)	− 119.00	
		+ 2,856.00
Total payment due		$4,888.50

RECEIPT OF GOODS

Sales terms such as "3/10, ROG" (**receipt of goods**) indicate that the discount period begins on the day the merchandise arrives. These terms are used especially when the goods are likely to be in transit for a long while, such as a shipment by cross-country rail freight or by boat from a foreign country. For instance, assume that an invoice dated December 15 has terms of 3/10, ROG and that the merchandise is received by the purchaser on January 9. The discount period begins January 9, the date of receipt, not December 15, the invoice date.

POSTDATING OR "AS OF"

Distributors may sometimes induce additional sales by guaranteeing that payment can be postponed without penalty. However, an invoice is dated and mailed on the same day as the order is shipped (which notifies the buyer that the order has been filled), and the distributor's ordinary sales terms (such as 2/10, 1/30, n/60) may already be printed on the invoice. Thus, another date must also be given to indicate how long the extended discount period applies, for example, "Date: January 12 **AS OF** March 15." This **postdating** means that the buyer may wait until March 15 to start applying the given sales terms. (The 2% discount could be taken any time on or before March 25.)

EXTRA

The **extra (X)** method also allows payment to be postponed and the cash discount still be taken. For example, terms of "3/10-80X" (or "3/10-80 extra," or "3/10–80 ex.") indicate that the discount is permitted if paid within 10 days plus 80, or a total of 90 days from the date of the invoice.

Extended cash discount periods, as described in the previous two paragraphs, are particularly popular in the sale of seasonal goods, such as lawnmowers or Christmas decorations. With delayed payments possible, retailers are willing to place orders sooner. The distributor can then spread shipments to all retailers over a wider period of time, which is easier for the wholesaler. The retailer may also benefit, since in many cases the merchandise can be sold before the cash discount period expires.

Example 3 An order arrived June 4. What is the last day on which a cash discount may be claimed, given the following conditions? (a) The invoice is dated April 27, with sales terms of 3/15, ROG. (b) The invoice is dated April 27 AS OF July 7, with sales terms of 3/15, n/30. (c) The invoice is dated April 27, with sales terms of 3/15–60X.

(a) The cash discount may be taken through June 19 (or 15 days after June 4, when the goods were received).
(b) The last day for the discount will be July 22 (or 15 days after the July 7 postdate).
(c) The last day for the discount will be July 11 (or 75 days after April 27).

On occasion it may not be possible to pay an entire invoice during the discount period. Whatever payment is made, however, may be considered to be a net payment determined by calculating the discount on some part of the debt. That is, given a 3% cash discount, each 97¢ paid reduces the amount due by $1. These partial-payment problems may be solved using a variation of the % Paid formula: % Paid × Credit toward account = Net partial payment. (Some firms grant the cash discount only if the account is paid in full.)

Example 4 An invoice dated August 3 for $4,500 has terms of 3/20, n/60. What payment must the buyer make to reduce the obligation by $1,000 if paid by August 23?

The $1,000 portion of the account is subject to the 3% discount, as illustrated previously.

				3% discount means that 97% was paid:
Portion paid off		$1,000	*or*	% Pd × Credit = N
Less (3% × $1,000)	−	30		0.97 × $1,000 = N
Net payment		$ 970		$970 = N

Thus, a payment of $970 will deduct $1,000 from the balance due. That is, if $970 of the $4,500 account is paid, $4,500 − $1,000 = $3,500 remains to be paid.

Example 5 Sales terms of 2/15, n/30 were given on an invoice for $380 dated February 12. If a $245 payment is made on February 27, by how much has the debt been reduced? What amount is still due?

A 2% discount means that 98% was paid:

$$\% \text{ Pd} \times \text{Credit} = N$$

$$0.98C = \$245$$

$$C = \frac{245}{0.98}$$

$$C = \$250$$

Since $250 has been deducted from the debt by a $245 payment, $380 − $250 = $130 remains to be paid.

SECTION 2 PROBLEMS

Determine the applicable discount rate, if any, for the following:

	SALES TERMS	DATE OF INVOICE	DATE GOODS RECEIVED	DATE PAID	APPLICABLE DISCOUNT
1.	a. 2/10, 1/15, n/30	Mar. 5	Mar. 18	Mar. 30	
	b. 3/10, 2/20, n/30	Feb. 12	Feb. 15	Feb. 22	
	c. 1/10, EOM	July 22	July 28	Aug. 10	
	d. 2/15, ROG	Aug. 3	Aug. 12	Aug. 27	
	e. 3/10, n/30	Sept. 4 AS OF Oct. 1	Sept. 10	Oct. 10	
	f. 2/10–30X	Dec. 8	Dec. 15	Jan. 17	

	SALES TERMS	DATE OF INVOICE	DATE GOODS RECEIVED	DATE PAID	APPLICABLE DISCOUNT
2. a.	3/10, 1/15, n/30	Sept. 8	Sept. 15	Sept. 30	
b.	2/10, 1/15, n/60	June 16	June 20	June 30	
c.	2/10, EOM	Aug. 26	Sept. 1	Sept. 10	
d.	1/10, ROG	May 1	May 15	May 25	
e.	3/15, n/30	Mar. 5 AS OF Aug. 5	Aug. 1	Aug. 20	
f.	2/10–60X	Nov. 13	Nov. 30	Jan. 12	

Determine the amount due on each invoice if paid on the date indicated.

	DATE OF INVOICE	INVOICE AMOUNT	DATE GOODS RECEIVED	SALES TERMS	DATE PAID	AMOUNT DUE
3. a.	Feb. 18	$ 824	Feb. 26	2/10, n/30	Mar. 5	
b.	Apr. 9	600	Apr. 15	3/10, 2/15, n/30	Apr. 24	
c.	May 23	325	May 28	2/10, EOM	June 10	
d.	June 15	780	June 20	1/10, PROX	July 15	
e.	Nov. 8 AS OF Dec. 10	450	Dec. 1	2/10, n/30	Dec. 20	
f.	Sept. 3	860	Oct. 15	3/10, ROG	Oct. 25	
g.	Mar. 10	550	Apr. 9	4/10–50X	May 9	
4. a.	Apr. 12	$1,200	Apr. 15	1/10, n/30	Apr. 30	
b.	June 8	4,700	June 12	3/10, 2/15, n/30	June 23	
c.	Aug. 25	800	Aug. 28	3/10, EOM	Sept. 10	
d.	Oct. 26	940	Oct. 31	1/10, PROX	Nov. 15	
e.	Dec. 18 AS OF Feb. 10	5,500	Jan. 5	2/10, n/30	Feb. 20	
f.	Mar. 12	6,300	Mar. 25	3/10, ROG	Apr. 4	
g.	July 1	2,600	Aug. 1	2/10–30X	Aug. 10	

Complete the following if partial payment was made on each invoice within the discount period.

	INVOICE AMOUNT	SALES TERMS	CREDIT TOWARD ACCOUNT	NET PAYMENT AMOUNT	AMOUNT STILL DUE
5. a.	$2,000	2/10, n/30	$500		
b.	860	1/10, n/30	300		
c.	670	2/10, n/60		$441	
d.	790	3/15, n/30		388	
e.	560	4/10, n/60			$200

	INVOICE AMOUNT	SALES TERMS	CREDIT TOWARD ACCOUNT	NET PAYMENT AMOUNT	AMOUNT STILL DUE
6. a.	$3,500	2/10, n/30	$1,750		
b.	4,800	1/10, n/30	2,000		
c.	880	3/10, n/30		$485	
d.	5,600	4/10, n/30		960	
e.	2,200	2/10, n/30			$1,000

7. Determine the amount due on an invoice that totals $383.20, including a prepaid freight charge of $33.20. Trade discounts of 20% and 20% and sales terms of 3/10, n/30 are shown on the invoice. Assume that the amount due is paid within the cash discount period.

8. Trade discounts of 30% and 10% are to be taken on an invoice that totals $2,546.45, including prepaid shipping charges of $46.45. Sales terms on the invoice are 3/10, n/30. What is the amount due if paid within the cash discount period?

9. The following items were purchased on a February 24 invoice. The merchandise was subject to trade discounts of 30% and 25%. The invoice contained sales terms of 2/10, n/30 and had prepaid freight charges of $12.95. How much should be paid on March 5?

8 dozen packages of paper towels at $8.00 per dozen
10 dozen bottles of liquid detergent at $18.00 per dozen
2½ dozen cans of furniture polish at $1.20 per can
3 dozen bottles of glass cleaner at $1.50 per bottle

10. The following merchandise was subject to trade discounts of 20% and 30%. The invoice dated July 8 contained sales terms of 2/15, n/60 and prepaid shipping charges of $50. How much should be remitted on July 23?

3 dozen cardigan sweaters at $216 per dozen
2 dozen turtleneck sweaters at $141 per dozen
½ dozen V-neck sweaters at $180 per dozen
4 dozen zip tunics at $240 per dozen

11. Three invoices from Cather Industries contained sales terms of 2/15, 1/20, n/30. The invoices totaled $850 on March 3, $1,000 on March 10, and $620 on March 15. If all the invoices are paid on March 30, how much is due?

12. A retailer has received three invoices from Tucker Enterprises, each containing terms of 3/10, 2/15, n/30. The invoices are $500 on July 6, $1,350 on July 11, and $1,800 on July 16. If all the invoices are paid on July 26, how much is due?

13. An invoice dated June 22 for $1,200 contains sales terms of 2/15, 1/20, n/30, PROX. On July 15, the buyer wishes to make a payment that will discharge a fourth of his obligation.

a. How much should he remit on July 15?

b. On July 20, the buyer is able to pay the remainder of the debt. How much is his second payment?

c. What total amount is paid to discharge the $1,200 obligation?

14. An invoice dated April 26 for $5,000 contains sales terms of 3/10, 1/15, n/30 EOM. On May 10, the retailer wishes to make a payment that will discharge one-half of the debt.

a. How much should be remitted on May 10?

b. On May 15, the retailer pays the remaining balance. How much is this second payment?

c. How much is paid to discharge the total $5,000 debt?

15. Creative Video Co. sent the Greenberg Connection an invoice dated August 21 AS OF September 15 for $870. Sales terms on the invoice were 3/15, n/60.

a. If Creative Video Co. received a check for $421.95 on September 30, how much credit should be given the Greenberg Connection?

b. How much does the Greenberg Connection still owe?

16. A software company sent Computer Upgrades, Inc. an invoice dated September 18 AS OF November 1 for $1,500. Sales terms were 2/10, n/30.

a. If the software company received a check for $490 on November 11, how much credit should be given to Computer Upgrades, Inc.?

b. How much does Computer Upgrades, Inc. still owe?

CHAPTER 12 GLOSSARY

AS OF (or Postdating). A later date shown on an invoice, at which time the given sales terms begin.

Cash discount. A discount given for early payment of an invoice or monthly statement.

Complement (of a percent). A percent such that the sum of it and a given percent equals 100%; 100% minus the given percent. (Their fractional or decimal equivalents are also complements, equaling 1.)

Consumer. An individual who buys a product for personal use.

End of month (EOM). Sales terms that start at the beginning of the following month. (Same as "proximo" terms.)

Extra (X). Sales terms in which the discount period extends for the specified additional number of days. (Example: 2/10–50X indicates a total of 60 days.)

Invoice. A bill giving a detailed list of goods shipped or services rendered, with an accounting of all costs.

List price. A suggested retail price; price paid by the consumer.

Net cost. The total (invoice) cost after applicable trade and cash discounts.

Net cost rate factor. A factor that multiplies times the list price to produce the net cost; the product of the "percents paid" in a trade discount.

Ordinary sales terms. Sales terms in which the cash discount period begins on the date shown on the invoice.

Postdated sales terms. (See "AS OF.")

Proximo (PROX). (See "End of month.")

Receipt of goods (ROG). Sales terms in which the cash discount period begins on the day the merchandise arrives.

Retailer. A business that sells directly to consumers.

Sales terms. A code on an invoice or monthly statement that specifies the rate and the allowable time period for a cash discount, if offered.

Series (chain) discounts. Two or more trade discounts, with each succeeding discount being applied to the balance remaining after the previous discount.

Single equivalent discount. A trade discount that produces the same net cost as does a group of series discounts.

Trade discount. A discount off the list (suggested retail) price of merchandise, given to obtain the wholesale price.

Wholesaler. A business that buys merchandise from the manufacturer and sells it to retail stores.

CHAPTER 13

MARKUP

OBJECTIVES

Upon completing Chapter 13, you will be able to:

1. Define and use correctly the terminology associated with each topic.
2. Given a percentage markup based on cost (C) or on selling price (S), use the formula "$C + M = S$" to find (C: Section 1: Examples 1–4; Problems 1–18; S: Section 2: Examples 1–3; Problems 1–18):
   a. Selling price
   b. Cost
   c. Dollar markup.
3. a. Determine the percent of markup based on cost (C) using the formula "$\underline{?}$% of $C = M$" (Section 1: Example 5; Problems 7, 8, 13, 14; Section 2: Example 3; Problems 1, 2, 7–14).
   b. Similarly, compute the percent of markup based on selling price (S) using the formula "$\underline{?}$% of $S = M$" (Section 1: Example 5; Problems 3, 4, 7–18; Section 2: Problems 9, 10).
4. Calculate the selling price factor ($\overline{Spf}$), and use "$\overline{Spf} \times C = S$" to find selling price (S) or cost (C) (C: Section 1: Example 6; Problems 3–6, 9, 10; S: Section 2: Example 2; Problems 1–4, 7, 8, 11–16).
5. a. Find the selling price required on perishable items so that the goods expected to sell will provide the desired profit on the entire purchase (Section 3: Example 1; Problems 1, 2, 5, 7–12).
   b. Similarly, determine the regular selling price when some of the items will be sold at a given reduced price (Section 3: Example 2; Problems 3, 4, 6, 13–16).

In accordance with the American economic philosophy, merchants offer their goods for sale with the intention of making a profit. If a profit is to be made, the merchandise must be marked high enough so that the cost of the goods and all selling expenses are recovered. On the other hand, merchants cannot mark their goods so as to yield an unreasonably high profit, or else they will not be able to meet the competition from other merchants in our free enterprise system. To stay within this acceptable range, therefore, it is important that business people have a thorough understanding of the factors involved and the methods used to determine the selling price of their merchandise. The first step in this direction is to become familiar with the terms involved.

The actual **cost** of a piece of merchandise includes not only its catalog price (often with trade and/or cash discounts deducted), but also charges for insurance, freight, and any other expenditures incurred during the process of getting the merchandise to the store.

The difference between this cost and the selling price is known as **gross profit.** Other names for gross profit are **markup, margin,** and **gross margin.**

Gross profit is not all clear profit, however, for out of it must be paid business expenses, or **overhead.** Examples of overhead are salaries, rent, depreciation, lights and water, heat, advertising, taxes, and insurance. Whatever remains after these expenses have been paid constitutes the clear or **net profit.** The net profit remaining is often quite small; indeed, expenses sometimes exceed the margin, in which case there is a loss. Another way of defining gross profit is to say that gross profit is overhead plus the desired net profit (assuming that the merchandise sells at the regular price).

There are many retail terms used interchangeably by business people that are confusing to business students. A term such as "cost" can mean the amount the retailer must pay to the wholesaler or manufacturer. It can also mean the amount the consumer pays to the retailer for the same goods, although, at this point in the distribution channel, it would be called the "selling price" by the retailer. Caution should be exercised to determine the position in the distribution of merchandise (manufacturer to retailer or retailer to consumer) before solving problems in Chapters 13 and 14.

Conceptually, the regular selling price of an item is determined as follows:

	Catalog list price	
	− Trade and/or cash discounts	
	+ Freight and other charges	
	= Net cost	(C)
Markup (M) {	+ Overhead	(OH)
	+ Desired net profit	(P)
	= Regular marked selling price	(S)

The following diagram illustrates the breakdown of selling price. Markup (or gross profit) may be computed as a percentage of cost, or it may be based on sales. But regardless of which method is used, the same relationship between cost and markup always exists as shown in the diagram; specifically, Cost plus Markup (or margin or gross profit) always equals Selling price. From this relationship we obtain our fundamental selling price formula:

$$C + M = S$$

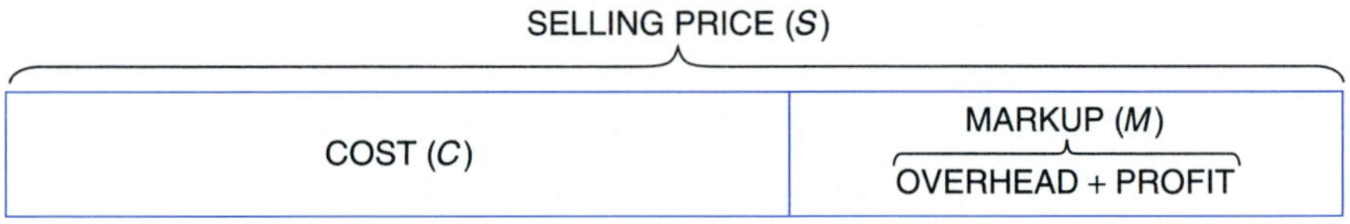

Observe that the dollar amount of markup (M) can be obtained in two ways:

$$S - C = M$$

or

$$\overline{OH} + P = M$$

where $\overline{OH}$ = overhead and P = profit.

Because the markup is composed of overhead and profit, we can express the basic "Cost + Markup = Selling price" formula in a slightly different manner, as follows:

$$C + (\overline{OH} + P) = S$$

SECTION 1

MARKUP BASED ON COST

Before attempting to calculate selling price, let us consider the following examples.

Example 1 Suppose that a furniture store bought a lamp for \$50 and sold it for \$70. Assume that the store's expenses (overhead) run 30% of cost. (a) How much markup (or margin or gross profit) did the store obtain? (b) What were the store's expenses related to this sale? (c) What amount of profit was made?

(a) Using the basic equation $C + M = S$ to find the markup,

$$C + M = S$$

$$\$50 + M = \$70$$

$$M = \$20$$

The markup (or margin or gross profit) on the sale was \$20.

(b) Overhead averages 30% of cost; thus,

$$30\% \times C = \overline{OH}$$

$$0.3(\$50) = \overline{OH}$$

$$\$15 = \overline{OH}$$

The furniture store's expenses for this sale were \$15.

(c) Since the overhead plus net profit make up the markup, then

$$\overline{OH} + P = M$$

$$\$15 + P = \$20$$

$$P = \$5$$

By using our previous diagram, we see that a net profit of \$5 was made on the lamp. The \$70 selling price would be broken down as shown below:

Net cost	(C)	=	\$50
+ Overhead	$(\overline{OH})$	=	15
+ Profit	(P)	=	5
= Selling price (S)		=	\$70

Example 2 The wholesale cost of a winter coat was \$80. A clothing store sold the coat for \$100. According to the store's last income statement, expenses have averaged 27% of sales income. (a) What was the markup on the coat? (b) How much overhead should be charged to this sale? (c) How much profit or loss was obtained on the coat?

(a) Starting with the basic equation as before,

$$C + M = S$$

$$\$80 + M = \$100$$

$$M = \$20$$

The selling price produced a markup of \$20.

(b) The expected overhead was 27% of the selling price; thus,

$$27\%S = \overline{OH}$$

$$0.27(\$100) = \overline{OH}$$

$$\$27 = \overline{OH}$$

Business expenses related to this sale would be $27.

(c) You can immediately realize that, if expenses were $27 while only $20 was brought in by the sale, there was obviously a $7 loss on the transaction. This calculation is shown in the following series of equations:

$$\overline{OH} + P = M$$

$$\$27 + P = \$20$$

$$P = \$20 - \$27$$

$$P = -\$7$$

The calculation produces a negative difference, and a negative profit (−$7) indicates a $7 loss on the sale.

Example 2 illustrates a vital fact about selling—just because an item sells for more than it had cost does not necessarily mean that any profit is made on the sale. Thus, it is essential that business people have the mathematical ability to take their expenses into consideration and price their merchandise to obtain the required markup.

Now let us consider some examples where the markup is based on cost.

Example 3 A merchant has found that his expenses plus the net profit he wishes to make usually run 40% of the cost of his goods. For what price should he sell a dress that cost him $50?

(a) We know that the markup equals 40% of the cost. Since 40% of $50 = $20, then

$$C + M = S$$

$$\$50 + \$20 = S$$

$$\$70 = S$$

(b) This same problem can be computed in another way. Since markup equals 40% of the cost ($M = 0.4C$), by substitution,

$$C + M = S$$
$$C + 0.4C = S$$
$$1.4C = S$$

But, since cost equals \$50,

$$1.4(\$50) = S$$
$$\$70 = S$$

It is customary in business to price merchandise as shown in Example 3(b), primarily because this is quicker than the method in Example 3(a), where markup was computed separately and then added to the cost (an extra step). In using method (b), you should make it a habit to *combine the* Cs *before substituting the dollar value of the cost.* Another advantage of method (b) is that it allows you to find the cost when the selling price is known. The following example illustrates how cost is found (as well as why the Cs should be combined before substituting for cost).

Example 4 A home improvement center marks up its merchandise 30% on cost. (a) What would be the selling price of a screen door that cost \$60? (b) What was the cost of a weed trimmer selling for \$65?

(a) Markup is 30% of cost, or $0.3C$.

$$C + M = S$$
$$C + 0.3C = S$$
$$1.3C = S$$
$$1.3(\$60) = S$$
$$\$78 = S$$

The door should sell for \$78 in order to obtain a gross profit (markup) of 30% on cost.

(b) Using the same 30% markup on cost when the selling price is \$65,

$$C + M = S$$
$$C + 0.3C = S$$
$$1.3C = S$$
$$1.3C = \$65$$
$$C = \$50$$

The weed trimmer cost \$50.

You can quickly see that markup based on cost is quite different from markup based on selling price. For instance, suppose that an article cost \$1.00 and sold for \$1.50. The \$0.50 markup is 50% of the cost, but only $33\frac{1}{3}\%$ of the selling price.

A merchant may wish to know what margin based on selling price is equivalent to a margin based on cost, or vice versa. To find the percent margin based on cost, one would use the formula

$$\underline{?}\,\% \text{ of } C = M$$

where C = cost and M = markup or margin. To determine the percent of margin based on selling price, you would use the formula

$$\underline{?}\,\% \text{ of } S = M$$

where S = selling price and M = margin.

Example 5 A circular saw listed for \$57 and required a \$3 freight charge. It sold for \$75. (a) What percent markup on cost did this represent? (b) What percent markup based on selling price did this represent?

The actual cost of the saw was \$57 + \$3 = \$60. The markup is then

$$C + M = S$$

$$\$60 + M = \$75$$

$$M = \$75 - \$60$$

$$M = \$15$$

(a) To find what percent of cost is markup,

$$\_\_\%C = M$$

$$\_\_\% \times \$60 = \$15$$

$$60x = 15$$

$$x = \frac{1}{4}$$

$$x = 25\% \text{ markup, based on cost}$$

(b) To find what percent of selling price is markup,

$$\_\_\%S = M$$

$$\_\_\% \times \$75 = \$15$$

$$75x = 15$$

$$x = \frac{1}{5}$$

$$x = 20\% \text{ markup, based on selling price}$$

Altogether, the components of this sale can be expressed as follows:

Catalog list price		= $57
+ Freight		= 3
= Net cost	(C)	= $60
+ Markup	(M)	= 15
= Selling price	(S)	= $75

To summarize, the $75 selling price is composed of a $60 cost plus a $15 markup. This $15 markup can be expressed as a percent either of the cost or of the selling price, as shown below. Notice that, for the given dollar markup, the percent based on cost (25%C) is larger than the percent based on selling price (20%S). This illustrates a general rule: The markup percent when based on cost is always larger than the markup percent when based on selling price. The diagram on the next page will help to visualize why: Since cost is smaller than selling price, the $15 represents a larger portion of the cost than it does of the selling price.

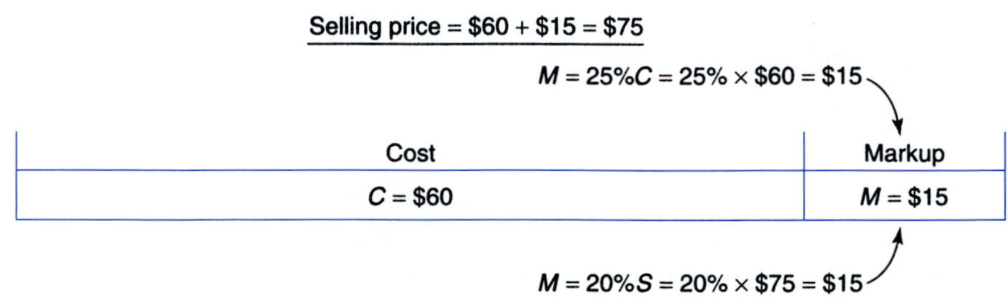

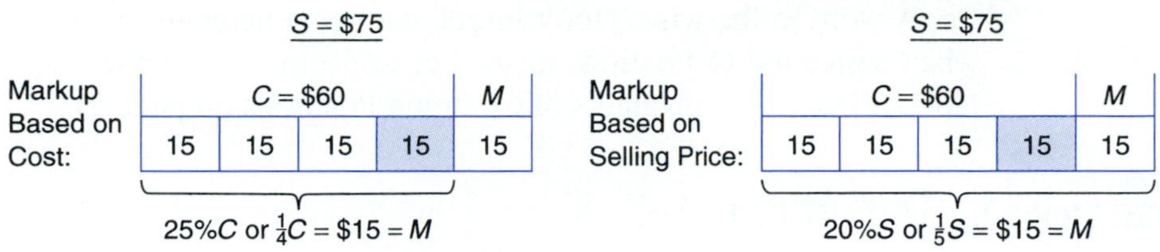

$\underline{S = \$75}$

Markup Based on Cost:	$C = \$60$				M
	15	15	15	15	15

$25\%C$ or $\frac{1}{4}C = \$15 = M$

$\underline{S = \$75}$

Markup Based on Selling Price:	$C = \$60$				M
	15	15	15	15	15

$20\%S$ or $\frac{1}{5}S = \$15 = M$

Note. In all cases where the same markup is used, the beginning calculations would be the same. Suppose that the markup is 25% on cost ($M = 0.25C$); then

$$C + M = S$$

$$C + 0.25C = S$$

$$1.25C = S$$

When the margin is 25% of cost, the basic calculation would always yield $1.25C = S$. This 1.25 is called the **selling price factor** ($\overline{Spf}$), a number that can be multiplied times the cost to obtain the selling price. This definition can be expressed by the formula

$$\overline{Spf} \times C = S$$

Thus, in actual practice, a retail store would not price each item by the complete process, starting with $C + M = S$. Rather, any store that uses a markup of 25% on cost would know that its selling price factor is 1.25. (The selling price factor of 1.25 could be expressed in any equivalent form: $1\frac{1}{4}$, $\frac{5}{4}$, or 125%.) Hence, the store would eliminate the first steps of the markup formula and would start with the selling price factor.

Example 6 Use $\overline{Spf} \times C = S$ to price an item that cost \$24, if the markup is 25% on cost.

To price an item that cost \$24, the store above would perform only the following calculation:

$$\overline{Spf} \times C = S$$

$$1.25C = S$$

$$1.25(\$24) = S$$

$$\$30 = S$$

The item would sell for \$30.

Note. After trying a few examples, you will observe that, *when markup is based on cost, the selling price factor can be found simply by adding 1 to the decimal or fractional equivalent of the markup.* (This is *not* true when markup is based on selling price.)

A word to the wise: Don't forget that many percents can be handled more easily when converted to fractions instead of to decimals. If review is needed, refer to the table of Percent Equivalents of Common Fractions on page 51.

SECTION 1 PROBLEMS

Answer the following questions about each problem. (1) How much markup does each sale include? (2) How much are the overhead expenses? (3) Is a profit or a loss made on the sale, and how much?

	COST	SELLING PRICE	OVERHEAD
1. a.	$80	$ 95	15% of cost
b.	55	73	40% of cost
c.	75	89	20% of cost
d.	33	40	25% of selling price
e.	72	96	15% of selling price
2. a.	$16	$ 20	20% of cost
b.	50	66	25% of cost
c.	30	35	30% of cost
d.	85	100	12½% of selling price
e.	56	64	15% of selling price

Find the missing information. (Compute the selling price before finding the markup.)

	PERCENT MARKUP ON COST	$\overline{Spf}$	COST	SELLING PRICE	MARKUP	PERCENT MARKUP ON SELLING PRICE
3. a.	40%		$90			
b.	20		60			
c.	12½		32			
d.	25			$15		
e.	60			64		
f.			54	66		
g.				36	9	
4. a.	25%		40			
b.	33⅓		60			
c.	20		20			
d.	50			96		
e.	10			55		
f.			36	42		
g.				48	18	

5. a. What would the selling price factor be for a markup of 42% on cost?
   b. What would the selling price be of an item costing $13?
   c. What dollar markup does this yield?
6. a. A store uses a markup of 20% on cost. What would its selling price factor be?
   b. What would the selling price be of an item costing $15?
   c. What dollar markup does this represent?
7. Men's socks cost $48 per one dozen pairs and sell for $5 each pair.
   a. What percent of cost is the markup?
   b. What percent of selling price is the markup?
8. Ballpoint pens cost $18 per dozen and sell for $2 each.
   a. What percent of cost is the markup?
   b. What percent of selling price is the markup?
9. a. What selling price factor is equivalent to a margin of 37½% on cost?
   b. Determine the selling price of each item that has the following cost: $8.00, $4.80, $6.40, and $16.40.
   c. What percent markup on selling price does this represent?
10. a. What selling price factor is equivalent to a gross profit of 40% on cost?
    b. Determine the selling price of each item having the following cost: $14, $9, $25, and $30.
    c. What percent markup on selling price does this represent?
11. a. After a markup of 30% on cost, an infant car seat sold for $78. What was the cost?
    b. A baby stroller selling for $85 was marked to obtain a gross profit of 25% on cost. Find the cost and the equivalent percent markup on retail sales price.
12. a. After a markup of 50% on cost, a bracelet sold for $135. What was the cost?
    b. A pair of earrings selling for $80 has been marked up so as to obtain a gross profit of 60% on cost. Find the cost and the equivalent markup on the retail sales price.
13. The catalog list price of a softball glove is $70. The Sports Center buys the gloves at a 50% trade discount and sells them at a 25% discount off the list price. Find
    a. The cost and the selling price of a glove
    b. The percent markup based on cost
    c. The percent markup based on selling price
14. The catalog suggested list price of a video game is $55. The Toy Chest buys the games at a 40% trade discount and sells them at a 10% discount off the suggested list price. Find
    a. The cost and the selling price
    b. The percent markup based on cost
    c. The percent markup based on selling price
15. A shipment of 15 smoke detectors had an invoice price of $115. Freight on the shipment was $20.
    a. At what price should each smoke detector be marked to obtain a 25% margin on cost?
    b. What percent margin on selling price does this represent?
16. Ten tennis rackets had an invoice price of $1,850 plus shipping charges of $100.
    a. What is the selling price of each racket if the markup on cost is 33⅓%?
    b. What is the markup percent on selling price?

17. A store buys 6 bottles of designer perfume for \$120 plus shipping charges of \$24. The store's markup is 60% on cost.
   a. What is the selling price per bottle of perfume?
   b. What percent markup on selling price does this represent?

18. Twelve golf clubs were purchased on a recent invoice by a sports shop for \$1,870 plus freight charges of \$50. The shop's markup is 25% on cost.
   a. What is the selling price per club?
   b. What percent markup on selling price does this represent?

SECTION 2 MARKUP BASED ON SELLING PRICE

Many business expenses are calculated as a percent of net sales. For example, sales representatives' commissions are based on their sales. Sales taxes, of course, are based on sales. When deciding how much to spend for advertising or research and development, a firm may designate a certain percent of sales. Many companies take inventory at the sales value. The income statement, which is one of two key reports used to indicate the financial condition of a company, lists the firm's sales, all its expenses, and its net profit; each of these items is then computed as a percent of net sales.

With so many other items calculated on the basis of sales, it is not surprising, then, that most businesses prefer to price their merchandise so that markup will be a certain percent of the selling price. Markup based on sales also offers a merchant the advantage of being able to refer to the daily cash register tape and immediately estimate the gross profit.

Example 1 Using a markup of 30% on sales, price an item that costs \$42.

As before, we start with the equation $C + M = S$. We know that the markup equals 30% of sales, or $M = 0.3S$. Thus, we substitute $0.3S$ for M in the formula $C + M = S$.

$$C + M = S$$

$$C + 0.3S = S$$

$$C + 0.3S - 0.3S = S - 0.3S$$

$$C = 0.7S$$

It is essential that you understand the preceding calculations. When $0.3S$ is substituted for M, there are then Ss on both sides of the equation. When an equation contains the same variable on both sides of the equals sign, the variables cannot be combined directly. So, to combine the Ss, "$0.3S$" must be subtracted from both sides of the equation. (When markup is based on selling price, it will always be necessary to subtract on both sides of the equation. This should be done before the actual dollar value is substituted for the cost.)

Now we are ready to find the selling price. The \$42 is substituted for the cost C:

$$C = 0.7S$$

$$\$42 = 0.7S$$

To find S, we must eliminate the 0.7. This is accomplished by multiplying both sides of the equation by the reciprocal, $\frac{1}{0.7}$:

$$\$42 = 0.7S$$

$$\frac{1}{0.7}(\$42) = 0.7S\left(\frac{1}{0.7}\right)$$

$$\$60 = S$$

Thus, the selling price is \$60. In this example, $\frac{1}{0.7}$ (or $\frac{10}{7}$ or 1.42857) is the selling price factor, because $\frac{1}{0.7}$ multiplied times the cost, \$42, produces the selling price, \$60. Notice that the \$60 retail price, which includes a markup of \$60 − \$42 = \$18, meets the stated requirement that markup be 30% of sales, as illustrated below:

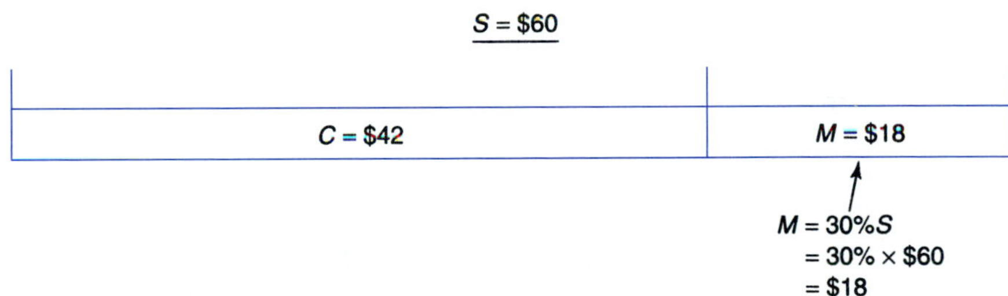

Sometimes it is more convenient and more accurate to convert the markup percent to a fraction than to a decimal. The following example illustrates such a problem.

Example 2 A department store prices its merchandise so as to obtain a gross profit of 33⅓% of the selling price. What selling price will be shown on a redwood picnic table that cost \$38?

The markup is to be 33⅓% of sales ($M = \frac{1}{3}S$). Recall that $S = \frac{3}{3}S$; thus,

$$C + M = S$$

$$C + \frac{1}{3}S = S$$

$$C + \frac{1}{3}S - \frac{1}{3}S = S - \frac{1}{3}S$$

$$C = \frac{2}{3}S$$

$$\frac{3}{2}C = \frac{2}{3}S\left(\frac{3}{2}\right)$$

$$\frac{3}{2}(\$38) = S$$

$$\$57 = S$$

The table would sell for \$57. The gross profit is \$57 − \$38 = \$19, which is 33⅓% of the \$57 selling price, as required.

Note. The value of the selling price factor ($\overline{Spf}$) when markup is based on retail selling price can be obtained as follows:

1. *Compute the complement of the given markup;* then
2. *Find the reciprocal of that complement.*

In Example 1, the given markup based on sales is 30%*S*. The complement is $\frac{1}{0.7}$; therefore, $\overline{Spf} = \frac{1}{0.7}$. Similarly in Example 2, the given markup is 33 1/3% of sales, or $\frac{1}{3}S$ The complement is $\frac{2}{3}$; thus, the selling price factor would be the reciprocal of the complement, or $\overline{Spf} = \frac{3}{2}$ or 1.5. (That is, $\overline{Spf} \times C = \frac{3}{2} \times \$38 = \$57$.)

Example 3 A set of patio furniture retailed for \$120. The store's expenses have run 19% of sales, and net profit has been 6% of sales. (a) What was the cost of the set? (b) What percent markup on cost does this represent?

(a) Recall that markup (gross profit) equals expenses plus net profit. Thus,

$$M = \overline{OH} + P$$

$$= 19\%S + 6\%S$$

$$M = 25\%S$$

Since $M = 25\%S$, then

$$C + M = S$$

$$C + 0.25S = S$$

$$C = 0.75S$$

$$= 0.75(\$120)$$

$$C = \$90$$

The markup was $120 − $90 = $30. The selling price of $120 would then be broken down as shown below:

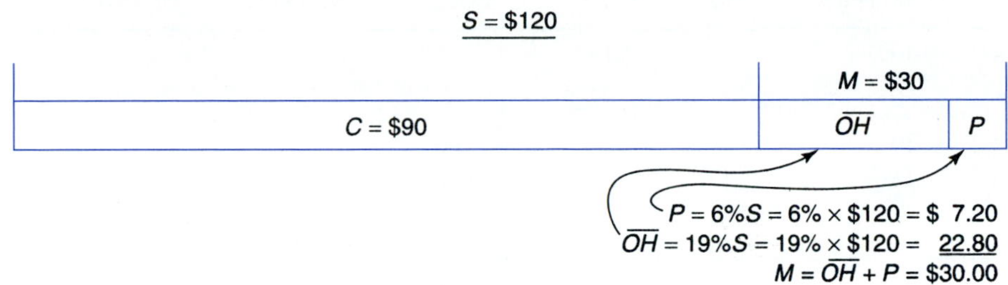

(b) Using the $90 cost and $30 markup, the markup percent (based on cost) is

$$\_\_\%C = M$$

$$\_\_\%(\$90) = \$30$$

$$90x = 30$$

$$x = \frac{30}{90}$$

$$= \frac{1}{3}$$

$$x = 33\frac{1}{3}\% \text{ markup on cost}$$

That is, $33\frac{1}{3}\%C = 33\frac{1}{3}\% \times \$90 = \$30$, which is the known markup.

SECTION 2 PROBLEMS

Complete the following.

	PERCENT MARKUP ON SELLING PRICE	$\overline{Spf}$	COST	SELLING PRICE	MARKUP	PERCENT MARKUP ON COST
1. a.	25%		$30			
b.	37½		35			
c.	33⅓			$42		
d.	50			12		
e.			18		$2	

	PERCENT MARKUP ON SELLING PRICE	$\overline{Spf}$	COST	SELLING PRICE	MARKUP	PERCENT MARKUP ON COST
2. a.	40%		$15			
b.	60		22			
c.	20			90		
d.	15			120		
e.			24		6	

3. a. If a markup of 30% on sales is used, what would the selling price factor be?
b. What would the selling price be for an article that cost $14?
c. What gross profit would be made?

4. a. What is the selling price factor for a markup of 35% on sales?
b. What would the selling price be for a package of batteries if the cost was $5.20?
c. What gross profit would be made on the sale?

5. A store uses a markup of 37½% on selling price. If this yields a gross profit of $15,
a. What was the selling price?
b. What was the cost?

6. A convenience store uses a markup of 60% on selling price. If this yields a gross profit of $5.40,
a. What was the selling price?
b. What was the cost?

7. a. Find the selling price factor corresponding to a 60% markup on retail sales.
b. Price items which cost $5.00, $6.40, $7.20, and $8.50.
c. What percent markup on cost does this represent?

8. a. What is the selling price factor for a 25% markup on retail sales?
b. Price items that cost $6, $9.90, $18, and $14.40.
c. What percent markup on cost does this represent?

9. A store purchased 50 blankets for a total cost of $2,215. It sold 20 blankets for $60 each, 16 for $54 each, 9 for $46 each, and 5 for $36 each.
a. How much gross profit was made on the entire transaction?
b. What percent markup on cost was made?
c. What percent markup on selling price was this?

10. A store purchased 70 toolboxes for a total cost of $912. It sold 45 of these toolboxes for $18, 10 for $15 each, and 15 for $12 each.
a. How much gross profit was made on the entire transaction?
b. What percent of cost was the markup?
c. What percent of selling price was the markup?

11. a. Would a markup of 50% on cost or 50% on selling price yield a larger gross profit?
b. Which would yield a larger margin, 20% on cost or 28% on selling price? (*Hint:* Compare selling price factors.)

12. a. Which would yield a larger gross profit, 30% on cost or 30% on selling price?
    b. Would 40% markup on cost or 31% on selling price yield a larger gross profit? (*Hint:* Compare selling price factors.)
13. Using a margin of 37½% on selling price, a store sold an article for $192. Determine its
    a. Selling price factor
    b. Cost
    c. Corresponding percent markup on cost
14. A store used a markup of 33⅓% on selling price and sold an item for $33. What was the
    a. Selling price factor?
    b. Cost?
    c. Corresponding percent markup on cost?
15. A home appliance manager wants to buy washing machines that can retail for $340. She must obtain a gross profit of 35% on selling price.
    a. What is the most she can pay for a washer that she plans to resell?
    b. What is the selling price factor?
16. The manager of an electronics department wants to purchase VCRs that will retail for $100. The department's required markup is 30% on selling price.
    a. What is the most that she can pay for a VCR that will be resold?
    b. What is the selling price factor?
17. A store's overhead is 25% of sales, and its net profit is 5% of sales. The store paid $42 less 16⅔% for an electric chain saw. At what price should the chain saw be priced?
18. A jewelry store expects its overhead to be 25% of sales, and its net profit is 50% of sales. At what price should it price a watch that cost the store $40 less 20%?

SECTION 3

MARKING PERISHABLES

A number of businesses handle products that perish in a short time if they are not sold. Examples of such businesses would be produce markets, bakeries, florists, dairies, and so forth. To a lesser extent, many other businesses would fall into this category, since styles and seasons change and the demand for a certain product diminishes. Clothing stores, appliance stores, and automobile dealers would be typical of this category.

These merchants know that not all of their goods will sell at their regular prices. The remaining items will have to be sold at a reduced price or else discarded altogether. Experienced business people know approximately how much of their merchandise will sell. They must, therefore, price the items high enough so that those which do sell will bring in the gross profit required for the entire stock.

The selling price on perishable merchandise is computed by the following basic procedure:

1. Determine the *cost* of the entire purchase.
2. Determine the *total selling price* of the entire purchase.
3. Determine the *amount of merchandise* that is expected to sell.
4. Express a *pricing equation:* Divide the total sales in item 2 by the amount of merchandise in item 3 to obtain the selling price (per item or per pound).

Example 1 A grocer bought 100 pounds of bananas at 45¢ per pound. Experience indicates that 10% of these will spoil. At what price per pound must the bananas be priced in order to obtain a margin of 50% on cost?

The bananas cost 100 pounds × 45¢, or \$45. Since the markup will be 50% on cost ($M = 0.5C$), then

$$C + M = S$$

$$C + 0.5C = S$$

$$1.5C = S$$

$$1.5(\$45) = S$$

$$\$67.50 = S$$

At least \$67.50 must be made from the sale of the bananas.

The grocer expects 10% or 10 pounds of bananas to spoil; thus, his needed \$67.50 in sales must come from the sale of the other 90 pounds. Now, if p = price per pound of bananas, then

$$90p = \$67.50$$

$$p = \frac{\$67.50}{90}$$

$$p = 75¢ \text{ per pound}$$

Note. When a fraction of a cent remains, retail merchants customarily mark the item up to the next penny even though the fraction may have been less than one-half cent. It should also be observed that in the preceding example, on all sales above the expected 90 pounds, the entire selling price will be additional profit.

Example 2 A delicatessen made 40 snack cakes at a cost of 30¢ each. About 15% of these will be sold as "day old" at 20¢ each. Find the regular price in order that the delicatessen can make 40% on cost.

The snack cakes cost 40 × 30¢ or $12 to bake, and the delicatessen uses a markup of 40% on cost ($M = 0.40C$). Thus,

$$C + M = S$$

$$C + 0.40C = S$$

$$1.4C = S$$

$$1.4(\$12) = S$$

$$\$16.80 = S$$

Now, 15% or 6 cakes will probably be sold at 20¢ each, which means that $1.20 will be earned at the reduced price. This leaves 34 cakes to be sold at the regular price. The regular sales and the reduced sales together must total $16.80. Thus, if p = regular price per cake,

$$\text{Regular} + \text{Reduced} = \text{Total}$$

$$34p + 6(\$0.20) = \$16.80$$

$$34p + \$1.20 = \$16.80$$

$$34p = \$15.60$$

$$p = \$0.46 \text{ per cake}$$

SECTION 3 PROBLEMS

Find the missing information.

	Quantity Bought	Cost per Unit	Total Cost	Markup on Cost	Required Sales	Percent to Spoil	Amt. to Sell	Selling Price
1. a.	70 lb.	$0.60		20%		10%		
b.	20 doz.	1.40		25		5		
c.	50	4.00		12½		4		
2. a.	100 lb.	$0.36		50		7%		
b.	125	6.00		40		4		
c.	110 doz.	1.80		30		10		

Complete the following, which involves part of the goods being sold at the regular price and the remaining amount at the reduced price.

	QUANTITY BOUGHT	COST PER UNIT	TOTAL COST	MARKUP ON COST	REQUIRED SALES	AMT. AT REGULAR PRICE	PERCENT AT REDUCED PRICE	AMT. AT REDUCED PRICE	REDUCED SELLING PRICE	REGULAR SELLING PRICE
3. a.	80 lb.	$0.40		20%			5%		$0.35	
b.	25	3.00		40			8		2.50	
4. a.	160 lb.	$0.90		50%			10%		$1.00	
b.	250	4.00		36			4		3.00	

5. Suppose that in Problem 1(a) the store actually sold all 70 pounds.

a. How much gross profit would the store earn?

b. What percent markup on cost would that be?

c. What percent markup on selling price would that be?

6. Suppose that in Example 2 the delicatessen actually sold 30 snack cakes at the regular price and the remainder at 30¢ each.

a. What gross profit would be obtained?

b. What percent markup on cost does that represent?

c. What percent markup on selling price does that represent?

7. A grocery store purchased 200 pounds of grapes at $0.60 per pound. Experience shows that 10% of the grapes will spoil. At what price per pound would the store make 15% markup on cost?

8. The Greenery Nursery purchased 500 pumpkins before Halloween at a cost of $2 each. From past experience, the manager knows that 10% of these will not sell. At what price per pumpkin must each be priced in order to obtain a margin of 55% on cost?

9. A market purchased 300 pounds of tomatoes at $0.72 per pound. The manager knows from past experience that 7% of the tomatoes will not sell. At what price per pound should the manager mark the tomatoes in order to receive a gross profit of 25% on cost?

10. A florist imports 40 orchids at a cost of $2.50 each. Past experience indicates that 15% of the orchids will not be sold. In order to make an 80% markup on cost, at what price per flower must the store sell each orchid?

11. A garden shop purchased 20 dozen cut daffodils at $6 per dozen. Past experience indicates that 5% of the daffodils will not sell. In order to make a 45% markup on cost, at what price per dozen should the flowers be marked?

12. A market purchased 400 dozen eggs at $0.60 per dozen. The produce manager knows that 20% of these eggs will be broken or will spoil. What must the marked price per dozen be if the store wants to make a 40% markup on cost?

13. The bakery department of a supermarket bakes 50 loaves of French bread at a cost of $0.55 a loaf. The bakery's markup is 60% of cost. Approximately 10% of the loaves will sell at a reduced price of $0.40 a loaf. What is the regular price per loaf?

14. A bakery bakes 120 cupcakes at a cost of $0.46 each. The baker's markup is 50% of cost. Approximately 20% of the cakes will sell at a reduced price of $0.48 each. What must the regular price per cake be?

15. The Ski King Shop purchased 75 insulated ski jackets at \$120 each. The shop has a markup of 35% on cost. About 12% of these jackets will sell at a reduced price of \$118 each. What must the regular list price of the jackets be?
16. A department store purchased 20 wheeled garment bags at \$150 per bag. The department's markup is 40% of cost. About 25% of the bags will be sold at a clearance price of \$180 each. What must the regular price of each bag be?

CHAPTER 13 GLOSSARY

Cost. The amount paid for merchandise, including the catalog price less any trade and/or cash discounts, plus charges for insurance, freight, and any other expenses incurred during the process of getting the merchandise to the store.

Gross profit. (See "Markup.")

Margin. (See "Markup.")

Markup (or gross profit or margin). The difference between cost and selling price of merchandise; the overhead and net profit combined:

$$C + M = S; \quad M = \overline{OH} + P$$

Markup is usually stated as a percent of cost:

$$\_\% \text{ of } C = M$$

or of selling price:

$$\_\% \text{ of } S = M$$

Net profit (P). The spendable profit remaining after cost of merchandise and overhead are deducted from selling price.

Overhead ($\overline{OH}$). Business expenses other than cost of merchandise.

Selling price factor ($\overline{Spf}$). A number (usually greater than 1) that multiplies times cost to produce selling price:

$$\overline{Spf} \times C = S$$

CHAPTER 14

MARKDOWN

OBJECTIVES

Upon completion of Chapter 14, you will be able to:

1. Define and use correctly the terminology associated with each topic.
2. Use "$C + M = S$," "%Pd × $L = N$," and "_____%$S = M$" to find (Section 1: Examples 1–4; Problems 1–16):
   a. Regular selling price (S_1)
   b. Reduced sales price (S_2)
   c. Percent of markup (at a sale price or regular price)
   d. Net cost (when trade discounts are offered).

   (The order in which the formulas are used depends on which information is given and which must be computed.)
3. a. Using "_____% of Original = Change," compute (Section 2: Examples 1, 4; Problems 1–4, 9, 10, 13–16):
      (1) Amount of markdown
      (2) Percent of markdown.
   b. When necessary, first use appropriate markup formulas to determine cost and regular selling price (Section 2: Examples 4, 5; Problems 1–4, 11–16).
4. When a markdown has occurred, find the total handling cost and use it to determine the amount and percent of (Section 2: Examples 2, 3, 5; Problems 1–18):
   a. Operating profit
   b. Breakeven
   c. Operating loss
   d. Absolute loss.

Essentially all retail concerns must sell at least some of their merchandise at reduced selling prices. Indeed, many firms must expect that some of their merchandise will not sell at all. A significant aspect of retail business thus involves pricing goods in anticipation of unsold items and computing markdowns to minimize losses. The following sections explain the mathematics required to compute reduced selling prices and markdowns.

SECTION 1

ACTUAL SELLING PRICE

Few, if any, retail merchants are able to sell their entire stock at its regular marked prices. Clearance sales and "storewide discounts" are common. In the accompanying Sears, Roebuck, and Co. ad, athletic shoes have been marked down 20% to 40% from the regular selling price. At the reduced selling price, or **actual selling price,** Sears will have a smaller markup than at its regular selling price. Recall that markup is composed of the overhead plus the profit. At the actual selling price, the business will try to recover its overhead cost and still make a profit. The following problems and examples involve markup when merchandise is sold at a sale or discount price.

sale 23.99
A. boys' Converse Crazed
Sizes 9-13 Full, 1-3.
Reg. 29.99
kids'
sale 34.99
D. women's Reebok Princess
Sizes 6-10M. Reg. 43.99
sale 39.99
E. men's Reebok Surreal
Sizes 7½-11, 12M.
Reg. 54.99
A
D
E
sale 47.99
B. women's adidas Fortitude
Sizes 6-10M. Reg. 59.99
B women's
C
men's
F
sale 47.99
F. men's New Balance 606
Sizes 7½-11, 12M.
Reg. 59.99
sale 39.99
C. women's Avia 1815
Sizes 6-10M. Reg. 49.99
save 20-40%
all athletic shoes
plus, save 20-40% on all men's casual and dress shoes, steel-toe and insulated work boots

Using the illustration from Chapter 13, Section 1 (page 340), and adding two additional lines, we see how the cost through the actual selling price flows:

	Cost (catalog list price)	
−	Trade and/or cash discount	
+	Freight and other charges	
=	Net cost	(C)
+	Overhead	$(\overline{OH})$
+	Desired net profit	(P)
=	Regular marked selling price	(S_1)
−	Consumer discount or markdown	$(\overline{MD})$
=	Actual selling price	(S_2)

Both the **regular** and the **reduced prices** of merchandise are commonly known by several different terms that are used interchangeably, as follows:

REGULAR PRICE (S_1)	ACTUAL SELLING PRICE (S_2)
Regular marked price	Reduced price
Regular selling price	Sale price
Marked price	Net selling price
List price	Price after reduction

When $M =$ _____$\%S$ is used for percent of markup based on sales, S may represent either the regular selling price (S_1) or the reduced selling price (S_2), depending on the information given. By using S_1 or S_2 in the formulas, confusion over which S is meant will be eliminated. There will be two Ms also, M_1 for markup at the regular price and M_2 for markup at the reduced or actual selling price. At a reduced price, markdown is represented by $\overline{MD}$. The **markdown** is the amount or percent by which the regular selling price is reduced to obtain the actual selling price (or "sale" price). The following illustration shows the breakdown of selling price (S) both for a regular marked price (S_1) and for a reduced sale price (S_2).

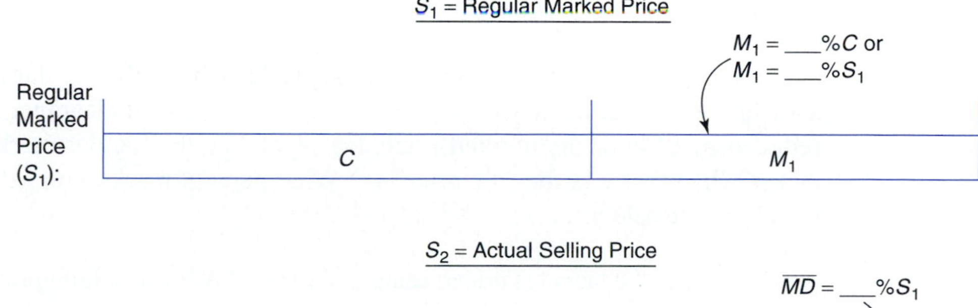

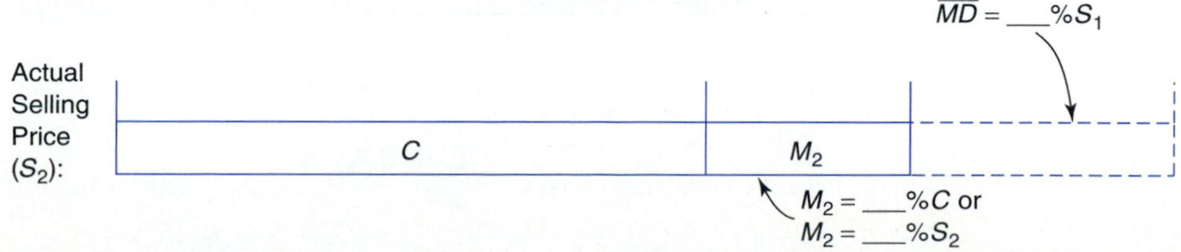

Example 1 The regular marked price of a suitcase is \$70, including a markup of 25% on *cost.* During a special sale, the case was reduced so that the store obtained a 12½% markup on the clearance sale price. What were (a) the cost and (b) the sale price?

(a) When $M_1 = 0.25$ of cost,

$$C + M_1 = S_1$$

$$C + 0.25C = S$$

$$1.25C = \$70$$

$$C = \$56$$

(b) When $M_2 = 0.125$ of the sale price, then

$$C + M_2 = S_2$$

$$C + 0.125S = S$$

$$C = S - 0.125S$$

$$C = 0.875S$$

$$\$56 = 0.875S$$

$$\$64 = S_2$$

The sale price was \$64, giving a margin of \$64 − \$56 = \$8. Notice that this \$8 markup is 12.5%S = 12.5% × \$64, as stipulated.

For situations in which a "sale" price is involved, you may often find it helpful to think through the entire pricing process from the time an item enters the store until it is sold "on sale."

Example 2 A markup of 40% on the selling price was used to obtain the regular marked price of a suede jacket that had cost \$48. During an end-of-season clearance, the jacket was reduced by 25% of the marked price. (a) What was the regular marked price of the jacket? (b) What was the sale price? (c) What percent markup on sales was made at the clearance sale price?

(a) First, the jacket is priced using a markup of 40% on selling price ($M_1 = 0.4S_1$):

$$C + M_1 = S_1$$

$$C + 0.4S = S$$

$$C = S - 0.4S$$

$$C = 0.6S$$

$$\$48 = 0.6S$$

$$\$80 = S_1$$

The regular marked price (or regular selling price) of the jacket was \$80.

(b) Later this regular price was reduced: A 25% discount means that 75% (or $\frac{3}{4}$) of the regular marked price (or list price) was paid:

$$\% \text{ Pd} \times \text{List} = \text{Net sales price } (S_2)$$

$$0.75(\$80) = S_2$$

$$\$60 = S_2$$

The clearance sale price (S_2) of the jacket was \$60.

(c) Before finding the percent markup based on the sale price, we must first determine the actual dollar markup at the sale price:

$$C + M = S \qquad\qquad \_\%S_2 = M_2$$

$$M = S - C \qquad\qquad \_\%(\$60) = \$12$$

$$M = \$60 - \$48 \qquad\qquad 60x = 12$$

$$M = \$12 \qquad\qquad x = 20\%$$

Thus, 20% of the sale price (20% × \$60 = \$12) was the gross profit.

Example 3 During a special promotion, Avery Jewelers marked a watch down 30%, selling it for \$42. Even at this reduced sale price, Avery still made a gross profit (markup) of 15% on sales. What percent on selling price would Avery have made if the watch had sold at its regular marked price?

Before we can use the formula $\_\%S = M$ to determine the *percent* markup on sales at the regular price, we must first know the regular markup (M_1) and the regular sales price (S_1). Thus, we must first calculate (a) the cost and (b) the regular marked price (S_1). From these we can determine (c) the regular markup (M_1). Then we can find (d) the percent markup on sales.

(a) At the clearance sale price of \$42, the markup is 15% on sales (or $M_2 = 0.15S_2$):

$$C + M_2 = S_2$$

$$C + 0.15S = S$$

$$C = S - 0.15S$$

$$C = 0.85(\$42)$$

$$C = \$35.70$$

The watch cost the jeweler \$35.70.

(b) The watch was reduced by 30% of the regular price to obtain the \$42 sale price:

$$\% \text{ Pd} \times L = \text{Net sale price } (S_2)$$

$$0.7L = \$42$$

$$L = \frac{\$42}{0.7}$$

$$L = \$60$$

(c) Since the watch was regularly priced at \$60, the regular markup (M_1) was \$60 − \$35.70 = \$24.30.

(d) Thus, at the regular price,

$$\_\_\_\%S_1 = M_1$$

$$\_\_\_\%(\$60) = \$24.30$$

$$60x = \$24.30$$

$$x = 40.5\%$$

The regular markup, based on sales, was 40.5% of the regular marked price ($40.5\%S = 40.5\% \times \$60 = \$24.30$). The regular price and the reduced (sale) price are illustrated in the following diagram.

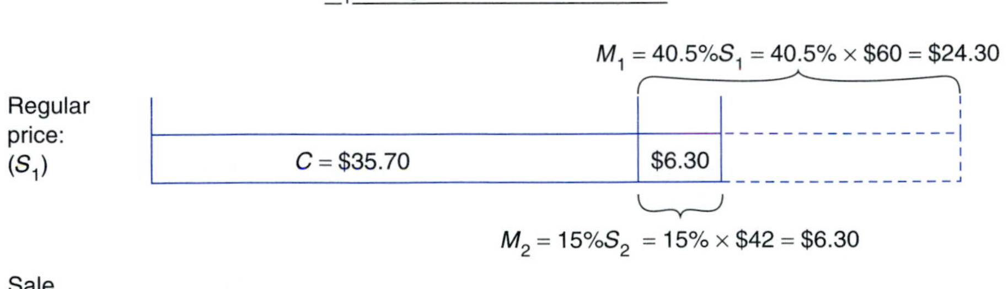

Many businesses deal in discounts every day. These merchants must price their goods very carefully, so that they can allow the customer a "discount" off the marked price and still make their desired markup on the sale. The discount price thus appears to be a "sale" price (and is computed similarly), although in reality this is the price at which the goods are always expected to sell; the business never expects to obtain the "regular marked price" for its merchandise.

Example 4 Caldwell Furniture paid a wholesale distributor \$750 less 20% and 20% for a dining room suite. Caldwell must obtain a markup of $33\frac{1}{3}\%$ on its actual selling price. At what price must the dining room suite be marked so that Caldwell can allow 20% and 10% off this marked price and still make its $33\frac{1}{3}\%$ markup?

This solution involves three parts: (a) to find the cost, (b) to determine what the actual selling price (the apparent "sale" price) will be, and (c) to determine what "regular" marked price, less discounts of 20% and 10%, will leave the actual (net) selling price that Caldwell needs.

(a) First, find the actual cost after discounts of 20% and 20%:

$$\% \text{ Pd} \times L = \text{Net cost}$$

$$(0.8)(0.8)(\$750) = N$$

$$0.64(750) = N$$

$$\$480 = N$$

The dining room suite cost Caldwell \$480 after trade discounts.

(b) A $33\frac{1}{3}\%$ markup on selling price (S_2) is used to determine the actual selling price (which will later be made to appear as a "sale" price):

$$C + M_2 = S_2$$

$$C + \frac{1}{3}S = S$$

$$C = S - \frac{1}{3}S$$

$$C = \frac{2}{3}S$$

$$\$480 = \frac{2}{3}S$$

$$\frac{3}{2}(480) = S$$

$$\$720 = S_2$$

Caldwell Furniture will actually sell the suite for \$720.

(c) Finally, we compute a "regular" marked price (S_1) which, after discounts of 20% and 10%, will leave the actual $720 net selling price (the apparent "sale" price):

$$\% \text{ Pd} \times L = \text{Net selling price}$$

$$(0.8)(0.9)L = \$720$$

$$0.72L = \$720$$

$$L = \$1{,}000$$

The marked price on the dining room suite will be $1,000 (although Caldwell knows that the suite will never be sold at that price). These marked and actual prices are illustrated in the following diagram.

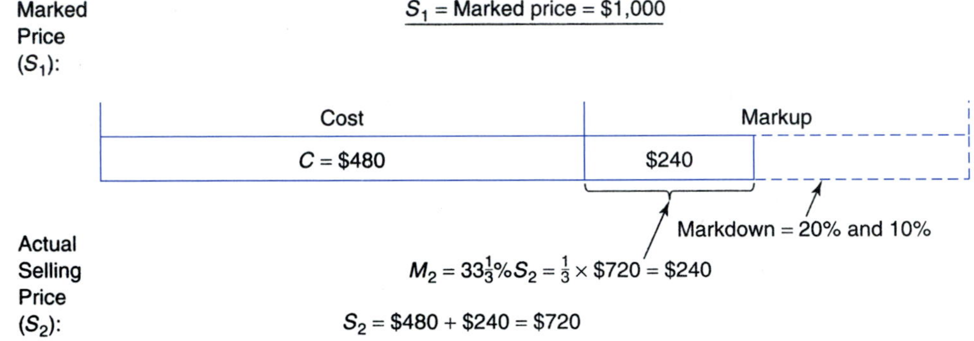

SECTION 1 PROBLEMS

Compute the missing items in the following problems.

	COST	PERCENT REGULAR MARKUP	REGULAR PRICE	PERCENT DISCOUNT	SALE PRICE	PERCENT SALE MARKUP
1. a.		40% of *C*	$ 70	X		20% of *S*
b.		33⅓% of *C*	64	X		16⅔% of *S*
c.	$ 28	30% of *S*		20%		?% of *S*
d.	30	25% of *S*		10		?% of *S*
e.		?% of *S*		10	$ 45	40% of *S*
f.		?% of *S*		30	175	15% of *S*
g.	$400 less 40% = ?	?% of *S*		33⅓/20		25% of *S*
h.	$200 less 20/20 = ?	?% of *S*		37½/20		20% of *S*

	COST	PERCENT REGULAR MARKUP	REGULAR PRICE	PERCENT DISCOUNT	SALE PRICE	PERCENT SALE MARKUP
2. a.		30% of *C*	$ 52	X		20% of *S*
b.		50% of *C*	900	X		25% of *S*
c.	$ 36	40% of *S*		25%		?% of *S*
d.	52	35% of *S*		20		?% of *S*
e.		?% of *S*		37.5	$ 30	10% of *S*
f.		?% of *S*		40	150	30% of *S*
g.	$ 40 less 30%	?% of *S*		30/20		50% of *S*
h.	$600 less 25/20	?% of *S*		20/20		10% of *S*

3. The regular price of a gold ring was $87, which included a 45% markup on cost. During a holiday sale, the ring was marked down so that the sale price produced a 25% gross profit on sales.

a. What was the cost of the ring?

b. What was the sale price?

4. A store sells a breadmaker for a regular price of $90 which produced a 25% gross profit on cost. During a sale, the price was reduced so that the sale price produced a 10% gross profit on sales.

a. What was the cost?

b. What was the sale price?

5. Steven's Furniture usually sells a dining suite with a markup of 37½% of the selling price. During a sale, the suite that costs $800 received a 25% reduction off its regular price.

a. What was the marked price?

b. How much was the sale price?

c. What percent of the sale price was the margin?

6. Julie's Interiors marks its furniture to obtain a markup of 33⅓% of the selling price. During a sale, a sofa that cost $424 received a 20% reduction off its regular price.

a. What was the regular marked price of the sofa?

b. What was the sale price?

c. What percent of the sale price was the markup?

7. An electric griddle cost a store $18. It was priced to obtain a markup of 55% on the regular selling price. Later, it was marked down 25% during a special sale. Find

a. The regular marked price

b. The sale price

c. The percent markup on sales at the special sale price

8. A canister vacuum cleaner that cost a department store $126 was marked to obtain a gross profit of 30% on regular selling price. During an end-of-year sale, it was marked 10% off the regular marked price.

a. What was the regular marked price?

b. What was the sale price?

c. What percent of the sale price was the markup?

9. After a markdown of 25%, a garage door opener sold for $90. At this sale price, the store made a 40% margin on sales. What had been
   a. The regular price of the garage door opener?
   b. The cost?
   c. The percent markup on the regular selling price?
10. At an after-holiday sale, a store marked a crystal clock down 15%, making the sale price $51. At this sale price, the store made a 20% gross profit on sales. Determine
   a. The regular price
   b. The cost
   c. The percent markup on the regular selling price
11. A cutlery set was marked down 12½%, making the sale price $63. At this sale price, the retailer made a 20% gross profit on sales. Determine
   a. The regular price of the cutlery
   b. The cost
   c. The percent markup on the regular selling price
12. Flannel dress pants were marked down 40%, making the sale price $45. At this sale price, the store made a 30% gross profit on sales. Determine
   a. The regular price of the pants
   b. The cost
   c. The percent markup on regular selling price
13. A portable compact disc player cost a merchant $140 less 30/25. After marking the player so that the merchant could allow a "40% discount," he still made a markup of 33⅓% on the actual selling price. What had been
   a. The cost?
   b. The actual selling price?
   c. The marked price?
14. A pair of skis cost a sport shop $200 less 40/10. After marking the skis so that the shop could allow a "20% discount," it still made a markup of 25% on the actual selling price. What was
   a. The cost?
   b. The actual selling price?
   c. The marked price?
15. A dress cost a store $50 less 30%. The store marked the dress so that it could allow a "12½% discount." At this sale price, the store still made a gross profit of 60% of the actual selling price. Determine
   a. The cost
   b. The sales price
   c. The marked price
16. CDs cost a retailer $15 less 50%. The retailer marked the CDs so that he could allow a "33⅓% discount." At this sale price, the store still made a gross profit of 40% of the actual selling price. Determine
   a. The cost
   b. The sales price
   c. The marked price

MARKDOWN VERSUS LOSS

In the previous section we have seen that merchandise must often be sold at a reduced price. This reduction may be expected and planned for well in advance, as in the case of seasonal clearance sales on clothing. Other reductions may be less predictable, as when the merchants' association decides to have a special citywide promotion or when the competition has unexpectedly lowered its price.

Regardless of what motivates a price reduction, there are guidelines that merchants generally follow in determining how much to mark down their merchandise. The wholesale cost plus the overhead on a sale is known as the **total handling cost.** Businesses prefer to price merchandise so that the selling price exceeds the total handling cost ($C + \overline{OH}$); thus, an **operating profit** exists. If the merchant's selling price only recovers the total handling cost, no profit will be made, but no loss will be made either. This critical value between profit and loss is often called the **breakeven point.**

If the merchant fails to recover the wholesale cost plus all the operating expenses, there has been an **operating loss.** If not even the wholesale cost of the article was recovered, the transaction has resulted in an **absolute** (or **gross**) **loss.** These various sales conditions are shown in the following diagram, which assumes a sale price lower than cost.

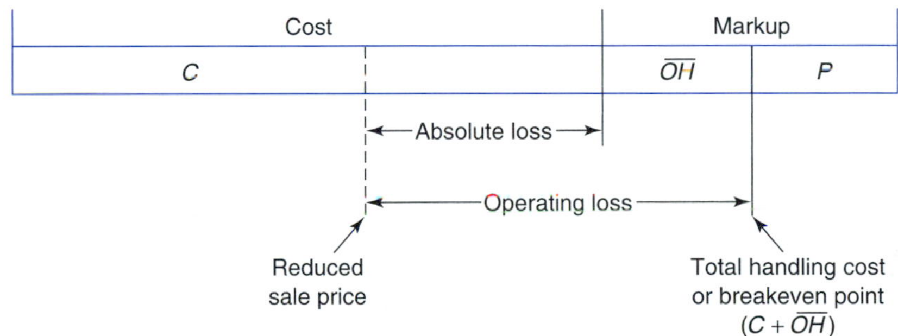

As described above, the following relationships* exist among actual selling (or sale) price ($\overline{ASP}$), total handling cost ($\overline{THC}$), and wholesale cost (C):

If $\overline{ASP} > \overline{THC}$, then a net profit exists.
If $\overline{ASP} = \overline{THC}$, then breakeven exists.
If $\overline{ASP} < \overline{THC}$, then an operating loss exists.
If $\overline{ASP} < C$, then an absolute loss exists.

We will be interested here in determining whether or not any profit was made following a markdown, as well as certain percentages related to markdown.

*The symbol “>” indicates “is greater than,” and “<” denotes “is less than.”

Example 1 A lamp that sold for \$120 was reduced to \$84. (a) What was the markdown on the lamp? (b) By what percent was the price reduced?

(a) The lamp was reduced by \$120 − \$84 = \$36.

(b) ___% of Original selling price = Change?

$$\_\_\_\%(\$120) = \$36$$

$$120x = 36$$

$$x = \frac{36}{120} \text{ or } \frac{3}{10}$$

$$x = 30\% \qquad \text{Reduction on selling price}$$

Example 2 The lamp in Example 1 had cost the furniture store \$75. The store's operating expenses run 20% of cost. Was an operating profit (net profit) or an operating loss made on the lamp?

The breakeven point or total handling cost = $C + \overline{OH}$. Since $\overline{OH}$ = 20% of cost and C = \$75, then

$$\begin{aligned} \text{Total handling cost} &= C + \overline{OH} \\ &= C + 0.2C \\ &= 1.2C \\ &= 1.2(\$75) \\ \overline{THC} &= \$90 \end{aligned}$$

In order to recover all operating expenses, the lamp's actual selling price (or sale price) should have been \$90. Since the lamp sold for less than this breakeven figure, the store's operating loss on the lamp was the difference between total handling cost and actual selling price:

Total handling cost	\$90
Actual selling price	− 84
Operating loss	\$ 6

Example 3 The regular marked price of a woman's suit was \$144. The suit cost \$120, and related selling expenses were \$18. The suit was later marked down by 25% during a clearance

sale. (a) What was the operating loss? (b) What was the absolute loss? (c) What was the percent of absolute loss (based on wholesale cost)?

The total handling cost ($C + \overline{OH}$) of the suit was \$120 + \$18 = \$138. The actual net selling price (or sale price) of the suit was

$$\% \text{ Pd} \times L = \text{Net}$$

$$0.75(\$144) = \text{Net}$$

$$\$108 = \text{Net selling price } (S_2)$$

(a) Since the suit sold for less than its total handling cost,

Total handling cost	\$138
Actual selling price	− 108
Operating loss	\$ 30

(b) The absolute or gross loss is the difference between the wholesale cost and the actual selling price:

Wholesale cost	\$120
Actual selling price	− 108
Absolute (or gross) loss	\$ 12

(c) Absolute loss or gross loss is always computed as some percent of the wholesale cost, as follows:

<u>What percent</u> <u>of</u> <u>wholesale cost</u> <u>was</u> <u>the absolute loss?</u>

$$\_\_\%(\$120) = \$12$$

$$120x = 12$$

$$x = \frac{12}{120} \text{ or } \frac{1}{10}$$

$$x = 10\% \text{ absolute (or gross) loss}$$

The operating and gross loss incurred in this sale may be visualized by referring to the following diagram.

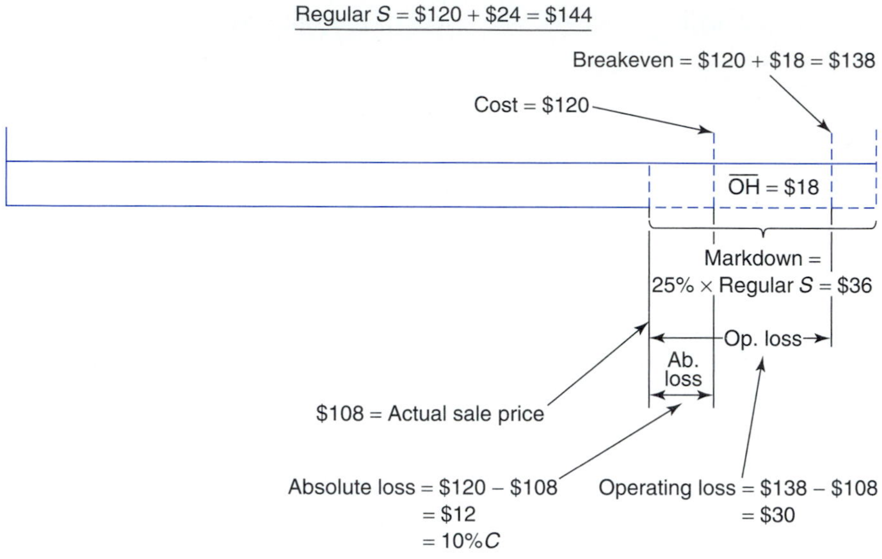

Example 4 A dress cost a department store \$40. The store's operating expenses are 35% on cost, and normal profit is 15% on cost. (a) What is the regular selling price of the dress? (b) What is the maximum percent of markdown that can be offered without taking an operating loss?

(a) The normal markup of overhead plus net profit is

$$M_1 = \overline{OH} + P$$

$$= 35\%C + 15\%C$$

$$M_1 = 50\%C$$

Therefore, using $M = 50\%\ C$ and a cost of \$40, the regular selling price is

$$C + M_1 = S_1$$

$$C + 0.5C = S$$

$$1.5C = S$$

$$1.5(\$40) = S$$

$$\$60 = S_1 \text{ (regular selling price)}$$

(b) The operating expenses are 35% or $\frac{7}{20}$ of cost; thus,

$$\text{Total handling cost} = C + \overline{OH}$$
$$= C + 0.35C$$
$$= 1.35C$$
$$= 1.35(\$40)$$
$$\overline{THC} = \$54$$

Regular selling price	\$60
Total handling cost	− 54
Maximum markdown	\$ 6

The lowest sale price the store could offer without taking any operating loss would be the breakeven point, \$54. This means that the dress could be marked down \$6 from its regular selling price. The percent of reduction would be

$$\underline{\quad}\% \text{ of Original} = \text{Change}$$
$$\underline{\quad}\%(\$60) = \$6$$
$$60x = 6$$
$$x = \frac{6}{60} \quad \text{or} \quad \frac{1}{10}$$
$$x = 10\% \text{ reduction in price}$$

Example 5 A variety store bought 200 pen sets at \$2 each. The regular price of \$2.50 included 15% on cost for overhead. After 180 sets were sold at the regular price, the remaining 20 sets were closed out at \$1 each. (a) How much net profit, operating loss, or absolute loss was made? (b) What percent of wholesale cost was it?

(a) Cost: $200 \times \$2 = \$400$

Total sales:	
$180 \times \$2.50 =$	\$450
$20 \times \$1.00 =$	+ 20
	\$470

Total handling cost:

$$\overline{THC} = C + \overline{OH}$$
$$= C + 0.15C$$
$$= 1.15C$$
$$= 1.15(\$400)$$
$$\overline{THC} = \$460$$

Total sales	\$470
Total handling cost	− 460
Net profit	\$ 10

(b) $\_\_\_\%\text{ of wholesale cost} = \text{Net profit}$

$$\_\_\_\%(\$400) = \$10$$

$$400x = 10$$

$$x = 2.5\% \quad \text{Net profit on cost}$$

SECTION 2 PROBLEMS

Complete the following.

	Regular Selling Price	Markdown Percent	Markdown Amount	Sale Price	Wholesale Cost	Overhead	$\overline{THC}$	Operating Profit or (Loss)
1. a.	$50	40%			$22	$ 4		
b.	30		$ 9		19	3		
c.	45			$35		8		$4
d.	18					7	$19	(3)
e.			6		50		59	(5)
f.		20		32		8		6
2. a.	$20	10%			$14	$ 3		
b.	15		$ 3		9	5		
c.	60			$36		7		$5
d.	80					10	76	(6)
e.			15		28		42	(9)
f.		50		12		4		2

	Regular Selling Price	Markdown Percent	Markdown Amount	Sale Price	Wholesale Cost	Overhead	$\overline{THC}$	Operating Loss	Gross Loss Amount	Gross Loss Percent
3. a.	$36		$ 6		$32	$3				
b.	60	40%			40			$10		
c.		20		$32		8	$43			
d.		30		63		4			$7	
4. a.	$40		$10		$35	$2				
b.	48	33⅓			40			$12		
c.		25		$42		6	51			
d.		20		72		8			$8	

5. A man's suit that regularly sold for $210 was advertised for sale at a 30% discount. The suit cost the department store $130, and operating expenses were 25% of cost. How much operating profit or loss was made on the sale of the suit?

6. A 30% discount was offered on a wrench set that regularly sold for $35. The set cost $23, and operating expenses were 15% of cost. How much operating profit or loss was made on this sale?

7. The price of a $40 pair of shoes was reduced by 25%. If the shoes cost $26, and operating expenses were 20% of cost, how much operating profit or loss was made on the sale?

8. A $45 perfume and body lotion gift set was reduced by 40%. The gift set cost the shop $25, and operating expenses were 20% of cost. How much operating profit or loss was made on the sale?

9. A $54 brass candlestick was sold for $45.
   a. What percent markdown did this represent?
   b. If the dealer had paid $50 less 20% and 15% for the candlestick, and operating expenses were $5.40, how much operating profit or loss was made?

10. A $120 dinnerware set was sold for $84.
    a. What percent markdown did this represent?
    b. If the store paid $80 less 10% and 20% for the dinnerware, and operating expenses were $10.40, how much operating profit or loss was made?

11. A store paid $140 for a solid wood kitchen table. The store's expenses are 20% of the regular selling price, and the expected net profit is 10% of regular selling price. During a special sale, the table was sold at a 50% discount.
    a. What was the regular price of the table?
    b. How much operating loss did the store experience?
    c. How much and what percent gross or absolute loss did the store suffer?

12. A jewelry store paid $90 for a blue topaz ring. The store's expenses are 10% of the regular selling price, and the expected net profit is 40% of sales. During a special sale, the ring was sold at a 60% discount.
    a. What was the regular price of the ring?
    b. What was the operating loss?
    c. How much and what percent of gross or absolute loss did the store suffer?

13. A garden shop purchases daylily bulbs for $7.50 per dozen, less 20% and 20%. The shop marks up the bulbs to provide for expenses of 30% on cost and a net profit of 20% on cost.
    a. What is the regular selling price per bulb?
    b. What is the maximum markdown that the shop could allow without taking an operating loss?
    c. What percent markdown would this be?

14. A computer store purchases blank disks for $7 per dozen less 33⅓/10. The marked price includes 40% on cost for expenses and 20% on cost for net profit.
    a. What does a single disk regularly sell for?
    b. What is the largest markdown that could be allowed on a disk without experiencing an operating loss?
    c. What percent markdown would this represent?

15. A cosmetic department purchases lipsticks for $90 per dozen, less 40/20. Expenses run 25% of cost, and net profit is expected to be 50% of cost.
    a. What should the regular price per lipstick be?
    b. What is the maximum markdown that the department could allow and still break even?
    c. What percent markdown would this represent?

16. A dozen candles are purchased for $30 less 20/20. Expenses run 25% of cost, and net profit is expected to be 10% of cost.
    a. What should the regular price per candle be?
    b. What is the maximum markdown that the shop could allow and still break even?
    c. What percent markdown would this represent (to the nearest tenth)?
17. A store purchased 200 Halloween masks for $2.80 each. The store sold 180 of the Halloween masks for $3.50 each. The last 20 of the masks were sold on Halloween Day for $2 each. The regular price included a markup of 15% on cost for overhead.
    a. What were the total wholesale cost, total handling cost, and total sales?
    b. Was a net profit, an operating loss, or a gross loss made on the 200 masks?
    c. How much was it, and what percent of wholesale cost did it represent (to the nearest tenth)?
18. A store paid $40 each for 50 35mm cameras. It sold 40 cameras for $55 each. The price for the last 10 cameras was $35 each. The regular markup for overhead was 20% on cost.
    a. What were the total wholesale cost, total handling cost, and total sales?
    b. Was a net profit, an operating loss, or a gross loss made on the 50 cameras?
    c. How much was it, and what percent of the wholesale cost did it represent?
19. A candy shop purchased 25 1-pound boxes of candy, paying $4 each. It sold 18 boxes for $6 each and 7 boxes for $5 each. The regular markup for overhead was 30% of cost.
    a. What were the total wholesale cost, total handling cost, and total sales?
    b. Was a net profit, an operating loss, or a gross loss made on the 25 boxes?
    c. How much was it, and what percent of the wholesale cost did it represent?
20. A department store purchased 50 packages of gift wrapping paper that cost $1.50 each. It sold 38 packages at $1.90 each and 12 at $1.60. The regular markup for overhead was 25% of cost.
    a. What were the total wholesale cost, total handling cost, and total sales?
    b. Was a net profit, an operating loss, or a gross loss made on the 50 packages?
    c. How much was it, and what percent of the wholesale cost did it represent (to the nearest tenth)?

SECTION 3

COMPREHENSIVE PROBLEMS FOR CHAPTERS 12, 13, AND 14

The following problems trace the merchandise through the pricing structure, which includes most or all of the following steps:

- Wholesalers and retailers normally purchase their merchandise at the list price less trade and cash discounts.
- Freight charges and any other charges related to the merchandise are added to the cost of the merchandise before the selling price is established.
- The total cost of the merchandise is marked up either on the cost or on the selling price.
- If the merchandise does not sell at this regular selling price, it will be marked down to a reduced selling price.

- The wholesaler or retailer wants to recover as much (if not all) of the total handling cost as possible:
  - If the actual selling price is greater than the total handling cost, there is an operating profit.
  - If the total handling cost is greater than the selling price, then an operating loss is experienced.
  - Sometimes, the actual selling price is below the wholesale cost; then a gross or an absolute loss occurs.

Work through these steps that merchants must use on a daily basis when marketing their merchandise, to determine how well they are succeeding. Knowing the details that are involved when marketing merchandise can make the difference between running a successful business or going into bankruptcy.

SECTION 3 PROBLEMS

1. The Dixon Sporting Goods purchased 80 life jackets at $20 each, less a 30% trade discount, from Elite Manufacturing Co. on April 5. The terms of the invoice were 2/10, 1/20, n/30. On April 15, the accounting department at Dixon sent a check for an amount that gave it credit for one-half the balance due. On April 25, the other half of the invoice was paid. Dixon has a markup of 20% on cost for operating expenses and 20% on cost for net profit. During May through July, 50 life jackets were sold at the regular marked prices. In August, the shop marked down the remaining merchandise by 40% for an after-season sale. As one of the assistants in the accounting department for Dixon, your job is to determine the following:
   a. The total cost of the life jackets before the cash discounts were taken
   b. The amount remitted on April 15
   c. The amount remitted on April 25
   d. The total cost and the individual cost per jacket after the cash discounts
   e. The regular list price of each jacket
   f. The reduced price of each jacket in August
   g. The total revenue received from the sale of the 80 life jackets
   h. The total amount of operating profit or operating loss
   i. The percent of cost that this profit or loss was (to the nearest tenth)
2. The Bailey Office Store purchased 60 labeling machines at $105 each, less a 25% trade discount, on May 10 from CAS Co. The terms of the invoice were 2/10, 1/20, n/60. On May 20, the Bailey accounting department sent a check for an amount that gave it credit for $\frac{1}{3}$ the balance due. On May 30, the remaining balance was paid. Bailey has a markup of 10% on cost for operating expenses and 20% on cost for net profit. During May through August, 45 labeling machines were sold at the regular marked price. During a September sale, the remaining merchandise was marked down by 10% and sold. It is your task to determine the following:
   a. The total cost of the labeling machines before the cash discounts were taken
   b. The amount remitted on May 20

c. The amount remitted on May 30
d. The total cost and the individual cost per labeling machine after the cash discounts
e. The regular list price of each machine
f. The reduced price of each machine in September
g. The total revenue received from the sale of the 60 labeling machines
h. The total amount of operating profit or operating loss earned after selling all 60 machines
i. The percent of cost that this profit or loss was (to the nearest tenth)

3. On August 1, the Marydale Department Store purchased directly from the manufacturer 100 chenille twin-size blankets at $125 each, less trade discounts of 15% and 20%. The credit terms on the invoice were 3/10 ROG. Marydale Department Store received the merchandise on September 2 and paid the manufacturer on September 12. An additional $35 freight charge was paid directly to Speedy Delivery Service on September 30. Marydale has a markup of 25% on cost for overhead and 15% on cost for net profit. During September through February, the store sold 75 blankets at the regular list price. In March, the remaining blankets were marked down 50% and sold. Determine the following:
   a. The amount remitted on September 12
   b. The total cost and the individual cost per blanket after the cash discount
   c. The regular list price of each blanket
   d. The reduced price of each blanket in March
   e. The total revenue received from the sale of the 100 blankets
   f. The amount of operating profit or operating loss
   g. The percent of cost that this profit or loss was (to the nearest tenth)
4. On October 15, Jackson Brothers purchased 1,500 gourmet popcorn tins for $3 each, less 20%, to be sold during the holiday season. The invoice contained credit terms of 2/10 EOM. The popcorn tins were received by Jackson Brothers on October 20, and the invoice was paid on November 10. A freight charge of $20 related to this purchase was paid directly to Qwik Delivery on October 21. The regular markup was 20% on cost for overhead and 15% on cost for net profit. During October through December, 1,000 tins were sold at the regular price. In January, the remaining popcorn tins were marked down 60% and sold. Determine the following:
   a. The amount remitted on November 10
   b. The total cost and the individual cost per tin after the cash discounts
   c. The regular list price of each tin
   d. The reduced price of each popcorn tin in January
   e. The total revenue received from the sale of the 1,500 popcorn tins
   f. The amount of operating profit or operating loss
   g. The percent of cost that this profit or loss was (to the nearest tenth)

CHAPTER 14 GLOSSARY

Absolute (or gross) loss. Loss resulting when the sale price of merchandise is less than cost: Absolute loss = Cost − Sale price.

Actual selling price. A "sale price" lower than the regular marked price; a reduced price. Actual selling price = Regular price − Markdown.

Breakeven point. A sale price that produces neither profit nor loss; same as Total handling cost.

Gross loss. (See "Absolute loss.")

Markdown. An amount or percent by which the regular selling price is reduced to obtain the actual selling price (or "sale" price).

Operating loss. Loss resulting when the sale price of merchandise is less than the total handling cost: Operating loss = Total handling cost − Sale price.

Operating profit. Profit resulting when the sale price of merchandise is greater than the total handling cost: Operating profit = Sales price − Total handling cost.

Reduced price. Selling price after a deduction or discount off the regular (higher) marked price; also called sale price, actual selling price, or net selling price.

Regular price. Selling price at which merchandise would normally be marked; also called regular marked price, regular selling price, marked price, or list price.

Total handling cost. The wholesale cost plus the related overhead on a sale ($C + \overline{OH}$).

CHAPTER 15

SIMPLE INTEREST

OBJECTIVES

Upon completing Chapter 15, you will be able to:

1. Define and use correctly the terminology associated with each topic.
2. a. Use the simple interest formula $I = Prt$ to find any variable, when the other items are given (Section 1: Examples 1, 3, 4; Problems 1, 2, 5–10, 13, 14).
   b. Use the amount formula $M = P + I$ to find the amount or interest (Examples 1, 2, 4; Problems 1–4, 7–10).
   c. Use the amount formula $M = P(1 + rt)$ to find the amount or principal (Examples 2, 5; Problems 3, 4, 11, 12).
3. a. Compute ordinary or exact interest using exact time (Section 2: Examples 1–4; Problems 1–6; Section 3: Examples 1–3; Problems 1–8, 13–16).
   b. Predict how a given change in rate or time will affect the amount of interest (Section 3: Problems 9–12).
4. Given information about a simple interest note, use any of the above formulas to compute the unknown (Section 4: Examples 1–3; Problems 1–20):
   a. Interest
   b. Maturity value
   c. Rate
   d. Time
   e. Principal.
5. Compute the present value of a simple interest note, either
   a. On the original day (Section 5: Examples 1, 2; Problems 1–4, 7–14), or
   b. On some other given day prior to maturity (Example 3; Problems 5, 6, 15–20).

The borrowing and lending of money is a practice that dates far back into history. Never, however, has the practice of finance been more widespread than it is today. Money may be loaned at a simple interest or a simple discount rate. The loan may be repaid in a single payment or a series of payments, depending upon the type of loan. When money is invested with a financial institution, compound interest is usually earned on the deposits. The following chapters explain the basic types of loans, important methods of loan repayments, and fundamental investment procedures. Before you proceed, a review of operations with parentheses found in Chapter 1 (page 5) would be helpful.

SECTION 1

BASIC SIMPLE INTEREST

Persons who rent buildings or equipment expect to pay for the use of someone else's property. Similarly, those who borrow money must pay rent for the use of that money. Rent paid for the privilege of borrowing another's money is called **interest.** The amount of money that was borrowed is the **principal** of a loan.

A certain percentage of the principal is charged as interest. The percent or **rate** is quoted on a yearly (per annum) basis, unless otherwise specified. The **time** is the number of days, months, or years for which the money will be loaned. In order to make rate and time correspond, time is always converted to years, since rate is always given on a yearly basis.

When **simple interest** is being charged, interest is calculated on the whole principal for the entire length of the loan. Simple interest is found using the formula

$$I = Prt$$

where I = interest, P = principal, r = rate, and t = time. The **amount** due on the ending date of the loan (also called **maturity value**) is the sum of the principal plus the interest. This is expressed by the formula

$$M = P + I$$

where M = amount (or maturity value or sum), P = principal, and I = interest.

Example 1 A loan of \$900 is made at 16% for 5 months. Determine (a) the interest and (b) the maturity value of this loan.

(a) $P = \$900$ $\qquad$ $I = Prt$

$r = 16\%$ or 0.16 $\qquad$ $= 900 \times 0.16 \times \frac{5}{12}$

$t = 5 \text{ months}^* = \frac{5}{12}$ year $\qquad$ $I = \$60$

*Keep in mind that, for periods less than 1 year, time must be expressed as a fraction of a year.

(b) $M = P + I$

$= \$900 + \$60$

$M = \$960$

The two formulas in Example 1 can be combined, allowing the maturity value to be found in one step. Since $I = Prt$, by substitution in the formula

$$\begin{aligned} M &= P + I \\ &\quad\;\;\downarrow \\ &= P + Prt \\ M &= P(1 + rt) \end{aligned}$$

If the formula $M = P(1 + rt)$ is used to calculate the amount, the interest may be found by taking the difference between maturity value and principal.

Example 2 Rework Example 1, using the maturity value formula $M = P(1 + rt)$. As before, $P = \$900$, $r = 0.16$, and $t = \frac{5}{12}$ year.

(a) $M = P(1 + rt)$

$= \$900\left(1 + 0.16 \times \dfrac{5}{12}\right)$

$= 900\left(1 + \dfrac{0.80}{12}\right)$

$= 900\left(\dfrac{1.80}{12}\right)$

$= 900(0.15)$

$M = \$960$

(b) $P + I = M$

$I = M - P$

$= \$960 - \$900$

$I = \$60$

Hint. When using the equation $M = P(1 + rt)$, remember that you must multiply $r \times t$ before adding 1, then multiply by P.

Calculator techniques . . . FOR EXAMPLE 2

When a formula contains parentheses, compute that value and place it into memory; then multiply by the number preceding the parentheses.

Note. To store the value in parentheses, you can perform the "1 [M+]" step either before or after you [M+] the "$r \times t$" calculation.

0.16 [×] 5 [÷] 12 [M+] → 0.0666666; 1 [M+] → 1

[MR] → 1.0666666 [×] 900 [=] → 959.99994

Rounded to the nearest cent, this equals $960. This technique will apply for many problems in this chapter and the next one.

Note. The "$r \times t$" can be computed as 5 [×] 16 [%] → 0.8 [÷] 12 [M+], if you prefer to use the [%] key rather than converting the 16% interest rate to its decimal equivalent.

Example 3 At what rate will $480 earn $28 in interest after 10 months?

$$P = \$480 \qquad I = Prt$$

$$r = ? \qquad \$28 = \$480r \times \frac{5}{6}$$

$$t = \frac{10}{12} \text{ or } \frac{5}{6} \text{ year} \qquad 28 = \frac{2{,}400}{6}r$$

$$I = \$28 \qquad 28 = 400r$$

$$r = \frac{28}{400} \text{ or } \frac{7}{100} \text{ or } 0.07$$

$$r = 7\%$$

Example 4 How long will it take at an 8% simple interest rate for $950 to amount to $988?

This problem may be worked using the maturity value formula $M = P(1 + rt)$. However, the simple interest formula $I = Prt$ is easier to work with and should be used whenever possible. Notice that the formula $I = Prt$ may be used to find a missing variable whenever the interest is given or can be found quickly.

In this case, the interest may be computed easily, since

$I = M - P$	$I = Prt$
$= \$988 - \$950$	$\$38 = \$950 \times 0.08t$
$I = \$38$	$38 = 76t$
$P = \$950$	$\frac{38}{76} = t$
$r = 8\%$	$t = \frac{38}{76}$ or $\frac{1}{2}$ or 0.5
$t = ?$	$t = \frac{1}{2}$ year or 6 months

Example 5 What principal would have a maturity value of \$636 in 8 months if the interest rate is 9%?

In this problem the interest is not given and is impossible to determine because the maturity value and principal are not both known. Thus, the amount formula $M = P(1 + rt)$ must be used to find the principal.

$P = \$?$	$M = P(1 + rt)$
$r = 9\%$ or 0.09	$\$636 = P\left(1 + 0.09 \times \frac{2}{3}\right)$
$t = \frac{8}{12}$ or $\frac{2}{3}$	$= P\left(1 + \frac{0.18}{3}\right)$
$M = \$636$	$= P(1 + 0.06)$
	$= P(1.06)$
	$\frac{\$636}{1.06} = P$
	$\$600 = P$

SECTION 1 PROBLEMS

Find the interest and the maturity value.

1. a. \$600 at 10% for 1 year b. \$600 at 10% for 6 months c. \$1,200 at 8% for 4 months
d. \$800 at 8.5% for 3 months e. \$2,400 at 9% for 9 months

2. a. \$700 at 9% for 1 year b. \$700 at 9% for 6 months c. \$1,400 at 8% for 3 months
d. \$2,400 at 6.5% for 9 months e. \$840 at 7% for 8 months

3. Find the interest and maturity value of Parts (a), (c), and (e) in Problem 1 using the maturity value formula $M = P(1 + rt)$.

4. Utilize the maturity value formula $M = P(1 + rt)$ to find both interest and maturity value for Parts (a), (c), and (e) in Problem 2.

5. What was the interest rate if the interest due on a 9-month loan of \$3,000 was \$180?

6. What was the interest rate on a note with a principal of \$2,000 for 6 months if the interest was \$90?

7. How many months will it take for a loan of \$1,500 at 7% to acquire a maturity value of \$1,587.50?

8. How many months would it take \$880 to amount to \$896.50 if an interest rate of 7.5% was charged?

9. If \$6,000 is worth \$6,300 after 6 months, what interest rate was charged?

10. What interest rate was charged on a note with a principal of \$1,600 and time of 3 months if the maturity value was \$1,636?

11. What principal will amount to \$9,680 after 18 months at 14%?

12. What principal will have a maturity value of \$14,880 after 36 months at 8%?

13. How long will it take for an investment to double at 16% simple interest? (*Hint:* Choose any principal; what will the interest be when the investment matures?)

14. At what simple interest rate will an investment double itself in 20 years? (*Hint:* Choose any principal; what will the interest be when the investment matures?)

SECTION 2

ORDINARY TIME AND EXACT TIME

The problems given in Section 1 were limited to those with time periods of even months or years. This is not always the case, as many loans are made for a certain number of days. Suppose that a loan is made on July 5 for 60 days; do we consider 60 days to be 2 months and consider the loan due on September 5, or do we take the 60 days literally and consider the amount due on September 3? Suppose that this loan made on July 5 is for a time of 2 months; would we consider the time to be $\frac{2}{12} = \frac{1}{6}$ year, or should we use the actual 62 days between July 5 and September 5? The answers to these questions depend upon whether ordinary time or exact time is being used.

When **ordinary time** is being used, each month is assumed to have 30 days. Thus, any 5-month period would be considered as $5 \times 30 = 150$ days. Similarly, 90 days would be considered as 3 months. In actual practice, however, ordinary time is seldom used, as we shall see later.

If **exact time** is being used, the specific number of days within the time period is calculated. Thus, 90 days would usually be slightly less than 3 months. Exact time may be calculated most conveniently using Table 14, The Number of Each Day of the Year, in the Tables Booklet. In this table, each day of the year is listed in consecutive order and assigned a number from 1 to 365. The following examples demonstrate use of the table.

Example 1 Using exact time, find (a) the due date of a loan made on April 15 for 180 days and (b) the exact time from March 10 to July 23.

(a) From the table of each day of the year, we see that April 15 is the 105th day. Thus,

$$\begin{array}{rr l} \text{April 15} = & 105 \text{ day} & \\ & +\underline{180 \text{ days}} & \\ & 285\text{th day} & \text{is October 12, the due date of the note using exact time} \end{array}$$

(b) From March 10 to July 23:

$$\begin{array}{rr l} \text{July 23} = & 204 \text{ day} & \\ \text{March 10} = & -\underline{\ \ 69 \text{ day}} & \\ & 135 \text{ days} & \text{between March 10 and July 23, using exact time} \end{array}$$

Example 2 A loan is taken out on June 1 for 5 months. Determine (a) the due date and (b) the number of days in the term of the loan, using both ordinary and exact time.

(a) Five months after June 1 is November 1. The loan (principal plus interest) will be due on November 1 regardless of which kind of time is used.*

(b) From June 1 to November 1:

Ordinary Time	*Exact Time*
5 months = 5 × 30	November 1 = 305 day
= 150 days	June 1 = −152 day
	153 days

Sometimes the time period of a loan includes a leap year. This does not affect the calculation of ordinary time, of course, since all months are considered to have 30 days. The leap year must be given consideration, however, when exact time is

**Note:* If the given number of months would end on a day that does not exist, the due date is the last day of the month. For example, a loan on March 31 for 8 months would be due on November 30, since there is no November 31.

being used. During a leap year, February 29 becomes the 60th day of the year; thus, from March 1 on, the number of each day is one greater than is shown in Table 14. For example, during a leap year, March 15 is the 75th day, instead of the 74th day as the table shows. This change in the number of the day must be made before the exact time is calculated.

A simple test allows you to determine whether any given year is a leap year: A leap year always falls on a year evenly divisible by 4, except for years that are even 100s. Century years are leap years *only* if they are divisible by 400, such as 1600 and 2000. The year 2002 is not evenly divisible by 4, so 2002 is not a leap year; but the year 2004 is evenly divisible by 4, so it is a leap year. The year 1900 is not evenly divisible by 400, so 1900 was not a leap year. United States Presidential elections also fall on leap years.

Example 3 Find the exact time from January 16, 2000, to May 7, 2000.

Since 2000 was a leap year, May 7 is the 128th day.

$$\begin{aligned} \text{May 7} &= \quad 128 \text{ day} \\ \text{January 16} &= -\underline{\ \ 16 \text{ day}} \\ & \quad\ \ 112 \text{ days} \quad \text{from January 16, 2000 to May 7, 2000} \end{aligned}$$

Example 4 Find the exact time between November 16 of one year and April 3 of the next (assuming that no leap year is involved).

We must find the remaining days in the present year and then add the days in the following year:

$$\begin{aligned} \text{Present year} &= \quad 365 \text{ days} \\ \text{November 16} &= -\underline{320 \text{ day}} \\ & \qquad 45 \text{ days remaining in present year} \\ \text{April 3} &= +\underline{\ \ 93 \text{ day}} \\ & \quad\ \ 138 \text{ days from November 16 to April 3} \end{aligned}$$

SECTION 2 PROBLEMS

If no year is given in the following problems, assume that it is not a leap year. Find the exact time from the first date to the second date.

1. a. March 6 to May 31
b. June 5 to September 15
c. July 10 to December 10
d. January 14 to November 28
e. February 12, 2000, to May 31, 2000
f. September 23, 2001, to April 1, 2002
g. July 4, 2003, to March 4, 2004
h. November 15, 1999, to June 6, 2000

2. a. November 9 to December 6
b. April 12 to October 18
c. May 27 to December 25
d. February 7 to November 23

e. January 16, 2000, to April 3, 2000
f. August 5, 2001, to March 12, 2002
g. June 25, 2003, to April 3, 2004
h. December 15, 2007, to May 1, 2008

Determine the due date, using exact time.

3. a. 30 days after August 21
b. 60 days after October 14
c. 120 days after June 18
d. 200 days after March 1
e. 150 days after July 17
f. 180 days after November 23

4. a. 60 days after September 3
b. 270 days after February 8
c. 90 days after June 12
d. 300 days after January 3
e. 180 days after May 1
f. 150 days after October 19

Find the due date and the exact time (in days) for loans made on the given dates.

5. a. February 18 for 6 months
b. May 24 for 3 months
c. January 26, 2002, for 9 months
d. March 30, 2006, for 4 months
e. January 13, 2004, for 5 months

6. a. August 7 for 4 months
b. May 26 for 2 months
c. January 5, 2000, for 8 months
d. July 18, 2003, for 4 months
e. February 3, 2008, for 3 months

SECTION 3

ORDINARY INTEREST AND EXACT INTEREST

It has been seen that the time of a loan is frequently a certain number of days. As stated previously, a time of less than one year must be expressed as a fraction of a year. The question then arises: When converting days to fractional years, are the days placed over a denominator of 360 or 365? It is this question with which ordinary and exact interest are concerned.

The term "interest" here refers to the number of days representing one year. Just as ordinary time indicated 30 days per month, **ordinary interest** indicates interest calculated using 360 days per year. Similarly, the term **exact interest** indicates interest calculated using 365 days per year. (A 366-day year is not often used in business, even for leap years.)

We have now discussed two ways of determining time (days in the loan period) and two types of interest (days per year). There are four possible combinations of these to obtain t for the interest formula:

I. Exact time over ordinary interest: $t = \dfrac{\text{Exact days}}{360}$

II. Exact time over exact interest: $t = \dfrac{\text{Exact days}}{365}$

III. Ordinary time over ordinary interest: $t = \dfrac{\text{Approximate days}}{360}$

IV. Ordinary time over exact interest: $t = \dfrac{\text{Approximate days}}{365}$

Type 1. The combination of exact time and ordinary interest is known as the **Bankers' Rule,** a name derived from the fact that banks (as well as other financial institutions) generally compute their interest in this way:

$$\frac{\text{Exact days}}{360}$$

Example 1 A \$730 loan is made on March 1 at 12% for 9 months. Compute the interest by the Bankers' Rule (that is, using exact time and ordinary interest).

Nine months after March 1 is December 1. The exact time is

$$\begin{aligned} \text{December 1} &= 335 \text{ day} \\ \text{March 1} &= \underline{\ 60 \text{ day}} \\ &\ 275 \text{ days} \end{aligned}$$

$P = \$730$ $\qquad I = Prt$

$r = 0.12$ $\qquad = \$730 \times 0.12 \times \dfrac{55}{72}$

$t = \dfrac{275}{360}$ or $\dfrac{55}{72}$ $\qquad = \dfrac{730 \times 0.12 \times 55}{72}$

$I = ?$ $\qquad I = \$66.92$ (exact time/ordinary interest)

Type II. The combination of exact time and exact interest is gaining in usage by financial institutions and has been used extensively in certain parts of the country during periods when interest rates have been quite high. This is partly to ensure that they do not exceed legal maximum (usury) rates on interest charges and partly because this method results in slightly less interest paid to depositors.

Another major application is by the Federal Reserve Bank and the U.S. government, which use this method when computing interest:

$$\frac{\text{Exact days}}{365 \text{ or } 366}$$

This exact-time/exact-interest method applies both for interest owed to the government (such as interest for late payment of taxes) and for interest paid by the government to purchasers of certain U.S. interest-bearing certificates. (For example, earnings on U.S. Treasury Notes are computed using a half-year variation of exact interest.)

Example 2 Compute Example 1 using type II (exact time and exact interest).

$$P = \$730 \qquad I = Prt$$

$$r = 0.12 \qquad = \$730 \times 0.12 \times \frac{55}{73}$$

$$t = \frac{275}{365} \text{ or } \frac{55}{73} \qquad = \frac{87.6}{73}$$

$$I = ? \qquad I = \$66.00 \quad \text{(exact time/exact interest)}$$

Type III. The combination of ordinary time and ordinary interest is equivalent to the calculations used in the first section of this chapter. (For example, any 6-month period is equivalent to 180 days or 0.5 year.) This method of interest calculation is not used commercially, but it is used in private loans between individuals:

$$\frac{\text{Approx. days}}{360} \text{ or } \frac{\text{Months}}{12}$$

Example 3 Rework Example 1 using ordinary time (9 months or 270 days) and ordinary interest (360 days or 12 months).

$$P = \$730 \qquad I = Prt$$

$$r = 0.12 \qquad = \$730 \times 0.12 \times 0.75$$

$$t = \frac{9}{12} \text{ or } \frac{270}{360} \qquad I = \$67.50 \quad \text{(ordinary time/ordinary interest)}$$

$$= \frac{3}{4} \text{ or } 0.75$$

$$I = ?$$

Type IV. The other possible combination of ordinary time and exact interest is not customarily used. (Example 1 reworked using 270 days over a 365-day year yields interest of $64.80.)

Comparing Examples 1 through 3 reveals that type I, the combination of *exact time over ordinary interest* (a 360-day year), produces the most interest on a loan. Since this combination results in the most interest due to the lender, it is therefore the method most often used historically—which accounts for its name, the Bankers' Rule. From this point on, our problems will emphasize this type I method (Bankers' Rule), with supporting use of the type II method.

SECTION 3 PROBLEMS

Find the ordinary interest (type I) and the exact interest (type II).

1. a. On $2,190 for 45 days at 8% b. On $2,920 for 60 days at 9%

2. a. On $6,500 for 90 days at 7% b. On $5,200 for 120 days at 6.5%

3. A loan of $8,000 was made on June 6 for 4 months at 14%. Calculate the interest on this loan using each type of combination of time and interest (types I, II, and III).

4. A 2-month note dated September 5 for $10,000 had an interest rate of 9%. Determine the interest due using the three combinations of time and interest (types I, II, and III).

Use the Bankers' Rule (type I) to compute the ordinary interest.

5. Compute ordinary interest on $3,000 for 90 days at the given rates. Observe carefully the relationship between rates and interest due.

a. 5% b. 10% c. 15% d. 7.5%

6. Find the ordinary interest on $6,000 for 60 days at each of the given rates. Observe the relationships among your answers.

a. 8% b. 4% c. 12% d. 6%

7. Find the ordinary interest on a $4,000 loan at 12% for each of the given time periods. Carefully observe the relationships among your answers.

a. 90 days b. 30 days c. 120 days d. 270 days

8. Compute the ordinary interest on a $2,400 loan at 7% for each of the given time periods. Carefully observe the relationships among your answers.

a. 30 days b. 90 days c. 45 days d. 180 days

9. The ordinary interest on a loan for a certain number of days at 8% was $36. Using what you have learned in Problem 5, compute the ordinary interest on this loan for the same time at

a. 16% b. 10% c. 12%

10. The ordinary interest on a loan for a certain number of days at 9% was $50. Using what you learned in Problem 6, compute the ordinary interest on this loan for the same time at

a. 18% b. 4.5% c. 13.5%

11. When interest was computed at a certain rate, the interest due on a 120-day loan was $40. Using your conclusions from Problem 7, determine the interest due on the same principal and at the same rate if the time periods were as follows:

a. 60 days b. 300 days c. 240 days

12. The ordinary interest on a loan at a certain rate was $80 for 30 days. Applying what you learned in Problem 8, determine the interest due on the same principal and at the same rate if the time periods were as follows:

a. 120 days b. 60 days c. 90 days

Use type II to compute the exact interest.

13. On April 5, a loan of $2,190 was taken out for 2 months. Find the exact interest due on the loan at
a. 10% b. 5% c. 15%

14. A 3-month note was taken out on July 10 for $1,095. Calculate the exact interest due on the note at
a. 12% b. 6% c. 15%

15. A loan for $1,460 was made at 9.5%. Determine the exact interest for
a. 150 days b. 180 days c. 270 days

16. A loan for $2,555 was made at 8%. Determine the exact interest for
a. 60 days b. 90 days c. 360 days

SECTION 4

SIMPLE INTEREST NOTES

A person who borrows money usually signs a written promise to repay the loan; this document is called a **promissory note,** or just a **note.** The money value that is specified on a note is the **face value** of the note. If an interest rate is mentioned in the note, the face value of this simple interest note is the **principal** of the loan. Interest is computed on the whole principal for the entire length of the loan, and the **maturity value** (principal plus interest) is repaid in a single payment on the **due date** (or maturity date) stated in the note. If no interest is mentioned, the face value is the maturity value of the note. The person who borrows the money is called the **maker** of the note, and the one to whom the money will be repaid is known as the **payee.**

The use of simple interest notes varies from state to state; the banks in some states issue simple interest notes, whereas others normally use another type of note. Figure 15-1

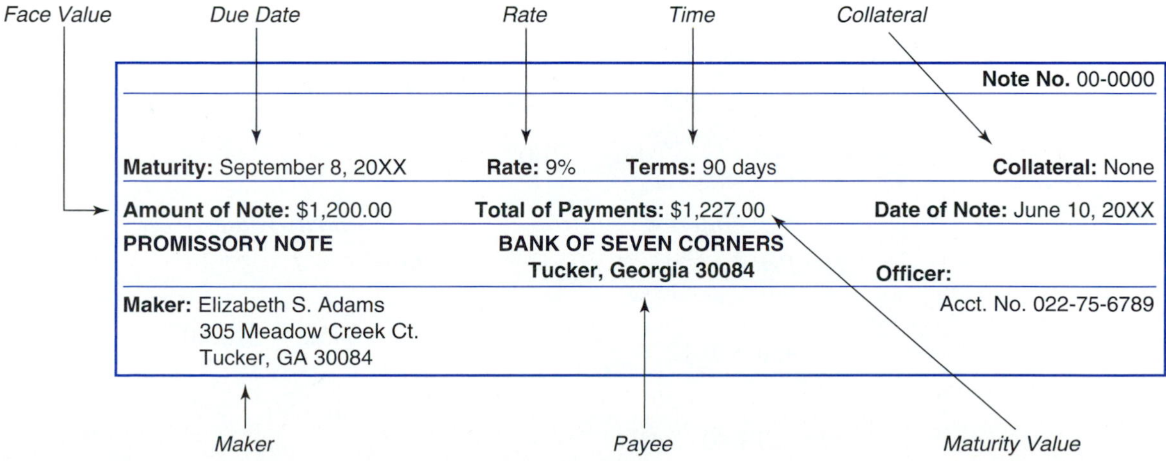

Figure 15-1 Interest-Bearing Note

shows the top portion of a simple interest bank note. Notes between individuals are usually simple interest notes also.

Before loaning money, a bank may require the borrower to provide **collateral**—some item of value that is used to secure the note. Thus, if the loan is not repaid, the bank is entitled to obtain its money by selling the collateral. (Any excess funds above the amount owed would be returned to the borrower.) Some items commonly used as collateral are cars, real estate, insurance policies, stocks and bonds, as well as savings accounts. Figure 15-2 illustrates a simple interest note in which a certificate of deposit is the collateral.

Most promissory notes are short-term notes—for 1 year or less. The most common time period for notes is one quarter (or 90 days). When a note matures after one quarter, the interest due must be paid, and the note may usually be renewed for another quarter at the rate then in effect for new loans.

In many localities, short-term loans from banks are almost exclusively either installment loans or simple discount loans. For an installment loan, the maturity value is calculated as for a simple interest loan; however, this amount is repaid in weekly or monthly payments rather than being repaid in one lump sum, as is the case for simple interest notes. (The simple discount loan and the installment loan are both discussed in detail in later topics.)

The bank is required by law to provide the borrower with information about the annual percentage rate and the finance charge. Some banks provide this information on the promissory note itself. Other banks provide a separate statement, the "Truth-in-Lending Disclosure Statement" (Figure 15-3).

Example 1 A note for \$450 was dated June 12 and repaid in 3 months with interest at 12%. Find (a) the due date and (b) the amount repaid.

(a) The due date was September 12.

(b) Using the Bankers' Rule, interest and maturity value were as follows:

Sept. 12 = 255 day
June 12 = 163 day
$t =$ 92 days or $\frac{92}{360}$ year

$P = \$450$
$r = 12\%$

Interest

$$I = Prt$$
$$= 450 \times 0.12 \times \frac{92}{360}$$
$$I = \$13.80$$

Maturity Value

$$M = P + I$$
$$= \$450 + \$13.80$$
$$M = \$463.80$$

Consumer Note — Cardinal Bank *Cardinal Bank*

William J. McDaniel — Borrower | May 1, 20XX — Date | Garden City — Originating Office

Co-Borrower | 4561-9121 — Account No. | B2-22314 — Note No.

3001 Dawes Avenue; Alexandria, VA 22311 — Borrower's Address

Five thousand and no/100 dollars — Loan Amount

Dollars($ 5,000.00)

In this Note, the words "you" and "your" mean the Borrower and any Co-Borrower. "We," "our," "us," and "the Bank" mean Cardinal Bank.

This note covers your loan with the Bank. When you sign it, each one of you is fully responsible for fulfilling all of the promises you make in this Note. By signing this Note, you acknowledge that you received a loan from us in the Loan Amount shown above and agree to pay to us at any of our offices, or at such place as the Bank may in writing designate, the Loan Amount shown above, plus or including interest, any other amounts due, upon the terms described below.

Terms of Note

☐ **Simple Interest Instalment Loan**
You will make a total of ____ monthly payments of principal and interest beginning on ________________________, 20__ and continuing on the same day of each succeeding month until this Note is paid in full. You will make ______ monthly payments of principal and interest equal to $__________ and then a final payment equal to the unpaid principal balance plus interest and any other amounts due.

☐ **Principal Plus Instalment Loan**
You will make a total of ______monthly payments of principal and interest beginning on ______________, 20__ and continuing on the same day of each succeeding month until this Note is paid in full. Each month for _____ months you will make a payment equal to $_________ in principal plus all interest accrued on the unpaid balance, and then a final payment equal to $_________ in principal plus all accrued interest any other amounts due.

☑ **Single Payment Loan** due on May 31, 20XX (30 days from the date of this note). You will make one payment in full of principal plus interest from the date of this Note until this Note is paid in full plus any other amounts due.

☑ New obligation ☐ Renewal — New disclosure required ☐ Renewal — Same term and no new disclosure required

Interest

Interest on an Instalment Loan will accrue on a 30/360 day basis On all other loan types, interest will accrue daily on the basis of a 360-day year. Interest will accrue at the stated interest rate on the unpaid balance from the date of this Note until paid in full. Interest will continue to accrue after maturity, whether by acceleration or otherwise, at the stated interest rate until this Note has been paid full. Subject to the above, the interest rate applicable to this Note (the "Rate") is:

☑ 9.0 % per annum, fixed for the term of this Note.

Collateral

We will have a security interest in the property described below (the "Collateral"). The Collateral is described more fully in a

☐ Security Agreement dated__________ ☐ Assignment on Deposit dated_________________

☐ Deed of Trust dated_____________ ☐ Credit Line Deed of Trust dated________________

☑ Other Certificate of Deposit for $5,000 at Cardinal Bank dated March 1, 20XX

By signing below, you agree to all the terms of this Note and acknowledge receipt of a completed copy.

William J. McDaniel (Seal) — Borrower's Signature

________________________ (Seal) — Co-Borrower's Signature

Figure 15-2 Note with Collateral

Truth-in-Lending Disclosure Statement *Cardinal Bank*

Creditor: Cardinal Bank Borrower: William J. McDaniel

Loan Type: Single Payment Loan Co-Borrower: ____________

Use of Terms: In this Disclosure, the words "you" and "your" mean any Borrower or Co-Borrower. "We," "our," and "us" mean the Creditor. "Schedule Payment" means any payment disclosed in this disclosure or specified in the Note to which this disclosure applies, whether such payment is a payment of principal and interest or a payment of interest only.

ANNUAL PERCENTAGE RATE The cost of your credit as a yearly rate	FINANCE CHARGE The dollar amount the credit will cost you	AMOUNT FINANCED The amount of credit provided to you on your behalf	TOTAL OF PAYMENTS Amount you will have paid after you have made all payments as scheduled
9.0%	$37.50	$5,000.00	$5,037.50

Payment Schedule:

Number of Payments	Amount of Payments	When Payments are Due
One	$5,037.50	May 31, 20XX

This obligation is payable upon demand.

Prepayment: If you pay off early, you will not have to pay a penalty and will not be entitled to rebate of any prepaid Finance charge.
Read all your loan documents for additional information about nonpayment, default, late charges, any required payment in full before the scheduled date, security interests and other matters pertaining to this credit transaction.

Acknowledgment of Receipt: The undersigned acknowledges receipt of a completed copy of this disclosure before entering into any agreement with Creditor concerning this credit transaction.

William J McDaniel 5/1/xx
Borrower Date Co-Borrower Date

Figure 15-3 Truth-in-Lending Disclosure Statement

Example 2 The maturity value of a 60-day simple interest note was \$976, including interest at 10%. What was the face value of the note?

$$M = \$976 \qquad M = P(1 + rt)$$

$$r = 0.10 \qquad \$976 = P\left(1 + 0.10 \times \frac{1}{6}\right)$$

$$t = \frac{60}{360} \text{ or } \frac{1}{6} \qquad 976 = P\left(1 + \frac{0.10}{6}\right)$$

$$P = ?$$

$$976 = P(1 + 0.016666)$$

$$976 = P(1.016666)$$

$$\frac{976}{1.016666} = P$$

$$\$960 = P$$

Calculator techniques . . . FOR EXAMPLE 2

Here, since you can't multiply the value in parentheses times *P*, you must divide by the parenthetical value (using [MR]) to find *P*:

0.1 [÷] 6 [M+] ⟶ 0.0166666; 1 [M+] ⟶ 1

976 [÷] [MR] [=] ⟶ 960.00006

Recall that the "1 [M+]" can be done first, if you prefer.

Example 3 A \$750 note drawn at 15% exact interest had a maturity value of \$795. What was the exact time of the note?

$$M = \$795$$
$$P = -\ 750$$
$$I = \$\ 45$$

$$r = \frac{15}{100}$$

$$t = ?$$

$$I = Prt$$

$$\$45 = \$750 \times 0.15t$$

$$45 = 112.5t$$

$$\frac{45}{112.5} = t$$

$$0.4 \text{ year} = t$$

$$t = 0.4 \text{ year} \times 365 \text{ days per year}$$

$$t = 146 \text{ days}$$

SECTION 4 PROBLEMS

Note. You should use exact days in all problems unless exact time is impossible to determine from the information given. For example, if a problem involves a 3-month loan but no date is included, it is impossible to compute exact time; thus, ordinary time would have to be used. However, if a problem states that a 3-month loan began on some specific date, then the exact days during that particular 3-month interval would be used.

Further, ordinary interest should normally be computed, using the Bankers' Rule (exact days over 360). However, a 365-day year should be observed when the loan problem specifies "exact interest."

1–2. Identify each part of the note in Figure 15-2 and the Truth-in-Lending Disclosure Statement in Figure 15-3 shown previously.

a. Face Value b. Maker c. Payee d. Date
e. Due date f. Principal g. Rate h. Time
i. Interest j. Maturity value

Complete the following simple interest problems, using the Bankers' Rule.

	Principal	Rate	Date	Due Date	Time	Interest	Maturity Value
3. a.	$4,800	9%	2/12	11/9			
b.	900	11	4/22		60 days		
c.	4,500	12	8/21		3 months		
d.		9		7/13	100 days	$20	
4. a.	$9,000	8%	5/15	6/4			
b.	1,800	9	7/12		120 days		
c.	3,600	10	6/20		3 months		
d.		6		3/21	60 days	$54	

In completing the following problems, use exact days if possible. Find the ordinary interest (360-day year) unless the exact interest (365-day year) is indicated.

5. A note dated January 24, 2003, was made for 4 months at 9%. If the principal of the note was $6,000, find the exact interest and the maturity value of the note.

6. A 3-month note dated January 5, 2002, for $2,920 had interest at 8%. Find the exact interest and the maturity value of the note.

7. A note dated October 5, 2000, read: "Two months from date, I promise to pay $400 with interest at 10.5%." What was the maturity value of the note?

8. The Hollister Co. received a 1-month note dated February 18, 2001, for merchandise sold. The principal of the note was $6,000 with interest at 12%. What maturity value will the Hollister Co. receive?

9. Find the maturity value of a 9%, 10-month note with a face value of $2,000.

10. A 6-month note is drawn for $9,000 with interest at 8%. What maturity value will be paid?

11. A 6-month note dated February 14, 2002, has a face value of $2,190 and a maturity value of $2,298.60. Determine the exact interest rate to be charged.
12. A note dated June 10, 2003, will be due in 5 months. If the face value of the note is $5,475, and if $5,727.45 will be paid to discharge the debt at maturity, what exact interest will be charged?
13. How long will it take $730 to earn $18 at 10% exact interest?
14. How long will it take $3,650 to earn $162 interest at a 9% exact interest rate?
15. What time is required for $400 to earn $36 at 12% interest?
16. What time is required for $2,000 to earn $30 at 6% interest?
17. Find the principal of a 180-day note made at 14% if the interest was $105.
18. The interest on a 120-day note was $48. If interest was computed at 9%, what was the principal of the note?
19. Find the face value of a 10% note dated September 7 that matures on December 6. The maturity value on the note is $3,075.
20. A 10% note dated April 15 matures on July 14. Find the face value of the note if the maturity value is $7,380.

SECTION 5

PRESENT VALUE

In Problems 19 and 20 of Section 4, we were told the maturity value of a note and asked to find what principal had been loaned. When the principal is being found, this is often referred to as finding the **present value.** Present value can also be explained by answering the question: What amount would have to be invested today (at the present) in order to obtain a given maturity value? This "X" amount that would have to be invested is the present value.

Present value at simple interest can be found using the same maturity value formula that has been used previously: $M = P(1 + rt)$. However, when present value is to be found repeatedly, the formula is usually altered slightly by dividing by the quantity in parentheses.

$$M = P(1 + rt)$$

$$\frac{M}{1 + rt} = \frac{P\cancel{(1 + rt)}}{\cancel{(1 + rt)}}$$

$$\frac{M}{1 + rt} = P$$

or

$$P = \frac{M}{1 + rt}$$

The formula can be used more conveniently in this form, since it is set up to solve for the principal or present value. However, Example 1 illustrates a present value problem solved with both versions of the formula, and either form may be used for this type of problem.

Example 1 If the maturity value of a 12%, 3-month note dated March 11 was \$618.40, find the present value (or principal, or face value) of the note.

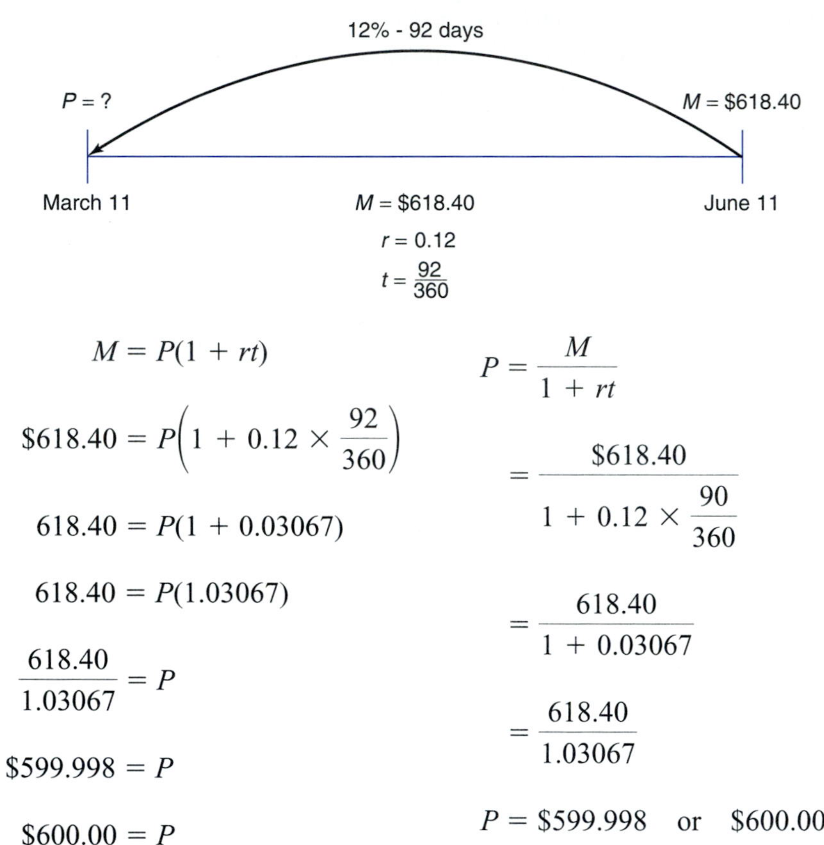

$$M = P(1 + rt) \qquad P = \frac{M}{1 + rt}$$

$$\$618.40 = P\left(1 + 0.12 \times \frac{92}{360}\right) \qquad = \frac{\$618.40}{1 + 0.12 \times \frac{90}{360}}$$

$$618.40 = P(1 + 0.03067) \qquad = \frac{618.40}{1 + 0.03067}$$

$$618.40 = P(1.03067) \qquad = \frac{618.40}{1.03067}$$

$$\frac{618.40}{1.03067} = P$$

$$\$599.998 = P \qquad P = \$599.998 \quad \text{or} \quad \$600.00$$

$$\$600.00 = P$$

Example 1 involves only one interest rate. Many investments, however, involve two interest rates. To illustrate present value at two interest rates, consider the following case.

Suppose that I find a place where I can earn 20% on my investment. The most I have been offered any other place is 12%, so I immediately invest \$1,000 in this "gold mine." On the way home, I meet a friend who offers to buy this investment for the same \$1,000 I just deposited. Would I sell the investment for \$1,000, knowing I could not earn 20% interest on any other investment I might make? No! My \$1,000 is really worth more than \$1,000 to me, because I would have to deposit more than \$1,000 elsewhere in order to earn the same amount of interest.

This case demonstrates that an investment is not always worth its exact face value; it may be worth more or less than its face value when compared to the average rate of interest being paid by most financial institutions. This typical, or average, interest rate is referred to as the **rate money is worth.** Thus, if most financial institutions are paying 8%, then money is worth 8%.

The rate money is worth varies considerably, depending on whether one is borrowing or depositing—and on the kind of deposit made. An ordinary **savings account,** where the money can be withdrawn at any time, earns less interest than a **certificate of deposit** (or **CD**). A CD usually requires a minimum deposit (often \$1,000), and the money must remain on deposit for a specified period of time (usually a 3-month minimum) in order to earn interest at the higher rate. CD rates increase when more money is deposited and/or longer time periods are established.

Another popular account, the **money-market account,** shares some characteristics with each of the others. The money-market account may also require a minimum deposit (often \$500), but it allows a limited number of withdrawals or checks (usually three per month) without penalty so long as the minimum balance remains met. Its rate of interest usually falls between that of the other two accounts.

As would be expected, institutions must charge higher rates for notes and other loans than they pay on deposits, or they could not cover expenses and return a profit to their owners.

If an investment is made at a rate that is not the prevailing rate money is worth, then the present value of the investment is different—either greater or less—than the principal. We therefore find what investment, made at the rate money is worth, would have the same maturity value that the actual investment has. The size of this "rate-money-is-worth investment" is the present or true value of the actual investment.

Example 2 A 90-day note for \$1,000 is drawn on November 12 at 8%. If money is worth 12%, find the present value of the note on the day it is drawn.

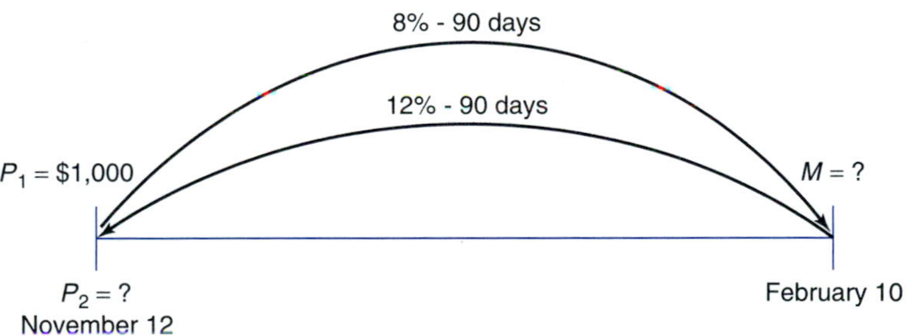

The solution to this kind of problem involves two steps: (a) Find the *actual maturity value* of the note, and (b) find the *present value at the rate money is worth* of that calculated amount.

(a) $P_1 = \$1,000$ $\qquad M = P(1 + rt)$

$r = 8\%$ $\qquad = \$1,000(1 + 0.08 \times 0.25)$

$t = \dfrac{90}{360}$ or 0.25 $\qquad = 1,000(1 + 0.02)$

$\qquad = 1,000(1.02)$

$M = ?$ $\qquad M = \$1,020$

(b) $M = \$1{,}020$

$r = 12\%$

$t = \frac{90}{360}$ or 0.25

$P_2 = ?$

$$P_2 = \frac{M}{1 + rt}$$

$$= \frac{\$1{,}020}{1 + 0.12 \times 0.25}$$

$$= \frac{1{,}020}{1 + 0.03}$$

$$= \frac{1{,}020}{1.03}$$

$$P_2 = \$990.29$$

The present value of the note on the day it was drawn was \$990.29. This means that the lender (payee) could have invested only \$990.29 at the rate money is worth (12%) and would achieve the same maturity value (\$1,020) that will be obtained on the \$1,000 note at only 8% interest. That is, \$990.29 invested at 12% would earn \$29.71 interest (\$990.29 + \$29.71 = \$1,020), producing the same \$1,020 maturity value as \$1,000 plus \$20 interest. Thus, the \$1,000 is really worth only \$990.29 to the lender.

When computing present value of an investment where two interest rates are involved, you should know beforehand whether the present value is more or less than the principal. In general, the present value is less than the principal when the investment rate is less than the rate money is worth. That is, when the actual rate is less, then the present value is less. The reverse is also true: If the actual rate is more than the rate money is worth, then the present value is more than the principal.

Previous examples have required finding the present value on the day a note was drawn. It is often desirable to know the worth of an investment on some day nearer the due date. The worth of an investment on any day prior to the due date is also called **present value.** The procedure for finding this present value is the same as that already discussed, except that in the second step the time as well as the rate will be different. It should be emphasized that present value is

1. Computed using the maturity value.
2. Computed for an exact number of days *prior* to the due date.
3. Computed using the rate money is worth.

Example 3 Find the (present) value on October 11 of a \$720, 4-month note taken out on July 10 at 15%, if money is worth 12%.

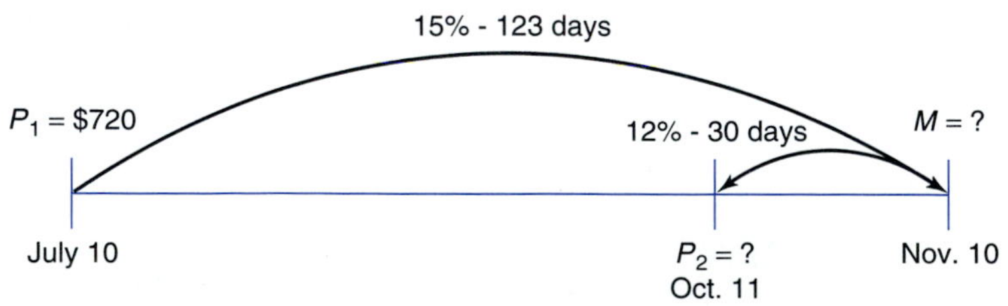

(a) $P_1 = \$720$ $\qquad I = Prt$ $\qquad M = P + I$

$r = 15\%$ $\qquad = \$720 \times 0.15 \times \frac{123}{360}$ $\qquad = \$720 + \$36.90$

$t = \frac{123}{360}$ $\qquad I = \$36.90$ $\qquad M = \$756.90$

$M = ?$

(b) $M = \$756.90$ $\qquad P_2 = \frac{M}{1 + rt}$

$r = 12\%$

$t = \frac{30}{360}$ or $\frac{1}{12}$

$P_2 = ?$

$$= \frac{\$756.90}{1 + 0.12 \times \frac{1}{12}}$$

$$= \frac{756.90}{1 + 0.01}$$

$$= \frac{756.90}{1.01}$$

$$P_2 = \$749.41$$

This means that if the note were sold on October 11, it would be sold for \$749.41; this money could then be invested at 12% interest, and on November 10 the maturity value of the new investment (\$749.41 plus \$7.49 interest) would also be \$756.90.

SECTION 5 PROBLEMS

Any convenient method may be used to find the amount in the following problems: $I = Prt$ and $M = P + I$, or $M = P(1 + rt)$.

Determine the present value, using the Bankers' Rule.

	Maturity Value	Rate	Time (Days)	Present Value
1. a.	$ 735	12%	150	
b.	2,484	14	90	
2. a.	$6,450	10%	270	
b.	2,020	8	45	

Complete the following, finding the present value on the day of the investment.

	Principal	Rate	Time (Days)	Maturity Value	Rate Money Is Worth	Present Value
3. a.	$6,000	10%	90		9%	
b.	8,000	12	120		15	
4. a.	$5,400	7%	180		8%	
b.	9,000	9	120		6	

Complete the following, finding the present value on the day indicated.

	Principal	Rate	Time (Days)	Maturity Value	Rate Money Is Worth	Days Before Maturity	Present Value
5. a.	$ 9,000	12%	300		10%	180	
b.	7,500	15	120		12	60	
6. a.	$30,000	10%	120		9%	60	
b.	50,000	8	270		11	180	

7. The maturity value was $4,062 on a 9% note dated July 15. If the note was due on September 15, what had the face value been?

8. On April 19, $20,425 was paid in settlement of a note dated January 19. If interest was charged at 8.5%, what was the face value of the note?

9. The maturity value of $6,165 was paid on April 15 to discharge a simple interest note dated January 15. If money is worth 11%, what was the present value of the note on the day it was made?

10. On October 6, $2,520 was paid at the maturity of a note dated April 9. If money is worth 10%, what was the present value of the note on the date it was made?

11. A 4-month note for $3,500 was drawn on February 8 at 15%. If money is worth 12%, find the present value of the note on the day it was issued.

12. A 2-month note for $4,500 was drawn on July 3 at 8%. If money is worth 9%, find the present value of the note on the day it was made.

13. A 4-month note dated February 12 is made at 11% for $6,000. What is the present value of the note on the day it was made if money is worth 12%?

14. Determine the present value of an 8%, 5-month note with a face value of $48,000 made on February 18 if money is worth 9%.

15. A 3-month note dated May 16 is made at 15% for $12,000. Find the present value of the note on July 17 if money is worth 12%.

16. A 7-month note for $18,000 was made at 10% on April 20. If money is worth 9%, what was the present value of the note on September 21?

17. A 4-month, 9% note dated January 17 had a face value of $2,500. Find the present value on April 17 if money is worth 7.5%.

18. On January 25, a 5-month note for \$6,000 was made at 6%. What was the present value of the note on May 26 if money is worth 7%?

19. A contractor is considering three payment options for a contract: \$30,000 on signing the contract today; \$30,700 in 3 months; or \$31,500 on the completion of the contract in 6 months. Money is worth 12%. Which is the best offer? (*Hint:* Compare the present value of each payment.)

20. The following offers were made on a house: \$155,000 cash; \$163,000 in 4 months; or \$165,000 in 9 months. If money is worth 9%, which is the best offer? (*Hint:* Compare the present value of each payment.)

CHAPTER 15 GLOSSARY

Amount. (See "Maturity value at simple interest.")

Bankers' Rule. The most commonly used method of interest calculation, where time is expressed in exact days over 360 days per year.

Certificate of deposit (CD). A savings investment that earns a higher interest rate than savings accounts because the money is committed for a set period of time (usually a 3-month minimum); may also require a minimum investment (often \$1,000).

Collateral. An item of value that is pledged by a borrower in order to insure that a note will be repaid or the equivalent value received.

Date (of a note). The date on which a note is drawn (or signed).

Due date (of a note). The date on which a note is to be repaid; the maturity date.

Exact interest. Interest calculated using 365 days per year.

Exact time. The time period of a loan expressed in the specific number of calendar days.

Face value. The money value specified on a promissory note; the principal of a simple interest note.

Interest. Rent paid for the privilege of borrowing money: $I = Prt$, where I = interest, P = principal, r = rate, and t = time.

Maker. A person who signs a note promising to repay a loan.

Maturity value (at simple interest). The total amount (principal plus interest) due on a simple interest loan; also called "amount" or "sum": $M = P + I$, or $M = P(1 + rt)$, where M = maturity value, P = principal, I = interest, r = rate, and t = time.

Maturity value (of a note). The amount to be repaid on the due date of a note (which may or may not include interest).

Money-market account. A combined savings/CD account that requires a minimum balance (often \$500) but allows a limited number of checks or withdrawals. Pays higher interest than regular savings, but lower than CDs.

Note. A written promise to repay a loan (either with or without interest, and with or without collateral).

Ordinary interest. Interest calculated using 360 days per year.

Ordinary time. The time period of a loan where each month is assumed to have 30 days.

Payee. The person (or bank or firm) to whom the maturity value of a note will be paid.

Present value. (1) Principal; also (2) the value of another investment which, if made (on any given date) at the rate money is worth, would have the same maturity value as has an actual, given investment:

$$P = \frac{M}{1 + rt}$$

where P = principal or present value, M = maturity value, r = rate, and t = time.

Principal. The original amount of money that is loaned or borrowed; the investment value on which interest is computed.

Promissory note. (See "Note.")

Rate. An annual percent at which simple interest is computed.

Rate money is worth. An average or typical rate currently being used at most financial institutions.

Savings account. A low-interest account where any amount may be deposited or withdrawn at any time; also called an open account.

Simple interest. Interest that is computed on the original principal of a loan for the entire time of the loan, and is then added to the principal to obtain the maturity value. (See also "Interest.")

Sum. (See "Maturity value at simple interest.")

Time. The duration of a simple-interest investment. (Time must be expressed as a fractional year, in order to correspond with the annual percentage rate at which interest is calculated.)

CHAPTER 16

BANK DISCOUNT

OBJECTIVES

Upon completion of Chapter 16, you will be able to:

1. Define and use correctly the terminology associated with each topic.
2. a. Determine the proceeds of a simple discount note, using the simple discount formulas (Section 1: Examples 1, 2; Problems 1–24):
   (1) $D = Mdt$
   (2) $p = M - D$.
   b. Also, apply $p = M(1 - dt)$ to compute
   (1) Proceeds (Section 1: Examples 2, 3; Problems 9, 10, 17–20)
   (2) Maturity value (Section 2: Example 1; Problems 1, 2, 5–10).
3. a. Determine the equivalent simple interest rate charged on a simple discount note (Section 1: Example 4; Problems 7, 8).
   b. Find the unknown discount rate or time of a simple discount note (Section 1: Example 5; Problems 13–16, 21, 22).
4. a. Rediscount a simple discount note (or a simple interest note) on a given day prior to the maturity date (Section 2: Examples 2, 3; Problems 3–6, 11–18).
   b. Also, determine the amount of interest made
   (1) By the original lender
   (2) By the second lender.
5. Compute the savings gained by signing a discount note on the last day allowed by invoice sales terms, in order to take advantage of the cash discount (assuming that the bank note will be repaid on the same due date specified by the sales terms) (Section 3: Example 1; Problems 1–10).
6. Explain the similarities and differences of simple interest and simple discount notes (Section 4: Problems 1–22).

In the preceding chapter we discussed the simple interest note. Some banks, however, use the simple discount note (illustrated in the following section) for short-term loans of 1 year or less. There are some similarities and several major differences between the two types which it is imperative that you understand. (The two notes will also be summarized in Section 4.)

SECTION 1

SIMPLE DISCOUNT NOTES

Recall that the face value of the simple interest note is the principal of the loan. Interest is computed on this principal and added to it to obtain the maturity value.

The **face value** of a discounted note, however, is the **maturity value**—the amount that will be repaid to the payee. As in the case of the simple interest note, the maturity value is repaid in a single payment at the end of the loan.

The interest on a discounted note is called **bank discount.** Bank discount is computed on the maturity value (face value) of the note and is then *subtracted* from the maturity value (as opposed to simple interest, which is computed on the *principal* and then *added* to the principal to *obtain* the maturity value). The amount remaining after the bank discount has been subtracted from the maturity value of the note is called the **proceeds.** The proceeds is the actual amount that the borrower will receive from the bank and that can then be spent. Because the bank discount is subtracted from the face value before the borrower receives any money, this is often referred to as paying "interest in advance."

To reemphasize: The distinguishing aspect of the simple discount note is that interest is charged on the amount that is to be paid back, rather than on the amount that was actually borrowed.

Figures 16-1 and 16-2 illustrate typical simple discount notes. Figure 16-1 is an unsecured or signature note; that is, the loan is made simply on the basis of the borrower's good credit standing. Banks often establish some maximum amount that they will lend on an unsecured note. Figure 16-2 illustrates a collateral note, which requires the maker (borrower) to secure the note with some item of value.

Bank discount is computed using the formula

$$D = Mdt$$

where D = discount, M = maturity value (face value) of the discount note, d = discount rate, and t = time. We will utilize the Bankers' Rule for most discount notes, although a 365-day year is sometimes used, as described later.

Proceeds p are then found using the formula

$$p = M - D$$

NOTE NO________ OFFICER APPROVAL________________

UNSECURED PROMISSORY NOTE

$ 2,400.00 January 30 19XX

For value received, the undersigned Borrower(s) jointly and severally promise to pay to North Star Savings Association (hereinafter called "NORTH STAR"), or order, at the main banking office of NORTH STAR in Richmond, VA, the sum of Two thousand four hundred and no/100------------Dollars in lawful money of the United States of America with principal and interest payable as stated in the schedule of payments.

SCHEDULE OF PAYMENTS Select one (terms not selected no not apply)

☒ **DISCOUNT NOTE**

The above stated amount shall be due and payable on March 31, 19XX, 60 days from the date of this Note, and includes interest in the amount of $40.00, discounted at the rate of 10 %.

☐ **SINGLE MATURITY NOTE WITH INTEREST FROM DATE**

The above stated amount shall be due and payable on ____________, 19____, ______ days from the date of this Note, plus interest in the amount of $____________ computed at the rate of ________%.

☐ **INSTALLMENT TERM NOTE**

The above stated amount shall be payable in ________ consecutive ________ installments of $__________ each, beginning on ____________, 19______, and one final installment or balloon payment of $__________ payable on ____________, 19______. The installment payment ☐ includes *or* ☐ is in addition to interest payable at the rate of ________%.

If final payment is a balloon payment, NORTH STAR makes no commitment to refinance same if not paid when due. If default be made on any payment of this Note, or any other obligation to NORTH STAR, the maturity of this Note may be accelerated and the entire balance shall become due and payable at the option of NORTH STAR. In the event of any default or failure to pay by maturity, interest thereafter shall be payable at the maximum contract rate then allowed by law. In the event that any payment shall be past due for 10 days or more, there will be a delinquency charge of the lesser of 12% of the unpaid payment or $25, which is in addition to any interest owed.

THIS NOTE IS SUBJECT TO ANY ADDITIONAL PROVISIONS, WARRANTIES, AND RIGHTS SET FORTH ON THE REVERSE SIDE HEREOF.

In Witness Whereof, the Borrower has caused this Note to be executed by its duly authorized officers (if a corporation), or has hereunto set hand and seal (if an individual) to be affixed hereto on the day and year first written above.

CORPORATE BORROWER:

____
Name of Corporation

By____________________
President

By____________________
Secretary

Corporate Seal: Seal

Address:____________________

INDIVIDUAL BORROWER(S):

A. Sample Borrower (Seal)

____________________(Seal)

____________________(Seal)

____________________(Seal)

Form 7126 NOTICE: SEE OTHER SIDE FOR IMPORTANT INFORMATION

Figure 16-1 Simple Discount Note

$ 1,500 July 15, 19X1

90 days after date, for value received and with interest discounted to maturity

at 12 percent per annum,

the undersigned promise(s) to pay to **NORTH CAROLINA NATIONAL BANK** or order

One thousand five hundred and no/100---Dollars

payable at any office of the North Carolina National Bank in North Carolina. Interest shall be computed on the basis of a 360 day year for the actual number of days in the interest period and interest shall accrue after maturity or demand, until paid, at the rate stated above.

To secure the payment of this note and liabilities as herein defined, the parties hereto hereby pledge and grant to said Bank (the word "Bank" wherever used herein shall include any holder or assignee of this note) a security interest in the collateral described as follows:

25 shares - IBM stock

and any collateral added thereto or substituted therefor, including shares issued as stock dividends and stock splits and dividends representing distribution of capital assets. The Bank is hereby authorized at any time to charge against any deposit accounts of any party hereto any and all liabilities whether due or not. The Bank may declare all liabilities due at once in the event any party hereto becomes subject to any proceedings for the relief of creditors including but not limited to proceedings under the Bankruptcy Act or otherwise, or if in the judgment of Bank the collateral decreases in value so as to render Bank insecure and Bank demands additional collateral which is not furnished, or if Bank at any time otherwise deems itself insecure. In the event the indebtedness evidenced hereby or liabilities as defined herein be collected by or through an attorney at law, the holder shall be entitled to collect reasonable attorneys' fees.

Upon failure to pay any liability when due, Bank may sell the collateral at public or private sale, for cash or on credit, as a whole or in parcels, without notice, and Bank may at any such sale purchase the collateral or any part thereof for its own account, and the proceeds of any such sale shall be applied first to the costs of such sale and the expenses of collection, including reasonable attorneys' fees, and then to the outstanding balance due on said liabilities, the application to be made in the manner and proportions as Bank elects. The Bank may forbear from realizing on the collateral or any part thereof, by sale or otherwise, all as the Bank may decide, and the liabilities of the parties hereto shall not be released, discharged or in any way affected by any such forbearance, nor shall any of the parties hereto have any rights or recourse against the Bank by reason of any action the Bank may take or omit to take under this note, by reason of any deterioration, waste, or loss of any of the collateral unless such deterioration, waste, or loss be caused by the willful act or willful failure to act of the Bank. Upon payment of this note the Bank may release the collateral but shall have the right to retain the same to secure any unpaid liabilities. Upon any transfer of this note and the collateral, the Bank shall be fully relieved of responsibility with reference thereto. "Liabilities" or "Liability" as herein used, shall include this note and all obligations of every kind of any party hereto in whatever capacity to Bank, now or hereafter existing, whether arising directly or acquired from others as collateral or otherwise, whether absolute or contingent, joint or several, joint and several, secured or unsecured, due or not due, direct or indirect, including, but not limited to, liabilities arising by operation of law, contractual or tortious, liquidated or unliquidated or otherwise.

All persons bound on this obligation, whether primarily or secondarily liable as principals, sureties, guarantors, endorsers or otherwise, hereby waive presentment, protest, notice of dishonor and of acceleration of maturity and any right to require the Bank to retain any collateral pledged as security for this note or any other liabilities and agree that any extension of time for payment with or without notice shall not affect their joint and several liabilities.

Witness our/my hand(s) and seal(s).

Address 2008 E. Corning Dr. Mason R. Weatherford (Seal)

NCNB 2158 Rev. 2/75 Due Oct. 13 No. 73885 Joan D. Weatherford (Seal)

Courtesy of North Carolina National Bank, Statesville

Figure 16-2 Collateral Discount Note

Let us consider the difference between the amount of money one would be able to spend if (s)he borrows at a discount rate versus the amount available to spend when borrowing at an interest rate. (The following discount computation illustrates Figure 16-2.)

Example 1 The face values of two notes are \$1,500 each. The discount rate and the interest rate are both 12%, and the time of each is 90 days.

Interest Note	*Discount Note*
$P = \$1{,}500$	$M = \$1{,}500$
$r = 12\%$	$d = 12\%$
$t = \frac{90}{360}$ or 0.25	$t = \frac{90}{360}$ or 0.25

$$I = Prt \qquad D = Mdt$$
$$= \$1{,}500 \times 0.12 \times 0.25 \qquad = \$1{,}500 \times 0.12 \times 0.25$$
$$I = \$45 \qquad D = \$45$$

$$M = P + I \qquad p = M - D$$
$$= \$1{,}500 + \$45 \qquad = \$1{,}500 - \$45$$
$$M = \$1{,}545 \qquad p = \$1{,}455$$

Notice that on the simple interest note, the borrower pays \$45 for the use of \$1,500. On the simple discount note, however, the borrower pays \$45 for the use of only \$1,455. Thus, based on the actual amount of money the borrower has available to spend, more interest is paid at a discount rate than at an interest rate. Or stated another way: The borrower is really paying slightly more than 12% on the money (s)he actually uses, since 12% of \$1,455 for 90 days would be only \$43.65 interest due, not the \$45 actually charged.

From this example we can conclude that, when a borrower obtains money at a discount rate, (s)he will always be paying somewhat more than the stated rate on the money actually received. In addition, the **Truth-in-Lending Law** enables the borrower to be more informed about the true rate being paid. This true rate is the **equivalent simple interest rate.** The law requires the lender to reveal both the amount of interest and the annual percentage rate charged on the money received, correct to the nearest $\frac{1}{4}$% or 0.25%. Banks have available tables or computer programs that enable them to determine this true rate very easily.

It will be of interest here to know the true rate required under the Truth-in-Lending Law for the examples above. Figure 16-3 shows the Truth-in-Lending Disclosure Statement that corresponds to the note illustrated in Figure 16-2 and Example 1. We see that the true annual percentage rate paid for the \$1,455 that the borrower received is actually 12.25% (correct to the nearest $\frac{1}{4}$% or 0.25%), rather than 12% (the discount rate). For the note in Figure 16-1, the true rate is 10.25%, correct to the nearest ¼% or 0.25%, compared to the stated 10% discount rate. This procedure is described later in the chapter.

The bank discount formulas given above can be combined into a single formula for finding proceeds. By substituting *Mdt* for *D,* we have

$$p = M - D$$
$$= M - Mdt$$
$$p = M(1 - dt)$$

Example 2 A \$3,000 note dated January 31 is to be discounted at 10% for 5 months. Find the bank discount and the proceeds using (a) $D = Mdt$ and (b) $p = M(1 - dt)$.

TRUTH IN LENDING DISCLOSURE STATEMENT
NORTH CAROLINA NATIONAL BANK
COMMERCIAL BANKING DEPARTMENT

DATE OF NOTE (Date on which finance charge begins to accrue): July 15, 19X1

BORROWER'S NAME: Mason R. & Joan D. Weatherford

AMOUNT OF NOTE: $ 1,500.00

FINANCE CHARGES			AMOUNT FINANCED		
Interest	2	$ 45.00	Paid to Borrower	11	$ 1,455.00
Fees	3	$	Credit Life Insurance Prem. (If any)	12	$ -
Other (specify):	4	$	Financing Statement Filing Fee (If any)	13	$ -
	5	$	Other Amounts Paid on Behalf of Borrower (specify):	14	$
	6	$		15	$
Total FINANCE CHARGE	7	$ 45.00		16	$
ANNUAL PERCENTAGE RATE	8	12.25 %	AMOUNT FINANCED	17	$ 1,455.00

No. of Payments	Due Dates or Periods	Amount of Regular Payment	Amount of Final Payment	Balloon Payment (If any) *	Total of Payments
18 1	19 October 13, 19X1	20 -	21 -	22 -	23 1,500.00

24 To secure the obligation borrower has granted NCNB a Security Interest in the following collateral (if applicable)

25 shares - IBM stock

25

NOTE: The Security Agreement will secure future or other indebtedness and will cover after-acquired property.

☐ If checked, the collateral includes the principal residence of the borrower, and two (2) copies of the Notice of Right to Rescind are attached hereto.

26 In the event of prepayment of the obligation before maturity, borrower shall receive a rebate of unearned Finance Charge as follows:

☐ Not applicable

27

ADDITIONAL INFORMATION

Property insurance, if required, may be purchased from any reputable insurer selected by the borrower.
If the obligation is collected by or through an attorney at law after maturity, borrower shall be required to pay all collection costs and reasonable attorney's fees.

28 *Conditions (if any) under which Balloon Payment may be refinanced if not paid when due

Receipt of the foregoing statement fully completed and any attachments referred to therein is hereby acknowledged.

29 Mason R. Weatherford
Borrower's Signature

Credit Life Insurance Election (if applicable):
I understand that Credit Life Insurance is not required as a condition of this loan but may be purchased through NCNB at a cost of $ ______ for the term of the loan. I hereby affirm my desire to purchase such insurance.

30 ______
Borrower's Signature for Credit Life Insurance

NCNB 2260 Rev 12/77

Courtesy of North Carolina National Bank, Statesville

Figure 16-3 Truth-in-Lending Disclosure Statement

The Bankers' Rule is applied below when computing bank discount, as follows:

$$\begin{aligned} \text{June } 30 &= 181 \text{ day} \\ \text{January } 31 &= -\underline{\ 31\text{day}\ } \\ \text{time} &= 150 \text{ days} \end{aligned} \qquad \begin{aligned} M &= \$3{,}000 \\ d &= 10\% \\ t &= \frac{150}{360} \text{ or } \frac{5}{12} \end{aligned}$$

(a) $D = Mdt$

$$= \$3{,}000 \times 0.10 \times \frac{5}{12}$$

$$D = \$125$$

$$p = M - D$$

$$= \$3{,}000 - \$125$$

$$p = \$2{,}875$$

(b) $p = M(1 - dt)$

$$= \$3{,}000\left(1 - 0.10 \times \frac{5}{12}\right)$$

$$= 3{,}000\ (1 - 0.041667)$$

$$= 3{,}000\ (0.958333)$$

$$p = \$2{,}874.999 \quad \text{or} \quad \$2{,}875$$

$$D = M - p$$

$$= \$3{,}000 - \$2{,}875$$

$$D = \$125$$

Although bank discount and proceeds may be found in either of the above ways, you would probably use method (a), since the computation would be easier.

It was shown in Example 1 that when interest and bank discount are equal, the borrower has less spendable money from a discounted note. In that example, however, the maturity values of the two notes were not the same. We will now consider a simple interest note and a simple discount note that both have the same maturity value, to compare the amounts that the borrower would have available to spend.

Example 3 A simple interest note and a simple discount note both have a maturity value of $1,000. If the notes both had interest computed at 15% for 120 days, find (a) the principal (or present value) of the simple interest note and (b) the proceeds of the discount note.

(a) *Simple Interest Note*

$M = \$1{,}000$

$r = 15\%$

$t = \dfrac{120}{360}$ or $\dfrac{1}{3}$

$P = ?$

(b) *Simple Discount Note*

$M = \$1{,}000$

$d = 15\%$

$t = \dfrac{120}{360}$ or $\dfrac{1}{3}$

$p = ?$

$$P = \frac{M}{1 + rt}$$
$$= \frac{\$1{,}000}{1 + 0.15 \times \frac{1}{3}}$$
$$= \frac{\$1{,}000}{1 + 0.05}$$
$$= \frac{\$1{,}000}{1.05}$$
$$P = \$952.38$$

$$p = M(1 - dt)$$
$$= \$1{,}000\left(1 - 0.15 \times \frac{1}{3}\right)$$
$$= \$1{,}000(1 - 0.05)$$
$$= \$1{,}000(0.95)$$
$$p = \$950$$

In this case, the maker of the note would have \$2.38 more to spend if the note were a simple interest note. If \$952.38 were invested at 15% for 120 days, the maturity value would be \$1,000.

As mentioned earlier, borrowing money by signing a simple discount note is more expensive than by signing a simple interest note. To determine the equivalent simple interest rate on the simple discount note in Example 3, use the simple interest formula, $I = Prt$.* The I is the discount or the difference between the maturity value and the proceeds. The proceeds will be substituted for the P in the formula, since the amount of money available to spend will be the proceeds of a discount note. The t will be the same, but r is not 15% since 15% is the discount rate on a maturity value of \$1,000, not 15% of the proceeds. The r will be the unknown.

Example 4 Determine the equivalent simple interest rate being charged on the simple discount note with a maturity value of \$1,000, a discount rate of 15%, and a time of 120 days (Example 3).

$$M = \$1{,}000$$
$$p = \$950$$
$$I = \$1{,}000 - \$950$$
$$= \$50$$
$$t = \frac{120}{360} \quad \text{or} \quad \frac{1}{3}$$

$$I = Prt$$
$$\$50 = \$950 \times r \times \frac{1}{3}$$
$$50 = \frac{950}{3}r$$
$$\left(\frac{3}{950}\right)50 = r$$
$$r = 15.79\% \quad \text{or} \quad 15.75\%$$

*Another formula for finding the equivalent simple interest rate is as follows: $r = \frac{D}{p} \times$ reciprocal of time. In Example 4, this is $\frac{\$50}{\$950} \times \frac{3}{1} = 0.1578947$ or 15.75%, correct to the nearest ¼%.

The rate of interest disclosed to the borrower would be 15.75%, correct to the nearest 0.25%.

As was true for simple interest notes, the use of a 365-day year has become more widespread with bank discounts. The trend began during a time of higher rates of interest and discount, which made the difference between a 360-day and a 365-day year more significant. In some parts of the country, bank discount is thus routinely calculated using an "exact" discount rate (exact days over 365 or 366). This exact discount method is also applied when a commercial bank borrows funds from a Federal Reserve bank.

Example 5 A bank received proceeds of \$14,150 on a 90-day discounted note from its Federal Reserve bank. The face value of the note was \$14,600 (using as collateral certain notes owed to the bank). What exact discount rate was charged?

$$\begin{aligned} M &= \$14{,}600 \\ p &= -\underline{14{,}150} \\ D &= \$450 \end{aligned}$$

$$t = \frac{90}{365}$$

$$d = ?$$

$$D = Mdt$$

$$\$450 = \$14{,}600 \times d \times \frac{90}{365}$$

$$450 = 3{,}600d$$

$$\frac{450}{3{,}600} = d$$

$$d = 0.125 \quad \text{or} \quad 12.5\%$$

Another type of investment that utilizes discount calculations is U.S. Treasury Bills, which may be purchased in either 13-week or 26-week categories. The price that an investor pays for a Treasury Bill (or "T-bill," as it is often called) is the proceeds remaining after the discount calculation. The face of the Treasury Bill is the amount that the investor will receive from the government when the bill matures. (It should be noted, however, that *discounts on Treasury Bills are calculated by the Bankers' Rule*— that is, $\frac{13 \times 7}{360}$ or $\frac{26 \times 7}{360}$ or $\frac{52 \times 7}{360}$.)

SECTION 1 PROBLEMS

1. Identify the parts of the discounted note shown in Figure 16-1.

a. Face value   b. Maker   c. Payee   d. Date

e. Due date   f. Rate   g. Time   h. Bank discount

i. Proceeds   j. Maturity value

2. Answer Parts (a) through (j) of Problem 1 as they pertain to the discounted note shown in Figures 16-2 and 16-3.

Supply the missing entries, using the Bankers' Rule.

	Maturity Value	Discount Rate	Date	Due Date	Time (Days)	Bank Discount	Proceeds
3. a.	$ 9,000	8%	10/5		30		
b.	7,600	15	2/25		270		
c.	5,000	10	5/14	8/12			
d.	3,000	9		9/3	60		
4. a.	$ 4,500	8%	5/15		90		
b.	6,200	9	4/10		120		
c.	9,000	10	9/20	10/20			
d.	10,000	11		11/14	180		

Use an exact bank discount (365-day year) to find the proceeds.

5. a. $3,650 at 9% for 30 days b. $7,300 at 11% for 120 days c. $1,095 at 15% for 90 days

6. a. $2,920 at 10% for 60 days b. $4,380 at 12% for 270 days c. $14,600 at 9% for 90 days

Use the Bankers' Rule and supply the missing parts. Round the equivalent simple interest rate correct to the nearest $\frac{1}{4}$%.

	Maturity Value	Discount Rate	Time (Days)	Bank Discount	Proceeds	Equivalent Simple Interest Rate
7. a.	$ 4,000	12%	90			
b.	8,000	15	150			
c.	7,200	13	120			
d.	5,400	10	60			
8. a.	$ 6,600	8%	120			
b.	12,000	10	270			
c.	10,000	9	90			
d.	5,400	7	300			

Complete the following, using the Bankers' Rule.

	MATURITY VALUE	DISCOUNT RATE	DATE	DUE DATE	TIME	PROCEEDS
9. a.	$4,320	14%	1/9		5 months	
b.	6,000	9	10/3		2 months	
c.	6,500	12	3/17		90 days	
d.	9,000	10	5/20		180 days	
10. a.	$7,500	8%	2/19		4 months	
b.	2,000	6	11/15		1 month	
c.	5,600	9	8/1		120 days	
d.	4,200	10	9/5		60 days	

11. A 90-day note for $10,000 was discounted by a bank at 11%. What amount will the borrower receive from the bank?

12. A mail-order firm signed a 60-day note with a face value of $4,000 and a discount rate of 9%. How much did the business receive from the bank?

13. The face value of an 8% note was $7,800. If the bank discount was $156, what was the time of the note?

14. The bank discount was $70 on a note with a face value of $3,500. If the note was discounted at 12%, what was the time of the note?

15. A borrower received $8,554 when his $9,400 note was discounted at the bank. If the note had a time of 270 days, what discount rate was charged?

16. The face value of a note was $15,000, and the proceeds were $14,550. What was the discount rate of this 120-day note?

17. A 3-month note having a maturity value of $6,000 was dated January 4. The bank charged a discount rate of 13%. What were the proceeds?

18. A 4-month note dated February 20 has a maturity value of $9,000. The bank charged a discount rate of 8%. What were the proceeds?

19. The maturity values of two 6-month notes are both $3,000. Compare the present value at 9% with the proceeds at a 9% discount.

20. The maturity values of two 4-month notes are both $6,000. Compare the present value at 12% interest with the proceeds at a 12% discount.

21. A bank used $54,750 of loans as collateral and received $53,940 when it discounted a note at the Federal Reserve bank. What exact discount rate was charged on the 60-day note?

22. A bank discounts a 30-day note at its Federal Reserve bank, receiving $39,886. Collateral for the note was $40,150 in loans owed to the bank. At what exact discount rate was the note computed?

23. A $10,000 U.S. Treasury Bill was purchased at a 10% discount rate for 26 weeks.

a. How much did the investor pay for the T-bill?

b. What was the maturity value?

c. How much interest was earned on the T-bill?

24. An investor purchased a $20,000 U.S. Treasury Bill at a 9% discount rate for 13 weeks.

a. How much was invested in the T-bill?

b. What was the maturity value?

c. How much interest was earned on the investment?

SECTION 2

MATURITY VALUE (AND REDISCOUNTING NOTES)

It often happens that the proceeds of a simple discount note are known and one wishes to find the maturity value of the note. When problems of this type are to be computed, it is customary to use an altered version of the basic discount formula $p = M(1 - dt)$. The formula is altered by dividing both sides of the equation by the expression in parentheses:

$$p = M(1 - dt)$$

$$\frac{p}{1 - dt} = \frac{M(\cancel{1 - dt})}{(\cancel{1 - dt})}$$

$$\frac{p}{1 - dt} = M \quad \text{or} \quad M = \frac{p}{1 - dt}$$

Thus, the formula is set up to solve for the maturity value, M. Example 1, however, shows a problem worked by both versions of the discount formula, and either form may be used to solve the assignment problems.

Example 1 The proceeds were \$290.80 on a 3-month note dated August 29 and discounted at 12%. What is the maturity value?

$p = \$290.80$

$d = 12\%$

$t = \frac{92}{360}$

$M = ?$

$$\begin{aligned} p &= M(1 - dt) \\ \$290.80 &= M\left(1 - 0.12 \times \frac{92}{360}\right) \\ &= M(1 - 0.030667) \\ &= M(0.969333) \\ \frac{\$290.80}{0.969333} &= M \\ \$300 &= M \end{aligned}$$

$$\begin{aligned} M &= \frac{p}{1 - dt} \\ &= \frac{\$290.80}{1 - 0.12 \times \frac{92}{360}} \\ &= \frac{\$290.80}{1 - 0.030667} \\ &= \frac{\$290.80}{0.969333} \\ M &= \$300 \end{aligned}$$

The maturity value is \$300 on November 29.

A promissory note, like a check or currency, is a negotiable instrument. That is, it can be sold to another party (a person, a business, or a bank), or it can be used to purchase something or to pay a debt.

If the payee of a discounted note (usually some financial institution) sells the note to another bank, the note has been **rediscounted.** In that case, the second bank also uses the maturity value of the note as the basis for determining the proceeds that it will pay for the note. It should be observed that the discount rates that financial institutions charge each other are somewhat lower than those that banks ordinarily charge businesses and individuals.

When a note is rediscounted, the actual amount of interest earned by the original lender is the difference between the rediscounted note's proceeds (p_2) and the original note's proceeds (p_1).

$$\begin{array}{rl} & \text{Proceeds}_2\ (p_2) \\ - & \underline{\text{Proceeds}_1\ (p_1)} \\ & \text{Net interest earned} \quad \text{(by original lender)} \end{array}$$

Example 2 City National Bank was the holder (payee) of a $1,200, 120-day note discounted at 12%. Thirty days before the note was due, City Bank rediscounted (or sold) the note to Capital Mortgage and Trust at 9%. (a) How much did City Bank receive for the note? (b) How much did City Bank actually make on the transaction?

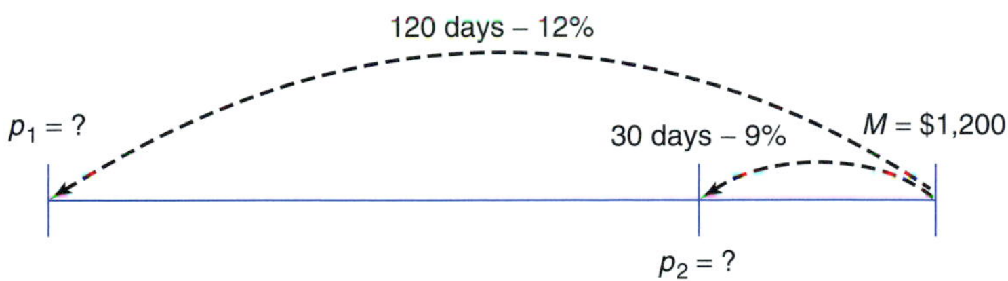

Original Note

$$M = \$1{,}200 \qquad D = Mdt$$

$$d = 12\% \qquad = \$1{,}200 \times 0.12 \times \frac{1}{3}$$

$$t = \frac{120}{360} \text{ or } \frac{1}{3} \qquad = \$1{,}200 \times 0.04$$

$$D = \$48$$

$$p_1 = ?$$

$$p = M - D$$

$$= \$1{,}200 - \$48$$

$$p_1 = \$1{,}152$$

The proceeds of the original note (p_1), the amount that City Bank loaned to the maker, were \$1,152.

Rediscounted Note

$M = \$1{,}200$

$d = 9\%$

$t = \frac{30}{360}$ or $\frac{1}{12}$

$p_2 = ?$

$$p_2 = M(1 - dt)$$
$$= \$1{,}200\left(1 - 0.09 \times \frac{1}{12}\right)$$
$$= \$1{,}200(1 - 0.0075)$$
$$= \$1{,}200(0.9925)$$
$$p_2 = \$1{,}191$$

(a) The proceeds of the rediscounted note (p_2) were \$1,191. That is, City Bank received \$1,191 when it sold the note to Capital Mortgage and Trust.

(b) City Bank made \$1,191 − \$1,152 = \$39 on the transaction. This is only \$9 less than the entire \$48 interest that City Bank would have earned if it had kept the note, although only $\frac{3}{4}$ of the time had elapsed.

DISCOUNTING SIMPLE INTEREST NOTES

It is also possible that the payee of a simple interest note may wish to sell the note at a bank before the due date. The amount that the bank would pay is determined as follows:

1. Find the maturity value of the simple interest note.
2. Discount the maturity value of the note using the bank's rate and the time *until the due date.* The proceeds remaining is the amount that the bank would pay for the note.
3. The interest that the original payee has made is the difference between what the note was sold for (the discounted proceeds) and the original amount loaned (the principal of the simple interest note):

Proceeds	
−Principal	
Net interest earned	(by original lender)

Example 3 A-1 Appliance Sales was the payee of a \$900, 80-day note with interest at 10%. Forty days before the due date, A-1's bank discounted the note at 9%. (a) How much did A-1 receive for the note? (b) How much did A-1 make on the transaction?

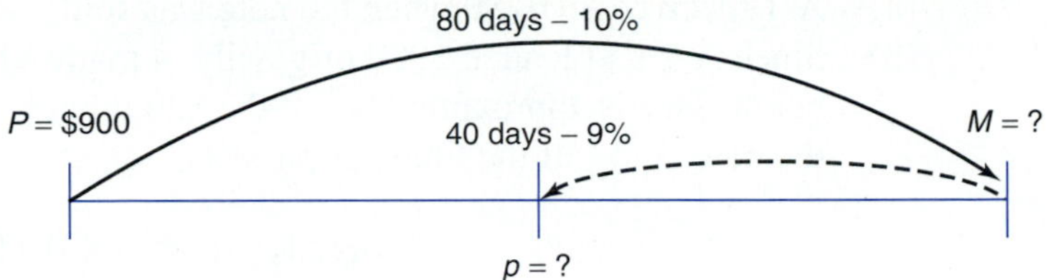

Simple Interest Note
(Finding Maturity Value)

$P = \$900 \qquad I = Prt$

$r = 10\% \qquad = \$900 \times 0.10 \times \frac{2}{9}$

$t = \frac{80}{360}$ or $\frac{2}{9} \qquad I = \$20$

$M = ?$

$$M = P + I$$

$$= \$900 + \$20$$

$$M = \$920$$

Discount Note
(Finding Proceeds)

$M = \$920 \qquad D = Mdt$

$d = 9\% \qquad = \$920 \times 0.09 \times \frac{1}{9}$

$t = \frac{40}{360}$ or $\frac{1}{9} \qquad D = \$9.20$

$p = ?$

$$p = M - D$$

$$= \$920 - \$9.20$$

$$p = \$910.80$$

(a) A-1 received $910.80 when the note was sold.

(b) Since A-1 had loaned $900 originally, it made $10.80 in interest, as shown below. This is approximately $\frac{1}{2}$ of the $20 interest that the maker will pay on the note, and $\frac{1}{2}$ of the time has passed.

Proceeds	$910.80
Principal	− 900.00
Net interest	$ 10.80

SECTION 2 PROBLEMS

Complete the following.

	Proceeds	Discount Rate	Date	Due Date	Time	Maturity Value
1. a.	$ 6,860	12%	Oct. 18		60 days	
b.	1,925	9	June 1		150 days	
c.	8,721	18	July 9		2 months	
d.	4,275	15	Feb. 25		4 months	
2. a.	$17,595	9%	Aug. 14		90 days	
b.	14,700	12	Sept. 7		60 days	
c.	5,220	10	Feb. 5		4 months	
d.	19,750	15	June 25		1 month	

Find the additional information.

	Original Note				Rediscounted Note			
	Maturity Value	Rate	Time (Days)	Proceeds	Rate	Time (Days)	Proceeds	Net Interest Earned
3. a.	$ 7,600	13%	270		14%	90		
b.	8,500	9	120		10	30		
c.	6,400	14	90		12	60		
4. a.	$10,000	7½%	120		6½%	45		
b.	12,000	9	150		8	60		
c.	9,000	10	300		12	120		

	SIMPLE INTEREST NOTE				DISCOUNTED NOTE			
	PRINCIPAL	RATE	TIME (DAYS)	MATURITY VALUE	DISCOUNT RATE	TIME (DAYS)	PROCEEDS	NET INTEREST EARNED
5. a.	$1,600	15%	90		12%	30		
b.	8,000	12	150		10	45		
6. a.	$4,000	9%	180		8%	45		
b.	1,500	6	60		7½	20		

7. The proceeds of a 4-month note dated January 10 were $3,800. If the discount rate was 15%, what was the face value of the note?

8. On November 16, the proceeds of a 1-month note were $15,840. If the bank discount had been computed at 12%, what was the face value of the note?

9. The maker of a 5-month note dated February 20 received proceeds of $8,470. If the discount rate had been 9%, how much did the maker pay at maturity?

10. The proceeds of a 2-month note dated March 15 were $6,381.18. If the discount rate was 9%, what was the maturity value of the note?

11. Cullman National Bank was the payee of a $20,000 note dated June 3. The note was made for 90 days and was discounted at 16%. On August 2, the note was rediscounted at Decatur Bank at 15%.

a. What amount had Cullman National Bank loaned the maker of the note?

b. How much had Cullman received when it rediscounted the note?

c. How much did Cullman make on the transaction?

d. How much more would Cullman have received if it had kept the note to maturity?

12. Tri State Bank was the payee of a $60,000 note dated August 16. The note was made for 120 days and was discounted at 10%. On November 14, the note was rediscounted at New Hope Bank at 9%.

a. How much did Tri State Bank loan to the maker of the note?

b. How much had Tri State received when it rediscounted the note?

c. How much did Tri State make on the transaction?

d. How much less interest was this than Tri State would have earned had it kept the note to maturity?

13. Professional Office Movers, Inc. was the payee of a $9,000, 120-day note dated March 16, with simple interest at 14%. On May 15, the business discounted the note at its bank at 12%.

a. How much did Professional Office Movers receive from the note?

b. How much interest did Professional Office Movers make on the note?

c. How much more interest would the business have made by keeping the note until maturity?

14. Case McCune, Inc. was the payee of an $18,000, 90-day note dated September 1 with simple interest at 7%. On November 10, Case McCune discounted the note at its bank at 9%.

a. How much did Case McCune receive from the bank?

b. How much did Case McCune make on the note?

c. How much more interest would the business have made by keeping the note to maturity?

15. Radford National Bank was the payee of a 120-day note dated June 25. The face value was $6,300, and the discount rate was 10%. On July 25, Radford National rediscounted the note at 9% at First Georgia State Bank.

a. What amount did Radford National lend to the maker of the note?

b. How much did Radford National receive when it rediscounted the note?

c. How much interest did Radford National earn from these transactions?

d. What interest did Radford National forego by not holding the note to maturity?

16. The Springfield National Bank was the payee of a 300-day note dated February 24. The face value of the note was $24,000, and the discount rate was 9%. On September 22, Springfield National Bank rediscounted the note at 8% at the Bank of Stafford.

a. What amount did Springfield National Bank lend to the maker of the note?

b. How much did Springfield receive when it rediscounted the note at the Bank of Stafford?

c. How much interest did Springfield earn from this transaction?

d. What interest did Springfield forfeit by not holding the note to maturity?

17. A 12%, 90-day simple interest note dated May 13 for $4,200 was payable at Sheraton National Bank. On June 22, Sheraton National Bank discounted the note at the Federal Reserve bank at an exact discount rate of 11%.

a. How much did Sheraton National Bank receive from the Federal Reserve bank?

b. How much interest did Sheraton National Bank earn from these transactions?

c. How much interest did Sheraton National Bank lose by not holding the original note to maturity?

18. Dickinson National Bank was the payee of a $30,000, 150-day note dated April 6, earning 10% interest. On July 20, the bank used the maturity value as collateral to arrange a note with its Federal Reserve bank at an exact discount rate of 8½%.

a. What amount did Dickinson National Bank receive from the Federal Reserve bank?

b. How much interest did Dickinson make on the funds it loaned?

c. How much interest did Dickinson forfeit by discounting the note?

SECTION 3

DISCOUNTING NOTES TO TAKE ADVANTAGE OF CASH DISCOUNT

In Chapter 12, we learned that many merchants offer cash discounts as inducements to buyers to pay their accounts promptly. These discounts typically are a full 1% to 2% in only 10 days. When businesses do not have sufficient cash available on the last day of a cash discount period, they often sign a bank note in order to take advantage of the cash discount. Although the bank charges interest on the note, the business still "comes out ahead" if it borrows for only a short term. Remember that a bank rate of 13.5% means 13.5% for an entire year; for a 20-day note, the actual interest charge would be only $\frac{3}{4}$%. Thus, the business would still save $1\frac{1}{4}$% by borrowing to take advantage of a 2% cash discount.

The business would save money by borrowing provided that it borrowed for a period of time short enough so that the bank discount it must pay is less than the cash discount. However, we will consider only the situation when the maximum amount will be saved: On the *last* day of the cash discount period, a note is discounted so that the proceeds exactly equal the amount required to take advantage of the cash discount. The note will be for only the time remaining until the net amount of the invoice would have been due.

Example 1 An invoice dated July 2 covers merchandise of $6,650 and freight of $30. Sales terms on the invoice are 2/10, n/30. A note is discounted at 9% in order to take advantage of the cash discount. (a) What will the face value of the note be? (b) How much will be saved by borrowing the necessary cash?

The last day of the discount period is July 12; the full net amount will be due 20 days later on August 1. Therefore, to obtain maximum savings, a discounted note will be signed on July 12 for 20 days.

(a)

	$6,650.00	Goods		$6,650.00	Goods
×	0.02		−	133.00	Cash discount
	$ 133.00	Cash discount		$6,517.00	
			+	30.00	Freight
				$6,547.00	Amount needed to take advantage of cash discount (proceeds needed)

$p = \$6{,}547$

$d = 9\%$

$t = \frac{20}{360}$ or $\frac{1}{18}$

$M = ?$

$$M = \frac{p}{1 - dt}$$

$$= \frac{\$6{,}547}{1 - 0.09 \times \frac{1}{18}}$$

$$= \frac{\$6{,}547}{1 - 0.005}$$

$$= \frac{\$6{,}547}{0.0995}$$

$$M = \$6{,}579.90$$

The required note will have a face (maturity) value of $6,579.90.

(b)	$6,650.00	Goods
+	30.00	Freight
	$6,680.00	Amount due August 1 if a note is not signed
−	6,579.90	Maturity value of note due August 1
	$ 100.10	Amount saved by discounting a note

SECTION 3 PROBLEMS

Complete the following. The notes will be discounted on the last day of the discount period to take advantage of the cash discounts.

	Amount of Invoice	Sales Terms	Amount Needed (Proceeds)	Time of Note (Days)	Discount Rate	Face Value of Note	Amount Saved
1. a.	$ 6,404.64	3/10, n/30			9%		
b.	7,614.80	2/15, n/30			12		
c.	4,242.86	2/15, n/60			8		
2. a.	$ 5,051.02	2/10, n/30			18%		
b.	8,206.19	3/15, n/30			12		
c.	10,193.30	3/15, n/60			9		

3. A bank note is discounted at 18% in order to take advantage of the cash discount on an invoice for $5,613.40. The terms of the invoice were 3/10, n/30.

a. What was the maturity value of the note?

b. How much was saved by borrowing to take advantage of the cash discount?

4. An invoice for $16,432.99 has sales terms of 3/15, n/30. A 9% note was discounted in order to take advantage of the cash discount.

a. What was the face value of the note?

b. How much was saved by borrowing to take advantage of the cash discount?

5. An invoice for $3,073.45 has terms of 3/15, n/30. A note is discounted at 15% in order to take advantage of the cash discount.

a. What will the face value of the note be?

b. How much will be saved by borrowing the necessary cash?

6. A business discounts a note at its bank at 9% in order to take advantage of a cash discount on an invoice for $20,153.06. The sale terms on the invoice were 2/10, n/60.

a. What was the maturity value of the note?

b. How much was saved by taking out the loan?

7. A note is discounted at 9% in order to take advantage of a cash discount on an invoice listing merchandise of $9,259.11 plus freight charges of $21.25. The cash discount is 4/10, n/50.

a. What will the face value of the note be?

b. How much will be saved by signing the note?

8. A note is discounted at 6% in order to take advantage of a cash discount on an invoice listing merchandise of $4,020.41 plus freight of $40. The cash discount is 2/30, n/60.
   a. What will the face value of the note be?
   b. How much will be saved by signing the note?
9. An invoice has terms of 2/10, n/30. Merchandise is listed for $2,715, and there is a freight charge of $18.30. If a bank note is discounted at 14%,
   a. What is the maturity value of the note?
   b. By borrowing the necessary cash, how much is saved?
10. A note is discounted at 12% in order to take advantage of a cash discount on an invoice listing merchandise for $5,093.75 and freight of $60. The cash discount shown on the invoice is 4/20, n/50.
   a. What is the face value of the note?
   b. How much is saved by borrowing?

SECTION 4

SUMMARY OF SIMPLE INTEREST AND SIMPLE DISCOUNT

It is essential that the student understand the identifying characteristics of both the simple interest and the simple discount notes as well as the differences between the two. Both kinds of *promissory notes* are

1. Generally for 1 year or less.
2. Repaid in a single payment at the end of the time.

As for their differences, the following are characteristic of the *simple interest note* (a sample of which is shown in Figure 16-4):

1. The face value is the principal of the note. The face value is the actual amount that was loaned.
2. Interest is computed on the face value of the note at the rate and for the time stated on the note.

Face value	(P)	$400
Interest rate	(r)	12%
Time	(t)	60 days
Interest	(I)	$8
Maturity value	(M)	$408

Figure 16-4 Simple Interest Note

3. The maturity value of the simple interest note is the sum of the face value plus interest.
4. A simple interest problem will contain one or more of these identifying characteristics: "*interest* rate of *x*%," "with *interest* at *x*%," a maturity value which exceeds the face value, "*amount.*"
5. The solution of simple interest problems requires one or more of the following formulas:

(a) $I = Prt$ (b) $M = P + I$

(c) $M = P(1 + rt)$ (d) $P = \dfrac{M}{1 + rt}$

The following are characteristic of the *simple discount note* (a sample of which is shown in Figure 16-5):

1. The face value of the simple discount note is the maturity value—the amount that will be repaid.
2. Discount notes may contain a rate quoted as a certain percent "discounted to maturity." The interest (called "bank discount") is computed on the face value of the note. (That is, the borrower pays interest on the amount that will be repaid, rather than on the amount actually borrowed.)
3. The actual amount that the borrower (maker) receives from the lender (payee) is called the "proceeds." The proceeds are found by subtracting the bank discount from the face value. This is often called paying interest in advance.
4. Simple discount problems may be distinguished from simple interest problems by one or more of the following: "*discount* rate of *x*%," "*discounted* at *x*%," "*proceeds,*" maturity value equals the face value, "interest in advance."

Face value	(*M*)	\$400
Discount rate	(*d*)	12%
Time	(*t*)	60 days
Discount	(*D*)	\$8
Proceeds	(*p*)	\$392
Maturity value	(*M*)	\$400

Figure 16-5 Simple Discount Note

5. The solution of simple discount problems requires one or more of the following formulas:

   (a) $D = Mdt$ (b) $p = M - D$

   (c) $p = M(1 - dt)$ (d) $M = \frac{p}{1 - dt}$

SECTION 4 PROBLEMS

Solve the following simple interest and simple discount problems.

1. A $5,000 note is signed for 45 days at 11%. Find the interest and the maturity value of the note.

2. Find the interest and the maturity value of a 120-day, 8.5% note with a face value of $4,500.

3. The face value of a $3,960 note is discounted for 40 days at 10%. What is the discount and what are the proceeds?

4. The face value of a 270-day was $9,000. If the discount rate is 11%, what is the discount and what are the proceeds?

5. Compare a simple interest note and a simple discount note. Both notes have face values of $1,600, interest charged at 15%, and a time of 120 days.

   a. How much interest will be charged on each note?

   b. How much will the maker receive on each note?

   c. What is the maturity value of each note?

6. A simple interest note and a simple discount note both have face values of $10,800. Both notes are for 90 days, and both have interest charged at 7%.

   a. Find the interest that will be paid on each.

   b. Find the amount the borrower actually receives.

   c. Find the maturity value of each note.

7. a. Determine the maturity value of a 60-day note that has proceeds of $2,352 if the discount rate is 12%.

   b. What is the face value of the note?

8. The proceeds of a 9% note for 30 days were $11,910. What was

   a. The maturity value?

   b. The face value?

9. A 180-day note with interest at 9.5% has a maturity value of $4,190.

   a. What is the principal of the note?

   b. What amount did the maker receive on the loan?

10. The maturity value is $15,375 on a 90-day note with interest at 10%.

   a. What is the face value of this note?

   b. How much interest did the maker receive on the loan?

11. The Regis Services Co. borrowed $7,000 from the bank by discounting a 90-day, 10% note. What equivalent simple interest rate (to the nearest 0.25%) was charged by the bank?

12. The Ritz Raines Co. discounted a 12.5%, 120-day note for $18,000 from Titone Telecom Co. What equivalent simple interest rate (to the nearest 0.25%) was charged by the bank?

13. Compare the present value at 15% interest with the proceeds at 15% discount on two notes with maturity values of $8,400 and terms of 120 days.

14. Compare the present value at 9% interest with the proceeds at 9% discount on two notes with maturity values of \$9,600 and terms of 30 days.

15. On August 22, a 4-month note with interest at 10% was made. The face value of the note was \$1,080. If most investments are earning 9%, what is the present value of the note on the day it was made?

16. On January 12, a 3-month note with interest at 12% was made. The face value of the note was \$11,000. What was the present value of the note on the day it was made if money is worth 11%?

17. On February 5, a 5-month note with a face value of \$5,200 was drawn at 12% interest. What is the present value of the note 90 days before maturity, if most investments are earning 10% interest?

18. On September 1, a 2-month note was signed for \$40,000 at 9% interest. What was the present value of the note 30 days before it was due, if money is worth 10.5%?

19. On January 8, Cook Construction Co. received a 6-month note for \$8,000 at 18% interest for a remodeling job. On April 9, Cook discounted the note at 15% at the bank.

a. How much did Cook Construction Co. receive from the bank?

b. How much did Cook Construction Co. make on the note?

c. How much interest did the bank make?

20. On October 9, Arvanitis Co. received a 2-month note for \$5,000 at 9% interest for merchandise sold. On November 9, Arvanitis discounted the note at 8% at its bank.

a. What amount did Arvanitis receive from the bank?

b. How much interest did Arvanitis make on the note?

c. How much interest did the bank earn?

21. American Bank was the payee of a \$6,000, 4-month note dated August 15 discounted at 9%. American Bank rediscounted the note at 8.5% at another bank on October 16.

a. What amount had American Bank loaned the maker?

b. How much did American Bank receive when it sold the note?

c. How much interest did American Bank earn?

d. How much interest did the second bank make?

22. Marbury Bank was the holder of a 3-month note dated June 18 with a discount rate of 12%. The face value was \$10,000. On July 20, Marbury Bank rediscounted the note at La Plata Bank at 11%.

a. How much did Marbury Bank lend to the maker of the note?

b. How much did Marbury Bank receive when it rediscounted the note?

c. How much interest did Marbury Bank earn?

d. How much interest did Marbury lose by not holding the note to maturity?

CHAPTER 16 GLOSSARY

Bank (or simple) discount. Interest that is computed on the maturity value of a loan and is then subtracted to obtain the proceeds; often called "interest in advance":

$$D = Mdt$$

where D = bank discount, M = maturity value, d = discount rate, and t = time.

Equivalent simple interest rate. The true rate of interest charged on a simple discount note; the interest rate which, if computed on the money the borrower actually received (the proceeds) and for the same time, would produce the same maturity value.

Face value. The money value specified on any promissory note; the maturity value of a simple discount note.

Maturity value. The total amount (Proceeds + Bank discount) due after a simple discount loan; the value on which simple discount is computed.

Proceeds. The amount remaining after the bank discount is subtracted from the maturity value; the amount actually received by a borrower:

$$p = M - D \quad \text{or} \quad p = M(1 - dt)$$

where p = proceeds, M = maturity value, D = bank discount, d = discount rate, and t = time.

Rediscounted note. A simple discount note that is sold for a value equal to the proceeds of the note at the buyer's discount rate.

Simple discount. (See "Bank discount.")

Truth-in-Lending Law. A federal law requiring lenders to disclose both their total finance charge and their corresponding annual percentage rate (based on the amount actually received by a borrower).

CHAPTER 17

Multiple Payment Plans

Objectives

Upon completion of Chapter 17, you will be able to:

1. Define and use correctly the terminology associated with each topic.
2. Apply the United States rule to
   a. Find the credit due for partial payments made on a note, and
   b. Find the balance due on the maturity date (Section 1: Example 1; Problems 1–6).
3. On an open-end charge account,
   a. Compute the interest due on adjusted monthly balances, and
   b. Determine the total amount required to pay off the account (Section 2: Example 1; Problems 1, 2, 5, 6).
   c. Compute the interest due on average daily balances, and
   d. Find the final balance (Example 2; Problems 3, 4, 7, 8).
4. Determine for an installment plan purchase (Section 3: Examples 1, 2; Problems 1, 2, 7–20, 23–26):
   a. The down payment and the finance charge
   b. The regular monthly or weekly payment
   c. The total cost on the time payment plan.
5. For loans or purchases repaid on an installment plan (Section 3: Examples 2–4; Problems 3, 4, 11–24):
   a. Compute "finance charge per $100" (or "rate × time").
   b. Consult a table to determine the annual percentage rate charged on an installment purchase.
6. Construct a table to verify that the annual percentage rate (Objective 5b) would actually result in the amount of interest charged (Section 3: Examples 3, 4; Problems 5, 6, 23–26, 28).
7. When prepayment occurs in an installment plan with set payments, determine the interest saved and the balance due
   a. Using the actuarial method (Section 4: Examples 1, 3; Problems 1, 2, 5–8)
   b. Using the rule of 78s (Example 2; Problems 3, 4, 9–12).

The simple interest and simple discount notes studied earlier are normally repaid in a single payment at the end of the time period. However, a borrower may prefer to repay part of a debt before the entire loan is due. The lender, in order to give the borrower proper credit for payments made prior to the maturity date, should apply the United States rule. Also, many firms, when extending credit, feel that the entire debt will more likely be repaid if the borrower is required to make regular installment payments that begin immediately, rather than a single payment that is not due for some time. Partial payments may thus be made under the **U.S. rule** or on basic **installment plans,** both of which are discussed in the following sections.

SECTION 1

UNITED STATES RULE

A person who has borrowed money on a note may sometimes wish to pay off part of the debt before the due date, in order to reduce the amount of interest paid. Such partial payments are not required (as they would be for an installment loan), but may be made whenever the borrower's financial condition permits. The United States rule, which we shall study, is used to calculate the credit that should be given for partial payments and to determine how much more is owed on the debt.

The **United States rule,** or **U.S. rule,** derives its name from the fact that this method has been upheld by the U.S. Supreme Court as well as by a number of state courts. It is the method used by the federal government in its financial transactions.* The Bankers' Rule should be used to determine "time" for the interest calculations in this section.

Basically, the U.S. rule is applied in the following manner:

1. Interest is computed on the principal from the first day until the date of the first partial payment.
2. The partial payment is used first to pay the interest due, and the remaining part of the payment is deducted from the principal.
3. The next time an amount is paid, interest is calculated on the adjusted principal from the date of the previous payment up to the present. The partial payment is then used to pay the interest due and to reduce the principal, as before. (This step may be repeated as often as additional payments are made.)
4. The balance due on the maturity date is found by computing interest due since the last partial payment and adding this interest to the unpaid principal.

*It might also be noted that interest calculations by the U.S. government use exact interest: t = exact days over 365 or 366. However, exact interest is not an essential element of the U.S. rule.

Example 1 On April 1, Mary Carrington Interiors took out a 90-day note for $1,500 bearing interest at 10%. On May 1, Carrington Interiors paid $700 toward the obligation, and on May 16, another $300. How much remains to be paid on the due date?

(a) Interest is computed on the original $1,500 from April 1 until the day of the first partial payment (May 1):

$$\begin{aligned} \text{May 1} &= 121 \text{ day} \\ \text{April 1} &= \underline{-\ 91 \text{ day}} \\ & \ \ 30 \text{ days} \end{aligned} \qquad \begin{aligned} I &= Prt \\ &= \$1{,}500 \times 0.10 \times \frac{1}{12} \end{aligned}$$

$$t = \frac{30}{360} \text{ or } \frac{1}{12} \qquad I = \$12.50$$

(b) The $700 is first applied to pay the interest, and the remaining amount ($700 − $12.50 = $687.50) is then deducted from the principal:

Principal	$1,500.00
Less: Payment to principal	− 687.50
Adjusted principal	$ 812.50

(c) Interest is computed on the $812.50 adjusted principal from May 1 until May 16. The $300 payment first pays this interest, and the remainder of the payment is then subtracted from the current principal:

$$\begin{aligned} \text{May 16} &= 136 \text{ day} \\ \text{May 1} &= \underline{-121 \text{ day}} \\ & \ \ 15 \text{ days} \end{aligned} \qquad \begin{aligned} I &= Prt \\ &= \$812.50 \times 0.10 \times \frac{1}{24} \end{aligned}$$

$$t = \frac{15}{360} = \frac{1}{24} \qquad I = \$3.39$$

Partial payment	$300.00
Less: Interest	− 3.39
Payment to principal	$296.61

Current principal	$812.50
Less: Payment to principal	− 296.61
Adjusted principal	$515.89

(d) No other payment is made until the maturity date; therefore, interest is computed on \$515.89 from May 16 until the 90-day period ends on June 30. The maturity value $(P + I)$ is the final payment.

$$\begin{aligned} \text{June 30} &= 181 \text{ day} \\ \text{May 16} &= \underline{-136} \text{ day} \\ & 45 \text{ days} \end{aligned} \qquad \begin{aligned} I &= Prt \\ &= \$515.89 \times 0.10 \times \frac{1}{8} \end{aligned}$$

$$t = \frac{45}{360} = \frac{1}{8} \qquad I = \$6.45$$

$$\begin{aligned} M &= P + I \\ &= \$515.89 + \$6.45 \\ M &= \$522.34 \end{aligned}$$

Thus, \$522.34 remains to be paid on June 30.

The total interest paid was \$12.50 + \$3.39 + \$6.45 = \$22.34. With no partial payments, interest on the note would have been $\$1{,}500 \times 10\% \times \frac{90}{360} = \$37.50$. Thus, \$37.50 − \$22.34 = \$15.16 was saved by making early payments.

The entire process required in the foregoing application of the U.S. rule is summarized as follows:

Original principal, 4/1		\$1,500.00
First partial payment, 5/1	\$700.00	
Less: Interest (\$1,500, 10%, 30 days)	− 12.50	
Payment to principal		− 687.50
Adjusted principal		\$ 812.50
Second partial payment, 5/16	\$300.00	
Less: Interest (\$812.50, 10%, 15 days)	− 3.39	
Payment to principal		− 296.61
Adjusted principal		\$ 515.89
Interest due (\$515.89, 10%, 45 days)		+ 6.45
Balance due, 6/30		\$ 522.34

Complications arise from the U.S. rule if the partial payment is not large enough to pay the interest due. The unpaid interest cannot be added to the principal, for then interest would be earned on interest; this would constitute compound interest, and the charging of compound interest on loans is illegal. This problem is solved simply by holding the payment or payments (without giving credit for them) until the sum of these partial payments is large enough to pay the interest due to that date. The procedure then continues as usual. (Note that the borrower does not save any interest unless the partial payment is sufficiently large to pay the interest due and reduce the principal somewhat.)

SECTION 1 PROBLEMS

Using the U.S. rule, determine the balance due on the maturity date of each of the following notes. (Use the Bankers' Rule.)

	NOTE			PARTIAL PAYMENTS	
	PRINCIPAL	RATE	TIME (DAYS)	AMOUNT	DAY
1. a.	$6,000	12%	150	$2,040	20th
				1,500	80th
b.	3,600	15	90	2,030	20th
				1,030	65th
c.	8,000	10	270	2,200	90th
				2,875	135th
2. a.	$9,000	10%	180	$1,225	90th
				2,000	135th
b.	7,200	8	120	3,288	30th
				2,000	105th
c.	5,000	9	240	3,150	120th
				950	180th

3. Carlton Interiors borrowed $10,000 by signing a 12%, 180-day note on April 5. The business made a $2,200 partial payment on June 4 and paid $3,240 on September 2. Determine the amount still owed at the end of the loan period.
4. Corcoran Hardware Co. borrowed $10,000 by signing a 9% note for 180 days on April 14. A payment of $2,150 was made toward the note on June 13, and $4,000 was paid on August 12. Find the balance due on the day the note matures.
5. On June 10, Perez Beauty Supplies borrowed $12,000 by signing an 18%, 90-day note. A partial payment of $2,120 was made on June 30, and a second partial payment of $5,225 was made on August 14. How much will be owed on the due date?
6. A & S Landscaping borrowed $12,000 by signing a 10% note dated June 15 for 150 days. The firm made a $2,100 partial payment on July 15 and another payment of $5,000 on October 13. What payment will be required when the note is due?

SECTION 2

INSTALLMENT PLANS: OPEN END

Undoubtedly, you already have some knowledge of that uniquely American institution, the **installment plan.** Almost all department stores, appliance stores, and furniture stores, as well as finance companies and banks, offer some form of installment plan whereby a customer may take possession of a purchase immediately and, for an additional charge, pay for it later by a series of regular payments. Americans pay exorbitant rates, as we shall see, for the privilege of buying on credit, but, because the dollar amounts involved are not extremely large, many people are willing to pay them.

The installment plan actually provides a great boost to the American economy, because many families buy things on the spur of the moment on the installment plan that they would never buy were they compelled to save the money required for a cash purchase. The lure of the sales pitch—"nothing down; low monthly payments," "budget terms," "buy now; no interest or payment for 60 days," "small down payment; 3 years to pay," "consolidate all your bills into one low, monthly payment"—has caused many families to feel that they will never miss that small extra amount each month. The result is that many families soon find themselves saddled with so many of these "small monthly payments" that they cannot make ends meet and are in danger of bankruptcy. The Family Service Agency or similar organizations in many cities provide inexpensive counseling to help such families learn to budget their incomes and "get back on their feet" financially.

The **Truth-in-Lending Law** (officially entitled Regulation Z) enables consumers to determine how much the privilege of credit buying actually costs, thereby allowing them to make intelligent decisions when faced with the question of when to buy on the installment plan. Indeed, the question has become "when to buy" rather than "whether to buy," because almost all Americans take advantage of time-payment plans at one time or another. The single item most often purchased in this way (other than homes) is the automobile, which is followed by other durable items such as furniture and appliances. Even realizing the expense of these plans, most people feel that the added convenience at times justifies the additional cost.

The high rate of interest charged by companies offering installment plans is not without justification. The merchant incurs numerous additional operating expenses as a result of the installment plan: the costs of investigating the customer's credit standing, discounting bank notes, buying extra insurance, paying cashiers and bookkeepers for the additional work, collection costs necessitated by buyers who do not keep up with payments, as well as "bad debt" loss from customers who never finish paying. And, of course, the merchant is entitled to interest, since the business has capital invested in the merchandise being bought on the time-payment plan. Thus, the installment plan is expensive for the seller as well as the buyer.

Basically, there are two types of installment plans: open-end accounts (this section) and accounts with a set number of payments (next section).

The first type of installment plan is called **open-end credit** because the time period of the credit account is not definite and the customer may receive additional credit before the first credit is entirely repaid. Under this plan, interest is computed each month on the unpaid balance (or sometimes on an average balance). Each payment is applied first toward the interest and then toward the balance due, in accordance with the U.S. rule of partial payments.

This plan is used by most department stores and other retail businesses offering "charge accounts" and "credit cards." The monthly interest rate is usually $1\frac{1}{4}\%$ (= 15% per year) to $1\frac{1}{2}\%$ (= 18% per year). Customer accounts at these stores usually vary quite a bit from month to month, because the account results from a number of small purchases rather than one large purchase. Many accounts of this type are never actually paid off, because new purchases continue to be made. These accounts are therefore often called **revolving charge accounts.** There is usually a limit to the amount a customer is entitled to charge, and a prescribed minimum monthly payment may be required according to the amount owed. (For instance, suppose the maximum balance that a charge account may reach is $500. The customer may be required to pay at least $15 per month if his balance is under $200 and pay at least $25 per month if he owes $200 or more.)

The following example illustrates the payment of a single purchase on an installment plan. This procedure would be used also to pay off an account balance when no other purchases are charged to the account.

Example 1 Installment Plan Having Monthly Rate

Mrs. Cohen's charge account at Butler-Stohr has a balance of $100. Interest of $1\frac{1}{2}\%$ per month is charged on unpaid customer accounts. Mrs. Cohen has decided to pay off her account by making monthly payments of $25. (a) How much of each payment is interest, and how much will apply toward her account balance? (b) What is the total amount Mrs. Cohen will pay, and how much of this is interest?

(a) The following schedule computes the interest and the payment that applies toward the balance due each month. "Interest" is found by taking $1\frac{1}{2}\%$ of the "balance due." This interest must then be deducted from the "monthly payment" in order to determine the "payment (that applies) toward the balance due." The "payment toward the balance due" is then subtracted from the "balance due" to find the "adjusted balance due," which is carried forward to the next line as the "balance due."

PAYMENT NUMBER	BALANCE DUE	$1\frac{1}{2}\%$ INTEREST PAYMENT	MONTHLY PAYMENT	PAYMENT TOWARD BALANCE DUE	ADJUSTED BALANCE DUE
1	$100.00	$1.50	$ 25.00	$ 23.50	$76.50
2	76.50	1.15	25.00	23.85	52.65
3	52.65	0.79	25.00	24.21	28.44
4	28.44	0.43	25.00	24.57	3.87
5	3.87	0.06	3.93	3.87	0.00
		$3.93	$103.93	$100.00	

(b) During the last month, Mrs. Cohen paid only the remaining balance due plus the interest due. She thus paid a total of $103.93 in order to discharge her $100 obligation, which included total interest of $3.93.

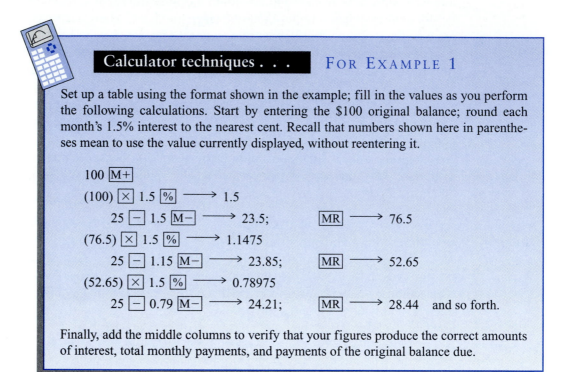

Calculator techniques . . . FOR EXAMPLE 1

Set up a table using the format shown in the example; fill in the values as you perform the following calculations. Start by entering the $100 original balance; round each month's 1.5% interest to the nearest cent. Recall that numbers shown here in parentheses mean to use the value currently displayed, without reentering it.

100 [M+]
(100) [×] 1.5 [%] ⟶ 1.5
25 [−] 1.5 [M−] ⟶ 23.5; [MR] ⟶ 76.5
(76.5) [×] 1.5 [%] ⟶ 1.1475
25 [−] 1.15 [M−] ⟶ 23.85; [MR] ⟶ 52.65
(52.65) [×] 1.5 [%] ⟶ 0.78975
25 [−] 0.79 [M−] ⟶ 24.21; [MR] ⟶ 28.44 and so forth.

Finally, add the middle columns to verify that your figures produce the correct amounts of interest, total monthly payments, and payments of the original balance due.

There are a number of variations in the way that interest may be computed on a revolving charge account. Example 1 illustrates interest computed on the *adjusted balance* remaining after payment was received during the month. By contrast, when the *previous balance* method is used, this month's interest is based on the balance due at the end of the previous month.

The most common interest method, the *average daily balance* method, computes a weighted average that is a combination of the two preceding methods. In this case, the previous month's balance is multiplied by the number of days until a payment is received. After that payment has been deducted, the adjusted balance is multiplied by the number of days remaining in the month (or other payment cycle). Their sum is then divided by 30 (or by 31 or other total days in the cycle) to obtain the average daily balance for the billing period, and interest is calculated on this average.

Ordinarily, interest is not charged on this month's "current" purchases until after the customer's statement is mailed. This is illustrated in the following example, which utilizes the average daily balance method. Example 2 also illustrates a more realistic revolving charge account, where each month's transactions include additional credit purchases as well as the required amount of payment.

Example 2 **Revolving Charge Account**

Michael Boyer's charge account at Wymer's had a \$50 balance at the beginning of the 6-month period illustrated below. Payments and additional charges occurred on the indicated dates of each 30-day month. Interest is computed by the average daily balance method at 2% per month. (No interest is charged on current purchases.) Each month's ending balance is indicated by an *. Calculation of three of the average balances is shown in footnotes. Notice also that a check confirms that the totals are correct. (Refer to the table on the following page for the monthly calculations of the revolving charge account.)

Calculator techniques . . . FOR EXAMPLE 2

There are many balances to keep up with in this problem, so ordinary addition/subtraction is suggested for most lines. Calculations using memory are suggested only to determine the average balance and its corresponding 2% interest charge. Here is the average-balance calculation of \$59.23 illustrated in footnote *b* of the associated table, with its resulting \$1.18 interest charge:

10 [×] 65.90 [M+] ⟶ 659; 20 [×] 55.90 [M+] ⟶ 1,118

[MR] ⟶ 1,777 [÷] 30 [=] ⟶ 59.233333 [×] 2 [%] ⟶ 1.1846666

		REVOLVING ACCOUNT					CURRENT ACCOUNT	
MO.	DATE	PREV. BAL.	PAYMT.	ADJ. BAL.	AVG. BAL.	2% INT. CHARGE	PURCHASES	CUR. BAL.
1	15	$ 50.00	$ 10	$ 40.00				$ 40.00
	20						$ 25	65.00
	30				$ 45.00[a]	$ 0.90		65.90*
2	5						30	95.90
	10	65.90	10	55.90				85.90
	14						20	105.90
	30				59.23[b]	1.18		107.08*
3	20	107.08	20	87.08				87.08
	22						15	102.08
	28						10	112.08
	30				100.41	2.01		114.09*
4	10	114.09	20	94.09				94.09
	18						15	109.09
	30				100.76	2.02		111.11*
5	8						25	136.11
	15	111.11	20	91.11				116.11
	20						5	121.11
	30				101.11	2.02		123.13*
6	2						24	147.13
	12	123.13	20	103.13				127.13
	22						15	142.13
	30				111.13[c]	2.22		144.35*
	Totals		$100			$10.35	$184	

*Denotes average daily balance.

[a] $\frac{(15 \times 50) + (15 \times 40)}{30} = \$45$

[b] $\frac{(10 \times 65.90) + (20 \times 55.90)}{30} = \$59.23$

[c] $\frac{(12 \times 123.13) + (18 \times 103.13)}{30} = \$111.13$

Check:

	$ 50.00	Beginning balance
	184.00	Purchases
+	10.35	Interest
	$244.35	
−	100.00	Payments
	$144.35	Ending balance

SECTION 2 PROBLEMS

For each account, complete a table (as in Example 1) that shows both the monthly interest due on the adjusted balance and the amount of each payment that applies to the balance due, until the account is paid in full.

	AMOUNT OF ACCOUNT	MONTHLY PAYMENT	INTEREST PER MONTH
1. a.	$175	$25	$1\frac{1}{4}\%$
b.	500	50	1
2. a.	$400	$50	$1\frac{1}{2}\%$
b.	500	75	1

Use the average daily balance method (as in Example 2) to compute 2% monthly interest for the following charge accounts. No interest is charged on current purchases. (Assume a 30-day billing cycle.) Compute total interest and final balance.

	MONTH	PREVIOUS BALANCE	PAYMENTS	PURCHASES	TOTAL INTEREST	FINAL BALANCE
3.	1	$ 300	$50 on 18th	$20 on 20th		
	2		$75 on 15th	$25 on 19th; $10 on 25th		
	3		$45 on 12th	$30 on 28th		
4.	1	$1,500	$40 on 12th	$50 on 16th		
	2		$150 on 15th	$25 on 20th; $10 on 25th		
	3		$400 on 5th; $200 on 20th	$60 on 12th		

5. Nicole Nelson charged a video camera for $300 to her account at Herman's Department Store. The store charges 1% per month interest on adjusted charge account balances. If she agrees to pay $50 monthly toward her account, how much interest will she pay?

6. Mai Nyugen's account balance at Daniel's Department Store is $400. The store charges 1¼% per month interest on adjusted charge account balances. If she agrees to pay $60 monthly toward her account, how much interest will she pay?

7. Sarah Castleberry's purchases and payments at Wilder Department Store are shown below. Interest at 1% is computed on her average daily balance but not on current purchases. A $10 payment is required when the preceding month's ending balance is under $100, and $20 for balances exceeding $100. Assuming that she pays the minimum amount each month and that each month has 30 days, find

a. The total payments b. The total interest c. The ending balance

MONTH	PREVIOUS BALANCE	DATE OF PAYMENT	PURCHASES
1	$85	15th	$30 on 20th
2		18th	$45 on 25th
3		12th	$25 on 21st

8. Brice Fulton's charge account reflects the following purchases. Interest at 1% is computed on his average daily balance (but not on current purchases). A $15 payment is required when the preceding month's ending balance is under $100, and $20 for balances exceeding $100. Assuming that he pays the minimum each month and that there are 30 days in each month, find

a. The total payments b. The total interest c. The ending balance

MONTH	PREVIOUS BALANCE	DATE OF PAYMENT	PURCHASES
1	$60	10th	$20 on 15th
2		5th	$80 on 10th
3		20th	$25 on 13th

SECTION 3 INSTALLMENT PLANS: SET PAYMENTS

A second type of installment plan—a set number of payments—is used to finance a single purchase (automobile, suite of furniture, washing machine, and so forth). In this case, the account will be paid in full in a specified number of payments; hence, finance charges for the entire time are computed and added to the cost at the time of purchase.

The typical procedure for an installment purchase is as follows:

1. The customer makes a **down payment** (sometimes including a trade-in) which is subtracted from the cash price to obtain the **outstanding balance.**
2. A **carrying charge** (or **time-payment differential**) is then added to the outstanding balance.
3. The resulting sum is divided by the number of payments (usually weekly or monthly) to obtain the amount of each installment payment.

This type of time-payment loan is also used by finance companies and by banks in their installment loan, personal loan, and/or automobile loan departments. The only variations in the plan as described above are that there is no down payment made (the amount of the loan becomes the outstanding balance) and that the interest charged corresponds to the carrying charge. Bank interest rates are usually lower, because banks will lend money only to people with acceptable credit ratings. Finance companies cater to persons whose credit ratings would not qualify them for loans at more selective institutions. Many states place essentially no restrictions on these small-loan agencies, allowing rates of 100% or more on small loans of 12 months or less. Their profits are not so high as their rates might indicate, however, for these companies have a high percentage of bad debts from uncollected loans.

As noted previously, high-powered advertising has often misled buyers into making financial commitments that were substantially greater than they realized beforehand. The Truth-in-Lending Law protects against false impressions by requiring that if a business mentions one feature of credit in its advertising (such as the amount of down payment), it must also mention all other important features (such as the number, amount, and frequency of payments that follow). If an advertisement states "only $20 down," for example, it must also state that the buyer will have to pay $10 a week for the next 2 years.

The principal accomplishment of the law is that it enables the buyer to know both the **finance charge** (the total amount of extra money paid for buying on credit) and the **annual percentage rate** to which this is equivalent. Some merchants may require a carrying charge, a service charge, interest, insurance, or other special charges levied only upon credit sales; the finance charge is the sum of all these items. The annual percentage rate, correct to the nearest quarter of a percent, must then be determined by one of the two methods approved in the law. Tables are available from the government which enable the merchant to determine the correct percentage rate without extensive mathematical calculations.

It is important to understand the distinction between the interest rates required under the law and simple interest rates such as we studied previously. Formerly, merchants often advertised a simple interest rate, and interest was computed on the original balance for the entire payment period. Consider the following, however: On a simple interest note, the borrower keeps all the money until the due date; thus, the borrower rightfully owes interest on the original amount for the entire period. On an installment purchase at simple interest, on the other hand, the buyer must immediately begin making payments. As the credit period expires, therefore, the buyer has repaid most of the obligation and then owes very little, yet the latter payments would still include just as much interest as the early payments did. This means that the buyer would be paying at a much higher interest rate than on the early payments. A simple interest rate would thus be deceiving; however, the Truth-in-Lending Law permits the buyer to know the true rate that is paid. (Examples in this section will illustrate the differences in these rates.) Since the rates required under the law are quite complicated to compute, the Federal Reserve System publishes tables that merchants can use to determine annual percentage rates. These tables contain rates corresponding to both monthly and weekly payments; however, this text will include tables only for monthly payments.

It should be pointed out that the Truth-in-Lending Law does not establish any maximum interest rates or maximum finance charges or otherwise restrict what a seller may charge for credit. Neither are there restrictions on what methods the seller may use to determine the finance charge. The law only requires that the merchant must fully inform buyers what they are paying for the privilege of credit buying. In addition to the finance charge and the annual percentage rate, other important information must also be itemized for the buyers, including penalties they must pay for not conforming to the provisions of the credit contract.

Note. The term "installment plan," as used in this section, describes high-interest payment plans that run for a limited period. (The maximum time is usually 4 years.) Chapter 20 includes the periodic payments made to repay home loans or other real estate loans, to repay long-term personal bank loans, or to repay other loans when a comparatively lower interest rate is computed only on the balance due at each payment (such as credit union loans or loans on insurance policies).

The following examples all pertain to installment plans of a set period of time, where the finance charge (carrying charge, interest, and similar items) is calculated at the time the contract is made.

Example 1 The Appliance Warehouse will sell a garden weeder for \$64 cash or on the following "easy-payment" terms: a $\frac{1}{4}$ down payment, a full 10% carrying charge, and monthly payments for 6 months. (a) Find the monthly payment. (b) What is the total cost of the weeder on the time-payment plan? (c) How much more than the cash price will be paid on the installment plan?

(a)

Cash price	\$64.00	
Down payment	− 16.00	($\frac{1}{4}$ of \$64)
Outstanding balance	\$48.00	
Carrying charge	+ 4.80	(10% of \$48)
Total of payments	\$52.80	

$$\frac{\$52.80}{6} = \$8.80 \quad \text{Monthly payment}$$

A payment of \$8.80 per month is required. Notice that the finance (carrying) charge was a full 10% of the outstanding balance (without multiplying by $\frac{1}{2}$ for the 6-month time period). The merchant may use this or any convenient method for determining the finance charge; however, this 10% rate is much lower than the rate the seller is required to disclose to the credit buyer. Since these payments cover only half a year, this rate is equivalent to simple interest at 20% per year. However, the rate that the merchant must reveal to the buyer is the rate that applies when interest is computed only on the balance due at the time each payment is made. Thus, the merchant must tell the buyer that an annual percentage rate of 33.5% is being charged. (Later examples will further illustrate these rates.)

Be aware of the correct procedure: If the finance charge is given as a rate (either a fraction or a percent), the finance charge is found by multiplying this rate times the outstanding balance remaining *after* the down payment has been made.

(b) The total cost of the weeder, when bought on the installment plan, is \$68.80, as follows:

Down payment	\$16.00
Total of monthly payments	+ 52.80
Total installment plan cost	\$68.80

(c) The \$4.80 carrying charge represents the additional cost to the credit customer:

Total installment plan price	\$68.80
Cash price	− 64.00
Time-payment differential	\$ 4.80

This \$4.80 finance charge, along with the 33.5% annual percentage rate, represents the two most important items the merchant must disclose to the credit buyer under the Truth-in-Lending Law, although several other things are also required.

Example 2 Gerald Wood bought a new color television that sold for \$675 cash. He received a \$35 trade-in allowance on his old set and also paid \$40 cash. The store arranged a 2-year payment plan, including a finance charge computed at regular 8% simple interest. (a) What is Wood's monthly payment? (b) What will the total cost of the TV be? (c) How much extra will he pay for the convenience of installment buying? (d) What annual percentage rate must be disclosed under the Truth-in-Lending Law?

(a) Wood's regular payment of \$29 per month is computed as follows:

Cash price	\$675	
Down payment	− 75	(\$35 trade-in + \$40 cash)
Outstanding balance	\$600	
Finance charge (I)	+ 96	$\left(\begin{aligned} I &= Prt \\ &= \$600 \times 0.08 \times 2\end{aligned}\right)$
Total of payments	\$696	

$$\frac{\$696}{24} = \$29 \text{ per month}$$

(b) The total cost of the TV will be \$771:

Down payment: Trade-in	\$ 35
Cash	40
Total of monthly payments	696
Total cost	\$771

(c) The \$771 total cost includes an extra \$96 interest (or finance or carrying) charge above the cash price.

(d) The annual percentage rate that the dealer must disclose is found using Table 14, Tables Booklet. First, the finance charge should be divided by the amount financed (the outstanding balance), and the result multiplied by 100. This gives the "finance charge per $100 of amount financed":

$$\frac{\text{Finance charge}}{\text{Amount financed}} \times 100 = \frac{\$96}{\$600} \times 100 = 0.16 \times 100 = \$16$$

Thus, the credit buyer pays $16 interest for each $100 being financed. Note that this result is also the total simple interest percent charged; that is, the "rate × time": 8% × 2 years = 16%. This relationship will always exist, so either method could be used to determine the table value.

To use Table 14, look in the left-hand column headed "Number of Payments" and find line 24, since this problem includes 24 monthly payments. Look to the right until you find the column containing the value nearest to the $16 we just computed. The percent at the top of that column will then be the correct annual percentage rate. In this example, the value nearest to our 16 is the entry 16.08. (In case a computed value is midway between two table values, the larger table value should be used.) Thus, the percent at the top of the column indicates that the annual percentage rate is 14.75%.

For practical purposes, this means that if interest were computed at 14.75% only on the remaining balance each month and the U.S. rule were used so that the $29 applied first to the interest due and then to reduce the principal, then 24 payments of $29 would exactly repay the $600 and the $96 interest.* Thus, it is the 14.75% rate that the merchant must state is being charged.

When simple interest is calculated on the *original balance* for the entire length of a loan, this rate is often called a **nominal interest rate.** The interest computed in this way is frequently called **add-on interest.** By contrast, an annual rate that is applied only to the *balance due at the time of each payment* is called an **effective interest rate.** The annual rates determined for the Truth-in-Lending Law by Table 14 are known as **actuarial rates,** which are almost identical to effective rates. It is this actuarial (or effective) rate that must be disclosed to the buyer by the merchant.

Example 3 (a) What actuarial (or effective) interest rate is equivalent to a nominal (or simple) interest rate of 9%, if there are six monthly installments? (b) How much simple interest

*Strictly speaking, there is a mathematical distinction between the annual percentage rates in Table 14 and the annual rates that one should use to apply the U.S. rule (as was done here), although either rates are acceptable under the Truth-in-Lending Law. The U.S. rule computes interest on the amount financed from the original date until the day payments are made. The rates in Table 14, however, are applied in the opposite direction: The present value on the original day of all the payments must equal the amount financed. Although this is a distinct difference in mathematical procedure, the difference in the actual percentage rates is insignificant for our purposes. Thus, since a table of rates applicable to the U.S. rule was not available, we shall use the rates from Table 14 in applying the U.S. rule.

would be required on a \$480 loan at a nominal 9% for 6 months? What is the monthly payment? (c) Verify that the actuarial rate would produce the same amount of interest, if six monthly payments were made.

(a) The value we look for in Table 14 is found by multiplying "rate × time":

$$9\% \times \frac{1}{2}\text{ year} = 4.5$$

This indicates that the interest due is 4.5% of the amount borrowed. It also means that the borrower pays \$4.50 interest for each \$100 borrowed.

Because this is a 6-month loan, we look on line 6 for the column containing the value nearest 4.50. The nearest value is 4.49. We thus find that a simple (nominal) interest rate of 9% for 6 months corresponds to an actuarial (effective) rate of 15.25% correct to the nearest $\frac{1}{4}$%. It is this actuarial rate* that is required under the Truth-in-Lending Law.

(b) A loan for \$480 at 9% simple interest for 6 months would require interest as follows:

$$\begin{aligned} I &= Prt \\ &= \$480 \times \underbrace{9\% \times 0.5} \\ &= \$480 \times \quad 4.5\% \qquad \text{[as in Part (a)]} \\ I &= \$21.60 \end{aligned}$$

or at \$4.50 per \$100 financed:

4.8	hundreds
×\$ 4.5	per \$100
\$21.60	interest

If the \$480 principal plus \$21.60 interest were repaid in six monthly payments, then

$$\frac{\$501.60}{6} = \$83.60 \text{ per month}$$

*A very similar rate can be computed without tables using the formula $e = \frac{2nr}{n+1}$, where e = effective rate, r = annual simple interest rate, and n = number of payments (either monthly or weekly). Thus, for Example 3, $e = \frac{2(6)(9)}{6+1} = 15.43\%$.

(c) The U.S. rule is used to verify the 15.25% actuarial rate. Each month, the interest due is subtracted from the \$83.60 payment. The remainder of the payment is then deducted from the outstanding principal to obtain the adjusted principal upon which interest will be calculated at the time of the next payment.

MONTH	$PRT = I$		PAYMENT TO PRINCIPAL
1	$\$480.00 \times 0.1525 \times \frac{1}{12} =$	\$ 6.10	\$ 77.50
2	$402.50 \times 0.1525 \times \frac{1}{12} =$	5.12	78.48
3	$324.02 \times 0.1525 \times \frac{1}{12} =$	4.12	79.48
4	$244.54 \times 0.1525 \times \frac{1}{12} =$	3.11	80.49
5	$164.05 \times 0.1525 \times \frac{1}{12} =$	2.08	81.52
6	$82.53 \times 0.1525 \times \frac{1}{12} =$	1.05	82.55
		\$21.58 +	\$480.02 = \$501.60

Thus, we see that a rate of 15.25%, applied only to the unpaid balance each month, would result in interest of \$21.58 over 6 months' time. This is \$0.02 less than the \$21.60 simple interest that would be due at a 9% nominal rate. (The difference results from rounding and from the use of Table C-00 in applying the U.S. rule. Recall that Table 14 is correct only to the nearest quarter of a percent.)

Calculator techniques . . . FOR EXAMPLE 3(C)

You could use the techniques shown for Example 1 of Section 2. However, since this "rate × time" calculation produces such a long decimal, you may prefer to keep that in memory, as shown here (rather than keeping the running balance in memory). Recall that $83.60 is the monthly payment. (Parentheses appear around values you use without reentering.)

.1525 [÷] 12 [M+] ⟶ 0.0127083

480 [×] [MR] [=] ⟶ 6.099984

83.60 [−] 6.10 [=] ⟶ 77.5; 480 [−] 77.5 [=] ⟶ 402.5

(402.5) [×] [MR] [=] ⟶ 5.11509075

83.60 [−] 5.12 [=] ⟶ 78.48; 402.5 [−] 78.48 [=] ⟶ 324.02

(324.02) [×] [MR] [=] ⟶ 4.117743366

83.60 [−] 4.12 [=] ⟶ 79.48; 324.02 [−] 79.48 [=] ⟶ 244.54

and so on. You will use this technique again for several subsequent examples.

Example 4 To repay a loan of $160 from IOU Finance Co. requires six payments of $28 each. Determine (a) the amount of interest that will be paid and (b) the annual (actuarial) percentage rate charged. (c) Verify that this annual percentage rate would actually result in the amount of interest charged.

(a) The six payments of $28 will total $168 altogether. The interest charged will thus be $8:

$ 28		Total of payments	$168
× 6		Amount of loan	− 160
$168		Interest (finance charge)	$ 8

(b) Before finding the actuarial rate, we first find the finance charge per $100 of amount financed:

$$\frac{\text{Finance charge}}{\text{Amount financed}} \times 100 = \frac{\$8}{\$160} \times 100 = 0.05 \times 100 = \$5.00$$

Then on line 6 of the table, we find that the value nearest to 5.00 is the entry 5.02. Thus, IOU Finance Co. must inform the borrower that (s)he is paying interest at an annual percentage rate of 17%.

(c) As before, the monthly payment applies first to the interest due, and the remainder of the payment is then deducted from the outstanding principal.

MONTH	$PRT = I$		PAYMENT TO PRINCIPAL
1	$\$160.00 \times 0.17 \times \frac{1}{12} = \$2.27$		$ 25.73
2	$134.27 \times 0.17 \times \frac{1}{12} = 1.90$		26.10
3	$108.17 \times 0.17 \times \frac{1}{12} = 1.53$		26.47
4	$81.70 \times 0.17 \times \frac{1}{12} = 1.16$		26.84
5	$54.86 \times 0.17 \times \frac{1}{12} = 0.78$		27.22
6	$27.64 \times 0.17 \times \frac{1}{12} = 0.39$		27.61
	$8.03	+	$159.97 = $168

Thus, we find that an effective rate of 17%, applied only to the unpaid balance each month, would require $8.03 in interest on a $160.00 loan. (This calculation contains a $0.03 rounding error.) The $8.00 interest on this loan would be equivalent to simple interest at the rate of only 10%, which illustrates well how misleading advertised installment rates could be before the Truth-in-Lending Law was passed.

Note. In verifying rates as in Part (c), the "rate × time" will be the same each month (in this case, $0.17 \times \frac{1}{12}$). In doing calculations of this kind, therefore, it is easier to find the product ($0.17 \times \frac{1}{12} = 0.01417$) and use this factor to multiply times the principal each month. (Be sure that you carry the factor out to enough places to obtain correct cents. Review Section 1, "Accuracy of Computation," in Chapter 1.)

As a general rule, the shorter the period of time, the higher installment rates tend to be. Customers seem to be willing to pay these very high rates, because the actual dollar amounts for shorter periods are not very large.

SECTION 3 PROBLEMS

Find the regular (monthly or weekly) payment required to pay for the following installment purchases.

	Cash Price	Down Payment Required	Finance Charge	Time Period	Regular Payment Required
1. a.	$ 500	$50	$20	10 months	
b.	960	25%	$30	12 weeks	
c.	645	$\frac{1}{3}$	10% simple interest	24 months	
2. a.	$1,000	$50	$25	5 months	
b.	1,500	10%	$50	20 weeks	
c.	1,100	$\frac{1}{4}$	8% simple interest	12 months	

For the following installment problems, determine the annual (actuarial) percentage rate charged.

	Amount Financed	Finance Charge	Months	Annual Percentage Rate
3. a.	$ 120	$20	9	
b.	300	$30	12	
c.	—	14% simple interest	36	
d.	—	12% simple interest	5	
4. a.	$ 900	$60	10	
b.	1,200	$90	12	
c.	—	10% simple interest	24	
d.	—	9% simple interest	6	

Complete a table similar to the one found on page 454 [Example 4(c)] to verify that the following annual percentage rates would produce the given interest.

	Principal	Interest	Monthly Payments	Months	Annual Rate	Total Interest Paid	Total Payment to Principal
5.	$800	$40	$140	6	17.00%		
6.	400	40	88	5	39.25		

7. A man's suit sells for $336 cash or for ⅓ down, a finance charge of $40, and the balance in 12 monthly installments. What is the monthly payment?

8. A five-piece dining set is priced at $1,400 cash or a ¼ down payment, a finance charge of $60, and the balance in 24 monthly installments. Compute the monthly payment.

9. A multimedia computer system can be purchased for $2,500 cash. On the installment plan, a 30% down payment, 18 monthly payments, and a finance charge of $50 are required. Compute the monthly payment.

10. A leather sofa is marked $1,600. The time-payment plan requires a 20% down payment, 18 monthly payments, and a finance charge of $70. What is the amount of each payment?

11. Four radial tires can be purchased for $520 cash. The tires can be purchased on the time-payment plan with a 25% down payment and 10 monthly payments that include a finance charge computed at 12% nominal (simple) interest.

a. What is the finance charge?

b. How much is the monthly payment?

c. What is the total cost of the tires on the installment plan?

d. Compute the annual percentage rate that must be disclosed under the Truth-in-Lending Law.

12. A couple purchased a freezer costing $1,000. The appliance may be purchased for a 10% down payment and 36 monthly payments. For convenience, the dealer computes the finance charge at 9% nominal (simple) interest.

a. What is the finance charge on the freezer?

b. How much is the monthly payment on the purchase?

c. Determine the total cost of the appliance on the installment plan.

d. What annual percentage rate must the dealer disclose under the Truth-in-Lending Law?

13. Monthly payments of $125 are required for 24 months to repay a loan of $2,600. Determine

a. The finance charge the borrower pays

b. The annual percentage rate the seller must reveal

14. Eighteen monthly payments of $70 are required to repay a $1,150 loan. Determine

a. The finance charge the borrower pays

b. The annual percentage rate charged on the loan

15. A $580 recliner chair was purchased by making payments of $35 for 18 months. Determine

a. The finance charge on the chair

b. The annual percentage rate charged on the installment plan

16. A $1,600 loan is repaid in 36 monthly payments of $57 each.

a. What is the finance charge on the loan?

b. At what annual percentage rate is interest paid?

17. A laser printer cost $1,400 cash. It can also be purchased for $200 down and 24 monthly payments of $56.

a. How much is the total cost of the printer when purchased on credit?

b. What is the finance charge on the purchase?

c. At what annual percentage rate was interest paid?

18. A stereo television costs $1,100 cash. If purchased on the installment plan, it can be purchased for $100 down and 12 monthly payments of $95 each. Find

a. The total cost of the television when purchased on credit

b. The finance charge on the purchase

c. The annual percentage rate the lender must disclose

19. The Smiths can purchase a lawn tractor for $1,300 by making a down payment of $250 and 36 monthly payments of $35. Determine

a. The total cost of the tractor on the installment plan

b. The finance charge

c. The annual percentage rate that must be disclosed to the Smiths

20. A diamond necklace can be purchased for \$2,000 by making a down payment of \$200 and 15 monthly payments of \$140 each.

a. What is the total cost of the necklace on the installment plan?

b. What is the finance charge for this time-payment plan?

c. What annual percentage rate must be disclosed?

21. Jerry's PC Services computes interest at 14% simple interest on installment purchases. Determine the annual percentage rate that the business must disclose under the Truth-in-Lending Law on the following monthly installment purchases.

a. A 12-month installment purchase

b. An 18-month installment purchase

c. A 9-month installment purchase

22. For simplicity, Stan's Furniture Mart computes interest at 9% simple interest. What annual percentage rate must it reveal under the Truth-in-Lending Law on the following monthly installment loans?

a. A 1-year loan

b. A 24-month loan

c. A 10-month loan

23. Decor Furniture Co. computes 9% nominal interest on an installment purchase of \$2,400 to be repaid in 6 monthly payments.

a. What amount of interest will be due?

b. What monthly payment is required?

c. What annual actuarial rate is charged?

d. Verify that this actuarial rate is correct.

24. a. What amount of interest would be due on a 3-month loan of \$800 if a nominal rate of 10% was charged?

b. How much will the monthly payments be?

c. What annual actuarial rate is charged on the loan?

d. Verify that this actuarial rate is correct.

25. Suppose that \$2,200 was borrowed under the conditions in Part (d) of Problem 3.

a. How much interest would be charged?

b. How much is the monthly payment?

c. Verify that the annual rate you found in Part (d) of Problem 3 would actually result in the correct amount of interest.

26. Assume that \$800 was borrowed at the rate and time of Part (d) of Problem 4.

a. How much interest would be charged?

b. What is the monthly payment?

c. Verify that the annual rate found in Part (d) of Problem 4 would result in the correct amount of interest.

27. Look for newspaper advertisements showing installment purchases. What terms are given? Do these merchants adhere to the requirements of the Truth-in-Lending Law?

28. Find an installment plan in a newspaper or other periodical. Verify the actuarial rate given (as in Problems 25 and 26).

SECTION 4

INSTALLMENT PLAN PREPAYMENTS

Like the borrower on a promissory note, the borrower on an installment plan may sometimes wish to pay off the remaining balance early, in order to avoid further payments and to save interest. In this case, the lender deducts from the remaining payments the interest that would otherwise have been due on the remaining principal.

Since lenders often do not keep schedules that show the remaining principal, however, the unearned interest must usually be determined by some computation. As noted earlier, the Truth-in-Lending Law places no restrictions on how the lender computes interest, so long as the borrower is clearly informed of the rate actually paid. The examples in this section illustrate the **actuarial method,** which utilizes the Truth-in-Lending tables, as well as the **rule of 78s,** a traditional method that computes a similar amount of unearned interest without tables.

Unearned interest is computed by the actuarial method using the formula

$$\frac{n \times \text{Pmt} \times \text{Value}}{100 + \text{Value}}$$

where n is the number of monthly installment payments *remaining* in the established time period, Pmt is the regular monthly payment, and Value is the value from the Truth-in-Lending table that corresponds to the annual (actuarial) percentage rate for n payments.

Example 1 Mason Jansen borrowed \$600 at 13% simple interest, to be repaid in 12 monthly payments of \$56.50. When making his eighth payment, Jansen also wishes to pay his remaining balance. (a) What annual (actuarial) percentage rate was charged on the loan? (b) How much interest will Jansen save, computed by the actuarial method? (c) What is the remaining balance on that day?

(a) As presented in the preceding section, the value we look for in the table is the product of "rate × time":

$$13\% \times 1 \text{ year} = 13 \quad \text{Corresponding to a 23.25\% annual percentage rate}$$

(b) Since 8 payments have been made, there are 4 payments remaining. Referencing the annual percentage rate table for 4 periods at 23.25%, we find a value of 4.89. Thus, the unearned interest by the formula is as follows:

$$\frac{n \times \text{Pmt} \times \text{Value}}{100 + \text{Value}} = \frac{4 \times \$56.50 \times 4.89}{100 + 4.89}$$

$$= \frac{\$1{,}105.14}{104.89}$$

$$= \$10.54$$

Jansen will save $10.54 interest by the actuarial method.

(c) Since the remaining payments total 4 × $56.50 = $226, Jansen's remaining balance is as follows:

$226.00	Total of remaining payments
− 10.54	Interest saved
$215.46	Balance due

A payment of $215.46 will terminate Jansen's installment loan. (This is in addition to the regular $56.50 payment paid on that day.)

The second method for determining unearned interest on installment plans, the rule of 78s, does not use the actuarial tables. The name of this method is derived from the sum of the digits for a 12-month period (12 + 11 + 10 + 9 + 8 + 7 + 6 + 5 + 4 + 3 + 2 + 1 = 78). Originally, when the concept of installment loans first began, most of the loans were for a period of just one year, or 12 monthly installments. Over the period of a 12-month installment loan, the borrower has the use of $\frac{12}{12}$ of the money during the first month, $1\frac{11}{12}$ of the money during the second month, and so forth. The lender, the bank, has more money at risk during the early part of the loan period than it does toward the end of the loan period. Thus, the bank claims that it has the right to earn a greater amount of interest in the earlier months of the loan than in the later months.

A formula is derived for allocating interest over the term of a 1-year loan. During the first month of the loan, when the borrower has use of $\frac{12}{12}$ of the principal, the bank is entitled to receive $\frac{12}{78}$ of the total interest; during the second month, the bank is entitled to receive $\frac{11}{78}$ of the principal, and so forth.

The steps to follow for the rule of 78s are illustrated in the following example.

Example 2 Refer to Mason Jansen's loan in Example 1, where $600 was repaid in 12 monthly payments of $56.50. The loan is paid in full after the 8th installment payment, with his final payment computed here using the rule of 78s.

Step 1 Determine the total amount of *interest* on the loan.

The installment payments for all 12 months total 12 × \$56.50 = \$678, which includes \$78 interest:

\$678	Total payments
− 600	Principal
\$ 78	Total interest

Step 2 Determine the *numerator* for the remaining period either by adding the years' digits or by applying the formula:

$$\frac{N(N+1)}{2}$$

There are four payments remaining:

$$\frac{4(4+1)}{2} = \frac{20}{2} = 10$$

Step 3 Determine the *denominator,* which is the sum of the months' digits for the entire loan period.

$$\frac{12(12+1)}{2} = \frac{156}{2} = 78$$

Step 4 Multiply the fraction $\frac{n}{d}$ times the total interest (step 1). The result is the amount of interest saved.

$$\frac{10}{78} \times \$78 = \$10 \text{ interest saved}$$

Step 5 Determine the total of the *remaining regular payments.*

Jansen has four remaining payments:

$$4 \times \$56.50 = \$226$$

Step 6 Subtract the amount of interest saved (step 4) from the total of the remaining installment payments (Step 5) to determine the total *payment necessary to pay off the loan.*

\$226.00	Total of remaining payments
− 10.00	Interest saved
\$216.00	Balance due

Jansen will owe \$216 under the rule of 78s (compared with \$215.46 using the actuarial method).

Example 3 Refer again to Jansen's \$600 loan in Examples 1 and 2, with monthly payments of \$56.50. Verify that a 23.25% annual percentage rate produces the correct balance remaining after the 8th payment.

MONTH	$PRT = I$	PAYMENT TO PRINCIPAL
1	$\$600.00 \times 0.2325 \times \frac{1}{12} = \$11.63$	\$ 44.87
2	$555.13 \times 0.2325 \times \frac{1}{12} = 10.76$	45.74
3	$509.39 \times 0.2325 \times \frac{1}{12} = 9.87$	46.63
4	$462.76 \times 0.2325 \times \frac{1}{12} = 8.97$	47.53
5	$415.23 \times 0.2325 \times \frac{1}{12} = 8.05$	48.45
6	$366.78 \times 0.2325 \times \frac{1}{12} = 7.11$	49.39
7	$317.39 \times 0.2325 \times \frac{1}{12} = 6.15$	50.35
8	$267.04 \times 0.2325 \times \frac{1}{12} = 5.18$	51.32
	215.72 0.00	215.72
	\$67.72	\$600.00

By this calculation, we obtain \$215.72 of the principal remaining after the 8th payment. This compares with \$215.46 by the actuarial method and \$216 by the rule of 78s. (Differences in the first two figures are due to rounding.)

The rule of 78s, although named on the basis of 12-month installments, also applies for installment periods of other lengths. For convenience, the sums of numbers representing other time periods are given in Table 17-1.

TABLE 17-1 SUM OF MONTHS' DIGITS

NUMBER OF MONTHS	SUM: 1 THROUGH LARGEST MONTH
6	21
9	45
10	55
12	78
15	120
18	171
24	300
36	666

Thus, if a 24-month installment plan is paid off 4 months early, $\frac{10}{300}$ of the total interest is saved. Similarly, if an 18-month plan is terminated 6 months early, $\frac{21}{171}$ of the interest is avoided.

As a reminder, for convenience in finding the sum of a series of consecutive integers for months 1 through *n,* the following formula may be applied (as illustrated for the sum $8 + 7 + 6 + 5 + 4 + 3 + 2 + 1 = 36$):

$$\frac{n(n+1)}{2} = \frac{8(9)}{2} = \frac{72}{2} = 36$$

SECTION 4 PROBLEMS

Based on the information given, determine the interest saved and the balance due according to the actuarial method.

	Months in Plan	Months Early	Monthly Payment	Annual Rate	Total Interest	Interest Saved	Balance Due
1. a.	24	6	$110.00	18.25%	$440.00		
b.	15	5	237.50	26.75	562.50		
2. a.	10	4	$860.00	16.0%	$600.00		
b.	12	5	46.67	21.5	60.00		

3. Repeat Problem 1 using the rule of 78s.

4. Repeat Problem 2 using the rule of 78s.

5. George Lunt purchased a diamond ring that had a cash price of $2,800. He purchased it on an installment plan specifying 24 monthly payments at 10% simple interest.

a. What amount of interest was computed for the 24-month plan?

b. What was his monthly payment?

c. What was the annual percentage rate that must be disclosed by the seller?

d. How much interest would be saved by the actuarial method if he paid the balance off after the 10th payment?

e. How much is the remaining balance after the 10th payment?

6. J. H. Bailey bought furniture for $3,300 under a finance plan of 12 monthly payments at 8% simple interest.

a. What is the finance charge for the plan?

b. What will his monthly payment be?

c. What annual percentage rate must the lender reveal?

d. As Bailey makes his 8th payment, he also wants to pay his remaining balance. How much interest will his prepayment save by the actuarial method?

e. What is his remaining balance?

7. Ruth Blake borrowed $7,500 at 11% simple interest to be repaid in 15 monthly payments. After making her 9th payment, she decided to pay her remaining balance. Determine

a. The interest that would be included in the 15-month plan

b. The monthly payment

c. The actuarial percentage rate charged
d. The interest saved under the actuarial method by paying the remaining balance early
e. The remaining balance after the 9th payment

8. Jim Oliver borrowed $3,000 at 8% simple interest to be repaid in 10 monthly payments. When making his 6th payment, Oliver decided to pay his remaining balance.

a. How much interest will be included in the 10 payments?
b. What will his monthly payment be?
c. What is the actuarial percentage rate charged?
d. How much interest will Oliver save by the actuarial method when he pays his remaining balance early?
e. What is his remaining balance?

9. Janet Hunter borrowed $3,600 by signing a 14% simple interest note requiring 9 monthly payments.

a. What is the total interest due on the 9-month note?
b. How much does Hunter pay each month?
c. After making her 4th payment, Hunter pays off her balance. How much interest did she save under the rule of 78s?
d. How much was required to pay her balance due?

10. Ned Welsh is buying a $5,000 high-definition television by making 12 monthly payments at 10% simple interest. After the 8th payment, he decided to pay off his balance.

a. What is the total interest included in the 12 payments?
b. How much does Welsh pay each month?
c. How much interest was saved under the rule of 78s?
d. What amount was required to pay the remaining balance due?

11. An oil painting costing $1,200 was purchased on an installment plan permitting 15 monthly payments at 12% simple interest.

a. How much interest was included in the 15 payments?
b. What was the monthly payment?
c. After making the 6th payment, the purchaser paid off the balance. How much interest was saved under the rule of 78s?
d. What was the remaining balance required to pay for the painting?

12. Steve Ali purchased a $1,200 health club membership under a plan requiring 24 monthly payments at 8% simple interest.

a. How much interest was included in the 24 payments?
b. How much did Ali pay each month?
c. Ali paid off the balance after making his 14th payment. How much interest did he save under the rule of 78s?
d. What is the remaining balance at that time?

CHAPTER 17 GLOSSARY

Actuarial (or effective) interest rate. The annual percentage rate required by the Truth-in-Lending Law; a rate computed on the balance due at the time of each installment payment.

Actuarial method. A method of determining the interest savings when a set installment plan is paid in full early, by applying an actuarial interest rate.

Add-on interest. Simple interest.

Annual percentage rate. The annual rate required to be disclosed by the Truth-in-Lending Law; an actuarial or effective rate.

Carrying charge. An extra charge paid for the privilege of buying on the installment plan; a time-payment differential or interest.

Down payment. An initial payment made at the time of purchase and deducted from the cash price before an installment payment is computed.

Effective interest rate. (See "Actuarial interest rate.")

Finance charge. The total amount of extra money paid for credit buying—carrying charge plus service charge, insurance, or any other required charges; the amount that must be disclosed under the Truth-in-Lending Law.

Installment plan. A payment plan that requires periodic payments for a limited time, usually at a relatively high interest rate; either (1) an open-end account or (2) an account with a set number of payments.

Nominal (or simple) interest rate. An interest rate computed on the original balance for the entire time, without regard to the decreasing balance due after each payment to an installment account.

Open-end credit. An account without a definite repayment period and to which additional credit may be added before the first credit is entirely repaid; a revolving charge account.

Outstanding balance. On an installment purchase, the portion of the cash price remaining after the down payment and before the carrying charge is added.

Revolving charge account. An open-end charge account offered by many retail businesses, where new charges may be added up to a set maximum, and minimum monthly payments (depending on the balance due) are made, with interest usually at $1\frac{1}{4}\%$ to $1\frac{1}{2}\%$ monthly computed on the unpaid balance.

Rule of 78s. A method of determining the interest savings when a set installment plan is paid in full early. The savings is that fraction of the total interest equal to the sum of the digits of the early months over the sum of the digits of the total months.

Simple interest rate. (See "Nominal interest rate.")

Time-payment differential. (See "Carrying charge.")

Truth-in-Lending Law. A federal regulation requiring sellers to inform buyers of (1) the finance charge and (2) the effective annual percentage rate paid on any credit account (installment purchase, installment loan, promissory note, etc.).

U.S. rule. A standard method used to give credit for partial payments; each payment is first used to pay the interest due, and the remainder is then deducted from the principal, to obtain an adjusted principal upon which future interest will be computed.

CHAPTER 18

COMPOUND INTEREST

OBJECTIVES

Upon completion of Chapter 18, you will be able to:

1. Define and use correctly the terminology associated with each topic.
2. Compute compound amount without use of a table for short periods (Section 1: Examples 1–4; Problems 1–10).
3. Compute compound interest and amount using the formula $M = P(1 + i)^n$ and using the compound amount table (Section 2: Examples 1–3; Problems 1–20).
4.
   a. Find compound interest and amount at institutions paying interest compounded daily from date of deposit to date of withdrawal (Section 3: Examples 1–4; Problems 1–10).
   b. Also, compare this with interest and amount that would be earned if interest were not compounded daily (Problems 3, 4).
   c. Use the formulas $I = P \times$ Dep. tab. and $I = W \times$ W/D tab. for computing interest compounded daily (Examples 1–4; Problems 7–10).
5. Compute present value at compound interest on investments made at either simple or compound interest. The formula $P = M(1 + i)^{-n}$ and the present value table will be used (Section 4: Examples 1–3; Problems 1–14).

It was previously noted that money invested with a financial institution earns compound interest. Compound interest is more profitable than simple interest to the investor, because, at compound interest, "interest is earned on interest." That is, interest is earned not only on the principal, but also on all interest accumulated since the original deposit. Most compound interest is calculated using prepared tables or computer programs. To be certain that you clearly understand compound interest, however, you should compute a few problems yourself.

Note. Pages 527–528 contain a summary of Chapters 18 through 20, which all contain topics involving interest at a compound rate. As you study the forthcoming chapters, you may also refer to the summary to identify the characteristics of each topic.

SECTION 1

COMPOUND INTEREST (BY COMPUTATION)

Recall that for a simple interest investment, interest is paid on the *original principal* only; at the end of the time, the maturity value is the total of the principal plus the simple interest. Now consider the following example.

Example 1 Suppose that Mr. A makes a \$1,000 investment for 3 years at 7% simple interest. Then

$$P = \$1{,}000 \qquad I = Prt \qquad M = P + I$$

$$r = 7\% \qquad = \$1{,}000 \times 0.07 \times 3 \qquad = \$1{,}000 + \$210$$

$$t = 3 \text{ years} \qquad I = \$210 \qquad M = \$1{,}210$$

Mr. A would earn \$210 interest on this investment, making the total maturity value \$1,210.

Now suppose that Ms. B invests \$1,000 for only 6 months, also at 7% interest. Then

$$P = \$1{,}000 \qquad I = Prt \qquad M = P + I$$

$$r = 7\% \qquad = \$1{,}000 \times 0.07 \times 0.5 \qquad = \$1{,}000 + \$35$$

$$t = 6 \text{ months or } 0.5 \text{ year} \qquad I = \$35 \qquad M = \$1{,}035$$

Ms. B would have \$1,035 at the end of this 6-month investment.

Assume that Ms. B then reinvests this \$1,035 for another 6 months at 7%; she would earn \$36.23 on her second investment, making the total amount \$1,071.23. If the total were then deposited for another 6 months, the interest would be \$37.49, and Ms. B would have \$1,108.72. If this procedure were repeated each 6 months until 3 years had passed, Ms. B would have made six investments and the computations would be as follows:

First 6 Months

$$I = Prt$$
$$= \$1{,}000 \times 0.07 \times 0.5$$
$$I = \$35$$
$$M = \$1{,}035$$

Second 6 Months

$$I = Prt$$
$$= \$1{,}035 \times 0.07 \times 0.5$$
$$I = \$36.23$$
$$M = \$1{,}071.23$$

Third 6 Months

$$I = Prt$$
$$= \$1{,}071.23 \times 0.07 \times 0.5$$
$$I = \$37.49$$
$$M = \$1{,}108.72$$

Fourth 6 Months

$$I = Prt$$
$$= \$1{,}108.72 \times 0.07 \times 0.5$$
$$I = \$38.81$$
$$M = \$1{,}147.53$$

Fifth 6 Months

$$I = Prt$$
$$= \$1{,}147.53 \times 0.07 \times 0.5$$
$$I = \$40.16$$
$$M = \$1{,}187.69$$

Sixth 6 Months

$$I = Prt$$
$$= \$1{,}187.69 \times 0.07 \times 0.5$$
$$I = \$41.57$$
$$M = \$1{,}229.26$$

Thus, after 3 years, Ms. B's original principal would have amounted to \$1,229.26. Since Mr. A had only \$1,210 after his single, 3-year investment, Ms. B made \$1,229.26 − \$1,210.00 or \$19.26 more interest by making successive, short-term investments.

The above example illustrates the idea of **compound interest:** Each time that interest is computed, the interest is added to the previous principal; that total then becomes the principal for the next interest period. Thus, money accumulates faster at compound interest because *interest is earned on interest* as well as on the principal.

Interest is said to be "compounded" whenever interest is computed and added to the previous principal. This is done at regular intervals known as **conversion periods** (or just **periods**). Interest is commonly compounded annually (once a year), semiannually (twice a year), quarterly (four times a year), or monthly. The total value at the end of the investment (original principal plus all interest) is the **compound amount.** The *compound interest* earned is the difference between the compound amount and the original principal. The length of the investment is known as the **term.** The quoted interest rate is always the nominal (or yearly) rate.

Before compound interest can be computed, (1) the term must be expressed as its total number of periods and (2) the interest rate must be converted to its corresponding rate per period.

Example 2 Determine the number of periods for each of the following investments: (a) 7 years compounded annually, (b) 5 years compounded semiannually, (c) 6 years compounded quarterly, and (d) 3 years compounded monthly.

In general, the number of periods is found in this way:

$$\text{Years} \times \text{Number of periods per year} = \text{Total number of periods}$$

Thus,

(a) $\begin{aligned}\text{7 years compounded annually} &= \text{7 years} \times \text{1 period per year}\\ &= \text{7 periods}\end{aligned}$

(b) $\begin{aligned}\text{5 years compounded semiannually} &= \text{5 years} \times \text{2 periods per year}\\ &= \text{10 periods}\end{aligned}$

(c) $\begin{aligned}\text{6 years compounded quarterly} &= \text{6 years} \times \text{4 periods per year}\\ &= \text{24 periods}\end{aligned}$

(d) $\begin{aligned}\text{3 years compounded monthly} &= \text{3 years} \times \text{12 periods per year}\\ &= \text{36 periods}\end{aligned}$

Example 3 Determine the rate per period for each investment: (a) 6% compounded annually, (b) 8.5% compounded semiannually, (c) 5% compounded quarterly, and (d) 4% compounded monthly.

Keep in mind that the stated interest rate is always the yearly rate. Thus, if the rate is 8% compounded quarterly, the rate per period is 2% (since 2% paid four times during the year is equivalent to 8% annually).

You can find rate per period as follows:

$$\frac{\text{Yearly rate}}{\text{Number of periods per year}} = \text{Rate per period}$$

(a) $\begin{aligned}\text{6\% compounded annually} &= \frac{6\%}{\text{1 period per year}}\\ &= \text{6\% per period (or 6\% each year)}\end{aligned}$

(b) 8.5% compounded semiannually $= \dfrac{8.5\%}{2 \text{ periods per year}}$

$= 4.25\%$ per period (or 4.25% each 6 months)

(c) 5% compounded quarterly $= \dfrac{5\%}{4 \text{ periods per year}}$

$= 1.25\%$ per period (or 1.25% each quarter)

(d) 4% compounded monthly $= \dfrac{4\%}{12 \text{ periods per year}}$

$= \frac{1}{3}\%$ per period (or $\frac{1}{3}\%$ each month)

Now let us compute a problem at compound interest.

Example 4 Find the compound amount and the compound interest for the following investments:

(a) $1,000 for 3 years at 8% compounded annually:

$$3 \text{ years} \times 1 \text{ period per year} = 3 \text{ periods}$$

$$\frac{8\%}{1 \text{ period per year}} = 8\% \text{ per period}$$

First period:	Principal	$1,000.00	
	Interest	+ 80.00	(8% of $1,000)
Second period:	Principal	$1,080.00	
	Interest	+ 86.40	(8% of $1,080)
Third period:	Principal	$1,166.40	
	Interest	+ 93.31	(8% of $1,166.40)
	Compound amount	$1,259.71	

Compound amount	$1,259.71
Less: Original principal	− 1,000.00
Compound interest	$ 259.71

(b) $1,000 for 1 year at 7% compounded quarterly:

$$1 \text{ year} \times 4 \text{ periods per year} = 4 \text{ periods}$$

$$\frac{7\%}{4 \text{ periods per year}} = 1.75\% \text{ per period}$$

First period:	Principal	$1,000.00	
	Interest	+ 17.50	(1.75% × $1,000)
Second period:	Principal	$1,017.50	
	Interest	+ 17.81	(1.75% × $1,017.50)
Third period:	Principal	$1,035.31	
	Interest	+ 18.12	(1.75% × $1,035.31)
Fourth period:	Principal	$1,053.43	
	Interest	+ 18.44	(1.75% × $1,053.43)
Compound amount		$1,071.87	

Compound amount	$1,071.87
Less: Original principal	− 1,000.00
Compound interest	$ 71.87

Note. Compare this $1,071.87 compound amount with that in Example 1, in which the compound amount after 1 year at 7% compounded *semiannually* was $1,071.23.

SECTION 1 PROBLEMS

Determine the number of periods and the rate per period for each of the following.

	RATE	COMPOUNDED	YEARS	NO. OF PERIODS	RATE PER PERIOD
1.	a. 4.5%	Annually	6		
	b. 9	Monthly	2		
	c. 7.5	Semiannually	5		
	d. 6	Quarterly	8		
	e. 5	Semiannually	9		
2.	a. 5.8%	Annually	8		
	b. 3	Monthly	4		
	c. 8	Semiannually	6		
	d. 8.5	Quarterly	4		
	e. 5.6	Semiannually	3		

Find the compound amount and the compound interest for each of the following.

3. $4,000 invested for 4 years at 7% compounded (a) annually and (b) semiannually.
4. $2,500 invested for 3 years at 5% compounded (a) annually and (b) semiannually.
5. $3,000 invested for 1 year at 5% compounded (a) semiannually and (b) quarterly.
6. $6,000 invested for 1 year at 8% compounded (a) semiannually and (b) quarterly.
7. $7,000 invested for 9 months at 6% compounded (a) quarterly and (b) monthly.

8. $4,500 invested for 6 months at 4.5% compounded (a) quarterly and (b) monthly.
9. Study carefully your answers to Parts (a) and (b) of the preceding problems. What conclusion seems to be indicated?
10. Suppose that you are given the principal, rate, and years of a compound interest problem. What else must you know in order to work the problem?

SECTION 2

COMPOUND AMOUNT (USING TABLES)

Example 1 in Section 1 illustrated the advantage of a compound interest investment over a simple interest investment. All financial institutions in the United States pay compound interest on savings. There are several factors that an investor should consider before opening an account, however, as various types of accounts are available, often at considerably different rates and under different conditions.

In an **open account,** the owner may deposit or withdraw funds at any time. The most common type of open account is the **statement account.** The owner of a statement account receives a statement periodically from the bank to show deposits, withdrawals, and interest credited to the account for the period. (Deposits to the bank savings account are made using deposit slips similar to those for checking accounts.) Today, financial institutions pay interest on savings accounts from the date of deposit to the date of withdrawal.

Funds deposited for a specific period of time may be used to purchase a **certificate of deposit** (commonly called a **CD**). CDs offer higher rates than open accounts, and their rates increase as funds are committed for longer periods of time. CDs are commonly purchased for 30 days, 90 days (one quarter), 6 months, and 1 year. Periods of 2 or 3 years are available but are not popular during times of low interest rates, as investors hope rates will rise before then. Minimum deposits of $1,000 or more are usually required for CDs. If funds are withdrawn early from a CD, the owner forfeits the higher interest rate and is usually paid only at the institution's open-account rate on the withdrawn amount. Some institutions offer a no-penalty CD at a slightly lower interest rate than the traditional CD, which allows the owner to withdraw funds early without being assessed a penalty. These CDs are attractive to investors since a low interest rate is not locked in on a long-term investment. Figure 18-1 illustrates a certificate of deposit.

Money-market accounts offer interest rates comparable to the rates of certificates of deposit. Unlike CDs, money can be withdrawn from the money-market account at any time without penalty, although normally a limit of only two or three checks per month can be written on the account without charge. The interest rate on the money-market account may change daily, weekly, or monthly, whereas the rate on the CD remains the same throughout the term.

Historically, the rates that financial institutions paid to depositors were controlled by federal regulations. The phasing out of those rates caused increased competition

First National Bank

Certificate of Deposit

May 02, 20X1
Issue Date

May 02, 20X2
Maturity Date

***Virginia A. Jenkins**
Name

111111111
Taxpayer Number

7000006670746
Account Number

12 months
Initial Term

$2,800.00
Amount

TWO THOUSAND EIGHT HUNDRED AND 00 CENTS
Amount Written Out

4.000%
Interest Rate

4.060%
Annual Percentage Yield

TO THIS ACCOUNT QUARTERLY
Payable

SEMINARY PLAZA
Branch Name

5476
RU#

Alexandria, VA
City/State

The Annual Percentage Yield assumes interest remains on deposit until maturity. A withdrawal will reduce earnings.

By signing this:

- You acknowledge the receipt of a copy of the Rules and Regulations For Savings Certificates and accept the terms described therein.
- You understand this time deposit is subject to such Rules and Regulations and as amended from time to time.
- You acknowledge that the Bank's statement of early withdrawal penalties or time deposits was called to your attention. If this Savings Certificate is designated as a No Penalty certificate, one penalty-free withdrawal of all or part of your deposit may be made after funds have been on deposit seven (7) calendar days.
- You understand that this Savings Certificate will renew automatically for like successive periods unless you redeem this certificate on the maturity date or within ten (10) calendar days beginning with the maturity date. Certificates which earn a fixed rate of interest will renew at the interest rate in effect on the maturity date. For variable rate certificates which a floor rate, the rate in effect on the maturity date will be the floor rate for the renewal term.
- If joint, this Savings Certificate shall be a (choose one):

( ) Joint account with survivorship (See Rules and Regulations)

( ) Joint account with no survivorship

Virginia A. Jenkins
Signature ____________ **Signature** ____________

Signature ____________ **Signature** ____________

Estate or Trust Account Certification

I hereby certify that I am the executor/executrix or administrator/administratrix or trustee of the Estate/Trust of ______________________ ("Estate"/"Trust"). I also hereby certify that all beneficiaries of the Estate/Trust, irrespective of any possible remainder interests or powers of appointments, are natural persons. In witness thereof, this ____________ day of ____________________, 20____.

Signature ____________ **Signature** ____________

Prepared By Edwin A. Jones **Authorized By** George Rodgers

Non-Transferable - Initial Deposit Receipt **Copy 1: CUSTOMER**

Figure 18-1 Certificate of Deposit

among financial institutions for investors' deposits, and the rates now fluctuate with the general economic climate. The 1980s and early 1990s experienced an extreme drop in interest rates overall, however, despite competition, and rates still remain low compared to earlier years. The standard by which interest rates are compared is known as the **prime rate**—the lowest rate that large financial institutions charge their "best" customers for loans. Investors earn somewhat less than the prime rate on deposits, and small borrowers pay considerably more than the prime rate when they take out loans or mortgages.

Problems in the preceding section of this chapter demonstrated the advantage of more frequent compounding. Thus, if two accounts offer 5.25% interest, but one compounds daily and the other compounds quarterly, the depositor would earn slightly more interest in the account compounded daily. (Most CDs have their interest compounded quarterly, even though the CD is for a longer period of time.)

While computing the problems in Section 1, it no doubt became obvious to you that this procedure can become quite long and tedious. Compound amount can also be found using the formula

$$M = P(1 + i)^n$$

where M = compound amount, P = original principal, i = interest rate per period, and n = number of periods.

Example 1 (a) Using the formula for compound amount, compute the compound amount and compound interest on \$1,000 invested for 3 years at 8% compounded semiannually.

$P = \$1,000 \qquad M = P(1 + i)^n$

$i = 4\%$ per period $\qquad = \$1,000(1 + 4\%)^6$

$n = 6$ periods $\qquad = 1,000(1 + 0.04)^6$

$M = ? \qquad M = 1,000(1.04)^6$

Recall that the exponent (here, "6") tells how many times the factor 1.04 should be written down before being multiplied. Thus, 1.04 should be used as a factor 6 times:

$$\begin{aligned} M &= \$1,000(1.04)^6 \\ &= \underbrace{1,000(1.04)(1.04)(1.04)(1.04)(1.04)(1.04)} \\ &= 1,000 \quad \times \quad (1.2653190\ldots) \\ M &= \$1,265.32 \end{aligned}$$

Therefore, the maturity value (compound amount) would be $1,265.32. The compound interest is $1,265.32 − $1,000, or $265.32.

Even using this formula, however, the calculation of compound amount would still be quite tedious if the number of periods (the exponent) were very large. The computation is greatly simplified through the use of a compound amount table—a list of the values obtained when the parenthetical expression $(1 + i)$ is used as a factor for the indicated numbers of periods.

Example 1 (cont.) (b) Using the compound amount formula and the table, rework Part (a).

$$P = \$1{,}000 \qquad M = P(1 + i)^n$$

$$i = 4\% \qquad \quad = \$1{,}000(1 + 4\%)^6$$

$$n = 6$$

To find compound amount, turn to Table 17, Amount of 1 (at Compound Interest); Tables Booklet. Various interest rates per period are given at the top left-hand margin of each page; find the page headed by 4%. The lines of the columns correspond to the number of periods, and these are numbered on both the right-hand and left-hand sides of the page. Go down to line 6 (for 6 periods) of the Amount column on the 4% page and there read "1.2653190185"; this is the value of $(1.04)^6$. Now

$$\begin{aligned} M &= P(1 + i)^n \\ &= \$1{,}000(1 + 4\%)^6 \\ &= 1{,}000(1.2653190185) \\ M &= \$1{,}265.32 \end{aligned} \qquad \begin{aligned} I &= M - P \\ &= \$1{,}265.32 \\ &\quad - 1{,}000.00 \\ I &= \$\ \ 265.32 \end{aligned}$$

Note. The value in the table includes the "1" from the parenthetical expression $(1 + i)$. It is *not* correct to add "1" to the tabular value before multiplying by the principal. (Students should now review Section 1, "Accuracy of Computation," in Chapter 1, which demonstrates how many digits from the table must be used in order to ensure an answer correct to the nearest penny.)

Example 2 Find the compound amount and the compound interest on $2,000 invested at 7% compounded quarterly for 6 years.

$P = \$2{,}000$ $\quad M = P(1 + i)^n$ $\quad I = M - P$

$i = 1.75\%$ $\quad = \$2000(1.75\%)^{24}$ $\quad = \$3{,}032.89 - 2{,}000.00$

$n = 24$ $\quad = 2{,}000(1.516443)$ $\quad I = \$1{,}032.89$

$= 3{,}032.886$

$M = \$3{,}032.89$

After 6 years, the $2,000 investment would thus be worth $3,032.89, of which $1,032.89 is interest.

Example 3 Ray Copeland opened a savings account on April 1, 20X1, with a deposit of $800. The account paid 4% interest compounded quarterly. On October 1, 20X1, Ray closed that account and added enough additional money to purchase a $1,000, 6-month CD earning interest at 6% compounded monthly. (a) How much more did Ray deposit on October 1? (b) What was the maturity value of his CD on April 1, 20X2? (c) How much total interest was earned?

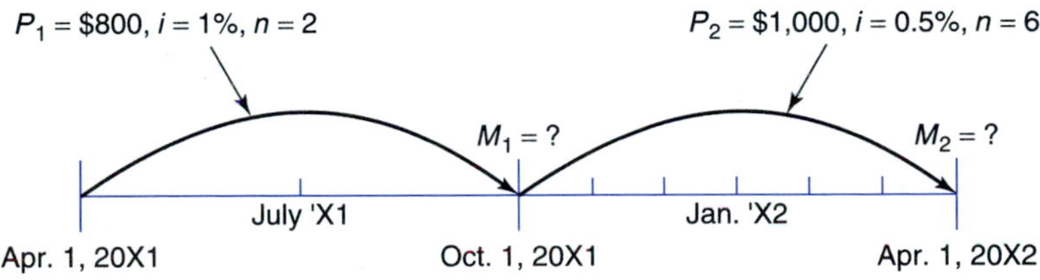

(a) Banking quarters usually begin on the first days of January, April, July, and October. This means that Ray's $800 deposit earned interest for two quarters. Hence,

$P_1 = \$800$ $\quad M_1 = P(1 + i)^n$ $\quad \$1{,}000.00 - 816.08$

$i = 1\%$ $\quad = \$800(1 + 1\%)^2$ $\quad \$\ 183.92$ Additional deposit

$n = 2$ $\quad = 800(1.020100)$

$= 816.080$

$M_1 = \$816.08$

(b) On October 1, Ray's savings account had a closing balance of $816.08. An additional $183.92 was required to purchase a $1,000 CD. There were then

6 months (2 quarters) until the CD matured on April 1, 20X2, during which time the $1,000 principal earned interest at 6% compounded monthly.

$$P_2 = \$1{,}000 \qquad M_2 = P(1 + i)^n$$

$$i = 0.5\% \qquad = \$1{,}000(1 + 0.5\%)^6$$

$$n = 6 \qquad = 1{,}000(1.030378)$$

$$M_2 = \$1{,}030.38$$

The maturity value of Ray's CD was $1,030.38 on April 1, 20X2.

(c) The total interest may be found in either of two ways:

1. By adding the interest paid on each of the principals.
2. By finding the difference between the final balance and the total of all deposits.

(1)

$$\begin{aligned} I &= M - P \\ &= \$816.08 - 800.00 \\ I_1 &= \$\ 16.08 \end{aligned} \qquad \begin{aligned} I &= M - P \\ &= \$1{,}030.38 - 1{,}000.00 \\ I_2 &= \$\ \ 30.38 \end{aligned}$$

$16.08	Interest_1
+ 30.38	Interest_2
$46.46	Total interest

(2)

$800.00	Deposit_1
+ 183.92	Deposit_2
$983.92	Total deposits

$1,030.38	Final balance
− 983.92	Total deposits
$ 46.46	Total interest

How Long Does It Take to Double an Investment?

For a sum of money invested at various interest rates compounded quarterly, the following table shows how long it takes to double (to the nearest quarter after the money doubles):

QUARTERLY RATE	YEARS TO DOUBLE	QUARTERLY RATE	YEARS TO DOUBLE	QUARTERLY RATE	YEARS TO DOUBLE
5%	14	$7\frac{1}{2}$%	$9\frac{1}{2}$	10%	$7\frac{1}{4}$
$5\frac{1}{2}$	$12\frac{3}{4}$	8	9	$10\frac{1}{2}$	$6\frac{3}{4}$
6	$11\frac{3}{4}$	$8\frac{1}{2}$	$8\frac{1}{4}$	11	$6\frac{1}{2}$
$6\frac{1}{2}$	11	9	8	$11\frac{1}{2}$	$6\frac{1}{4}$
7	10	$9\frac{1}{2}$	$7\frac{1}{2}$	12	6

SECTION 2 PROBLEMS

1–6. Rework Problems 3–8 of Section 1 (pages 470–471) using the compound amount formula $[M = P(1 + i)^n]$.

Compute the following compound amounts and compound interest, using Table 17.

	PRINCIPAL	RATE	COMPOUNDED	YEARS
7. a.	$ 500	5%	Monthly	3
b.	1,700	6	Quarterly	5
c.	3,200	7	Semiannually	6
d.	4,500	8	Monthly	4
e.	8,100	9	Quarterly	7
8. a.	$1,400	6%	Monthly	5
b.	2,500	5	Quarterly	4
c.	5,200	4	Semiannually	3
d.	9,000	7	Monthly	6
e.	800	8	Quarterly	2

9. Compute compound amount and interest on the following deposits at 5% compounded monthly for 3 years.
a. $500 b. $1,000 c. $2,000 d. What is your conclusion regarding the principal?

10. Compute compound amount and interest on the following investments at 7% compounded monthly for 4 years.
a. $800 b. $1,600 c. $3,200 d. What is your conclusion regarding the principal?

11. Calculate compound amount and interest on $1,000 drawing interest compounded quarterly for 2 years at the following rates.
a. 4% b. 8% c. 16% d. Does doubling the rate exactly double the interest?

12. Determine compound amount and interest on $1,000 earning interest compounded quarterly for 5 years at the following rates.
a. 3% b. 6% c. 12% d. Does doubling the rate exactly double the interest?

13. Determine compound amount and interest on $1,000 at 6% compounded semiannually for the following years.
a. 2 years b. 4 years c. 8 years d. Does the interest exactly double when the time is doubled?

14. Compute compound amount and compound interest on $1,000 at 5% compounded semiannually for
a. 3 years b. 6 years c. 12 years d. Does the interest exactly double when the time is doubled?

15. On January 1, 20X6, Patricia Humble opened a savings account with a deposit of $1,000. Her bank paid 4% interest compounded quarterly.

a. What was the value of Humble's account on July 1, 20X6?

b. On that day, Humble made a deposit sufficient to bring the value of the account to $1,200. What amount did she deposit?

c. How much was her account worth on July 1, 20X7?

d. How much total interest did her account earn?

16. On January 2, 20X4, Simon Ross opened a savings account with a deposit of $2,200. His bank paid 6% interest compounded quarterly.

a. What was the value of his account on July 2, 20X4?

b. On that day, Ross made a deposit sufficient to bring the value of the account to $2,400. What amount did he deposit on that date?

c. How much was his account worth on July 2, 20X5?

d. How much total interest did Ross earn?

17. On March 1, Elena Ticer purchased a $1,000, 90-day CD earning interest at 5% compounded monthly.

a. What was the maturity value of the CD?

b. On June 1, Ticer added enough additional money to purchase a $1,400, 1-year CD that paid interest at 6% compounded quarterly. How much was her additional deposit?

c. What was the maturity value of this 1-year CD?

d. How much total interest did Ticer earn?

18. On July 6, Laura Christensen purchased a $2,000, 6-month CD that paid interest at 7% compounded semi-annually.

a. What was the maturity value of the CD?

b. On January 6, Laura added enough money to purchase a $2,100, 1-year CD that paid interest at 7% compounded quarterly. How much was her additional deposit?

c. What was the maturity value of this 1-year CD?

d. How much total interest did she earn?

19. On May 1, Ed Grant purchased a $1,000, 6-month CD that earned interest at 5% compounded monthly.

a. What was the maturity value of the CD?

b. On November 1, Grant added additional funds to purchase a $1,400, 2-year CD paying 7% compounded quarterly. How much additional deposit was made?

c. What was the maturity value of this 2-year CD?

d. How much interest was earned on the two CDs?

20. On September 10, Jean Murphy purchased a $600, 6-month CD that earned interest at 4% compounded quarterly.

a. What was the maturity value of the CD?

b. On March 10, Murphy added enough money to purchase a $700, 2-year CD paying 5% compounded semiannually. How much additional deposit was made?

c. What was the maturity value of the 2-year CD?

d. How much interest was earned on both CDs?

INTEREST COMPOUNDED DAILY

Financial institutions also pay "daily interest" (interest compounded daily) on all open accounts where deposits or withdrawals may be made at any time. That is, financial institutions pay interest for the exact number of days that money has been on deposit.

Interest on deposits is compounded daily, but to eliminate excessive bookkeeping, most institutions enter interest in the depositor's account only once each quarter. For this reason, a daily interest table contains factors for one quarter. In accordance with the practice followed by most savings institutions, it is a 90-day quarter; deposits made on the 31st of any month earn interest as if they were made on the 30th. Recall that quarters begin in January, April, July, and October.

INTEREST ON DEPOSITS

As pointed out in the preceding section, interest rates on deposits in savings accounts and regular CDs have remained low in recent years, despite the termination of federal regulations controlling them.

Our study will be limited to 4.5% interest compounded daily, which approximates the historic passbook interest rate. Interest is found by multiplying the principal (deposit) times the appropriate value from Table 15, Interest from Day of Deposit, Tables Booklet. That is,

$$\text{Interest} = \text{Principal} \times \text{Deposit table}$$

which might be abbreviated

$$I = P \times \text{Dep. tab.}$$

In the daily interest deposit table, there is a column for each of the three months of the quarter; each column (month) contains entries for 30 days. To use the table, you look for the date on which a deposit was made; the factor beside that date is the number to be multiplied by the amount of deposit in order to obtain the interest (provided the money remained on deposit until the quarter ended).

Example 1 Find the interest that would be earned at 4.5% compounded daily, if $1,000 were deposited in a savings and loan association on July 17.

July is the first month of the quarter; therefore, we refer to the "1st Month" column of the deposit table and to the 17th line under that heading:

$$\begin{aligned} I &= P \times \text{Dep. tab.} \\ &= \$1{,}000(0.0092923) \\ &= 9.2923 \\ I &= \$9.29 \end{aligned}$$

When the quarter ends, interest of $9.29 will be added to the depositor's account, bringing the total to $1,009.29.

Example 2 Richard Chiles has an account in a financial institution where deposits earn interest at 4.5% compounded daily. When the quarter began on January 1, Chiles's account contained \$1,000. During the quarter, he made deposits of \$200 on February 7 and \$300 on March 13. (a) How much interest will the account earn during the quarter? (b) What will the balance be in Chiles's account at the end of the quarter?

For any deposits made after a quarter begins, interest must be computed separately for each deposit. The total interest for the quarter is the sum of all interest for the various deposits.

(a) Chiles's initial balance of \$1,000 will earn interest for the entire quarter:

$$I = P \times \text{Dep. tab.}$$

$$= \$1{,}000(0.0113128)$$

$$= 11.3128$$

$$I = \$11.31$$

The \$200 deposit on February 7 would earn interest from the 7th day of the second month:

$$I = P \times \text{Dep. tab.}$$

$$= \$200(0.0067724)$$

$$= 1.35448$$

$$I = \$1.35$$

Interest on the \$300 deposit made March 13 (the third month) is

$$I = P \times \text{Dep. tab.}$$

$$= \$300(0.0022524)$$

$$= 0.67572$$

$$I = \$0.68$$

The transactions for the entire quarter would thus be summarized as follows:

Principal		*Interest*		
\$1,000		\$11.31		
200		1.35		
+ 300		+ 0.68		
\$1,500	+	\$13.34	= \$1,513.34	Balance, end of quarter

The account would earn total interest of \$13.34 during the quarter.

(b) Chiles's account would have a balance of \$1,513.34 at the end of the quarter.

INTEREST ON WITHDRAWALS

When withdrawals are involved, interest may be calculated in the following manner:

1. Subtract the withdrawals from the opening principal. Compute interest on this remaining balance for the entire quarter.
2. Determine the interest that would be earned on the withdrawn funds until the date they were withdrawn. This interest is found as follows, using Table 16, Interest to Day of Withdrawal, Tables Booklet:

   $$\text{Interest} = \text{Withdrawal} \times \text{Withdrawal table}$$

   which may be abbreviated

   $$I = W \times \text{W/D tab.}$$

   (The daily interest withdrawal table is similar to the daily interest deposit table in that it is divided into the 3 months of the quarter, and you use the factor beside the appropriate date.)
3. Total interest for the quarter is the sum of the various amounts of interest found in steps 1 and 2.

Example 3 On July 1, a savings account in a credit union contained \$1,250. A withdrawal of \$250 was made on September 15. (a) How much interest did the account earn for the quarter, if interest was paid at 4.5% compounded daily? (b) What was the ending balance in the account?

(a) Since \$250 was withdrawn from the \$1,250 account, only \$1,000 earned interest for the entire quarter. In Example 2 we found that the interest on \$1,000 for one quarter is \$11.31.

Next, we compute interest on the $250 between the beginning of the quarter and September 15, when the funds were withdrawn. Since September is the third month of the quarter, we use the 15th line in the "3rd Month" column in the withdrawal table:

$$
\begin{aligned}
I &= W \times \text{W/D tab.} \\
&= \$250(0.0094185) \\
&= 2.3546 \\
I &= \$2.35
\end{aligned}
$$

The $250 will earn interest of $2.35 before it is withdrawn on September 15. Total interest for the quarter is thus

Interest	
$11.31	
+ 2.35	
$13.66	Total interest for the quarter

(b) The balance in this account at the end of the quarter was

Opening balance	$1,250.00	
Withdrawal	− 250.00	
	$1,000.00	
Interest	+ 13.66	
	$1,013.66	Balance, end of quarter

Example 4 Compute (a) the interest and (b) the balance at the end of the quarter after the following transactions, when daily interest is compounded at 4.5%.

Balance	April 1	$1,800
Withdrawal	May 18	200
Deposit	June 14	400
Withdrawal	June 21	200

(a) The account would earn interest on $1,800 for the entire quarter, if there had been no withdrawals. Since there were two withdrawals of $200, however, only $1,800 − $400, or $1,400, will earn interest for the whole quarter:

Withdrawals			
May 18	$200	Opening balance	$1,800
June 21	200	Withdrawals	− 400
	$400	Principal earning interest for entire quarter	$1,400

Interest for the quarter is thus computed as follows:

$1,400 principal for entire quarter:

$I = P \times \text{Dep. tab.}$

$= \$1{,}400(0.0113128)$

$= 15.8379$

$I = \$15.84$

$400 deposit on June 14 (third month):

$I = P \times \text{Dep. tab.}$

$= \$400(0.0021271)$

$= 0.8508$

$I = \$0.85$

The following interest was earned prior to the two withdrawals:

$200 withdrawal on May 18 (second month):

$I = W \times \text{W/D tab.}$

$= \$200(0.0060177)$

$= 1.2035$

$I = \$1.20$

$200 withdrawal on June 21 (third month):

$I = W \times \text{W/D tab.}$

$= \$200(0.0101758)$

$= 2.0352$

$I = \$2.04$

Thus, total interest for the quarter is

	Interest
Interest on funds on deposit for entire quarter	$15.84
Interest on $400 deposited June 14	0.85
Interest on $200 withdrawn May 18	1.20
Interest on $200 withdrawn June 21	2.04
Total interest for quarter	$19.93

(b) The balance after the quarter ended would be

Deposits	*Withdrawals*		
$400	$200	Opening balance	$1,800.00
	200	Deposits	+ 400.00
	$400		$2,200.00
		Withdrawals	− 400.00
			$1,800.00
		Interest	+ 19.93
		Balance, end of quarter	$1,819.93

The savings account would contain $1,819.93 after the quarter ended.

SECTION 3 PROBLEMS

The following problems are all for interest paid at 4.5% *compounded daily. Find (1) the amount of interest and (2) the balance at the end of the quarter for the following deposits.*

1. a. $1,000 on May 3 b. $800 on July 6 c. $4,000 on September 16 d. $600 on October 8

2. a. $1,000 on February 5 b. $350 on April 10 c. $2,300 on March 15 d. $700 on October 19

Compute the interest that would be earned for one quarter (1) at 4.5% *compounded daily and (2) at* 4.5% *compounded quarterly on the given principals. The factor for* 4.5% *compounded quarterly for one quarter is* 0.01125.

3. a. $7,500 b. $4,500

4. a. $5,000 b. $600

Find (1) the total amount of interest and (2) the balance after the quarter ended for the following accounts.

OPENING BALANCE	DEPOSIT
5. a. $700 on October 1	$ 500 on October 22 $ 300 on November 15 $ 400 on December 5
b. $600 on January 1	$ 200 on January 30 $ 400 on February 13 $ 300 on March 9
6. a. $1,000 on April 2	$ 200 on April 26 $ 500 on May 3 $ 400 on June 28
b. $1,500 on October 1	$ 900 on November 1 $1,000 on November 20 $ 600 on December 3

Compute (1) the interest for the quarter and (2) the balance when the quarter ended for the following problems.

	OPENING BALANCE	WITHDRAWALS
7.	a. $10,000 on April 1	$1,000 on May 4
	b. $ 5,500 on July 1	$ 100 on August 10 $ 400 on September 1
8.	a. $ 6,800 on October 5	$ 800 on December 5
	b. $ 2,400 on January 10	$ 200 on February 8 $ 500 March 20

Transactions for an entire quarter are given below. Calculate (1) the total interest each account would earn and (2) the balance in each account after the quarter ends.

	OPENING BALANCE	DEPOSITS	WITHDRAWALS
9.	a. $5,600 on October 1	$ 500 on October 30	$1,000 on November 11
	b. $4,900 on January 1	$ 800 on February 5 $1,000 on March 12	$ 600 on January 15 $ 400 on March 20
	c. $5,100 on April 1	$ 400 on April 8 $ 100 on May 9	$ 200 on April 27 $ 900 on June 20
10.	a. $7,000 on July 1	$ 300 on September 5	$ 600 on August 13
	b. $3,600 on October 1	$1,000 on November 1 $1,000 on December 1	$ 600 on October 6 $ 200 on December 19
	c. $6,000 on January 1	$ 800 on February 14 $ 500 on March 3	$1,000 on January 20 $ 600 on March 30

SECTION 4

PRESENT VALUE (AT COMPOUND INTEREST)

It often happens that someone wishes to know how much would have to be deposited now (at the present) in order to obtain a certain maturity value. The principal that would have to be deposited is the **present value.**

The formula for obtaining present value at compound interest is a variation of the compound amount formula. The formula is rearranged to solve for P rather than M, by dividing both sides of the equation by the parenthetical expression $(1 + i)^n$:

$$M = P(1 + i)^n$$

$$\frac{M}{(1 + i)^n} = P$$

or, reversing the order,

$$P = \frac{M}{(1 + i)^n}$$

Therefore, present value could be found by dividing the known maturity value M by the appropriate value from the compound amount table (Table 17) that we have already been using. However, present value tables have been developed which allow present value to be computed by multiplication.

Just as $\frac{12}{3} = 12 \cdot \frac{1}{3}$, so

$$\frac{M}{(1 + i)^n} = M \cdot \frac{1}{(1 + i)^n}$$

The entries in Table 20, Present Value, Tables Booklet, are the quotients obtained when the numbers from the compound amount table are divided into 1. These quotients (the present value entries) can then be multiplied by the appropriate maturity value to obtain the present value.

Since present value is usually computed using multiplication, the formula is commonly written in a form that indicates multiplication and uses a negative exponent. Recall that a negative exponent indicates that the factor actually belongs in the opposite part of the fraction. $\left(\text{That is, } 3x^{-2} \text{ means } \frac{3}{x^2}.\right)$ Thus,

$$P = \frac{M}{(1 + i)^n}$$

is usually written

$$P = M(1 + i)^{-n}$$

The following examples will illustrate that present value may be found using either the compound amount table (Table 17) or the present value table (Table 20). As a general rule, however, you should use the present value table when computing present value at compound interest.

Example 1 An investment made for 6 years at 9% compounded monthly is to have a maturity value of $1,000. Determine the present value using (a) the compound amount table (Table 17) and (b) the present value table (Table 20). Also, (c) find how much interest will be included.

(a) $M = \$1{,}000$ $\quad P = \dfrac{M}{(1+i)^n}$

$i = 0.75\%$ $\quad = \dfrac{\$1{,}000}{(1+0.75\%)^{72}}$

$n = 72$ $\quad = \dfrac{\$1{,}000}{1.712553}$

$P = ?$ $\quad = 583.924$

$P = \$583.92$

(b) $P = M(1+i)^{-n}$

$= \$1{,}000(1+0.75\%)^{-72}$

$= 1{,}000(0.583924)$

$P = \$583.92$

(c) $I = M - P$

$= \$1{,}000.00 - 583.92$

$I = \$\ 416.08$

Thus, \$583.92 invested now at 9% compounded monthly will earn \$416.08 interest and mature to \$1,000 after 6 years.

Example 2 Charles and Sue Baker would like to have \$6,000 in 4 years for a down payment on a condominium. (a) What single deposit would have this maturity value if CDs earn 7% compounded quarterly? (b) How much of the final amount will be interest?

(a) $M = \$6{,}000$ $\quad P = M(1+i)^{-n}$

$n = 16$ $\quad = \$6{,}000\left(1 + 1\frac{3}{4}\%\right)^{-16}$

$i = 1\frac{3}{4}\%$ $\quad = 6{,}000(0.7576163)$

$P = ?$ $\quad = 4{,}545.698$

$P = \$4{,}545.70$

(b) $I = M - P$

$= \$6{,}000.00 - 4{,}545.70$

$I = \$1{,}454.30$

If the Bakers purchase a \$4,545.70 CD now, it will earn interest of \$1,454.30 during 4 years at 7% compounded quarterly and will have a \$6,000 maturity value.

As was true of simple interest problems, present value (or present worth) may on occasion differ from principal. The actual amount of money that is invested is always the principal; it may or may not be invested at the interest rate currently being paid by most financial institutions. The present value of the investment is the amount that would have to be invested at the rate money is worth (the rate being paid by most financial institutions) in order to obtain the same maturity value that the actual investment will have. If an investment is sold before its maturity date, it should theoretically be sold for its present value at that time.

Example 3 Arthur Levy made a $12,000 real estate investment that he expects will have a maturity value equivalent to interest at 8% compounded monthly for 5 years. If most savings institutions are currently paying 6% compounded quarterly on 5-year CDs, what is the least amount for which Levy should sell his property?

This is a two-part problem. We must first determine the maturity value of the $12,000 investment. Then, using the rate money is actually worth, we must compute the present worth of the maturity value obtained by step 1.

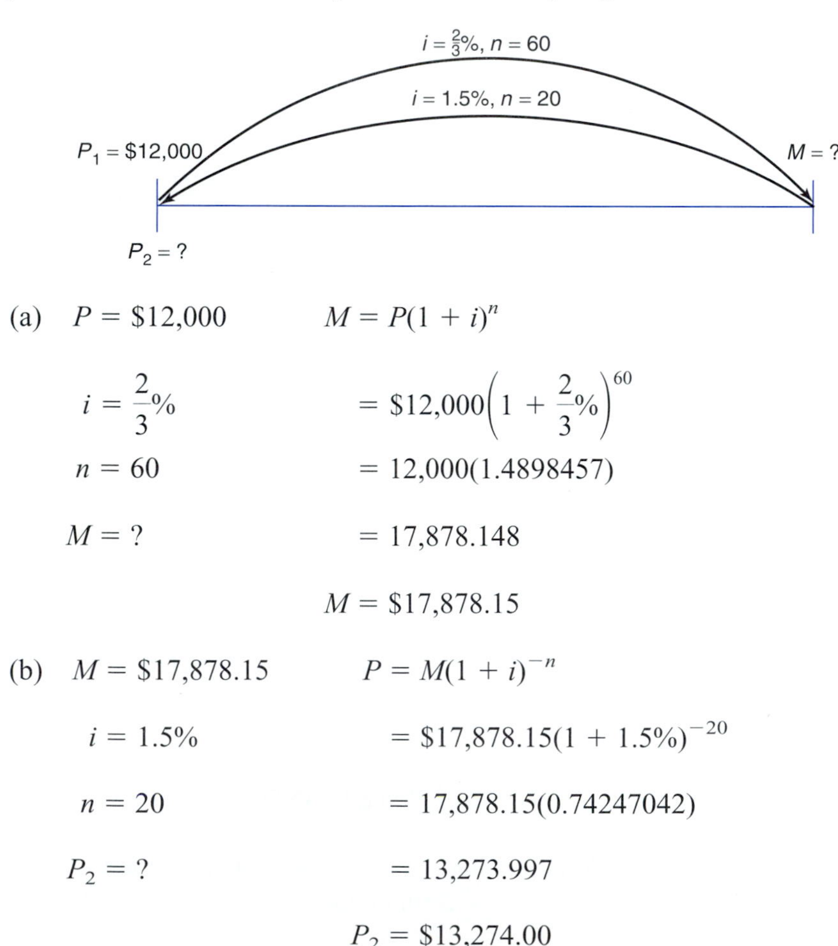

(a) $P = \$12{,}000$ $\qquad M = P(1 + i)^n$

$i = \frac{2}{3}\%$ $\qquad = \$12{,}000\left(1 + \frac{2}{3}\%\right)^{60}$

$n = 60$ $\qquad = 12{,}000(1.4898457)$

$M = ?$ $\qquad = 17{,}878.148$

$$M = \$17{,}878.15$$

(b) $M = \$17{,}878.15$ $\qquad P = M(1 + i)^{-n}$

$i = 1.5\%$ $\qquad = \$17{,}878.15(1 + 1.5\%)^{-20}$

$n = 20$ $\qquad = 17{,}878.15(0.74247042)$

$P_2 = ?$ $\qquad = 13{,}273.997$

$$P_2 = \$13{,}274.00$$

Levy should not sell the $12,000 property for less than $13,274, because he would have to purchase a CD of this value, paying interest at 6% compounded quarterly, in order to have $17,878.15 in 5 years.

Calculator Tip. For most calculators, the value of i that you use from the interest tables must be rounded to a maximum of 7 decimal places (or a maximum of 8 digits altogether). For instance, in Part (b) of example 3, you would use (0.7424704). In most problems, your result will be within a cent or two of the correct answer. Frequently you will obtain exactly the same amount, to the nearest cent, which happens here.

SECTION 4 PROBLEMS

Compute present value at compound interest, using the data given in Problems 1 and 2. Also determine the amount of interest that will be earned before maturity.

	MATURITY VALUE	RATE	COMPOUNDED	YEARS
1. a.	$ 3,200	4%	Semiannually	2
b.	2,500	5	Quarterly	5
c.	6,800	7	Monthly	4
d.	1,800	5	Semiannually	1
e.	4,400	6	Quarterly	3
2. a.	$ 1,400	5%	Quarterly	5
b.	40,000	8	Semiannually	3
c.	5,200	6	Monthly	6
d.	10,000	7	Quarterly	2
e.	8,700	5	Monthly	4

3. Bill and Beth Reno want to have $5,000 in savings in 1 year for a family vacation. If they can make an investment that pays 6% compounded monthly,

a. How much should the Renos invest now in order to have this maturity value?

b. How much interest will their investment earn?

4. The Fannon Group wants to have $25,000 in savings in 2 years for a renovation project.

a. What single deposit must be made now in a 2-year CD that earns 8% interest compounded quarterly?

b. How much interest will its investment earn?

5. The Capitol Plumbing Co. plans to increase the size of its warehouse in 2 years. The business will need $75,000 for the project.

a. What single deposit must be made now in an investment earning 8% interest compounded quarterly?

b. How much interest will the investment earn?

6. Furman Industries will require $30,000 in 2 years. If it can make an investment at 7% compounded monthly,

a. What single deposit made now will yield this amount?

b. How much interest will this investment earn?

7. Dominion Rentals, Inc. expects to remodel its offices in 6 years at a cost of $100,000. If the company can make an investment paying 9% compounded semiannually,
   a. What single deposit made now will produce this maturity value?
   b. How much interest will be included?
8. Dan's Boating Center expects to expand its operations in 5 years at a cost of $200,000. It can make an investment at 9% compounded quarterly.
   a. What single deposit made now will produce this amount?
   b. How much interest will be earned?
9. A $7,000 investment is made at 6% compounded quarterly for a 4-year period.
   a. What is the maturity value of the investment?
   b. If money is generally worth 5% compounded monthly, what is the present value of the investment?
10. A $9,000 investment is made at 5% compounded quarterly for a 5-year period.
   a. What is the maturity value of the investment?
   b. If money is generally worth 6% compounded semiannually, what is the present value of the investment?
11. Maggie Clifford is the payee of a 2-year, 11% simple interest note for $6,000. Most financial institutions are currently paying 7% compounded monthly.
   a. How much will the maturity value of the note be?
   b. How much should Clifford have received if she had sold the note on the day it was originated?
12. Pablo Metz is the payee of a 3-year, 9% simple interest note for $10,000. Money invested in a 3-year CD is worth 7% compounded monthly.
   a. How much will be due when the note matures?
   b. If Metz had sold the note on the day it was drawn, how much would he have received?
13. A $5,000, 1-year CD was purchased at a 9% simple interest rate.
   a. What maturity value will the CD have?
   b. If inflation is increasing at a rate equivalent to 6% compounded monthly (the rate money is worth), what was the value of the CD on the day it was purchased?
14. A 6% simple interest rate was earned on a 1-year CD for $1,000. Savings accounts are earning 5% compounded quarterly.
   a. What will the maturity value of the CD be?
   b. Using the savings account rate, what is the present value of the CD?

CHAPTER 18 GLOSSARY

Certificate of deposit (CD). A savings contract that offers a slightly higher rate when a (usually $1,000 minimum) deposit is made for a specified time period. (An early withdrawal would forfeit the higher rate.)

Compound amount. The total value at the end of a compound interest investment (original principal plus all interest).

Compound interest. Interest computed periodically and then added to the previous principal, so that this total then becomes the principal for the next interest period. "Interest earned on (previous) interest." The difference between final maturity value and original principal.

Conversion period. A regular time interval at which interest is compounded (or computed and added to the previous principal). Interest is commonly compounded daily, monthly, quarterly, semiannually, or annually.

Money-market account. A savings contract similar to a CD that pays higher interest rates than open accounts, permits the withdrawal of funds at any time, and allows up to three checks per month to be written.

Open account. A savings account upon which deposits or withdrawals may be made at any time.

Period. (See "Conversion period.")

Present value. The amount of money that would have to be deposited (on some given date prior to maturity) in order to obtain a specified maturity value.

Prime rate. The interest rate banks charge on loans to their "best" customers.

Statement account. An open account for which the bank sends periodic statements showing transactions that have occurred since the last statement.

Term. The length of time for a compound interest investment.

CHAPTER 19

ANNUITIES

OBJECTIVES

Upon completion of Chapter 19, you will be able to:

1. Define and use correctly the terminology associated with each topic.
2. Determine amount and compound interest earned on an annuity, using the procedure $M = \text{Pmt.} \times \text{Amt. ann. tab.}_{\overline{n}|i}$ and the amount of annuity table (Section 1: Example 1; Problems 1–10).
3. **a.** Compute present value of an annuity, using the procedure $\text{P.V.} = \text{Pmt.} \times \text{P.V. ann. tab.}_{\overline{n}|i}$ and the present value of annuity table (that is, determine the original value required in order for one to withdraw the given annuity payments) (Section 2: Examples 1, 2; Problems 1–12).
   b. Also, determine the total amount received and the interest included.
4. Use the same procedure (Objective 3) to determine (Section 2: Example 3, Problems 13–16):
   a. The total amount paid for a real estate purchase
   b. The equivalent cash price.

Notice that the compound interest problems in the previous chapter basically involved making a *single deposit* which remained invested for the entire time. There are few people, however, who have large sums available to invest in this manner. Most people must attain their savings goals by making a series of regular deposits. This leads to the idea of annuities.

An **annuity** is a series of payments (normally equal in amount) that are made at regular intervals of time. Most people think of an annuity as the regular payment received from an insurance policy when it is cashed in after retirement. This is one good example, but there are many other everyday examples that are seldom thought of as annuities. Besides savings deposits, other common examples are rent, salaries, Social Security payments, installment plan payments, loan payments, insurance payments—in fact, any equal payments made at regular intervals of time.

There are several time variables that may affect an annuity. For instance, some annuities have definite beginning and ending dates; such an annuity is called an **annuity certain.** Examples of an annuity certain are installment plan payments or the payments from a life insurance policy converted to an annuity of a specified number of years.

If the beginning and/or ending dates are uncertain, the annuity is called a **contingent annuity.** Monthly Social Security retirement benefits and the payments on an ordinary life insurance policy are examples of contingent annuities for which the ending dates are unknown, because both will terminate when the person dies. If a person provides in a will that following death a beneficiary is to receive an annuity for a fixed number of years, this is a contingent annuity for which the beginning date is uncertain. A man with a large estate might provide that his surviving wife receive a specified yearly income for the remainder of her life and that the balance then be donated to some charity; this contingent annuity would then be uncertain on both the beginning and ending dates.

Another factor affecting annuities is whether the payment is made at the beginning of each time interval (such as rent and insurance premiums, which are normally paid in advance) or at the end of the period (such as salaries and Social Security retirement benefits). An annuity for which payments are made at the beginning of each period is known as an **annuity due.** When payments come at the end of each period, the annuity is called an **ordinary annuity.**

We shall be studying *investment annuities*—annuities that earn compound interest (rather than rent or installment payments, for example, which earn no interest). An annuity is said to be a **simple annuity** when the date of payment coincides with the conversion date of the compound interest.

The study of contingent annuities requires some knowledge of probability, which is not within the scope of this text. Thus, since our purpose is just to give you a basic introduction to annuities, our study of annuities will be limited to simple, ordinary annuities certain—annuities for which both the beginning and ending dates are fixed, and for which the payments are made on the conversion date at the end of each period.

SECTION 1

AMOUNT OF AN ANNUITY

The **amount of an annuity** is the maturity value that an account will have after a series of equal payments into it. As in the case of other compound interest problems, amount of an annuity is usually found by using the appropriate table for the appropriate rate per period and the number of periods. Thus, we shall compute amount of an annuity by using Table 18, Amount of Annuity, Tables Booklet, and the following procedure:

$$\text{Amount} = \text{Payment} \times \text{Amount of annuity table}_{\overline{n}|i}$$

where n = number of periods and i = interest rate per period. For simplicity, the procedure* might be abbreviated as

$$M = \text{Pmt.} \times \text{Amt. ann. tab.}_{\overline{n}|i}$$

The total amount of deposits is found by multiplying the periodic payment times the number of periods. Total interest earned is then the difference between the maturity value and the total deposits made into the account.

Example 1 (a) Determine the amount (maturity value) of an annuity if Leonard Feldman deposits \$100 each quarter for 6 years into an account earning 8% compounded quarterly. (b) How much of this total will Feldman deposit himself? (c) How much of the final amount is interest?

(a) (Notice that deposits are made each quarter to coincide with the interest conversion date at the bank; we would not be able to compute the annuity if they were otherwise.)

$$\begin{aligned} \text{Pmt.} &= \$100 & M &= \text{Pmt.} \times \text{Amt. ann. tab.}_{\overline{n}|i} \\ n &= 24 & &= \$100 \times \text{Amt. ann. tab.}_{\overline{24}|2\%} \\ i &= 2\% & &= 100(30.42186) \\ & & &= 3{,}042.186 \\ & & M &= \$3{,}042.19 \end{aligned}$$

Feldman's account will contain \$3,042.19 after 6 years.

*Many texts give the procedure for amount of an annuity in the form $M = Pm_{\overline{n}|i}$, where M = maturity value, P = payment, and $m_{\overline{n}|i}$ indicates use of the amount of annuity table. Strictly speaking, $M = Pm_{\overline{n}|i}$ is also a procedure, not a formula. The actual formula is $M = P \times \dfrac{(1+i)^n - 1}{i}$.

(b) There will be 24 deposits of $100 each, totaling 24 × $100 or $2,400.
(c) The interest earned during this annuity period is

Amount of annuity	$3,042.19
Total deposits	− 2,400.00
Interest	$ 642.19

SECTION 1 PROBLEMS

Using the amount-of-an-annuity procedure, find the maturity value that would be obtained if one makes the payments given below. Also determine the total deposits and the total interest earned. Assume that all problems are ordinary annuities, where payments are made at the end of the period, as in Table 18.

	Periodic Payment	Rate	Compounded	Years	Maturity Value	Total Deposits	Total Interest Earned
1. a.	$ 500	6%	Quarterly	7			
b.	800	7	Quarterly	3			
c.	300	8	Semiannually	8			
d.	1,500	5	Quarterly	6			
e.	1,100	6	Monthly	3			
2. a.	$1,500	7%	Semiannually	4			
b.	2,000	6	Quarterly	5			
c.	600	5	Monthly	4			
d.	1,800	8	Quarterly	8			
e.	50	9	Monthly	4			

3. Erin Anderson deposited $500 each quarter for 2 years into a savings account that paid 7% compounded quarterly.
   a. What was the value of her account after 2 years?
   b. How much had actually been deposited?
   c. How much interest was earned?

4. Kara Marino deposited $300 each quarter for 4 years into her savings account that paid 4% compounded quarterly.
   a. What was the value of her account after 4 years?
   b. How much did Ms. Marino invest?
   c. How much interest did she earn?

5. Paul Smith Mattress Co. invested $500 each month into an account that earned 6% compounded monthly.
   a. How much will the company's account be worth in 5 years?
   b. How much of the total will the company have deposited?
   c. How much will be interest?

6. Adams Aquamarine Sales, Inc. invested $1,000 semiannually into an account that earned 7% compounded semiannually.

a. How much was the company's account worth after 8 years?

b. How much of this amount was invested by the company?

c. How much interest was earned on the account?

7. Strickland Printing Co. invested $6,000 each year for 5 years into an account earning 8% compounded annually.

a. How much will the company invest in 5 years?

b. What will the maturity value of the annuity be?

c. How much interest will the annuity earn?

8. Kelly Carpets deposited $5,000 each year for 10 years into an account earning 6% annually.

a. How much did the business invest?

b. What was the maturity value of this account?

c. How much interest was earned during the 10-year period?

9. Jones Telecommunications, Inc. invested $8,000 each quarter for 5 years into an account paying 6% compounded quarterly.

a. How much did the company invest?

b. What will the amount in the account be after 5 years?

c. How much interest will the maturity value include?

10. Booker Brothers, Inc. invested $800 each month into an account earning 5% compounded monthly for 2 years.

a. How much did the business invest?

b. What was the maturity value of the investment?

c. How much interest was earned?

SECTION 2

PRESENT VALUE OF AN ANNUITY

Observe that our study of amount of an annuity involved starting with an empty account and making payments into it so that the account contained its largest amount at the end of the term. The study of present value of an annuity is exactly the reverse: The account contains its largest balance at the beginning of the term, and someone receives payments from the account until it is empty. This balance which an account must contain at the beginning of the term is the **present value** or **present worth of an annuity.**

Rather than studying the actual formula for present value of an annuity, we will use an informal procedure as we did for amount of an annuity. By consulting Table 21, Present Value of Annuity, Tables Booklet, present value can be computed as follows:

$$\text{Present value} = \text{Payment} \times \text{Present value of annuity table}_{\overline{n|}\, i}$$

where, as before, n = number of payments and i = interest rate per period. The procedure* can be abbreviated

$$\text{P.V.} = \text{Pmt.} \times \text{P.V. ann. tab.}_{\overline{n}|i}$$

The total amount to be received from an annuity is found by multiplying the payment times the number of periods. As long as there are still funds in the account, it will continue to earn interest; thus, even though the balance of the account is declining during the term of the annuity, the account will still continue to earn some interest until the final payment is received. The total interest that the account will earn is found by subtracting the beginning balance (the present value) from the total payments to be received.

Example 1 Elaine Shaw wishes to receive a $100 annuity each quarter for 6 years while attending college and graduate school. Her account earns 8% compounded quarterly. (a) What must the (present) value of Elaine's account be when she starts college? (b) What total amount will Elaine actually receive? (c) How much interest will these annuity payments include?

(a) Pmt. = $100 $\quad$ $\text{P.V.} = \text{Pmt.} \times \text{P.V. ann. tab.}_{\overline{n}|i}$

$n = 24$ $\quad$ $= \$100 \times \text{P.V. ann. tab.}_{\overline{24}|2\%}$

$i = 2\%$ $\quad$ $= 100(18.91393)$

$= 1{,}891.393$

$\text{P.V.} = \$1{,}891.39$

Elaine's account must contain $1,891.39 when she enters college.

(b) She will receive 24 annuity payments of $100 each, for a total of $2,400.

(c) The account will earn interest of

Total payments	$2,400.00
Present value	− 1,891.39
Interest	$ 508.61

Thus, from an account containing $1,891.39, Elaine may withdraw $100 each quarter until a total of $2,400 has been withdrawn. During this time, the declining fund will have earned $508.61 interest. After the final payment is received, the balance in her account will be exactly $0.

*The procedure for present value of an annuity is often indicated as $A = Pa_{\overline{n}|i}$, where A = present value, P = payment, and $a_{\overline{n}|i}$ indicates use of the present-value-of-annuity table.

Example 2 Grover Campbell would like to receive an annuity of $5,000 semiannually for 10 years after he retires; he will retire in 18 years. Money is worth 6% compounded semiannually. (a) How much must Campbell have when he retires in order to finance this annuity? (b) What single deposit made now would provide the funds for the annuity? (c) How much will Campbell actually receive in payments from the annuity? (d) How much interest will the single deposit earn before the annuity ends?

This is a two-part problem: (1) to find the beginning balance (present value) required for the 10-year annuity, and (2) to find what single deposit made now (present value at compound interest) would produce a maturity value equal to the answer obtained in step 1.

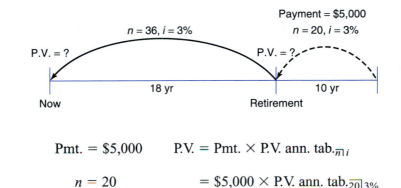

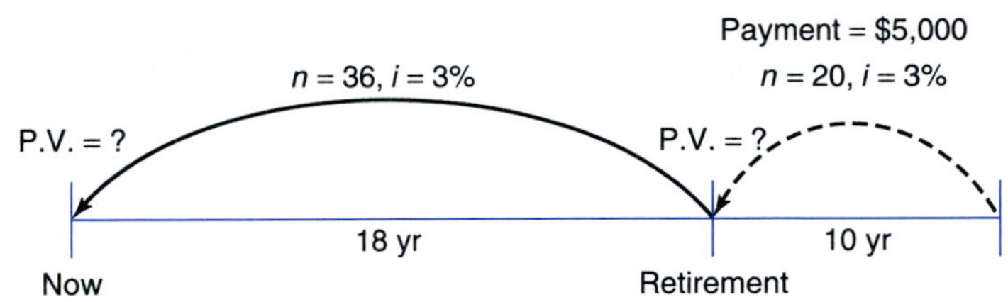

(a) Pmt. = $5,000 $\quad$ P.V. = Pmt. × P.V. ann. tab.$_{\overline{n}|i}$

$n = 20$ $\quad$ = $5,000 × P.V. ann. tab.$_{\overline{20}|3\%}$

$i = 3\%$ $\quad$ = 5,000(14.877475)

= 74,387.375

P.V. = $74,387.38

The account must contain $74,387.38 at Campbell's retirement, in order for him to receive $5,000 each 6 months for 10 years.

(b) We next find what single deposit, made now, would reach a maturity value of $74,387.38 when Campbell retires in 18 years.

$M = \$74{,}387.38 \quad P = M(1 + i)^{-n}$

$n = 36 \quad = 74{,}387.38(1 + 3\%)^{-36}$

$i = 3\% \quad = 74{,}387.38(0.3450324)$

$= 25{,}666.056$

$P = \$25{,}666.06$

A single deposit of \$25,666.06 now will be worth \$74,387.38 after 18 years at 6% compounded semiannually. Notice that $i = 0.3450324251$ was rounded to 7 decimal places in accordance with the Calculator Tip on p. 489.

(c) Campbell will receive 20 payments of \$5,000 each, for a total of \$100,000 during the term of his annuity.

(d) The total interest is the difference between what he will actually receive from the annuity (\$100,000) and the original amount invested (\$25,666.06):

Total received	\$100,000.00
Principal invested	− 25,666.06
Interest	\$ 74,333.94

The original investment of \$25,666.06 would produce interest of \$74,333.94 before the annuity expires. (Campbell will receive nearly 4 times the original amount invested.)

Equivalent Cash Price. Keep in mind that "present value" often refers to the value of an investment on the first day of the term. (It always refers to the value on some day prior to the maturity date.) This fact will help clarify the following example.

Example 3 A homebuyer made a down payment of \$20,000 and will make payments of \$7,500 semiannually for 25 years. Money is worth 7% compounded semiannually. (a) What would the equivalent cash price of the house have been? (b) How much will the buyer actually pay for the house?

(a) The "cash price" would have been the cost on the original day. This value on the beginning day of the term indicates that present value is required.

Cash price = Down payment + Present value of periodic payments

$$\text{Pmt.} = \$7{,}500 \qquad \text{P.V.} = \text{Pmt.} \times \text{P.V. ann. tab.}_{\overline{n}|\,i}$$

$$n = 50 \qquad = \$7{,}500 \times \text{P.V. ann. tab.}_{\overline{50}|\,3\frac{1}{2}\%}$$

$$i = 3\frac{1}{2}\% \qquad = 7{,}500(23.455618)$$

$$= 175{,}917.13$$

$$\text{P.V.} = \$175{,}917.13$$

Thus,

Down payment	\$ 20,000.00
Present value of periodic payments	+ 175,917.13
Cash price	\$195,917.13

(b) The buyer will actually make a down payment of $20,000 plus 50 payments of $7,500 each, for a total cost of $395,000:

Down payment	$ 20,000
Total of periodic payments	+ 375,000
Total cost	$395,000

The equivalent cash price was $195,917.13 because that sum, invested in an account earning 7% semiannual interest, would have the same maturity value ($1,094,182.80) as if the $20,000 down payment and each semiannual payment were deposited at the same rate. Notice that the buyer's total cost ($395,000) will be about 2 times the equivalent cash price. This is not an uncommon ratio for the payment of home loans. This subject is discussed in more detail in Section 2 on "Amortization" in the next chapter.

SECTION 2 PROBLEMS

Hint. Remember that the basic problems of amount and present value at compound interest involve a *single* deposit for some period of time, whereas the annuity problems involve a *series* of payments.

Using the present-value-of-annuity procedure, determine (1) the present value required in order to receive each of the annuities given below, (2) the total amount that each annuity would pay, and (3) how much interest would be included.

	Payment	Rate	Compounded	Years	Present Value	Total Payment	Total Interest
1. a.	$1,000	15%	Semiannually	8			
b.	1,500	9	Monthly	3			
c.	3,000	8	Quarterly	10			
d.	1,700	12	Quarterly	6			
e.	2,600	6	Semiannually	5			
2. a.	$1,000	8%	Semiannually	15			
b.	1,600	10	Monthly	3			
c.	2,400	9	Quarterly	8			
d.	3,500	7	Quarterly	2			
e.	1,200	5	Semiannually	6			

(1) Calculate the amount of an annuity (as in Section 1) and the interest included if one makes deposits as given below. Also, (2) determine the present value (as in Section 2) required in order to receive the given annuity payments, and determine the total interest that would be included in the annuity.

	PAYMENT	RATE	COMPOUNDED	YEARS	AMT. OF ANNUITY	TOTAL INTEREST	PRESENT VALUE OF ANNUITY	TOTAL INTEREST
3. a.	$6,000	8%	Quarterly	8				
b.	300	7	Semiannually	12				
c.	4,900	6	Monthly	4				
4. a.	$5,000	10%	Quarterly	7				
b.	8,000	9	Semiannually	10				
c.	900	7	Monthly	5				

5. Bill Jefferson wishes to purchase an annuity that will pay $450 quarterly for 3 years. If the current interest rate is 6% compounded quarterly,

a. How much would be required to finance the annuity?

b. What is the total amount Jefferson will receive from the annuity?

c. How much total interest will the annuity include?

6. Ngroc Tran wants to receive a monthly annuity of $250 while she attends a 1-year college prep course. If the current interest rate is 5% compounded monthly,

a. What amount is required in her account to finance the annuity?

b. What is the total amount she will receive from the annuity?

c. How much interest will her investment earn?

7. Sam Wallace wants to receive an annuity of $600 per quarter for a 2-year period. If his investment account pays 5% quarterly,

a. How much must Wallace have in his account to finance the annuity?

b. How much will he receive in total from the annuity?

c. How much interest will his investment earn?

8. Lucia Whelan wants to receive an annuity of $150 each month during a 5-year period. If her investment account pays 5% compounded monthly,

a. How much must Whelan have in her account to finance the annuity?

b. How much will she receive in total from the annuity?

c. How much interest will her investment earn?

In Problems 9–12, round your solution to each part to the nearest $100 *for use in succeeding parts.*

9. Eric Walton wants to receive a $700 annuity semiannually for 4 years after he retires. He can invest his money at 8% compounded semiannually.

a. What amount must be on deposit when he starts receiving the annuity?

b. Walton will retire in 10 years. What single deposit today must be made to provide the funds for the annuity?

c. How much interest will his deposit earn until the annuity ends?

10. Don Murphy wants to receive $2,500 quarterly for 6 years after he retires. He can invest his money at 7% compounded quarterly.

a. What amount must be on deposit when he starts receiving the annuity?

b. Murphy will retire in 10 years. What single deposit must he make today to provide the funds for the annuity?

c. How much interest will his deposit earn until the annuity ends?

11. Jenny Miller's daughter plans to attend college in 6 years. Miller wants to provide an annuity of $350 monthly during her daughter's 4 years in college. Her money can be invested at 6% compounded monthly.

a. What amount must be on deposit when the daughter enters college in order to receive this annuity?

b. How much would Miller have to deposit today in order to provide for the annuity?

c. How much total interest will this deposit earn during the 10-year period?

12. The Richardson Center wants to provide an annuity of $10,000 semiannually for a 3-year period. Money can be invested at 6% compounded semiannually. If the center plans for the annuity to begin in 2 years,

a. What amount must be on deposit at the beginning of the 3-year period?

b. How much would the Richardson Center have to deposit today in order to finance the annuity?

c. How much total interest would the center earn?

13. Cavett Enterprises purchased property for $25,000 down and monthly payments of $3,000 for 6 years.

a. What was the total cost to the business?

b. What cash price deposited at 10% compounded monthly would yield the seller of the property the same maturity value (s)he would have after 6 years of depositing Cavett's payments?

14. Liebert-Negra, Inc. purchased land for $50,000 down and monthly payments of $6,000 for 5 years.

a. What was the total cost of the property?

b. What cash price deposited at 7% compounded monthly would yield the seller of the land the same maturity value (s)he would have after 5 years of depositing Liebert-Negra's payments?

15. A local area network (LAN) system was purchased by making a $5,000 down payment and quarterly payments of $2,000 for 3 years.

a. What was the total cost of the LAN system?

b. What would the equivalent cash price have been, if money were worth 9% compounded quarterly?

16. A business purchased office furniture by making an $8,000 down payment and quarterly payments of $5,000 for 2 years.

a. What was the total cost of the furniture?

b. What would the equivalent cash price have been if money were worth 8% compounded quarterly?

CHAPTER 19 GLOSSARY

Amount (of an annuity). The maturity value of an account after a series of equal (annuity) payments into it.

Annuity. A series of equal payments made at regular intervals of time (either with or without interest).

Annuity certain. An annuity with definite beginning and ending dates.

Annuity due. An annuity paid (or received) at the beginning of each time interval. (Example: Rent or insurance premiums.)

Contingent annuity. An annuity for which the beginning and/or ending dates are uncertain.

Ordinary annuity. An annuity paid (or received) at the end of each time period. (Example: Salaries or stock dividends.)

Present value (of an annuity). The beginning balance an account must contain in order to receive a given annuity from it.

Simple annuity. An annuity in which the date of payment coincides with the conversion date of the compound interest.

CHAPTER 20

SINKING FUNDS AND AMORTIZATION

OBJECTIVES

Upon completion of Chapter 20, you will be able to:

1. Define and use correctly the terminology associated with each topic.
2. Using the procedure Pmt. = $M \times$ S.F. tab.$_{\overline{n}|i}$ and the sinking fund table, determine (Section 1: Example 1; Problems 1, 2, 5–10):
   a. The regular payment required to finance a sinking fund
   b. The total amount deposited
   c. The interest earned.
3. Prepare a sinking fund schedule to verify that the payments (Objective 2) will result in the required maturity value (Section 1: Example 2; Problems 13, 14).
4. Given the quoted price of a bond and its interest rate, determine its (Section 1: Examples 3, 4; Problems 3, 4, 11, 12):
   a. Purchase price
   b. Premium or discount
   c. Annual interest
   d. Current yield.
5. Using the procedure Pmt. = P.V. $\times$ Amtz. tab.$_{\overline{n}|i}$ and the amortization table, compute (Section 2: Examples 1, 2; Problems 1–14):
   a. The periodic payment required to amortize a loan
   b. The total amount paid
   c. The interest included.
6. Prepare an amortization schedule to verify that the payments (Objective 5) pay off the loan correctly (Section 2: Example 3; Problems 15, 16).
7. Explain the characteristics of the six basic types of compound interest and annuity problems (Section 3: Problems 1–24).

Far-sighted investors may wish to establish a savings plan whereby regular deposits will achieve a specified savings goal by a certain time. An account of this type is known as a **sinking fund.**

In the loan plans studied in previous chapters, either simple interest or simple discount was charged. For large loans that take many years to repay, however, ordinary simple interest would not be profitable to lenders, as later examples will show. Thus, long-term loans must be repaid in a series of payments that include all interest due since the previous payment. This loan repayment procedure is known as **amortization.**

Sinking funds and amortization are the subjects of the following sections; both topics involve finding a required periodic payment (or annuity).

SECTION 1

SINKING FUNDS AND BONDS

Section 1 of Chapter 19, on the amount of an annuity, dealt with making given periodic payments into an account (at compound interest) and finding what the maturity value of that account would be. It frequently happens that businesses know what amount will be needed on some future date and are concerned with determining what periodic payment (annuity) would have to be invested in order to obtain this amount.

When a special account is established so that the maturity value (equal periodic deposits plus compound interest) will exactly equal a specific amount, such a fund is called a **sinking fund.** Sinking funds are often used to finance the replacement of machinery, equipment, facilities, and so forth, or to finance the redemption of bonds.

Bonds are somewhat similar to promissory notes in that they are written promises to repay a specified debt on a certain date. Both notes and bonds earn simple interest. The primary difference is that interest on bonds is typically paid periodically, whereas the interest on a note is all repaid on the maturity date. Large corporations, municipalities, and state governments usually finance long-term improvements by selling bonds. (A term of 10 years or more is typical.) We shall be concerned first with setting up a sinking fund to provide the *face value* (or **par value** or redemption value) of the bonds at maturity, as well as sinking funds to provide significant maturity values for other purposes.

SINKING FUND PROCEDURES

The objective of a sinking fund problem, then, is to determine what periodic payment, invested at compound interest, will produce a given maturity value. Since amount-of-annuity problems also involve building a fund that would contain its largest amount at maturity, we can see that sinking fund problems are a variation of amount-of-annuity problems. The difference is that in this case we are concerned with finding the periodic payment of the annuity rather than the maturity value. (The periodic payment to a sinking fund is often called the **rent.**)

The period payment to a sinking fund can be found using the amount-of-annuity procedure $M = \text{Pmt.} \times \text{Amt. ann. tab.}_{\overline{n}|i}$ and solving for Pmt. Since use of this procedure would involve long and tedious division, however, you will normally find the periodic payment by using Table 19, Sinking Fund, Tables Booklet, and the following procedure:

$$\text{Payment} = \text{Maturity value} \times \text{Sinking fund table}_{\overline{n}|i}$$

where n = number of payments and i = interest rate per period. This procedure* may be abbreviated

$$\text{Pmt.} = M \times \text{S.F. tab.}_{\overline{n}|i}$$

Example 1 Brandon County issued bonds totaling \$1,000,000 in order to build an addition to the courthouse. The county commissioners set up a sinking fund at 9% compounded quarterly in order to redeem the 5-year bonds. (a) What quarterly rent must be deposited to the sinking fund? (b) How much of the maturity value will be deposits? (c) How much interest will the sinking fund earn?

(a) $M = \$1{,}000{,}000$ $\text{Pmt.} = M \times \text{S.F. tab.}_{\overline{n}|i}$

$n = 20$ $= \$1{,}000{,}000 \times \text{S.F. tab.}_{\overline{20}|2\frac{1}{4}\%}$

$i = 2\frac{1}{4}\%$ $= 1{,}000{,}000(0.04014207)$

$\text{Pmt.} = \$40{,}142.07$

A payment of \$40,142.07 must be deposited into the sinking fund quarterly in order to have \$1,000,000 at maturity.

(b) There will be 20 deposits of \$40,142.07 each, making the total deposits 20 × \$40,142.07, or \$802,841.40.

(c) The interest earned is thus

Maturity value	\$1,000,000.00
Total deposits	− 802,841.40
Interest	\$ 197,158.60

*The procedure for finding the payment to a sinking fund is often given as $P = M \times \frac{1}{m_{\overline{n}|i}}$, where P = payment, M = maturity value, and $\frac{1}{m_{\overline{n}|i}}$ indicates use of the sinking fund table. The procedure $P = M \times \frac{1}{m_{\overline{n}|i}}$ is a variation of the amount-of-annuity procedure $M = Pm_{\overline{n}|i}$.

When a sinking fund is in progress, businesses often keep a **sinking fund schedule.** The schedule shows how much interest the fund has earned during each period and what the current balance is. The schedule also verifies that the periodic payments will result in the desired maturity value.

Example 2 Prepare a sinking fund schedule to show that semiannual payments of $949 for 2 years will amount to $4,000 when invested at 7% compounded semiannually.

The periodic "interest" earned is found by multiplying the previous "balance at end of period" by the periodic interest rate i (example: $\$949 \times 3\frac{1}{2}\% = \$33.22$).

PAYMENT	PERIODIC INTEREST ($i = 3\frac{1}{2}\%$)	PERIODIC PAYMENT	TOTAL INCREASE	BALANCE AT END OF PERIOD
1	$ 0	$ 949.00	$ 949.00	$ 949.00
2	33.22	949.00	982.22	1,931.22
3	67.59	949.00	1,016.59	2,947.81
4	103.17	949.00	1,052.17	3,999.98
Totals	$203.98	$3,796.00		

$3,999.98 Final balance

Notice that the final balance in the sinking fund will be $0.02 less than the $4,000 needed, because of rounding.

BOND PROCEDURES

As described at the beginning of this topic, bonds are similar to promissory notes. The issuers of bonds (usually governmental bodies or large corporations) must pay periodic interest on the face (par) value at a rate specified on the bonds. For *registered bonds,* the bond owners are recorded, and checks are mailed directly. For *coupon bonds,* the owner collects the interest by clipping an attached coupon and submitting it (to the company or specified trustee) on the indicated date. A registered bond is shown in Figure 20-1.

If the Brandon County bonds in Example 1 pay 7% semiannual interest, then each 6 months the county must pay the bond owners interest that collectively totals (by $I = Prt$)

$$\$1,000,000 \times \frac{7}{100} \times \frac{1}{2} = \$35,000 \quad \text{Semiannual interest}$$

This $35,000 interest (the total for all bonds) is in addition to the semiannual payment Brandon County makes to the sinking fund in order to redeem the bonds at maturity.

Most bonds are issued with a par (redemption) value of $1,000 each. Buyers are willing to pay more or less than this face value, depending on the issuer's financial

REGISTERED REGISTERED

NOT MORE THAN 1,000,000 DOLLARS

NUMBER R 00000

DOLLARS $1,000.00

9% NOTE DUE 2010

SEE REVERSE FOR CERTAIN DEFINITIONS

CUSIP 459200 AF 8

INTERNATIONAL BUSINESS MACHINES CORPORATION

International Business Machines Corporation, a corporation duly organized and existing under the laws of the State of New York (therein called the "Company," which term includes any successor corporation under the Indenture referred to on the reverse hereof), for value received, hereby promises to pay to

John B. Vernon

, or registered assigns the principal sum of One thousand and no/00 DOLLARS

at the office or agency of the Company in the Borough of Manhattan, The City and State of New York, on May 1, 1998, in such coin or currency of the United States of America as at the time of payment shall be legal tender for the payment of public and private debts, and to pay interest, semi-annually on May 1 and November 1 of each year, on said principal sum at said office or agency, in like coin or currency, at the rate of 9% per annum, from the May 1 or the November 1, as the case may be, next preceding the date of this Note to which interest has been paid, unless the date hereof is a date to which interest has been paid, in which case from the date of this Note, or unless no interest has been paid on the Notes, in which case from May 1, 1988, until payment of said principal sum has been made or duly provided for. Notwithstanding the foregoing, if the date hereof is after April 15 or October 15, as the case may be, and before the following May 1 or November 1, this Note shall bear interest from such May 1 or November 1; provided, however, that if the Company shall default in the payment of interest due on such May 1 or November 1, then this Note shall bear interest from the next preceding May 1 or November 1 to which interest has been paid, or, if no interest has been paid on the Notes, from May 1, 1988. The interest so payable on any May 1 or November 1 will be paid to the person in whose name this Note is registered at the close of business on such April 15 or October 15, as the case may be, next preceding such May 1 or November 1, unless the Company shall default in the payment of interest due on such interest payment date, in which case such defaulted interest shall be paid to the person in whose name this Note is registered at the close of business on a special record date for the payment of such defaulted interest established by notice to the registered holders of Notes not less than 15 days preceding such special record date. Payment of interest may, at the option of the Company, be made by check mailed to the registered address of the person entitled thereto.

Reference is made to the further provisions of this Note set forth on the reverse hereof, which shall have the same effect as though fully set forth at this place.

Unless the certificate of authentication hereon has been executed by the Trustee by manual signature, this Note shall not be entitled to any benefit under said Indenture, or be valid or obligatory for any purpose.

In Witness Whereof, the Company has caused this instrument to be duly executed under its corporate seal.

Dated

International Business Machines Corporation

TRUSTEE'S CERTIFICATE OF AUTHENTICATION

THIS IS ONE OF THE NOTES REFERRED TO IN THE WITHIN-MENTIONED INDENTURE

THE BANK OF NEW YORK, AS TRUSTEE

BY SPECIMEN

AUTHORIZED SIGNATORY

©IBM

ATTEST: BY

SECRETARY CHAIRMAN OF THE BOARD

Courtesy of International Business Machines Corporation

Figure 20-1 Registered Bond

condition and whether the bond's interest rate is higher or lower than the prevailing market rate. Thus, if a financially sound company issues bonds at a higher-than-normal interest rate, the bonds may sell at a **premium** (more than face value). Low-interest bonds would probably sell at a **discount** (less than redemption value). If sold for face value, the bonds are sold at **par.** Generally, as interest rates increase, bond prices decrease, and as interest rates decrease, bond prices increase.

Bonds are listed on a sales exchange by company name, interest rate, and redemption date, as shown in the illustration below. For instance, "IBM 6⅜03" indicates that IBM bonds pay 6⅜% interest and mature in 2003. The optional "s" in "SouBell 6s04" is merely a plural form of reference. The initials "cv" in "PacSci 7¾03cv" indicate that these bonds are convertible to common stock under certain terms. The current yield of "6.5" is read as 6.5%. The volume indicates the number of bonds sold per $1,000 of face value. The "328" thus stands for $328,000. Bonds are quoted as a percentage of face value, which also indicates whether they are selling at a premium or a discount. Thus, the close figure "98½" or $985 is the bond price at the close of the previous business day. The net change, "⅜" or $3.75, is the change in the price of the bond from the previous trading day. The "vj" preceding a bond name indicates that trading has been suspended by the Securities and Exchange Commission, and that the company is in bankruptcy, receivership, or being reorganized under the Bankruptcy Act.

As this book went to press, the exchanges were listing bonds as whole numbers and fractions. The directors of the New York Stock Exchange voted in 1997 to trade stocks and bonds in dollars and cents (decimals instead of fractions) by the year 2000. This ends a 200-year tradition of quoting prices in fractions. All other major foreign stock markets, currency markets, and U.S. mutual funds use the decimal system.

CORPORATION BONDS

BONDS	CURRENT YIELD	VOLUME	CLOSE	NET CHANGE
vjColuG $10\frac{1}{4}$11	. . .	50	$145\frac{1}{8}$	. . .
IBM $6\frac{3}{8}$03	6.5	328	$98\frac{1}{2}$	$+\frac{3}{8}$
Kroger 9s02	8.8	90	$101\frac{7}{8}$	. . .
MobilCp $8\frac{3}{8}$01	7.8	35	108	+1
PacSci $7\frac{3}{4}$03cv	7.2	5	108	−1
Revlon $10\frac{7}{8}$10	11.0	474	$98\frac{1}{2}$	$-1\frac{1}{2}$
SouBell 6s04	6.5	6	$92\frac{3}{4}$	$+1\frac{3}{4}$

Example 3 Three $1,000 bonds are quoted as shown, producing the indicated selling price and premium (or discount):

QUOTE	FACE VALUE	SELLING PRICE	PREMIUM/DISCOUNT
104	$1,000	104% × $1,000 = $1,040	$40 premium
97	1,000	97% × 1,000 = 970	30 discount
$108\frac{1}{2}$	1,000	$108\frac{1}{2}$% × 1,000 = 1,085	85 premium

The buyer of a bond normally also pays the previous owner for interest accrued (earned) since the last interest payment. This amount will be reimbursed to the new owner, who will receive full interest at the end of the period. A small (insignificant for our purposes) brokerage commission is also added to a bond's quoted price when the bond is purchased.

Bond purchases are made to provide income for the investor. Certain government bonds pay tax-free income, which makes these bonds desirable even at reduced interest rates. When bonds sell at a discount or a premium, the stated interest rate will differ from the **current yield** (the rate of interest based on the amount actually paid).

Example 4 Two $1,000 bonds each pay 6% simple interest annually. The quoted price of bond A was 98, and of bond B 114. Find the current annual yield on each.

Annual interest on each bond, by $I = Prt$, is

$$\$1{,}000 \times 0.06 \times 1 = \$60$$

Bond A cost $980 and bond B was $1,140. Thus,

$$\text{Current yield} = \frac{\text{Annual interest}}{\text{Market or purchase price}}$$

$$\text{Bond A} = \frac{\$60}{\$980} = 6.12\% \text{ current yield}$$

$$\text{Bond B} = \frac{\$60}{\$1{,}140} = 5.26\% \text{ current yield}$$

SECTION 1 PROBLEMS

Using the sinking fund table, determine the periodic payment necessary to finance each sinking fund. Also calculate the total deposits and the total amount of interest contained in the maturity value.

	MATURITY VALUE	RATE	COMPOUNDED	YEARS
1. a.	$ 30,000	7%	Quarterly	5
b.	50,000	8	Semiannually	6
c.	75,000	6	Monthly	7
d.	100,000	5	Quarterly	4
e.	200,000	9	Monthly	3
2. a.	$ 80,000	8%	Quarterly	10
b.	500,000	9	Semiannually	8
c.	275,000	10	Monthly	5
d.	60,000	7	Quarterly	12
e.	140,000	12	Monthly	2

For each bond below, assume that the face value is $1,000. Compute the additional information related to each. Express current yield as a percent correct to 2 decimal places.

	PRICE QUOTE	PURCHASE PRICE	PREMIUM (P) OR DISCOUNT (D)	INTEREST RATE	ANNUAL INTEREST	CURRENT YIELD
3. a.	101			$6\frac{1}{4}$%		
b.	88			$5\frac{1}{8}$		
c.	100			$9\frac{3}{4}$		
d.	$95\frac{1}{2}$			$8\frac{1}{2}$		
e.	$106\frac{1}{4}$			10		
4. a.	106			$5\frac{1}{2}$%		
b.	94			5		
c.	100			$7\frac{3}{4}$		
d.	$85\frac{1}{2}$			$6\frac{1}{4}$		
e.	$104\frac{1}{4}$			8		

5. The Osteopathic Medical Center plans to remodel its offices in 4 years. The directors expect to need $75,000.
   a. What quarterly payment must be invested in a sinking fund invested at 9% compounded quarterly?
   b. What part of the maturity value will be deposits?
   c. How much interest will the fund draw?

6. Griffin Publishing plans to spend $250,000 in 2 years for new equipment. To finance the purchase, the firm will establish a sinking fund at 7% compounded quarterly.
   a. What periodic payment must Griffin make to the sinking fund each quarter?
   b. What will the total of these payments be?
   c. How much interest will the fund contain at maturity?

7. The First Edition Book Store wants $25,000 to upgrade its computer system in 2 years. The store can set up a sinking fund at 7% compounded semiannually.
   a. What semiannual payment must be made to finance the fund?
   b. What total amount will the store invest?
   c. How much interest will the account accumulate?

8. Goldstein Center plans to renovate its facilities in 5 years at a cost of $1,000,000. It can set up a sinking fund at 6% compounded semiannually.
   a. What periodic payment will be made?
   b. What will the total of these payments be?
   c. What will the total amount of interest earned be?

9. The city of Boise sold bonds totaling $1,000,000 to finance an addition to a recreation center. The bonds mature in 8 years. To provide for redemption of the bonds, a sinking fund was established at 9% compounded monthly.
   a. What will the monthly payment be to the fund?
   b. How much will the city invest?
   c. How much interest will the mature fund include?

10. The city of Plymouth sold bonds totaling $1,000,000 to finance the building of a new school. The bonds will mature in 7 years. If a sinking fund is established earning 8% compounded monthly,

a. How much must be deposited each month to finance the fund?

b. How much will the city invest?

c. How much of the maturity value will be interest?

11. Austin Enterprises, Inc. sold 500 bonds of $1,000 par value at 104. The bonds pay 8¼% interest annually to the bondholders.

a. What was the cost of each bond?

b. What was the premium or discount?

c. What total amount did Austin receive from the sale?

d. How much interest is paid on each bond each year?

e. What is the current yield for each bond?

f. How much total interest must the corporation pay to all bondholders each year?

12. Montgomery Software sold 5,000 bonds of $1,000 par value at 96. The bonds pay 10.25% interest annually to the bondholder.

a. How much did each bond cost?

b. What was the premium or discount?

c. What total amount did Montgomery receive from the sale?

d. How much interest is paid on each bond each year?

e. What is the current yield to the bondholder?

f. How much total interest must Montgomery pay to all bondholders each year?

Find the periodic payment necessary to finance each of the following sinking funds. Then prepare a sinking fund schedule to verify that these periodic payments will result in the desired maturity value. Directions for using an Excel® spreadsheet are given below. These problems are on the student data disk that accompanies this text.

SPREADSHEET DIRECTIONS FOR PROBLEM 13a.		✓
1.	Review the terminology on page 84. Place a check mark in the last column when each task is accomplished.	
2.	Start the Excel software program. Insert the student data disk in *Drive A*.	
3.	Click the *File* button; then click the *Open* button.	
4.	Change *Look-in:* to 3½″ Floppy (A:). Open the *Chapter 20* folder. Open *Problem 13a.*	
5.	Click in Cell B4 and key "0"; tab to C4 and key "932.08"; tab to D4 and key "932.08"; tab to E4 and key "932.08."	
6.	Click Cell C4 which selects the cell. Use the fill handle of C4 and drag down through C11. This copies the periodic payment to each of these cells.	
7.	Click Cell B5 and key "=E4*$B$2." Tab. This formula multiplies the rate times the balance at the end of the period.	

	SPREADSHEET DIRECTIONS FOR PROBLEM 13a.	✓
8.	Click Cell D5 and key "=B5+C5." Tab. This formula adds the periodic interest to the periodic payment.	
9.	Click Cell E5 and key "=D5+E4." Tab or Enter.	
10.	To copy the formula in B5 to the range B6:B11, use the fill handle of B5. Click B5, and drag the fill handle down through B11. Temporarily, only Cell B6 will contain a valid value. The other cells in Column B will change as values in other columns change. Note: The colon (:) is used in the expression B6:B11 to specify a range of cells.	
11.	Click Cell D5. Use the fill handle to drag the formula down through Cell D11. Only Cell D6 contains a valid value, but the other values will change as the other cells are completed.	
12.	Click Cell E5. Use the fill handle to drag the formula down through Cell E11. Now, cells in columns B, D, and E should be completed.	

	SPREADSHEET DIRECTIONS FOR PROBLEM 13b.	✓
1.	Review the first four steps of Problem 13a. Place a check mark in the last column when each task is accomplished.	
2.	Click Cell B4. Key "0." Tab.	
3.	In Cell C4, key "3547.92." Repeat this number in Cells D4 and E4.	
4.	Click Cell C4. Use the fill handle and drag down through Cell C8.	
5.	Click Cell B5. Key "=E4*$B$2." Tab twice.	
6.	In Cell D5, key "=B5+C5." Tab.	
7.	In Cell E5, key "=E4+D5." Enter.	
8.	Click Cell B5 and drag the fill handle down through Cell B8. Only Cell B6 will display a valid value at this time.	
9.	Click Cell D5 and drag the fill handle down through Cell D8.	
10.	Click Cell E5 and drag the fill handle down through Cell E8. All cells should display valid values.	

	SPREADSHEET DIRECTIONS FOR PROBLEM 14a.	✓
1.	Review the first four steps of Problem 13a. Place a check mark in the last column when each task is accomplished.	
2.	Click Cell B4 and key "0." Tab.	
3.	Click Cell C4 and key "2408.18." Repeat this number in Cells D4 and E4.	
4.	Click Cell C4. Use the fill handle and drag down through C7.	

SPREADSHEET DIRECTIONS FOR PROBLEM 14a.		✓
5.	Click Cell B5. Key "=E4*$B$2." Tab twice.	
6.	In Cell D5, key "=B5+C5." Tab.	
7.	In Cell E5, key "=E4+D5." Enter.	
8.	Click Cell B5 and drag the fill handle down through B7. Only Cell B6 will display a valid value at this time.	
9.	Click Cell D5 and drag the fill handle down through D7.	
10.	Click Cell E5 and drag the fill handle down through E7.	

SPREADSHEET DIRECTIONS FOR PROBLEM 14b.		✓
1.	Review the first four steps of Problem 13a. Place a check mark in the last column when each task is accomplished.	
2.	Click Cell B4 and key "0." Tab.	
3.	In Cell C4, key "8387.75." Repeat this number in Cells D4 and E4.	
4.	Click Cell C4. Use the fill handle and drag down through C9.	
5.	Click Cell B5 and key "=E4*$B$2." Tab twice to Cell D5.	
6.	In Cell D5, key "=B5+C5." Tab.	
7.	In Cell E5, key "=E4+D5." Enter.	
8.	Click Cell B5. Use the fill handle and drag down through B9. Only Cell B6 will display a valid value at this time.	
9.	Click Cell D5. Use the fill handle and drag down through D9.	
10.	Click Cell E5. Use the fill handle and drag down through E9. All cells should display valid values.	

	MATURITY VALUE	RATE	COMPOUNDED	YEARS
13. a.	$ 8,000	8%	Quarterly	2
b.	20,000	6	Annually	5
14. a.	$10,000	5%	Semiannually	2
b.	60,000	7	Annually	6

SECTION 2

AMORTIZATION

BANK LOANS

In our study of simple interest, it was pointed out that simple interest notes are usually for short periods of time—a year or less. Example 1 in Section 1 of Chapter 18 (p. 466) was given to illustrate the advantage to the depositor of compound interest over simple interest. The same example can be used to illustrate that simple interest notes for long periods are not profitable to bankers. Suppose that the investors in that example are replaced by bankers who are each lending \$1,000. Banker A lends \$1,000 at 7% simple interest for 3 years, for which he would receive \$210 interest.

Banker B, on the other hand, lends \$1,000 at 7% for only 6 months. When the borrower repays the loan after 6 months, banker B then lends both the principal and the interest for another 6 months. This process is repeated for 3 years; so, in effect, banker B earns compound interest on the \$1,000 she originally lent.

That is, banker B would earn \$229.26 total interest, or \$19.26 more than banker A. You can easily understand that banks would prefer to invest their money in successive short-term loans rather than tying it up in single long-term loans. If carried to its logical conclusion, the result would be that no one could borrow money for long-term projects such as building a home or starting a business.

Therefore, a borrower who receives a large bank loan makes periodic, partial payments. That is, at regular intervals, payments are made that include both a payment to reduce the principal and also the interest due on the principal still owed. This procedure is the same as paying on the installment plan at an annual percentage rate required by the Truth-in-Lending Law (that is, an effective interest rate—interest paid only on the balance due). At the same time, it enables the bank, in effect, to earn compound interest (by reloaning the interest just repaid).

Note. It should be emphasized that the interest rates associated with amortization are effective rates; that is, the interest rate each period is applied to the outstanding balance only. To put it another way: The borrower receives credit for the principal that has been repaid, and interest is paid only on the amount still owed.

Since the bulk of the debt exists at the beginning of the time period, and the problem is to determine what periodic payment is necessary to discharge the debt, amortization is thus a variation of the present-value-of-annuity problem already studied. The debt itself is the present (beginning) value of the annuity. The payment may be found by using the present-value-of-annuity procedure, P.V. = Pmt. × P.V. ann. tab.$_{\overline{n}|i}$, and solving for Pmt.

This procedure, however, necessitates dividing by the number from the present-value-of-annuity table. To simplify the calculations, financial institutions usually use Table 22, Amortization, Tables Booklet, which is an amortization table for regular payments scheduled each period. This table can be used in the following procedure:

$$\text{Payment} = \text{Present value} \times \text{Amortization table}_{\overline{n}|i}$$

where, as before, n = number of periods and i = interest rate per period. The procedure* may be abbreviated

$$\text{Pmt.} = \text{P.V.} \times \text{Amtz. tab.}_{\overline{n}|i}$$

The total amount paid to amortize a loan is found by multiplying the periodic payment by the number of payments. The total amount of interest is then the difference between the total paid and the original loan (present value).

Example 1 Marge and Gary Thomas bought a \$30,000 residential lot by making a \$3,000 down payment and making equal semiannual payments for 4 years. (a) How much was each payment if they had a 9% loan? (b) What was the total of their payments? (c) How much interest did the payments include? (d) How much did they pay altogether for the lot?

(a) Following the \$3,000 down payment, the principal (or present value) of the loan was \$27,000.

$$\begin{aligned} \text{P.V.} = \$27{,}000 \qquad \text{Pmt.} &= \text{P.V.} \times \text{Amtz. tab.}_{\overline{n}|i} \\ n = 8 \qquad\qquad &= \$27{,}000 \times \text{Amtz. tab.}_{\overline{8}|4\frac{1}{2}\%} \\ i = 4\tfrac{1}{2}\% \qquad\qquad &= \$27{,}000(0.1516097) \\ \text{Pmt.} &= \$4{,}093.46 \end{aligned}$$

A payment of \$4,093.46 each 6 months would repay both the principal and the interest due on a \$27,000 loan.

(b) There were 8 payments of \$4,093.46 each, making a total (principal plus interest) of \$32,747.68.

(c) Interest would be computed each 6 months on the balance still owed (balance times $4\frac{1}{2}\%$). The total interest would be

Total payments	\$32,747.68
Principal	− 27,000.00
Interest	\$ 5,747.68

*This procedure may also be indicated by $P = A \times \frac{1}{a_{\overline{n}|i}}$, where P = payment, A = present value, and $\frac{1}{a_{\overline{n}|i}}$ denotes use of the amortization table. (This table is often entitled "Annuity Whose Present Value is 1.") The form $P = A \times \frac{1}{a_{\overline{n}|i}}$ is a variation of the present-value-of-annuity procedure $A = Pa_{\overline{n}|i}$.

(d) The total cost of the lot, including the down payment, was thus

Total cost of loan	$32,747.68
Down payment	+ 3,000.00
Total cost of lot	$35,747.68

MORTGAGES

Monthly payments on home mortgages are probably the most familiar amortization payment. Since our amortization table (Table 22) contains only 100 periods, it can be used to determine monthly payments only for loans with terms no longer than $8\frac{1}{3}$ years. However, tables similar to the one below are used by real estate agents and mortgage bankers to determine monthly mortgage payments required at various prevailing rates on home loans of 20, 25, or 30 years.

MONTHLY PAYMENT PER $1,000 OF MORTGAGE[a]

RATE	20 YEARS	25 YEARS	30 YEARS
8%	$ 8.37	$7.72	$7.34
$8\frac{1}{4}$	8.53	7.89	7.52
$8\frac{1}{2}$	8.68	8.06	7.69
$8\frac{3}{4}$	8.84	8.23	7.87
9	9.00	8.40	8.05
$9\frac{1}{4}$	9.16	8.57	8.23
$9\frac{1}{2}$	9.33	8.74	8.41
$9\frac{3}{4}$	9.49	8.92	8.60
10	9.66	9.09	8.78
11	10.33	9.81	9.53

[a]Monthly payments including principal and interest.

The following example illustrates the cost of purchasing a home. As in Example 1, these mortgage payments include interest computed only on the balance due.

Example 2 The Townsends purchased a $98,000 home by making an $8,000 down payment and signing a 10% mortgage with monthly payments for 25 years. (a) Determine the Townsends' monthly payment. (b) What was the total of their payments, and how much of this was interest? (c) What was the total cost of their home?

(a) After their $8,000 down payment, the mortgage principal was $90,000. Their payment was thus

$ 9.09	per $1,000 at 10% for 25 years
× 90	
$818.10	monthly payment on $90,000 mortgage

(b) Twelve payments per year for 25 years equal 300 payments required to repay the mortgage. Their total payments and interest included would be

Monthly payment	$ 818.10
	× 300
Total payments	$245,430
Principal	− 90,000
Interest	$155,430

(c) Including their down payment, the Townsends' $98,000 home cost them

Total cost of loan	$245,430
Down payment	+ 8,000
Total cost of home	$253,430

The **fixed-rate mortgage** has been the traditional home mortgage in the United States. This mortgage has a fixed or set rate of interest, such as 8%, for a fixed time period, such as 20, 25, or 30 years. During the late 1970s and early 1980s, a period of rapid increases in interest rates, the fixed-rate mortgage lost popularity to the **adjustable-rate mortgage (ARM)** or **renegotiable-rate mortgage.** Typically, the interest rate on these mortgages was below market rates initially but then rose after a few years to a maximum rate that remained in force thereafter.

By the mid-1980s, interest rates decreased substantially throughout the economy, and fixed-rate mortgages once more became standard on new loans. As interest rates continued to fall through the 1990s, many people who had taken out the adjustable-rate mortgages refinanced their homes at the lower interest rate, thereby reducing their monthly payments.

At the time this edition went to press, conventional fixed-rate mortgages remained popular, and mortgage interest rates were approximately 8% (7.71%) for an average 30-year fixed-rate mortgage.

AMORTIZATION SCHEDULES

Persons making payments to amortize a loan are often given an **amortization schedule,** which is a period-by-period breakdown showing how much of each payment goes toward the principal, how much is interest, the total amount of principal and interest that has been paid, the principal still owed, and so forth. Figure 20-2 illustrates an actual amortization schedule of a $50,000 loan at 10% interest for 10 years, which will be repaid in monthly payments of $633.38. Observe that a total of $76,005.46 will be repaid altogether—the $50,000 principal plus $26,005.46 interest.

Mortgage Amortization Schedule						
Principal	Payment Amount	Periodic Rate	APR	Number of Payments	Total Amount Paid	Total Interest Paid
50000.00	633.38	0.0075	9%	120	76005.60	26005.60

Payment	Principal Owed	Interest	Payment to Principal
1	50000.00	375.00	258.38
2	49741.62	373.06	260.32
3	49481.30	371.11	262.27
4	49219.03	369.14	264.24
5	48954.79	367.16	266.22
6	48688.58	365.16	268.22
7	48420.36	363.15	270.23
8	48150.13	361.13	272.25
9	47877.88	359.08	274.30
10	47603.58	357.03	276.35
11	47327.23	354.95	278.43
12	47048.80	352.87	280.51
22	44147.07	331.10	302.28
23	43844.80	328.84	304.54
24	43540.25	326.55	306.83
25	43233.42	324.25	309.13
37	39366.96	295.25	338.13
38	39028.83	292.72	340.66
39	38688.17	290.16	343.22
40	38344.95	287.59	345.79
41	37999.16	284.99	348.39
42	37650.77	282.38	351.00
60	30913.44	231.85	401.53
61	30511.91	228.84	404.54
62	30107.37	225.81	407.57
63	29699.80	222.75	410.63
64	29289.16	219.67	413.71
65	28875.45	216.57	416.81
66	28458.64	213.44	419.94
67	28038.70	210.29	423.09
68	27615.61	207.12	426.26
69	27189.35	203.92	429.46
70	26759.89	200.70	432.68
101	11722.42	87.92	545.46
102	11176.96	83.83	549.55
103	10627.41	79.71	553.67
104	10073.73	75.55	557.83
105	9515.90	71.37	562.01
106	8953.89	67.15	566.23
107	8387.67	62.91	570.47
108	7817.20	58.63	574.75
109	7242.44	54.32	579.06
113	4900.01	36.75	596.63
114	4303.38	32.28	601.10
115	3702.28	27.77	605.61
116	3096.66	23.22	610.16
117	2486.51	18.65	614.73
118	1871.78	14.04	619.34
119	1252.43	9.39	623.99
120	628.45	4.71	628.67

Figure 20-2 Schedule of Direct Reduction Loan

Example 3 Prepare a simplified amortization schedule for Example 1, which found that a semiannual payment of \$4,093.46 would amortize a \$27,000 loan at 9% for 4 years.

The "interest" due each period is found by multiplying the "principal owed" by the periodic interest rate i. (Example: \$27,000 × $4\frac{1}{2}$% = \$1,215.) The next period's principal is found by subtracting the "payment to principal" from the "principal owed" (example: \$27,000 − \$2,878.46 = \$24,121.54).

PAYMENT	PRINCIPAL OWED	INTEREST ($i = 4\frac{1}{2}\%$)	PAYMENT TO PRINCIPAL (\$4,093.46 − INTEREST)
1	\$27,000.00	\$1,215.00	\$ 2,878.46
2	24,121.54	1,085.47	3,007.99
3	21,113.55	950.11	3,143.35
4	17,970.20	808.66	3,284.80
5	14,685.40	660.84	3,432.62
6	11,252.78	506.38	3,587.08
7	7,665.70	344.96	3,748.50
8	3,917.20	176.27	3,917.19
	Totals	\$5,747.69	\$26,999.99
	Total cost of loan	\$32,747.68	

Since interest is rounded to the nearest cent each period, the totals on an amortization schedule like this are often off by a few cents. In actual practice, the final loan payment is usually slightly different from the regular payment in order to have the total amounts come out exact.

The discussion at the beginning of this section pointed out that long-term simple interest notes (with a single payment at maturity) would not be profitable to bankers. We now see that such notes would also be unreasonably expensive to borrowers. Thus, effective interest with periodic payments works to the advantage of both the borrower and the lender. If financial institutions could charge simple interest and also receive periodic payments—as is frequently done on installment purchases and loans (although the Truth-in-Lending Law does require that the equivalent effective annual percentage rate be disclosed)—this would be the most profitable arrangement for the lender; however, this is not done on long-term loans from banking institutions.

EFFECTIVE VS. SIMPLE INTEREST

The advantage to the borrower of effective interest over simple interest is obvious from the following example, which compares 12% simple interest ($I = Prt$) with effective interest at 12% compounded semiannually ($i = 6\%$) on a $25,000 loan for 10, 20, and 30 years.

Effective versus Simple Interest on a $25,000 Loan at 12%

TERM (YEARS)	TOTAL EFFECTIVE INTEREST ($i = 6\%$)	TOTAL SIMPLE INTEREST (12%)	EXTRA INTEREST AT SIMPLE RATE
10	$18,592.20	$30,000	$11,407.80
20	41,461.60	60,000	18,538.40
30	67,813.40	90,000	22,186.60

Observe that on the 10-year loan, there would be over 60% more interest at the simple rate. The percentage difference decreases with longer terms, but on the 30-year loan there would still be 45% more interest at the simple rate. The simple interest alone on the 30-year loan would be 3.6 times as much as the principal that was borrowed. It is easy to see that no one could afford to buy a home or start a business if it were necessary to borrow money at simple interest.

SECTION 2 PROBLEMS

Using the amortization table, 22, find the payment necessary to amortize each loan. Also compute the total amount that will be repaid, and determine how much interest is included.

	PRINCIPAL (P.V.)	RATE	COMPOUNDED (PAID)	TERM (YEARS)
1. a.	$300,000	8%	Semiannually	15
b.	150,000	9	Quarterly	10
c.	80,000	7	Monthly	5
d.	500,000	10	Semiannually	7
e.	20,000	9	Monthly	8
2. a.	$ 60,000	6%	Semiannually	5
b.	900,000	5	Quarterly	12
c.	280,000	8	Monthly	7
d.	550,000	9	Semiannually	8
e.	90,000	5	Monthly	3

Using the table of Monthly Payments per $1,000 of Mortgage (page 518), determine the monthly payment required for each mortgage below. Also find the total amount of those payments and the amount of interest included.

	MORTGAGE PRINCIPAL	RATE	TERMS
3. a.	$200,000	8%	30
b.	300,000	$8\frac{3}{4}$	20
c.	60,000	$9\frac{1}{4}$	25
d.	90,000	9	20
4. a.	$120,000	$8\frac{3}{4}$%	30
b.	75,000	$8\frac{1}{2}$	20
c.	360,000	9	25
d.	500,000	$9\frac{3}{4}$	30

5. Paul Martin borrowed $30,000 to build an addition to his home. He will repay the 9% loan with monthly payments for 3 years.

a. What is Martin's monthly payment?

b. How much total interest will be paid?

6. Gerald Prince borrowed $150,000 in order to start his own business. His loan was made at 8% with monthly payments for 4 years.

a. What is his monthly payment?

b. How much total interest will Prince pay?

7. Wooten Veterinarian Supply signed a $60,000 loan to purchase a new inventory system. The company will make quarterly payments at 8% interest for 7 years.

a. What quarterly payment is necessary to discharge the loan?

b. How much interest will be included in the payments?

8. WDH, Inc. borrowed $1,000,000 to purchase new factory equipment. Quarterly payments will be made for 5 years in order to amortize the 7% loan.

a. What is the amount of each payment?

b. How much total interest will be charged?

9. The Dumas Co. purchased real property for a building site for $145,000. It paid $25,000 down and financed the balance at 8% quarterly for 5 years.

a. What quarterly payment was necessary to repay the loan?

b. How much interest was included in the payments?

c. What was the actual total cost of the property?

10. Mr. and Mrs. Stephenson purchased a $108,000 townhouse by paying $8,000 down and signing a 9% mortgage with quarterly payments for 10 years.

a. How much is each quarterly payment?

b. How much interest will the Stephensons pay?

c. What will the total cost of the townhouse be?

11. A new automated distribution system cost Davis Delivery Co. $175,000. Davis paid $25,000 down and financed the remainder at 7% monthly for 5 years.

a. What monthly payment is required to repay the loan?

b. How much interest will Davis pay?

c. What will the total cost of the system be?

12. Plant additions to Scott Technologies, Inc. cost $650,000. The firm paid $20,000 down and financed the balance with monthly payments for 2 years at 8%.

a. What monthly payment was necessary to repay the loan?

b. How much interest was included in the payments?

c. What was the actual total cost to Scott Technologies?

13. Jane and Tom Brown purchased a $245,000 home by making a $15,000 down payment. Their mortgage rate is 9½%.

a. Compare the monthly payment required for a 20-year mortgage versus a 25-year mortgage.

b. How much would each mortgage cost altogether?

c. What would the total cost of the house be in each case?

d. How much more would the house cost with the 25-year mortgage?

14. Dr. and Mrs. Rupert purchased a $415,000 home by making a $40,000 down payment. The interest rate was 8¾%.

a. Compare monthly payments of 25 years to 30 years.

b. What would the total payments be in each case?

c. Including down payment, how much would the house cost altogether in each case?

d. Determine how much extra the Ruperts would pay under the 30-year mortgage.

Find the periodic payment required for each of the following loans. Then verify your answer by preparing an amortization schedule similar to the one in Example 3, showing how much of each payment is interest and how much applies toward the principal. If you wish to use the Excel® software to complete the amortization schedules, directions are given below. The student data disk contains these problems.

SPREADSHEET DIRECTIONS FOR PROBLEM 15a.		✓
1.	Review the terminology on page 84. Place a check mark in the last column when each task is accomplished.	
2.	Start the Excel software program. Insert the student data disk in *Drive A*.	
3.	Open *Chapter 20* by clicking the *File* button and then the *Open* button.	
4.	Change *Look-in:* to 3½″ Floppy (A:). Open the *Chapter 20* folder. Open *Problem 15a.*	
5.	Click Cell C5. Key "5000." Tab twice.	
6.	In Cell E5, key "=C5*$C$3." Tab twice.	
7.	In Cell G5, key "=$B$3-E5." Tab.	
8.	Click Cell C6. Key "=C5-G5." Tab.	

SPREADSHEET DIRECTIONS FOR PROBLEM 15a.		✓
9.	Click Cell C6. Use the fill handle and drag down through C9. None of the cell values is valid at this time.	
10.	Click Cell E5. Use the fill handle and drag down through E9. Again, the cell values are not valid.	
11.	Click Cell G5. Use the fill handle and drag down through G9. All the cells should contain valid values.	
12.	Click Cell C10. Key "=C9-G9." Tab twice.	
13.	In Cell E10, click the *Auto Sum* (Σ) button on the Tool Bar twice. The total interest paid is displayed. Tab twice.	
14.	In Cell G10, click the *Auto Sum* button twice. The total of the payments to the principal is displayed. The totals are off a few cents due to rounding each month.	

SPREADSHEET DIRECTIONS FOR PROBLEM 15b.		✓
1.	Review the terminology on page 84. Place a check mark in the last column when each task is accomplished.	
2.	Start the Excel software program. Insert the student data disk in *Drive A.*	
3.	Open *Chapter 20* by clicking the *File* button and then the *Open* button.	
4.	Change *Look-in:* to 3½″ Floppy (A:). Open the *Chapter 20* folder. Open *Problem 15b.*	
5.	Click Cell C5. Key "6000." Tab twice.	
6.	In Cell E5, key "=C5*$C$3." Tab twice.	
7.	In Cell G5, key "=$B$3-E5." Tab.	
8.	Click Cell C6. Key "=C5-G5." Tab.	
9.	Click Cell C6. Use the fill handle and drag down through C10. None of the cell values is valid at this time.	
10.	Click Cell E5. Use the fill handle and drag down through E10. Again, the cell values are not valid.	
11.	Click Cell G5. Use the fill handle and drag down through G10. All the cells should contain valid values.	
12.	Click Cell C11. Key "=C10-G10." Tab twice.	
13.	In Cell E11, click the *Auto Sum* (Σ) button on the Tool Bar twice. The total interest paid is displayed. Tab twice.	
14.	In Cell G11, click the *Auto Sum* button twice. The total of the payments to the principal is displayed. The totals are off a few cents due to rounding each month.	

Spreadsheet directions for problem 16a.		✔
1.	Review the terminology on page 84. Place a check mark in the last column when each task is accomplished.	
2.	Start the Excel software program. Insert the student data disk in *Drive A*.	
3.	Open *Chapter 20* by clicking the *File* button and then the *Open* button.	
4.	Change *Look-in:* to 3½″ Floppy (A:). Open the *Chapter 20* folder. Open *Problem 16a.*	
5.	Click Cell C5. Key "10000." Tab twice.	
6.	In Cell E5, key "=C5*$C$3." Tab twice.	
7.	In Cell G5, key "=$B$3-E5." Tab.	
8.	Click Cell C6. Key "=C5-G5." Tab.	
9.	Click Cell C6. Use the fill handle and drag down through C12. None of the cell values is valid at this time.	
10.	Click Cell E5. Use the fill handle and drag down through E12. Again, the cell values are not valid.	
11.	Click Cell G5. Use the fill handle and drag down through G12. All the cells should contain valid values.	
12.	Click Cell C13. Key "=C12-G12." Tab twice.	
13.	In Cell E13, click the *Auto Sum* (Σ) button on the Tool Bar. The total interest paid is displayed. Tab twice.	
14.	In Cell G13, click the *Auto Sum* button twice. The total of the payments to the principal is displayed.	

Spreadsheet directions for problem 16b.		✔
1.	Review the terminology on page 84. Place a check mark in the last column when each task is accomplished.	
2.	Start the Excel software program. Insert the student data disk in *Drive A*.	
3.	Open *Chapter 20* by clicking the *File* button and then the *Open* button.	
4.	Change *Look-in:* to 3½″ Floppy (A:). Open the *Chapter 20* folder. Open *Problem 16b.*	
5.	Click Cell C5. Key "12000." Tab twice.	
6.	In Cell E5, key "=C5*$C$3." Tab twice.	
7.	In Cell G5, key "=$B$3-E5." Tab.	
8.	Click Cell C6. Key "=C5-G5." Tab.	
9.	Click Cell C6. Use the fill handle and drag down through C16. None of the cell values is valid at this time.	

SPREADSHEET DIRECTIONS FOR PROBLEM 16b.		✓
10.	Click Cell E5. Use the fill handle and drag down through E16. Again, the cell values are not valid.	
11.	Click Cell G5. Use the fill handle and drag down through G16. All the cells should contain valid values.	
12.	Click Cell C17. Key "=C16-G16." Tab twice.	
13.	In Cell E16, click the *Auto Sum* (Σ) button on the Tool Bar. The total interest paid is displayed. Tab twice.	
14.	In Cell G16, click the *Auto Sum* button twice. The total of the payments to the principal is displayed. The totals are off a few cents due to rounding each month.	

	Principal (P.V.)	Rate	Compounded (Paid)	Term (Years)
15. a.	$ 5,000	7%	Annually	5
b.	6,000	8	Semiannually	3
16. a.	$10,000	6%	Quarterly	2
b.	12,000	8	Monthly	1

SECTION 3 REVIEW

The following criteria may help enable you to distinguish between the basic types of problems studied in Chapters 18 through 20.

1. Determine whether the problem is a compound interest problem or an annuity problem: If a series of *regular payments* is involved, it is an *annuity;* otherwise, it is a compound interest problem.
2. Then decide whether amount or present value is required: In general, if the question in the problem refers to the *end* of the time period, it is an *amount* problem; if the question relates to the *beginning* of the time period, it is a *present value* problem.

Note. This review includes only the basic problems of each type so that students can confidently identify each without its being listed under a section heading. This is not intended to be a complete review of the entire unit; your personal review should also include the variations of the basic problems studied in each section. Table 20-1 identifies each basic type of problem and the corresponding procedure for each.

COMPOUND INTEREST	ANNUITY
The basic problem involves a *single* deposit invested for the entire time.	The problem involves a *series* of *regular* payments (or deposits).
1. *Amount* at compound interest: A single deposit is made, and you wish to know how much it will be worth at the *end* of the time. $M = P(1 + i)^n$	1. *Amount* of an annuity: Regular deposits are *made,* and you wish to know the value of the account at the *end* of the time. $M = \text{Pmt.} \times \text{Amt. ann. tab.}_{\overline{n}\rvert i}$
2. *Present value* at compound interest: You wish to find what single deposit must be invested at the *beginning* of the time period in order to obtain a given maturity value. $P = M(1 + i)^{-n}$	2. *Present value* of an annuity: Regular payments are to be *received,* and you want to find how much must be on deposit at the *beginning* from which to withdraw the payments. $\text{P.V.} = \text{Pmt.} \times \text{P.V. ann. tab.}_{\overline{n}\rvert i}$
	If the problem involves *finding a periodic payment,* one of the following applies:
	3. *Sinking fund:* You are asked to find what regular payment must be made to *build up* an account to a given amount. $\text{Pmt.} = M \times \text{S.F. tab.}_{\overline{n}\rvert i}$
	4. *Amortization:* You are asked to determine what regular payment must be made to *discharge* a debt. $\text{Pmt.} = \text{P.V.} \times \text{Amtz. tab.}_{\overline{n}\rvert i}$

SECTION 3 PROBLEMS

1. Assume that $5,000 is the principal of an investment made at 9% compounded quarterly for 7 years. Find

a. The compound amount

b. The compound interest

2. Assume that $20,000 is the principal of an investment made at 10% compounded monthly for 4 years. Find

a. The compound amount

b. The compound interest

3. Assume that $30,000 is the maturity value of an investment made at 8% compounded semiannually for 8 years. Compute

a. The present value

b. The compound interest

4. Assume that $6,000 is the maturity value of an investment made at 5% compounded monthly for 3 years. Compute

a. The present value

b. The compound interest

5. Assume that $200,000 is the desired maturity value of a sinking fund established at 8% compounded quarterly for 8 years. Find

a. The periodic payment required to finance the fund

b. The total of the deposits to the fund

c. The interest included in the maturity amount

6. Assume that $175,000 is the desired maturity value of a sinking fund established at 7% compounded monthly for 5 years. Find

a. The periodic payment to the fund

b. The total deposits

c. The total interest included in the maturity value

7. Assume that $25,000 is the principal (present value) of a loan. The loan will be financed at 6% semiannually for 4 years. What is

a. The periodic payment

b. The total amount paid to discharge the debt

c. The total interest paid

8. Assume that $350,000 is the principal (present value) of a loan. The loan will be financed at 8% quarterly for 6 years. What is

a. The period payment

b. The total amount paid to discharge the debt

c. The total interest paid

9. Assume that quarterly payments of $100,000 each are paid into an investment earning 5% compounded quarterly for 5 years.

a. How much is the maturity value?

b. How much will be deposited?

c. How much interest will be included?

10. Assume that monthly payments of $10,000 each are paid into an investment earning 7% compounded monthly for 3 years.

a. What is the maturity value?

b. How much will be deposited?

c. How much interest will be included?

11. Assume that a person wishes to receive a payment of $600 each month for 4 years. If the account earns 6% compounded monthly for 4 years,

a. What amount must be on deposit when the annuity begins?

b. How much interest will be earned over the life of the annuity?

12. Assume that a person wishes to receive a payment of $2,000 quarterly for 2 years. If the account earns 7% compounded quarterly,

a. What amount must be on deposit when the annuity begins?

b. What is the total interest earned over the life of the annuity?

13. Sam Levett wants to have $15,000 in 4 years to make a down payment on a house. His savings can earn 6% compounded monthly.

a. How much must he deposit now to achieve his goal?

b. How much of the $15,000 will be interest?

14. Betty Dickens finishes graduate school in 3 years. She would like to have $40,000 at that time to start a business. Money is worth 7% compounded monthly.

a. How much must she invest now in order to achieve her savings goal?

b. How much interest will her investment earn?

15. BBB Trucking Co. is investing $7,000 each quarter to purchase a new refrigerated truck in 4 years. The investment earns 7% compounded quarterly.

a. What amount will the company have after 4 years?

b. How much of this amount will be interest?

16. Connor Furniture Distributors is investing $10,000 each quarter to enlarge its warehouse in 5 years. Their investment earns 8% compounded quarterly.

a. What amount will the account contain in 5 years?

b. How much total interest will the account earn?

17. Lock Manufacturing Co. borrowed $50,000 to purchase furniture for its new headquarters. The company signed a 10%, 3-year loan, making semiannual payments.

a. What was the semiannual payment?

b. How much interest does the company pay on the loan?

18. Jason Quinn borrowed $12,000 to consolidate and pay off his debts. He signed a 9%, 5-year loan, making semiannual payments.

a. What is his semiannual payment?

b. How much total interest will he pay on the loan?

19. Instructor George Hamid plans to return to graduate school for 2 years, and he would like to receive $400 at the end of each month during this time. He can invest his money at 6% compounded monthly.

a. What amount must Hamid's account contain when he returns to school?

b. How much will he receive while he is in school?

c. What part of these payments will be interest?

20. To supplement his income while attending 4 years of college, Jim Vernon would like to receive $300 at the end of each month during this time. He can invest his money at 5% compounded monthly.

a. How much must Vernon's account contain at the beginning of college in order to provide these payments?

b. How much will he receive before the fund is exhausted?

c. How much interest will he earn?

21. Beltway Equipment Co. sold bonds worth $800,000 at maturity in order to expand its operations. Beltway will make deposits each 6 months for 10 years to provide the funds to redeem the bonds. It can invest money at 9% compounded semiannually.

a. What semiannual payment will be required to finance the fund?

b. How much of the $800,000 will be actual deposits?

c. How much interest will the deposits earn?

22. The Stewart family wants to travel to Europe in 2 years. They anticipate a need for $20,000. Money is currently worth 7% compounded monthly.

a. How much must the Stewarts save each month to reach their goal?

b. How much of the $20,000 will be their deposits?

c. How much interest will the account contain?

23. Maria Rodman purchased a $1,000 CD that pays 5% quarterly when held for 3 years.
    a. How much will the CD be worth at maturity?
    b. How much interest will the certificate earn?
24. Hannah Priest invested $4,500 in a savings account which earns 6% compounded quarterly.
    a. How much will the account be worth in 4 years?
    b. How much interest will she earn?

CHAPTER 20 GLOSSARY

Adjustable-rate mortgage (ARM). A mortgage in which the interest rate may change annually, but the amount and time are fixed for the entire loan period.

Amortization. The process of repaying a loan (principal plus interest) by equal periodic payments.

Amortization schedule. A listing of the principal and interest included in each periodic loan payment, as well as a statement of the balance still owed.

Bond. A written promise to repay a specified debt on a certain date, with periodic interest to be paid during the term of the bond. Bonds are typically sold by large corporations and by city and state governments when they borrow money.

Current yield. A rate of interest based on the amount actually paid for a bond (rather than its stated rate based on the par value).

Discount. The amount by which the purchase price of a bond is less than its par value.

Fixed-rate mortgage. A mortgage in which the amount, time, and interest rate are fixed at the beginning of the loan and do not change.

Par value. Face value or redemption value of a bond. The value upon which interest is computed.

Premium. The excess amount above par value that is paid for a bond.

Renegotiable-rate mortgage. (See "Adjustable-rate mortgage.")

Rent. The periodic payment to a sinking fund.

Sinking fund. An annuity account established so that the maturity value (equal periodic deposits plus interest) will exactly equal a specific amount.

Sinking fund schedule. A periodic listing of the growth (in principal and interest) of a sinking fund.

APPENDIX A

ARITHMETIC

OBJECTIVES

Upon completing Appendix A, you will be able to:

1. Define and use correctly the terminology associated with each topic.
2.
   a. Identify the place value associated with each digit in a whole number or a decimal number (Section 1: Problems 1, 2; Section 4: Problem 1).
   b. Pronounce and write out the complete number (Section 1: Problems 3, 4; Section 4: Problem 2).
3. Perform accurately the arithmetic operations (addition, subtraction, multiplication, and division) using whole numbers, common fractions, and decimal fractions (Section 2: Examples 1, 2; Problems 1–16; Section 3: Examples 1–4; Problems 6–10; Section 4: Examples 3–9; Problems 5–9).
4. Manipulate common fractions accurately (Section 3):
   a. Reduce (Problems 1, 5–10)
   b. Change to higher terms (least common denominator) (Problems 2, 3, 6, 7)
   c. Convert mixed numbers to improper fractions, and vice versa (Problems 4, 5)
   d. Convert common fractions to their equivalent decimal value (Section 4: Example 1; Problem 3).
5. Manipulate decimal fractions accurately (Section 4)
   a. Convert to common fractions (including decimals with fractional remainders) (Example 2; Problem 4).
   b. Multiply and divide by powers of 10 by moving the decimal point (Examples 5, 6, 8, 9; Problem 5).

Although calculators or computers are used in businesses to perform most arithmetic computations, it is still advantageous for everyone to have a certain facility with computation. The following topics are presented to give you an opportunity to develop or practice the necessary skills.

SECTION 1

READING NUMBERS

The Hindu-Arabic number system that we use is known as the "base 10" system. This means that it is organized by 10 and powers of 10. Any number can be written using combinations of 10 basic characters called **digits:** 0, 1, 2, 3, 4, 5, 6, 7, 8, and 9. When combined in a number, each digit represents a particular value according to its rank in the order of digits and to the position it holds in the number (the latter is known as its **place value**). The following chart shows what place value (and power of 10) is represented by each place in a number (the number 1,376,049,528 is illustrated). Larger place values are on the left, and smaller values are toward the right.

Note. Each power of 10 is indicated (or abbreviated) by an **exponent,** which is a small number written after the 10 in a raised position. The exponent indicates the number of tens that would have to be multiplied together in order to obtain the unabbreviated place value. Thus, in the expression 10^4, the exponent "4" indicates that the complete number is found by multiplying together four 10s: $10^4 = 10 \times 10 \times 10 \times 10 =$ 10,000. It should also be observed that the exponent indicates the number of zeros that the complete number contains. That is, 10^5 is equal to a "1" followed by 5 zeros:

$$10^5 = 10 \times 10 \times 10 \times 10 \times 10 = 100{,}000$$

BILLIONS			MILLIONS			THOUSANDS			HUNDREDS		
Hundred billions: 10^{11}	Ten billions: 10^{10}	Billions: 10^9	Hundred millions: 10^8	Ten millions: 10^7	Millions: 10^6	Hundred thousands: 10^5	Ten thousands: 10^4	Thousands: 10^3	Hundreds: 10^2	Tens: 10^1	Ones: 10^0
		1	3	7	6	0	4	9	5	2	8

The "2" in the number in the chart, because of the place it occupies, represents "two 10s" or 20. Similarly, the "4" represents "four 10,000s" or 40,000. The last three digits in the chart, "528" are read "five hundred twenty-eight."

For convenience in reading, large numbers are usually separated with commas into groups of three digits, starting at the right. Each three-digit group is then read as the appropriate number of hundreds, tens, and units followed by the family name of the groupings. Thus, the digits "376" above are read "three hundred seventy-six million." The entire number, written with commas, would appear 1,376,049,528 and would be read "one billion, three hundred seventy-six million, forty-nine thousand, five hundred twenty-eight." Commas in a number have no mathematical significance and are used merely to separate the family groupings and indicate the point at which you should pronounce the family name.

When preparing to read a large number, you should never have to name every place value (smallest to largest) in order to determine the largest value; rather, you should point off the groups of three, reading the family names as you proceed, until the largest family grouping is reached. Thus, the largest value of the number 207,XXX,XXX,XXX is quickly identified, reading right to left:

```
 B    M    T    H
207,XXX,XXX,XXX
```

The 207, in the billions grouping, thus represents "two hundred seven billion," and succeeding values would be read in decreasing order.

In reading the value of a number, it is incorrect to use the word "and" except to designate the location of a decimal point. The number 16,000,003 is pronounced "sixteen million, three."

SECTION 1 PROBLEMS

In the number 3,851,469,720, identify the place value of each digit.

1. a. 1 b. 2 c. 3 d. 4 e. 0

2. a. 5 b. 6 c. 7 d. 8 e. 9

Write in words the values of the following numbers:

3. a. 633,520,481 b. 25,543,128 c. 150,286,413 d. 6,046,125 e. 812,344,601,022

4. a. 842,416,375 b. 2,655,120,688 c. 48,136,472 d. 8,108,027 e. 762,300

SECTION 2

WHOLE NUMBERS

The simplest set of numbers in our number system is those used to identify a single object or a group of objects. These are the **counting numbers** or **natural**

numbers or positive **whole numbers.** The counting numbers, along with zero and the negative whole numbers (whole numbers less than zero), comprise the set known as the **integers.** Earliest "mathematics" consisted simply of using the integers to count one's possessions—sheep, cattle, tents, wives, and so on. As civilization advanced and became more complex, the need for more efficient mathematics also increased. This led to the perfection of the four arithmetic **operations:** addition, subtraction, multiplication, and division. (This review includes operations only with nonnegative numbers.)

ADDITION

This operation provides a shortcut that eliminates having to count each item consecutively until the total is reached. The names of the numbers in an operation of addition are as follows:

Addend	23
Addend	+16
Sum	39

Addition functions under two basic mathematical laws, expressed as (1) the commutative property of addition and (2) the associative property of addition. The **commutative property of addition** means that the *order* in which *two* addends are taken does not affect the sum. More simply, if two numbers are to be added, it does not matter which is written down first. The commutative property is usually written in general terms (terms which apply to all numbers) as

$$\text{CPA:} \quad a + b = b + a$$

The commutative property of addition can be illustrated using the numbers 2 and 3:

$$2 + 3 \stackrel{?}{=} 3 + 2$$

$$5 = 5$$

The **associative property of addition** applies to the *grouping* of *three* numbers; it means that, when three numbers are to be added, the sum will be the same regardless of which two are grouped together to be added first. The associative property is expressed in the following general terms:

$$\text{APA:} \quad (a + b) + c = a + (b + c)$$

The associative property can be verified as follows. Suppose that the numbers 2, 3, and 4 are to be added. By the associative property,

$$(2 + 3) + 4 \stackrel{?}{=} 2 + (3 + 4)$$

$$5 \quad + 4 \stackrel{?}{=} 2 + \quad 7$$

$$9 = 9$$

By applying both the commutative and associative properties of addition, one can verify other groupings or the addition of more addends.

An excellent way in which you can speed addition is by adding in groups of numbers that total 10. You should become thoroughly familiar with the combinations of two numbers that total 10 (1 + 9; 2 + 8; 3 + 7; 4 + 6; 5 + 5) and should be alert for these combinations in problems. For example,

$$\begin{array}{r} 3 \\ 7 \\ 6 \\ +\ \underline{4} \end{array}$$

This should not be added as "3 plus 7 is 10, plus 6 is 16, plus 4 is 20." Rather, you should immediately recognize the two groups of 10 and add "10 plus 10 is 20," as

$$\begin{array}{rl} 3 & \\ 7 & \} \ 10 \\ 6 & \\ +\ \underline{4} & \} \ 10 \\ 20 & \end{array}$$

Most problems do not consist of such obvious combinations, but the device may frequently be employed if you stay alert for it. For instance,

$$\begin{array}{rl} 8 & \\ 4 & 10 \ (8+2) \\ 2 & \\ +\ \underline{6} & 10 \ (4+6) \\ 20 & \end{array}$$

$$\begin{array}{r} 4 \\ 5 \\ 9 \\ +\ \underline{5} \\ 23 \end{array}$$

(added "4 plus 10 is 14, plus 9 is 23" or "4 plus 9 is 13, plus 10 is 23")

$$\begin{array}{r} 9 \\ 6 \\ 1 \\ 4 \\ +\ \underline{6} \\ 26 \end{array}$$

(added "10 plus 6 is 16, plus 10 is 26")

$$\begin{array}{r} 8 \\ 6 \\ 4 \\ 3 \\ 5 \\ +\ \underline{7} \\ 33 \end{array}$$

(added "8 plus 10 is 18, plus 10 is 28, plus 5 is 33")

This technique, like many others, can be carried to extremes. It is intended to be a timesaver; it will be, provided that you stay alert for combinations adding up to 10 when they are conveniently located near each other. However, a problem may take longer to add if you waste time searching for widely separated combinations.

Also, it is easy to miss a number altogether if you are adding numbers from all parts of the problem rather than adding in nearly consecutive order. Therefore, you should use the device when convenient but should not expect it to apply to every problem.

If a first glance at a problem reveals that the same digit appears several times in a column, the following device is useful: Count the number of times the digit appears and multiply this number by the digit itself; the remaining digits are then added to this total. For example,

$$\begin{array}{r} 7 \\ 4 \\ 7 \\ 7 \\ 8 \\ +\underline{\ 7} \\ 40 \end{array}$$

(added "four 7s are 28, plus 4 is 32, plus 8 is 40")

As you probably recall, addition problems may be checked either by adding upward in reverse order or simply by re-adding the sum.

Horizontal addition problems seem harder because the numbers being added are so far apart. More mistakes are made in horizontal problems because a digit from the wrong place is often added. However, many business forms require horizontal addition. The following hint may make such addition easier:

1. A right-handed person should use the left index finger to cover all the higher-place digits in the first number, over to the digit that is currently being added. (Unfortunately, left-handed persons will have to forego this aid.)
2. As digits from succeeding numbers are added, place the pencil point beneath the digit currently being added. Mentally repeat the current subtotal frequently as you proceed.

For example, while the ten's-place digits are being added in the following problem, the left index finger should cover the digits "162" of the first number. The pencil point should be placed under the 5, 1, 2, and 7 as each is added.

16,2)83 + 22,054 + 40,611 + 7,927 + 19,670 = <u>. . . 45</u>

SUBTRACTION

This operation is the opposite of addition. (Since subtraction produces the same result as when a negative number is added to a positive, subtraction is sometimes defined as an extension of addition.) The parts of a subtraction operation have the following names:

Minuend	325
Subtrahend	−198
Difference or Remainder	127

A subtraction problem can be checked mentally by adding together the subtrahend and difference; if the resultant sum equals the minuend, the operation is verified.

You can easily determine that neither the commutative nor the associative property applies to subtraction. For example, it makes a great deal of difference whether we find \$17 − \$8 or \$8 − \$17. And any example will show that the associative property does not hold:

$$(7 - 5) - 2 \stackrel{?}{=} 7 - (5 - 2)$$

$$2 \quad - 2 \stackrel{?}{=} 7 - \quad 3$$

$$0 \neq 4$$

As you know, subtraction often necessitates "borrowing," or regrouping of the place values. For example, the number 325 in our first example of subtraction—which represents three 100s, two 10s, and five 1s—had to be regrouped into two 100s, eleven 10s, and fifteen 1s. Both groupings equal 325, as the following example shows:

Original	*Regrouped*		*Subtraction*
300	200	(two 100s)	(two 100s)
20	110	(eleven 10s)	(eleven 10s)
5	15	(fifteen 1s)	(fifteen 1s)
325	325		

$$\begin{array}{r} \overset{2}{\cancel{3}}\ \overset{11}{\cancel{2}}\ {}^{1}5 \\ -1\ \ 9\ \ 8 \\ \hline 1\ \ 2\ \ 7 \end{array}$$

You are also aware that addition problems often contain many numbers, whereas subtraction problems are done only two numbers at a time. Subtraction problems could just as well contain more numbers were it not for the fact that regrouping becomes so complicated. Calculators, which can regroup repeatedly when necessary, can perform a series of subtractions before indicating a total.

MULTIPLICATION

This operation was developed as a shortcut for addition. Suppose that a student has three classes on each of five days; this week's total classes could be found by addition:

M	T	W	T	F	TOTAL
3 +	3 +	3 +	3 +	3 =	15

This approach to the problem would become quite tedious, however, if we wanted to know the total receipts at a grocery store from the sale of 247 cartons of soft drinks at $2.49 each.

Thus, multiplication originated when persons learned from experience what sums to expect following repeated additions of small numbers, and the actual addition process then became unnecessary. Perfection of the operation led to the knowledge that any two numbers, regardless of size, could be multiplied if one had memorized the products of all combinations of the digits 0–9. The frequent occurrence in business situations of the number 12 makes knowledge of its multiples quite valuable also.

The numbers of an operation of multiplication have the following names:

Multiplicand (or **Factor**)	18
Multiplier (or **Factor**)	× 4
Product	72

MULTIPLICATION TABLE

The following table lists the products of the digits 2–9, as well as the numbers 10–12. You should be able to recite instantly all products up through the multiples of 9. A knowledge of products through the multiples of 12 is also very useful.

	2	3	4	5	6	7	8	9	10	11	12
2	4	6	8	10	12	14	16	18	20	22	24
3	6	9	12	15	18	21	24	27	30	33	36
4	8	12	16	20	24	28	32	36	40	44	48
5	10	15	20	25	30	35	40	45	50	55	60
6	12	18	24	30	36	42	48	54	60	66	72
7	14	21	28	35	42	49	56	63	70	77	84
8	16	24	32	40	48	56	64	72	80	88	96
9	18	27	36	45	54	63	72	81	90	99	108
10	20	30	40	50	60	70	80	90	100	110	120
11	22	33	44	55	66	77	88	99	110	121	132
12	24	36	48	60	72	84	96	108	120	132	144

There are three properties affecting the operation of multiplication: (1) the commutative property of multiplication, (2) the associative property of multiplication, and (3) the distributive property.

The **commutative property of multiplication,** like the commutative property of addition, applies to the *order* in which *two* numbers are taken. The commutative property asserts that either factor can be written first, and the product will be the same. In general terms, the commutative property of multiplication is written

$$\text{CPM:} \quad a \cdot b = b \cdot a$$

It can be illustrated using the factors 4 and 7:

$$4 \cdot 7 \stackrel{?}{=} 7 \cdot 4$$

$$28 = 28$$

The **associative property of multiplication** is similar to the associative property of addition in that it applies to the *grouping* of any *three* numbers. That is, if three numbers are to be multiplied, the product will be the same regardless of which two are multiplied first. The associative property of multiplication is expressed in general terms as

$$\text{APM:} \quad (a \cdot b)c = a(b \cdot c)$$

Using the numbers 2, 5, and 3, the associative property is verified as follows:

$$(2 \cdot 5)3 \stackrel{?}{=} 2(5 \cdot 3)$$

$$(10)3 \stackrel{?}{=} 2(15)$$

$$30 = 30$$

The third property applies to the combination of multiplication and addition. It is the **distributive property,** sometimes called the **distributive property of multiplication over addition.** This property means that the product of a factor times the sum of two numbers equals the sum of the individual products. Stated differently, this means that if two numbers are added and their sum is multiplied by some other number, the final total is the same as if the two numbers were first separately multiplied by the third number and the products then added. The distributive property is expressed in general terms as

$$\text{DP:} \quad a(b + c) = ab + ac$$

Using 2 for the multiplier and the numbers 3 and 5 for the addends, the distributive property can be verified as follows:

$$2(3 + 5) \stackrel{?}{=} (2\cdot3) + (2\cdot5)$$

$$2(8) \stackrel{?}{=} \quad 6 + \quad 10$$

$$16 = 16$$

Note. The distributive property has frequent application in business. One of the best examples is sales tax. When several items are purchased, the sales tax is found by adding all the prices and then multiplying only once by the applicable sales tax percent. This is much simpler than multiplying the tax rate by the price of each separate item and then adding to find the total sales tax, although the total would be the same by either method.

While other methods are sometimes used, the easiest, quickest, and most reliable check of a multiplication problem is simply to repeat the multiplication. (If convenient, the factors may be reversed for the checking operation.)

Students often waste time and do much unnecessary work when *multiplying by numbers that contain zeros.* When multiplying by 10, 100, 1,000, etc., write the product immediately, simply by affixing to the original factor the same number of zeros as contained in the multiplier. Thus,

$$54 \times 1\underline{0} = 54\underline{0}; \quad 2{,}173 \times 1\underline{00} = 217{,}3\underline{00}; \quad \text{and} \quad 145 \times 1{,}\underline{000} = 145{,}\underline{000}$$

When multiplying numbers ending in zeros, write the factors with the zeros to the right of the problem as it will actually be performed. After multiplication with the other digits has been completed, affix all these zeros to the basic product, as in Example 1.

Example 1 (a) 148 × 60 (b) 6,700 × 52

```
(a)    148 ¦
     ×   6 ¦ 0
     ---------
     8,88 ¦ 0

(b)     5 2 ¦
     ×  6 7 ¦00
     ----------
       36 4 ¦
      312   ¦
     ----------
      348,4 ¦00
```

(c) 130 × 1,500

```
      13 ¦ 0
   ×  15 ¦ 00
   ----------
      65 ¦
     13  ¦
   ----------
    195, ¦ 000
```

Note here that all three zeros are affixed to the product

When a multiplier contains zeros within the number, many people write whole rows of useless zeros in order to be certain that the significant digits remain properly aligned. These unnecessary zeros should be omitted; the other digits will still be in correct order if you practice this basic rule of multiplication: On each line of multiplication, the first digit to be written down is placed directly beneath the digit in the multiplier (bottom factor) being multiplied at the time.

Example 2 (a) 671 × 305

Right		*Inefficient*
671		671
× 305		× 305
3 355		3 355
201 3	(because 300 × 1 = 300)	0 00
204,655		201 3
		204,655

(b) 4,382 × 3,004

4,382	
× 3,004	
17 528	
13 146	(because 3,000 × 2 = 6,000)
13,163,528	

DIVISION

This operation is the reverse of multiplication. The numbers in an operation of division are identified as follows:

	4	**Quotient**
Divisor	18)77	**Dividend**
	72	
	5	Remainder

That is,

$$\text{Divisor}\,\overline{)\text{Dividend}}^{\ \text{Quotient}} \quad \text{or} \quad \text{Dividend} \div \text{Divisor} = \text{Quotient}$$

Division is neither commutative, associative, nor distributive, as can be determined by brief experiments. Obviously, $4\overline{)16}$ yields quite a different quotient from $16\overline{)4}$. An example to test the associative property reveals different quotients:

$$(36 \div 6) \div 3 \stackrel{?}{=} 36 \div (6 \div 3)$$

$$6 \quad \div 3 \stackrel{?}{=} 36 \div \quad 2$$

$$2 \neq 18$$

Similarly, a test of the distributive property reveals that it does not apply either:

$$60 \div (4 + 6) \stackrel{?}{=} (60 \div 4) + (60 \div 6)$$

$$60 \div \quad 10 \quad \stackrel{?}{=} \quad 15 \quad + \quad 10$$

$$6 \neq 25$$

Division by one-digit divisors should be carried out mentally and the quotient written directly; this method is commonly called *short division.* You may find it helpful to write each remainder in front of the next digit as the operation progresses. For example,

$$7\overline{)9422} \qquad \text{written} \qquad \begin{array}{r} 1\ \ 3\ \ 4\ \ 6 \\ 7\overline{)9^{2}4^{3}2^{4}2} \end{array}$$

Division involving divisors of two or more digits is usually performed by writing each step completely. Division performed in this manner is known as *long division.* The following general rules apply to both long and short division and cover aspects of division where mistakes are often made:

1. The first digit of the quotient should be written directly above the last digit of the partial dividend which the divisor goes into. For example, when dividing $25\overline{)17628}$, you should write the first digit of the quotient above the 6:

$$\begin{array}{r} 7 \\ 25\overline{)17628} \\ \underline{175} \end{array}$$

2. The amount remaining after each succeeding step in division must always be smaller than the divisor. Otherwise, the divisor would have divided into the dividend at least one more time. If a remainder is larger than the divisor, the

previous step should be repeated to correct the corresponding digit in the quotient. For example, if the first step had been

$$\begin{array}{r} 6 \\ 25\overline{)17,628} \\ \underline{15\;0} \\ 2\;6 \end{array} \longleftarrow \text{Remainder is more than 25}$$

the quotient should be corrected:

$$\begin{array}{r} 7 \\ 25\overline{)17,628} \\ \underline{17\;5} \\ 1 \end{array} \longleftarrow \text{Remainder is less than 25}$$

3. Succeeding digits of the dividend should be brought down one at a time. For example,

$$\begin{array}{r} 7 \\ 25\overline{)17,628} \\ \underline{17\;5}\downarrow \\ 12 \end{array}$$

4. Each time a digit is brought down, a new digit must be affixed to the quotient directly over the digit that was just brought down. Thus, a zero is affixed to the quotient when a brought-down digit does not create a number large enough for the divisor to divide into it at least one whole time. For example,

$$\begin{array}{r} 70 \\ 25\overline{)17,628} \\ \underline{17\;5} \\ 12 \end{array}$$

The next digit is then brought down immediately and the division process continued as follows:

$$\begin{array}{r} 705 \\ 25\overline{)17,628} \\ \underline{17\;5} \\ 128 \\ \underline{125} \\ 3 \end{array}$$

5. The final remainder may be written directly after the quotient as follows:

$$\begin{array}{r} 705 \quad \text{R3} \\ 25\overline{)17{,}628} \\ \underline{17\,5} \\ 128 \\ \underline{125} \\ 3 \end{array}$$

Division problems can be checked by multiplying the divisor times the quotient and then adding the remainder (if any). This result should equal the original dividend. For example, for the above division problem,

$$\begin{array}{r} 705 \\ \underline{\times \quad 25} \\ 17{,}625 \\ \underline{+ \quad\; 3} \\ 17{,}628 \end{array}$$

SECTION 2 PROBLEMS

The following problems were especially designed to provide practice in adding by combinations that total 10.

1.

a.	b.	c.	d.	e.	f.	g.	h.	i.
7	4	9	6	4	6	2	32	64
2	8	7	5	5	1	5	54	23
3	6	1	2	1	9	3	48	33
+8	+1	5	4	8	2	6	66	75
		+5	+8	+9	+2	+3	+25	+11

2.

a.	b.	c.	d.	e.	f.	g.	h.	i.
3	6	8	9	5	7	3	13	48
2	4	2	4	5	9	2	25	16
7	5	6	3	8	1	4	77	52
+8	+5	3	1	4	3	6	65	84
		+7	+6	+2	+8	3	+32	+25
						3		
						+4		

Add, using time-saving techniques when possible. Check your answers.

3.

a.	b.	c.	d.	e.	f.	g.	h.	i.
6	8	6	4	23	42	12	88	67
6	7	7	4	14	48	16	14	13
5	2	5	8	53	44	46	76	44
4	3	5	3	63	46	56	54	27
+1	+4	3	4	42	+41	17	32	63
		+2	6	33		+66	+51	+54
			+8	+56				

4. a. 123 + 986 + 290 + 451　b. 802 + 463 + 623 + 440 + 716　c. 528 + 73 + 265 + 939 + 80　d. 7,132 + 3,040 + 6,355 + 8,098　e. 3,613 + 4,872 + 7,520 + 5,638 + 2,516

f. 186 + 4,351 + 265 + 68 + 3,081　g. 61,046 + 42,844 + 36,545 + 80,622　h. 44,315 + 34,045 + 25,438 + 66,055 + 47,681　i. $5,808.25 + 49.18 + 219.15 + 2,363.75 + 7.19 + 340.93

5. a. 41 + 39 + 22 + 16 + 88
b. 78 + 24 + 45 + 52 + 63
c. 154 + 238 + 448 + 261 + 311

6. a. 63 + 41 + 22 + 18 + 73
b. 240 + 313 + 467 + 243 + 324
c. 1,206 + 1,344 + 2,281 + 3,468 + 2,615

Subtract. Check by adding.

7. a. 36 − 28　b. 863 − 199　c. 3,468 − 1,547　d. 6,002 − 5,679

8. a. 43 − 17　b. 655 − 458　c. 5,044 − 3,950　d. 1,824 − 469

Multiply.

9. a. 466 × 10　b. 64 × 100　c. 553 × 100　d. 56 × 200　e. 1,800 × 46　f. 25,000 × 83

10. a. 312 × 100　b. 88 × 1,000　c. 65 × 30　d. 525 × 300　e. 3,300 × 74　f. 2,700 × 44

11. a. 76 × 24　b. 714 × 53　c. 1,203 × 317　d. 6,890 × 254

12. a. 47 × 35　b. 669 × 31　c. 3,365 × 842　d. 5,724 × 864

Divide. Check using multiplication.

13. a. $6\overline{)3{,}588}$　b. $8\overline{)5{,}968}$　c. $36\overline{)648}$　d. $48\overline{)1{,}248}$　e. $63\overline{)18{,}333}$

14. a. $5\overline{)13{,}545}$　b. $3\overline{)38{,}232}$　c. $537\overline{)13{,}425}$　d. $627\overline{)183{,}711}$　e. $52\overline{)13{,}416}$

15. The following table shows the sales made by various departments of a discount store during one week. Find the total sales of the store for each day and the total sales of each department for the entire week. (This final total serves as a check of your addition: The sum of the weekly totals for each register must equal the sum of the daily totals for the entire store. That is, the sum of the totals in the right-hand column must equal the sum of the totals across the bottom.)

THE DISCOUNT MART
Sales for Week of August 4

DEPT.	MONDAY	TUESDAY	WEDNESDAY	THURSDAY	FRIDAY	SATURDAY	DEPT. TOTALS
#1	$345.68	$406.29	$323.76	$386.91	$459.85	$481.86	$
#2	562.09	438.88	475.45	424.18	482.37	519.12	
#3	445.21	459.16	398.06	437.75	493.46	418.25	
#4	293.34	302.28	285.37	356.11	374.04	380.84	
#5	322.66	289.55	314.42	349.23	352.58	334.17	
#6	488.49	419.37	392.76	407.33	469.87	493.66	
Daily Totals	$	$	$	$	$	$	$

16. Complete the following sales invoice. Multiply the cost of each item by the number of items purchased. Add the totals in the final column to find the total cost of the purchase.

OFFICE SUPPLY HOUSE

Invoice

NO. PURCHASED	DESCRIPTION	COST PER ITEM	TOTAL
14	X-22 pencil sharpener	$ 8.12	$
12	X-25 pencil sharpener	10.55	
25	B-13 ream bond stationery	4.95	
15	K-46 ream fax paper	2.95	
26	R-12 ream computer paper	2.85	
6	TC-32 multimedia drawer	8.95	
72	C-28 pencil	.10	
46	L-4 ballpoint pen	.55	
8	RF-10 color inkjet printer	215.00	
5	P-7 computer speaker	24.95	
3	A-IV calculator	8.38	
2	HD-5 CD-ROM drive	142.95	
7	RP-34 notebook computer	2,142.00	
		Total	$

SECTION 3

COMMON FRACTIONS

The system of fractions was developed to meet the need for measuring quantities that are not whole amounts. The parts of a fraction are identified by the following terms:

A **common fraction** is an indicated division—a short method of writing a division problem. That is, if \$3 is to be divided among five persons, each person's share is indicated as

$$\$3 \div 5 \qquad \text{or} \qquad \$\frac{3}{5}$$

Thus, the fraction line indicates division (the numerator is to be divided by the denominator), and the entire fraction represents each person's share. We now see why common fractions are often known as "rational numbers." A **rational number** is any number that can be expressed as the quotient of two integers. The fraction $\frac{3}{5}$ is a rational number because it is expressed as the quotient of the integers (whole numbers) 3 and 5. The entire fraction represents the quotient—the value of 3 divided by 5.

A common fraction is called a **proper fraction** if its numerator is smaller than its denominator (for example, $\frac{4}{7}$) and if the indicated division would result in a value less than 1. The numerator of an **improper fraction** is equal to or greater than the denominator (for example, $\frac{7}{7}$ or $\frac{15}{7}$) and thus has a value equal to or greater than 1.

REDUCING FRACTIONS

Two fractions are equal if they represent the same quantity. When writing fractions, however, it is customary to use the smallest possible numbers. Thus, it is often necessary to *reduce* a fraction—to change it to a fraction of equal value but written with smaller numbers.

The fraction $\frac{3}{3}$ represents $3 \div 3$ or 1. Obviously, any number written over itself as a fraction equals 1 $\left(\frac{x}{x} = 1\right)$. We also know from multiplication that 1 times any number equals that same number ($1 \cdot x = x$). These facts are the mathematical basis for reducing fractions. A fraction can be reduced if the same number is a factor of both the numerator and the denominator. For example,

$$\frac{6}{15} = \frac{2 \times 3}{5 \times 3} \qquad \begin{matrix}\text{(the factors of 6 are 2 and 3)} \\ \text{(the factors of 15 are 5 and 3)}\end{matrix}$$

But since $\frac{3}{3} = 1$, then

$$\frac{6}{15} = \frac{2 \times 3}{5 \times 3} = \frac{2}{5} \times \frac{3}{3} = \frac{2}{5} \times 1 = \frac{2}{5}$$

Hence, $\frac{2}{5}$ represents the same quantity as $\frac{6}{15}$.

In actual practice, you reduce a fraction by mentally testing the numerator and denominator to determine the *greatest* factor that they have in common (that is, to determine the largest number that will divide into both evenly). The reduced fraction is then indicated by writing these quotients in fraction form:

$\frac{20}{15}$ (Think: 5 divides into 20 *four* times and into 15 *three* times) Write: $\frac{20}{15} = \frac{4}{3}$

Sometimes, however, the greatest common factor may not be immediately obvious. When the greatest common factor is not apparent, a fraction may be reduced to its lowest terms by performing a series of reductions:

$$\frac{60}{84} = \frac{2}{2} \times \frac{30}{42} = \frac{30}{42} = \frac{2}{2} \times \frac{15}{21} = \frac{15}{21} = \frac{3}{3} \times \frac{5}{7} = 1 \times \frac{5}{7} = \frac{5}{7}$$

The end result will be the same whether the reduction is performed in one step or in several steps. It should be noted that if a series of reductions is used, the product of these several common factors equals the number that would have been the greatest single common factor. For example, the fraction $\frac{60}{84}$ above was reduced by 2, 2, and 3. The product of these factors is 12, which is the greatest common factor:

$$\frac{60}{84} = \frac{2}{2} \times \frac{2}{2} \times \frac{3}{3} \times \frac{5}{7} = 1 \times 1 \times 1 \times \frac{5}{7} = \frac{5}{7} \quad \text{or} \quad \frac{60}{84} = \frac{12}{12} \times \frac{5}{7} = 1 \times \frac{5}{7} = \frac{5}{7}$$

Therefore, you should divide by the largest number that you recognize as a factor of both the numerator and the denominator. However, you should always examine the result to determine whether it still contains another common factor.

CHANGING TO HIGHER TERMS

Suppose that we wish to know whether $\frac{5}{6}$ or $\frac{7}{9}$ is larger. To compare the size of fractions, as a practical matter we must rewrite them using the same denominator. The process of rewriting fractions so that they have a common denominator (which is usually larger than either original denominator) is called *changing to higher terms.*

The product of the various denominators will provide a common denominator. However, it is customary to use the **least common denominator,** which is the smallest number that all the denominators will divide into evenly. Thus, the least (or lowest) common denominator is often a number that is smaller than the product of all the denominators. In fact, one of the given denominators may be the least common denominator.

When finding the least common denominator, you should first consider the multiples of the largest given denominator. If this method does not reveal the least common denominator quickly, then you should find the product of the denominators. Before using this product as the common denominator, though, first consider $\frac{1}{2}$, $\frac{1}{3}$, or even $\frac{1}{4}$ of this product to determine if that might be the lowest common denominator. If not, then the product itself probably represents the lowest common denominator.

Consider the fractions, $\frac{2}{3}$, $\frac{1}{6}$, and $\frac{3}{8}$. To find the least common denominator, examine the multiples of 8 (the largest denominator). The multiples of 8 are 8, 16, 24, 32, 40, etc. Neither 8 nor 16 is divisible by 3 or 6. However, 24 is divisible by both 3 and 6 (and, of course, by 8). Hence, 24 is the least common denominator of $\frac{2}{3}$, $\frac{1}{6}$, and $\frac{3}{8}$.

Next, consider the fractions $\frac{5}{6}$, $\frac{3}{7}$, and $\frac{5}{8}$. The product of the denominators is $6 \times 7 \times 8$, or 336. However, $\frac{1}{2}$ of 336, or 168, represents the smallest number that 6, 7, and 8 will all divide into evenly. Thus, 168 is the least common denominator. Also consider the fractions $\frac{1}{6}$, $\frac{2}{7}$, and $\frac{2}{9}$. The product of these denominators is $6 \times 7 \times 9$, or 378. One-half of 378 is 189, which is not divisible by 6. However, $\frac{1}{3}$ of 378 is 126, which is divisible by 6, 7, and 9 all. Hence, 126 is the least common denominator.

Changing to higher terms is based on the same principles as reducing fractions, but applied in reverse order. That is, the procedure is based on the facts that any number over itself equals 1 $\left(\frac{x}{x} = 1\right)$ and that any number multiplied by 1 equals that same number ($x \cdot 1 = x$). Thus, once the least common denominator has been determined, you must then find how many times the given denominator divides into the common denominator. This result is used as both the numerator and the denominator to create a fraction equal to 1, which is then used to convert the fraction to higher terms. For example, suppose that we have already determined that 12 is the least common denominator for $\frac{3}{4}$ and $\frac{1}{6}$. To convert $\frac{3}{4}$ to higher terms, think "4 divides into 12 *three* times." Thus, $\frac{3}{3}$ is used to change $\frac{3}{4}$ to higher terms:

$$\frac{3}{4} = \frac{3}{4} \times 1 = \frac{3}{4} \times \frac{3}{3} = \frac{3 \times 3}{4 \times 3} = \frac{9}{12}$$

We have multiplied $\frac{3}{4}$ by a value, $\frac{3}{3}$, equal to 1. Therefore, we know that the value of the fraction remains unchanged and that $\frac{3}{4} = \frac{9}{12}$. Similarly, 6 divides into 12 *two* times. Thus, $\frac{2}{2}$ is used to change $\frac{1}{6}$ to higher terms:

$$\frac{1}{6} = \frac{1}{6} \times 1 = \frac{1}{6} \times \frac{2}{2} = \frac{1 \times 2}{6 \times 2} = \frac{2}{12}$$

Hence, $\frac{1}{6} = \frac{2}{12}$.

In actual practice, it is unnecessary to write out this entire process. Rather, you mentally divide the denominator into the lowest common denominator and multiply that quotient times the numerator to obtain the numerator of the converted fraction. Therefore, to change $\frac{3}{4}$ to twelfths, think: "4 divides into 12 *three* times; that *three* times the numerator 3 equals 9; thus, $\frac{3}{4} = \frac{9}{12}$." Or, when changing $\frac{1}{6}$ to twelfths, think: "6 divides into 12 *two* times; *two* times 1 equals 2; therefore, $\frac{1}{6} = \frac{2}{12}$."

Now let us return to $\frac{5}{6}$ and $\frac{7}{9}$ to determine which is larger.

$$\frac{5}{6} = \frac{15}{18} \qquad \text{and} \qquad \frac{7}{9} = \frac{14}{18}$$

Thus, comparing the fractions on the same terms, $\frac{15}{18}$ is larger than $\frac{14}{18}$. Hence, $\frac{5}{6}$ is larger than $\frac{7}{9}$.

ADDITION AND SUBTRACTION

The parts of each operation have the same terminology that they did with integers, as follows:

Addend	$\frac{3}{7}$	Minuend	$\frac{3}{7}$
Addend	$+\frac{2}{7}$	Subtrahend	$-\frac{2}{7}$
Sum	$\frac{5}{7}$	Difference or remainder	$\frac{1}{7}$

Before common fractions can be either added or subtracted, the fractions must first be written with a common denominator. The numerators are then added or subtracted (whichever is indicated). The result is then written over the *same* common denominator. (Be certain that this fact is firmly impressed in your mind, for mistakes are frequently made at this point.) The answer should then be reduced, if possible. Consider the following examples.

Example 1

(a)
$$\begin{array}{rcr} \frac{3}{5} & = & \frac{9}{15} \\ +\frac{2}{3} & = & \frac{10}{15} \\ \hline & & \frac{19}{15} \end{array}$$

(b)
$$\begin{array}{rcr} \frac{3}{4} & = & \frac{9}{12} \\ \frac{2}{3} & = & \frac{8}{12} \\ +\frac{5}{6} & = & \frac{10}{12} \\ \hline & & \frac{27}{12} = \frac{9}{4} \end{array}$$

(c)
$$\frac{1}{4} + \frac{3}{8} + \frac{1}{6} = \frac{6}{24} + \frac{9}{24} + \frac{4}{24}$$
$$= \frac{6 + 9 + 4}{24}$$
$$\frac{1}{4} + \frac{3}{8} + \frac{1}{6} = \frac{19}{24}$$

(d)
$$\begin{array}{rcr} \frac{9}{16} & = & \frac{9}{16} \\ -\frac{3}{8} & = & -\frac{6}{16} \\ \hline & & \frac{3}{16} \end{array}$$

(e)
$$\begin{array}{rr} \dfrac{5}{7} = & \dfrac{15}{21} \\ -\dfrac{1}{3} = & -\dfrac{7}{21} \\ \hline & \dfrac{8}{21} \end{array}$$

(f)
$$\begin{aligned} \frac{1}{2} - \frac{1}{6} &= \frac{3}{6} - \frac{1}{6} \\ &= \frac{3-1}{6} \\ &= \frac{2}{6} \\ \frac{1}{2} - \frac{1}{6} &= \frac{1}{3} \end{aligned}$$

MULTIPLICATION AND DIVISION

As would be expected, the terms previously used to identify the parts of multiplication and division with whole numbers also apply to those operations with fractions.

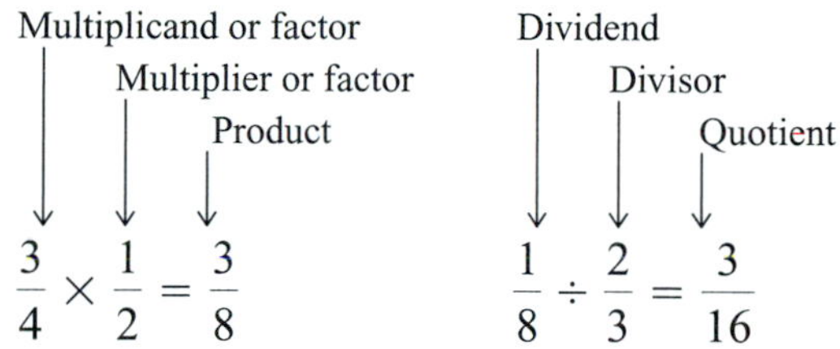

When multiplying fractions, we multiply the various numerators together to obtain the numerator of the product and then multiply the denominators to obtain the denominator of the product. Where possible, the product should be reduced.

$$\frac{2}{5} \times \frac{2}{3} = \frac{2}{5} \times \frac{2}{3} = \frac{2 \times 2}{5 \times 3} = \frac{4}{15}$$

When one fraction (dividend) is to be divided by another (divisor), the divisor (second fraction) must first be inverted (that is, the numerator and denominator must be interchanged). Then proceed as in multiplication. When a fraction is inverted, the new fraction is called the **reciprocal** of the original fraction. Hence, the reciprocal of $\frac{3}{5}$ is $\frac{5}{3}$. Thus, it is often said that to divide by a fraction, we "multiply by the reciprocal."

$$\frac{3}{4} \div \frac{2}{3} = \frac{3}{4} \times \frac{3}{2} = \frac{3 \times 3}{4 \times 2} = \frac{9}{8}$$

(To divide by $\frac{2}{3}$, *multiply* by the reciprocal $\frac{3}{2}$.)

Note. With common fractions, the number properties apply exactly as they did for whole numbers. That is, the *commutative* and *associative* properties apply for both addition and multiplication. Then the *distributive* property of multiplication over addition illustrates how the two operations interact. You might pick some fractions and try confirming that these properties do work, using the examples from the previous section as a guide. (As before, none of these properties holds for subtraction or division.)

You may be accustomed to "canceling" when multiplying or dividing fractions. When you use cancellation, this is actually just reducing the problem rather than waiting and reducing the answer. We know that the value of a fraction remains unchanged when it is reduced. Thus, cancellation is permissible because it is a reduction process and does not change the value of the answer.

Consider the problem $\frac{3}{4} \times \frac{1}{3}$. If this problem is multiplied in the standard manner, the product would be $\frac{3}{12}$, which reduces to $\frac{1}{4}$. The following operations illustrate how the problem itself can be reduced before the answer is obtained:

$$\frac{3}{4} \times \frac{1}{3} = \frac{3 \times 1}{4 \times 3}$$

Since multiplication is commutative, we can reverse the order of the numbers:

$$\begin{aligned} \frac{3 \times 1}{4 \times 3} &= \frac{1 \times 3}{4 \times 3} \\ &= \frac{1}{4} \times \frac{3}{3} \\ &= \frac{1}{4} \times 1 \end{aligned}$$

$$\frac{3 \times 1}{4 \times 3} = \frac{1}{4}$$

Thus, $\frac{3}{4} \times \frac{1}{3} = \frac{1}{4}$. However, you would not usually reduce by rewriting the problems to make the 3s form the fraction $\frac{3}{3}$ and thereby reduce to 1. Rather, seeing that there is one 3 in the numerator and another in the denominator, you would cancel the 3s (or reduce $\frac{3}{3}$ to 1) as follows:

$$\frac{3}{4} \times \frac{1}{3} = \frac{\overset{1}{\cancel{3}}}{4} \times \frac{1}{\underset{1}{\cancel{3}}} = \frac{1}{4}$$

Two numbers can be canceled even though they are widely separated in the problem, provided that one is in the numerator and the other in the denominator.

$$\frac{4}{5} \times \frac{3}{5} \times \frac{7}{8} = \frac{\overset{1}{\cancel{4}}}{5} \times \frac{3}{5} \times \frac{7}{\underset{2}{\cancel{8}}} = \frac{1}{5} \times \frac{3}{5} \times \frac{7}{2} = \frac{21}{50}$$

Now consider the division problem $\frac{4}{5} \div \frac{8}{15}$. If this problem were divided before it is reduced, the operation would be as follows:

$$\frac{4}{5} \div \frac{8}{15} = \frac{4}{5} \times \frac{15}{8}$$

$$= \frac{60}{40}$$

$$\frac{4}{5} \div \frac{8}{15} = \frac{3}{2}$$

The problem could also be reduced prior to the actual division:

$$\frac{4}{5} \div \frac{8}{15} = \frac{4}{5} \times \frac{15}{8}$$

$$= \frac{4}{5} \times \frac{5 \times 3}{4 \times 2}$$

$$= \frac{4 \times 5 \times 3}{5 \times 4 \times 2}$$

Because multiplication is commutative and associative, we may rearrange the numbers:

$$= \frac{5 \times 4 \times 3}{5 \times 4 \times 2}$$

$$= \frac{5}{5} \times \frac{4}{4} \times \frac{3}{2}$$

$$= 1 \times 1 \times \frac{3}{2}$$

$$\frac{4}{5} \div \frac{8}{15} = \frac{3}{2}$$

Hence, $\frac{4}{5} \div \frac{8}{15} = \frac{3}{2}$ by either method. In actual practice, however, neither method is normally used. Rather, after the divisor has been inverted, cancellation would be used to reduce the fractions. That is, the implied fraction $\frac{4}{8}$ would be reduced to $\frac{1}{2}$, and $\frac{15}{5}$ would be reduced to $\frac{3}{1}$ as follows:

$$\begin{aligned}\frac{4}{5} \div \frac{8}{15} &= \frac{4}{5} \times \frac{15}{8} \\ &= \frac{\overset{1}{\cancel{4}}}{\underset{1}{\cancel{5}}} \times \frac{\overset{3}{\cancel{15}}}{\underset{2}{\cancel{8}}} \\ &= \frac{1}{1} \times \frac{3}{2}\end{aligned}$$

$$\frac{4}{5} \div \frac{8}{15} = \frac{3}{2}$$

Thus, we see that cancellation is a method of reducing fractions and therefore produces the same answers as ordinary reduction. However, cancellation may be used only under the following conditions: Cancellation may be performed only in a multiplication problem, or in a division problem *after* the divisor has been inverted; and a number in the *numerator* and a number in the *denominator* must contain a common factor (that is, must comprise a fraction that can be reduced). The numbers being reduced are slashed with a diagonal mark; the quotient obtained when dividing by the common factor is written above the number in the numerator and below the number in the denominator. These reduced values are then used in performing the multiplication.

MIXED NUMBERS

A number composed of both a whole number and a fraction is known as a **mixed number** (for example, $3\frac{1}{2}$). Multiplication and division involving mixed numbers is easier if the mixed number is first converted to a common fraction.

Mixed numbers can be converted to fractions by applying the same basic principles that we have used to reduce fractions or to change to higher terms—namely, that any number over itself equals 1 $\left(\frac{x}{x} = 1\right)$ and that multiplying any number by 1 equals that same number ($x \cdot 1 = x$). You will recall that a whole number can be written in fraction form using a denominator of 1. (That is, $5 = \frac{5}{1}$. This is correct because $\frac{5}{1}$ means $5 \div 1$, which equals the original 5.) Now consider the mixed number $3\frac{1}{2}$.

$$\begin{aligned}3\frac{1}{2} &= 3 + \frac{1}{2} \\ &= (3 \times 1) + \frac{1}{2}\end{aligned}$$

$$= \left(3 \times \frac{2}{2}\right) + \frac{1}{2}$$

$$= \left(\frac{3}{1} \times \frac{2}{2}\right) + \frac{1}{2}$$

$$= \frac{6}{2} + \frac{1}{2}$$

$$3\frac{1}{2} = \frac{7}{2}$$

Thus, we see that $3\frac{1}{2}$ can be converted to the common fraction $\frac{7}{2}$ by multiplying the whole-number part by a fraction equal to 1 and then adding the fraction from the mixed number. (The "fraction equal to 1" is always composed of the denominator from the fraction in the mixed number over itself—in this case, $\frac{2}{2}$. Thus, the mixed number $4\frac{2}{5}$ would be converted by substituting $\frac{5}{5}$ for 1.) Any mixed number always equals an improper fraction.

The familiar procedure used to convert a mixed number to a common fraction is to "multiply the denominator times the whole number and add the numerator." This total over the original denominator (from the fraction part) is the value of the common fraction that is equivalent to the mixed number. Thus,

$$3\frac{1}{2} = 3 \overset{+}{\underset{\times}{\rightleftarrows}} \frac{1}{2} = \left\{ \begin{array}{l} \quad 3 \\ \times\ \underline{2} \text{ halves} \\ \quad 6 \text{ halves} + 1 \text{ half} = \dfrac{7}{2} \end{array} \right.$$

If the whole number is so large that the product is not immediately obvious, the multiplication can be performed digit by digit and the numerator added in the same way as are numbers that have been "carried." For example,

$16\frac{3}{8} = \frac{?}{8}$ Think: "8 × 6 is 48, plus 3 is 51."

Write 1 and carry the 5. $\left(16\frac{3}{8} = \frac{\ldots 1}{8}\right)$

Then, "8 × 1 is 8, plus 5 is 13." So

$$16\frac{3}{8} = \frac{131}{8}$$

Example 2 demonstrates how mixed numbers are handled in multiplication and division problems.

Example 2

(a) $$4\frac{1}{2} \times 3\frac{1}{3} = \frac{9}{2} \times \frac{10}{3} = \frac{\overset{3}{\cancel{9}}}{\underset{1}{\cancel{2}}} \times \frac{\overset{5}{\cancel{10}}}{\underset{1}{\cancel{3}}} = \frac{3}{1} \times \frac{5}{1} = \frac{15}{1}$$

$$4\frac{1}{2} \times 3\frac{1}{3} = 15$$

(b) $$5\frac{2}{5} \div 3\frac{3}{5} = \frac{27}{5} \div \frac{18}{5} = \frac{27}{5} \times \frac{5}{18} = \frac{\overset{3}{\cancel{27}}}{\underset{1}{\cancel{5}}} \times \frac{\overset{1}{\cancel{5}}}{\underset{2}{\cancel{18}}} = \frac{3}{1} \times \frac{1}{2}$$

$$5\frac{2}{5} \div 3\frac{3}{5} = \frac{3}{2}$$

(c) $$4\frac{7}{8} \div 1\frac{11}{16} = \frac{39}{8} \div \frac{27}{16} = \frac{39}{8} \times \frac{16}{27} = \frac{\overset{13}{\cancel{39}}}{\underset{1}{\cancel{8}}} \times \frac{\overset{2}{\cancel{16}}}{\underset{9}{\cancel{27}}} = \frac{13}{1} \times \frac{2}{9}$$

$$4\frac{7}{8} \div 1\frac{11}{16} = \frac{26}{9}$$

Notice in division that a mixed-number divisor should be converted to an improper fraction *before* the divisor is inverted.

In certain instances, it is necessary to change an improper fraction to a mixed number. This operation simply reverses the former procedure. Suppose that $\frac{19}{8}$ is to be converted to a mixed number. The numerator can first be broken down into a multiple of the denominator plus a remainder. The familiar properties of 1 $\left(\text{that is, } \frac{x}{x} = 1 \text{ and } x \cdot 1 = x\right)$ are then applied as before:

$$\frac{19}{8} = \frac{2(8) + 3}{8} = \frac{2(8)}{8} + \frac{3}{8}$$

$$= \left(\frac{2}{1} \times \frac{8}{8}\right) + \frac{3}{8}$$

$$= \left(\frac{2}{1} \times 1\right) + \frac{3}{8}$$

$$= 2 + \frac{3}{8}$$

$$\frac{19}{8} = 2\frac{3}{8}$$

Thus, $\frac{19}{8} = 2\frac{3}{8}$. In actual practice, however, the procedure used to convert improper fractions to mixed numbers applies the fact that a fraction is an indicated division. That is, $\frac{19}{8}$ means $19 \div 8$, which equals 2 with a remainder of 3. Hence, $\frac{19}{8} = 2\frac{3}{8}$.

Addition may be performed using mixed numbers (without changing them to improper fractions). Although it is generally considered acceptable to leave answers as improper fractions, it is not correct to leave a mixed number that includes an improper fraction. The improper fraction should be changed to a mixed number and this amount added to the whole number to obtain a mixed number that includes a proper fraction (reduced to lowest terms). Thus, if the sum of an addition problem is $7\frac{14}{6}$, this result must be converted as follows:

$$7\frac{14}{6} = 7 + \frac{14}{6}$$

$$= 7 + 2\frac{2}{6}$$

$$= 9\frac{2}{6}$$

$$7\frac{14}{6} = 9\frac{1}{3}$$

Example 3 shows cases of addition with mixed numbers.

Example 3

(a)

$$\begin{array}{r} 7\frac{5}{9} \\ +\ 3\frac{1}{9} \\ \hline 10\frac{6}{9} = 10\frac{2}{3} \end{array}$$

(b)

$$\begin{array}{rcl} 6\frac{5}{8} & = & 6\frac{15}{24} \\ +\ 9\frac{2}{3} & = & 9\frac{16}{24} \\ \hline & & 15\frac{31}{24} = 16\frac{7}{24} \end{array}$$

(c)
$$\begin{array}{rl} 16\frac{2}{3} = & 16\frac{4}{6} \\ +17\frac{5}{6} = & 17\frac{5}{6} \\ \hline & 33\frac{9}{6} = 34\frac{3}{6} = 34\frac{1}{2} \end{array}$$

Subtraction may also be performed using mixed numbers. You will recall that subtraction usually requires regrouping (or borrowing). This procedure, when applied to fractions, means that improper fractions must often be created where they did not exist.

Consider the problem $7\frac{3}{8} - 4\frac{7}{8}$. Although the mixed number $4\frac{7}{8}$ is smaller than $7\frac{3}{8}$, the fraction $\frac{7}{8}$ cannot be subtracted from $\frac{3}{8}$. In order to perform the operation, we must regroup the $7\frac{3}{8}$. Mixed numbers are regrouped by borrowing one unit from the whole number, converting this unit to a fraction equal to 1 (the common denominator over itself), and adding this fractional "1" to the fraction of the mixed number. Using this procedure, we regroup $7\frac{3}{8}$ as follows:

$$\begin{aligned} 7\frac{3}{8} &= 7 + \frac{3}{8} \\ &= 6 + 1 + \frac{3}{8} \\ &= 6 + \frac{8}{8} + \frac{3}{8} \\ &= 6 + \frac{11}{8} \\ 7\frac{3}{8} &= 6\frac{11}{8} \end{aligned}$$

Thus, $7\frac{3}{8}$ equals $6\frac{11}{8}$. In actual practice, the procedure of borrowing 1 from the whole number and adding it to the existing fraction may be done mentally, with only the result being written. The preceding subtraction problem and other problems are illustrated in Example 4.

Example 4

(a)
$$\begin{array}{r} 5\frac{5}{6} \\ -1\frac{1}{6} \\ \hline 4\frac{4}{6} = 4\frac{2}{3} \end{array}$$

(b)
$$\begin{array}{rl} 7\frac{3}{8} = & 6\frac{11}{8} \\ -4\frac{7}{8} = & -4\frac{7}{8} \\ \hline & 2\frac{4}{8} = 2\frac{1}{2} \end{array}$$

(c)
$$\begin{array}{rcrcr} 15\frac{1}{3} & = & 15\frac{4}{12} & = & 14\frac{16}{12} \\ -\ 8\frac{3}{4} & = & -\ 8\frac{9}{12} & = & -\ 8\frac{9}{12} \\ \hline & & & & 6\frac{7}{12} \end{array}$$

Notice that the fractions in mixed numbers are first changed to fractions with a common denominator before regrouping for subtraction, as illustrated by (c).

SECTION 3 PROBLEMS

1. Reduce the following fractions to lowest terms (that is, reduce as much as possible).

a. $\frac{16}{24}$ b. $\frac{14}{6}$ c. $\frac{28}{36}$ d. $\frac{48}{84}$ e. $\frac{34}{51}$

f. $\frac{75}{45}$ g. $\frac{57}{76}$ h. $\frac{108}{90}$ i. $\frac{189}{321}$ j. $\frac{168}{288}$

2. Convert the following fractions to fractions with least common denominators.

a. $\frac{1}{3}$ and $\frac{3}{4}$ b. $\frac{1}{2}$ and $\frac{4}{7}$ c. $\frac{4}{9}$ and $\frac{2}{3}$ d. $\frac{3}{8}$ and $\frac{1}{16}$

e. $\frac{5}{8}$ and $\frac{1}{6}$ f. $\frac{1}{5}$, $\frac{1}{4}$, and $\frac{2}{3}$ g. $\frac{2}{7}$, $\frac{1}{6}$, and $\frac{1}{2}$ h. $\frac{7}{10}$, $\frac{2}{5}$, and $\frac{1}{3}$

i. $\frac{2}{9}$, $\frac{3}{4}$, and $\frac{5}{6}$ j. $\frac{3}{5}$, $\frac{1}{4}$, and $\frac{5}{8}$

3. Compare the following fractions by converting to a common denominator. Then arrange the original fractions in order of size, starting with the smallest.

a. $\frac{3}{7}$, $\frac{3}{4}$, $\frac{5}{14}$, $\frac{1}{2}$, $\frac{5}{8}$ $\frac{5}{7}$ b. $\frac{7}{12}$, $\frac{5}{6}$, $\frac{3}{4}$, $\frac{7}{9}$, $\frac{5}{8}$, $\frac{13}{24}$

4. Change the following mixed numbers to improper fractions.

a. $3\frac{1}{4}$ b. $6\frac{5}{8}$ c. $8\frac{2}{7}$ d. $5\frac{2}{9}$ e. $12\frac{2}{5}$ f. $20\frac{1}{3}$

g. $26\frac{3}{7}$ h. $32\frac{1}{2}$ i. $18\frac{5}{6}$ j. $9\frac{2}{11}$ k. $15\frac{2}{7}$ l. $7\frac{7}{12}$

5. Convert the following improper fractions (and improper mixed numbers) to mixed numbers. Reduce if possible.

a. $\frac{17}{6}$ b. $\frac{46}{5}$ c. $\frac{34}{4}$ d. $\frac{24}{7}$ e. $\frac{49}{3}$ f. $\frac{37}{12}$

g. $\frac{92}{6}$ h. $\frac{54}{4}$ i. $4\frac{5}{3}$ j. $8\frac{16}{5}$ k. $9\frac{7}{3}$ l. $5\frac{9}{2}$

6. Find the sums, reducing when possible.

a. $\frac{1}{4} + \frac{2}{5}$ b. $\frac{1}{6} + \frac{3}{8}$ c. $\frac{3}{5} + \frac{2}{3} + \frac{1}{10}$ d. $\frac{1}{4} + \frac{1}{8} + \frac{5}{6}$ e. $\frac{7}{12} + \frac{4}{9} + \frac{1}{4} + \frac{1}{6}$ f. $\frac{5}{6} + \frac{2}{3} + \frac{4}{9} + \frac{3}{18}$

g. $6\frac{2}{5} + 5\frac{1}{2}$ h. $3\frac{3}{4} + 6\frac{1}{9}$ i. $7\frac{3}{8} + 7\frac{2}{5}$ j. $22\frac{5}{6} + 14\frac{4}{5}$

7. Subtract, reducing when possible.

a. $\frac{5}{6} - \frac{1}{3}$ b. $\frac{3}{4} - \frac{1}{5}$ c. $\frac{7}{8} - \frac{3}{7}$ d. $\frac{7}{12} - \frac{1}{2}$ e. $\frac{2}{3} - \frac{3}{5}$

f. $15\frac{1}{4} - 3\frac{1}{3}$ g. $24\frac{1}{6} - 5\frac{3}{4}$ h. $33 - 12\frac{1}{6}$ i. $56 - 40\frac{5}{7}$ j. $28 - 7\frac{1}{2}$

8. Multiply, reducing whenever possible.

a. $\frac{2}{7} \times \frac{5}{6}$ b. $\frac{3}{8} \times \frac{7}{12}$ c. $\frac{5}{9} \times \frac{9}{10}$ d. $\frac{14}{20} \times \frac{5}{28}$ e. $\frac{4}{15} \times \frac{3}{24}$

f. $\frac{6}{7} \times 2\frac{1}{2}$ g. $6\frac{2}{3} \times \frac{1}{5}$ h. $4\frac{1}{5} \times \frac{4}{7}$ i. $12\frac{1}{4} \times 5\frac{1}{7}$ j. $3\frac{3}{8} \times 3\frac{5}{9}$

9. Multiply, reducing whenever possible.

a. $\frac{4}{14} \times \frac{7}{15} \times \frac{3}{16}$ b. $\frac{3}{8} \times \frac{5}{6} \times \frac{7}{10}$ c. $4\frac{1}{2} \times \frac{2}{3} \times 8\frac{1}{6}$ d. $2\frac{5}{8} \times 2\frac{2}{7} \times 4\frac{2}{3}$

10. Divide, reducing to lowest terms.

a. $\frac{2}{9} \div \frac{2}{3}$ b. $\frac{4}{9} \div \frac{4}{5}$ c. $\frac{3}{8} \div \frac{6}{15}$ d. $\frac{5}{6} \div \frac{3}{18}$ e. $\frac{6}{7} \div \frac{18}{21}$

f. $\frac{3}{5} \div \frac{6}{11}$ g. $\frac{9}{10} \div \frac{1}{2}$ h. $3\frac{1}{5} \div \frac{2}{5}$ i. $\frac{3}{8} \div 2\frac{1}{2}$ j. $1\frac{1}{3} \div 7\frac{1}{5}$

SECTION 4

DECIMAL FRACTIONS

As the developing number system became more sophisticated, a method was devised for expressing parts of a whole without writing a complete common fraction; this is the system of **decimal notation.** Decimal values are actually abbreviated fractions.

That is, the written number is only the numerator of an indicated fraction. The denominator of the indicated fraction is some power of 10. This denominator is not written but is indicated by the number of digits to the right of the decimal point. For example, 0.7 indicates $\frac{7}{10}$, or 0.43 indicates $\frac{43}{100}$, without the 10 or 100 being written. Thus, since decimal values represent fractional parts of a whole, they are known as **decimal fractions.**

Decimal fractions are an extension of our whole number (and integer) system, both because they are based on powers of 10 and also because the value of any digit is determined by the position it holds in the number. The following chart shows the value and power of 10 that is associated with each place in a decimal number. (Some whole-number places are included to show that decimal values are an extension of whole-number place values.) Notice that the names for decimal place values repeat, in reverse order, the names of integer place values—except that the letters "-ths" are added to indicate that these are decimal-place values.

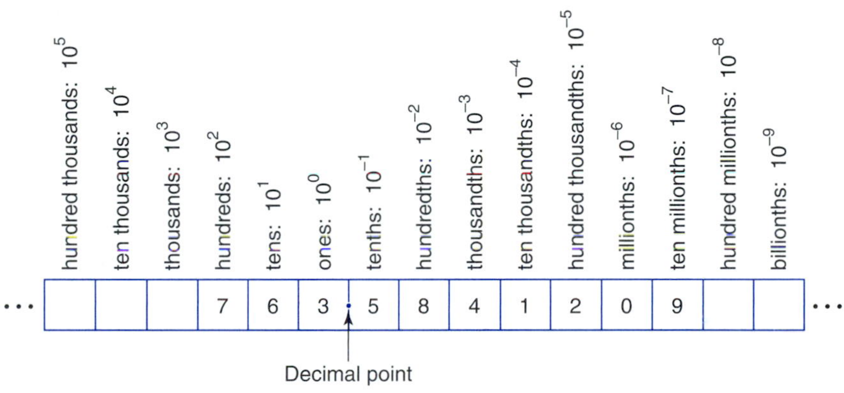

A decimal point (or period) is used in a number to show where the whole amount ends and the fractional portion begins. A whole number is usually written without a decimal point, but every integer is understood to have a decimal point after the units place. Thus, 15 could be written "15." (or "15.0," or "15.00," etc.) and would be just as correct.

You will recall from our study of integers that 10^2 represents 100, and the digit 7 in the chart above thus has the value 7×100, or 700. Similarly, 10^{-2} represents $\frac{1}{100}$, and the digit 8 has the value $8 \times \frac{1}{100}$, or $\frac{8}{100}$. In the same manner, the digit 5 represents "5 tenths," or $\frac{5}{10}$; the digit 4 represents "4 thousandths," or $\frac{4}{1000}$. Hence, any digit in the first place following a decimal represents that number divided by 10 $\left(\frac{x}{10}\right)$; any number in the second place represents that number over 100 $\left(\frac{y}{100}\right)$; any number in the third decimal place represents that number over 1,000 $\left(\frac{z}{1000}\right)$; and so on.

As pointed out previously, a decimal is actually an abbreviated fraction. Thus, the decimal fraction 0.475 (the zero is placed before the decimal point purely for clarity) represents a common fraction with a numerator of 475 and (since the decimal contains three digits to the right of the decimal point) a denominator of 1,000. Therefore, the decimal fraction 0.475 could be written $\frac{475}{1000}$. This illustrates why decimal fractions, like common fractions, are known as **rational numbers,** since they can be expressed as the quotient of two integers.

Decimal fractions are spoken (or written) by saying the digits as if they represented a whole number and then saying the place value associated with the farthest place to the right of the decimal point. Thus,

NUMBER	SPOKEN (OR WRITTEN) FORM
0.9	"Nine *tenths*"
0.14	"Fourteen *hundredths*"
0.007	"Seven *thousandths*"
0.3024	"Three thousand twenty-four *ten-thousandths*"
0.00508	"Five hundred eight *hundred-thousandths*"
25.6	"Twenty-five and six *tenths*"
2.03	"Two and 3 *hundredths*"
4,070.035	"Four thousand seventy and thirty-five *thousandths*"

Notice that zeros between the decimal point and the first significant digit do not affect the spoken form. (For example, 0.007 is said "*seven* thousandths.") Also, the word "and" is spoken (or written) solely to indicate the location of the decimal point. (Thus, 2.03 is pronounced "two *and* three hundredths.") Zeros to the right of a decimal fraction (with no other digits following) do not affect its value. Thus, 0.4, 0.40, 0.400, and 0.4000 are equal in value, because

$$\frac{4}{10} = \frac{40}{100} = \frac{400}{1000} = \frac{4,000}{10,000}$$

CONVERTING COMMON FRACTIONS TO DECIMAL FRACTIONS

Manual computation can often be accomplished efficiently using common fractions, although at times common fractions become cumbersome. Most computation in modern offices, however, is done on electronic machines, which are equipped to handle decimal values rather than common fractions. Thus, business efficiency requires a knowledge not only of common fractions and decimal fractions separately, but also of how to convert each form to the other.

You will recall that a common fraction is an indicated division. Hence, $\frac{2}{5}$ means $2 \div 5$. This principle is the basis for converting common fractions to decimal fractions. That is, a common fraction is converted to a decimal fraction by dividing the denominator into the numerator:

$$\frac{n}{d} = n \div d = d\overline{)n}$$

This procedure applies to improper as well as proper fractions. If a mixed number is to be converted to a decimal number, the fractional part of the mixed number is converted in this same manner, and the decimal equivalent is then affixed to the whole number of the mixed number. Example 1 illustrates these procedures.

Example 1

(a) $\frac{2}{5} = 5\overline{)2.0}$ = .4 = 0.4
$\quad\underline{2\,0}$

(b) $\frac{3}{40} = 40\overline{)3.000}$ = .075 = 0.075
$\quad\underline{2\,80}$
$\quad 200$
$\quad\underline{200}$

(c) $\frac{9}{8} = 8\overline{)9.000}$ = 1.125 = 1.125

(d) $2\frac{8}{25}$: $\frac{8}{25} = 25\overline{)8.00}$ = .32 = 0.32

so $2\frac{8}{25} = 2.32$

The student is no doubt aware that many fractions do not have an exact decimal equivalent. This will always be the case (unless the fraction represents a whole number) if the denominator is 3 or is divisible by 3, or if the denominator is a prime number larger than five. (A **prime number** is any number that is divisible by no numbers except itself and 1—for example, 7, 11, 17, 23, 31, to name but a few.)

You may not always recognize whether a fraction has an exact decimal equivalent. The *recommended procedure for converting fractions* is first to divide to the hundredths' place (two decimal places). If further inspection reveals that the quotient will come out to an exact digit after one more division, then include the third decimal place. If it is obvious that the quotient would not be exact even with three decimal places, then stop dividing after two decimal places, and express the remainder as a fraction. (Students using calculators may round quotients to the nearest thousandth, or the third decimal place. Rounding is reviewed in Chapter 1.) Consider the following examples.

Example 1 (cont.)

(e) $\frac{3}{8} = 8\overline{)3.000}$ = .375 = 0.375

(f) $\frac{13}{200} = 200\overline{)13.000}$ = .065 = 0.065

(g) $\frac{3}{7} = 7\overline{)3.00}$ = $.42\frac{6}{7} = 0.42\frac{6}{7}$ (or 0.429)

(h) $\frac{16}{29} = 29\overline{)16.00}$ = $.55\frac{5}{29} = 0.55\frac{5}{29}$ (or 0.552)

CONVERTING DECIMAL FRACTIONS TO COMMON FRACTIONS

The standard procedure for converting a decimal fraction to a common fraction is to pronounce the value of the decimal, write this value in fraction form, and then reduce this fraction. This is shown in Example 2.

Example 2

(a) 0.2 = 2 tenths $= \frac{2}{10} = \frac{1}{5}$

(b) 0.12 = 12 hundredths $= \frac{12}{100} = \frac{3}{25}$

(c) 0.336 = 336 thousandths $= \frac{336}{1{,}000} = \frac{42}{125}$

(d) 0.0045 = 45 ten-thousandths $= \frac{45}{10{,}000} = \frac{9}{2{,}000}$

If a decimal fraction also contains a common fraction, the conversion process begins in a similar manner but requires additional steps. First, we rewrite the given decimal as a fraction, and then we convert the numerator to an improper fraction. Next, we multiply by an appropriate fraction equal to 1 $\left(\frac{x}{x} = 1\right)$ and then reduce. The following examples show how $0.83\frac{1}{3}$ and $0.31\frac{1}{4}$ are converted to fractions.

Example 2 (cont.)

(e)
$$0.83\frac{1}{3} = \frac{83\frac{1}{3}}{100} = \frac{\frac{250}{\cancel{3}}}{100} = \frac{\frac{250}{3}}{100} \times \frac{\cancel{3}}{3} = \frac{250}{300}$$
$$0.83\frac{1}{3} = \frac{5}{6}$$

(f)
$$0.31\frac{1}{4} = \frac{31\frac{1}{4}}{100} = \frac{\frac{125}{4}}{100} = \frac{\frac{125}{\cancel{4}}}{100} \times \frac{\cancel{4}}{4} = \frac{125}{400}$$
$$0.31\frac{1}{4} = \frac{5}{16}$$

As noted previously, many computations can be done more easily by using fractional equivalents for decimal numbers, especially if the decimals are as complicated as those above. To save time, however, you should be able to recognize when a decimal is the equivalent of one of the common fractions most often used—without having to actually convert it. These equivalents are listed on page 51, and you should memorize all that you do not already know.

ADDITION AND SUBTRACTION

Addition and subtraction with decimals are very similar to addition and subtraction with integers. The same names apply to the parts of each problem:

Addend	0.36	**Minuend**	0.73
Addend	+0.58	**Subtrahend**	−0.46
Sum	0.94	**Difference** or **Remainder**	0.27

As in addition and subtraction with whole numbers, decimal numbers must also be arranged so that all the digits of the same place are in the same column. That is, all the tenth's places must be in a vertical line, all the hundredth's places in one column, and so forth. In so arranging decimal numbers, the decimal points will always fall in a straight line. Therefore, in actual practice, we arrange the numbers by aligning the decimal points.

The actual addition or subtraction of decimal fractions is performed in exactly the same way as with integers; the decimal points have no effect on the operations. The decimal point in the answer is located directly under the decimal points of the problem.

It was pointed out earlier that appending zeros to the right of a decimal number does not alter its value (that is, $3.7 = 3.70 = 3.700$, and so on). Thus, if the numbers of a problem contain different numbers of decimal places, you may wish to add zeros so that all will contain the same number of decimal places. This is particularly helpful in subtraction problems where the subtrahend contains more decimal places than the minuend.

Example 3 (a) Add: $0.306 + 9.8 + 14.65 + 6.775 + 24 + 0.09$. (b) Subtract: $18.6 - 5.897$.

(a)
```
    0.306
    9.8
   14.65
    6.775
   24.
 +  0.09
  -------
   55.621
```

(b)
```
   18.600
 −  5.897
  -------
   12.703
```

Notice that zeros were inserted in the minuend to facilitate the subtraction.

MULTIPLICATION

The standard terminology is used to identify decimal numbers in a multiplication problem:

Multiplicand or **Factor**	0.3
Multiplier or **Factor**	× 0.4
Product	0.12

Decimal numbers are written down for multiplication in the same arrangement that would be used if there were no decimal points. The actual multiplication of digits is also performed in the same way as for integers. The only variation comes in the placing of the decimal point in the product. To determine how this is done, consider the problem 12.4 × 3.13:

$$\begin{array}{r} 12.4 \\ \times\ 3.13 \\ \hline 372 \\ 124 \\ 372 \\ \hline 38812 \end{array}$$

Observe that this multiplication problem is approximately 12 × 3; therefore, the product must be approximately 36. This indicates that the decimal point must be located between the 8s. Thus,

$$\begin{array}{r} 12.4 \\ \times\ 3.13 \\ \hline 372 \\ 124 \\ 372 \\ \hline 38.812 \end{array}$$

The correct position of the decimal point in every multiplication product could thus be determined by approximating the answer and locating the decimal point accordingly. However, you would notice that the product invariably contains the same number of decimal places as the total number of decimal places in the factors. Therefore, as a time-saving device, it is customary to count the total number of decimal places in the factors and place the decimal point accordingly, rather than to estimate the product. (In the preceding problem, 12.4 contained one decimal place, and 3.13 contained two decimal places; thus, the product contains three decimal places—38.812.) This example can also be explained as $\frac{124}{10} \times \frac{313}{100} = \frac{38{,}812}{1{,}000}$. Observe that the denominator of 1,000 indicates three decimal places.

Sometimes basic multiplication produces a number containing fewer digits than the total number of decimal places required for the product. In this case, you must insert a zero or zeros between the decimal point and the first significant digit so that the product will contain the correct number of decimal places. Consider the example 0.04 × 0.2:

$$\begin{array}{r} 0.04 \\ \times\ 0.2 \\ \hline 8 \end{array}$$

The factors require a three-digit product. Thus, two zeros must be inserted after the decimal point:

$$\begin{array}{r} 0.04 \\ \times\ \ 0.2 \\ \hline 0.008 \end{array}$$

You can readily understand that 0.008 is the correct answer, because 0.04 × 0.2 is equal to $\frac{4}{100} \times \frac{2}{10} = \frac{8}{1{,}000}$.

Some additional examples are given below.

Example 4

(a)
$$\begin{array}{r} 41.8173 \\ \times\ \ \ 2.36 \\ \hline 2509038 \\ 1254519 \\ 836346 \\ \hline 98.688828 \end{array}$$

(b)
$$\begin{array}{r} 0.0547 \\ \times\ \ \ 3.6 \\ \hline 3282 \\ 1641 \\ \hline 0.19692 \end{array}$$

(c)
$$\begin{array}{r} 7.38 \\ \times\ 0.004 \\ \hline 0.02952 \end{array}$$

(d)
$$\begin{array}{r} 0.24 \\ \times\ 0.03 \\ \hline 0.0072 \end{array}$$

Multiplication by powers of 10 presents a special case. Study the operations in Example 5.

Example 5

(a)	(b)	(c)	(d)
25.13	65.8	78	3.9471
× 10	× 100	× 100	× 1,000
251.30	6,580.0	7,800	3,947.1000

Observe that each product in Example 5 contains the same digits as the multiplicand (along with some zeros appended) and that each decimal point has been moved to the right the same number of places as there were zeros in the power of 10. By applying this information, you should never actually have to multiply by a power of 10. Instead, you can obtain the product simply by moving the decimal point as many places to the right as there are zeros in the power of 10. This is shown in Example 6.

Example 6

(a) $25.13 \times 10 = 25_x1.3$

$25.13 \times 10 = 251.3$

(b) $65.8 \times 100 = 65_x80.$

$65.8 \times 100 = 6{,}580$

(c) $78 \times 100 = 78_x00.$

$78 \times 100 = 7{,}800$

(d) $3.9471 \times 1{,}000 = 3_x947.1$

$3.9471 \times 1{,}000 = 3{,}947.1$

DIVISION

Decimal numbers in a division problem are identified by the terms previously learned:

```
               5.3     Quotient
Divisor   1.5)7.95     Dividend
```

The basic operation of division is performed in the same manner for decimal numbers as for whole numbers. The decimal points do not affect the division process itself, but only the location of the decimal point in the quotient. To understand how the decimal points are handled, consider the following example:

```
        51
3.02)15.402
     15 10
       302
       302
```

Notice that this division is approximately 3)15; therefore, the quotient must be approximately 5. This indicates that the decimal point in the quotient must be located after the 5.

```
       5.1
3.02)15.402
     15 10
       302
       302
```

It would be possible to position the decimal point in every quotient by first estimating the result and then placing the decimal point accordingly. However, this requires additional time and effort; hence, some quick, easy method for determining the correct location of the quotient decimal is necessary. You are no doubt familiar with the method commonly used to position the decimal quickly; that is, to "move the decimal point to the end of the divisor and move the decimal in the dividend the same number of places." Let us see why this procedure is permissible.

It was observed that the decimal point of the quotient is always located directly above the decimal point of the dividend if the divisor is a whole number. For example,

$$\begin{array}{r} 2\ 13 \\ 12\overline{)25.56} \\ \underline{24} \\ 1\ 5 \\ \underline{1\ 2} \\ 36 \\ \underline{36} \end{array}$$

Since this problem is approximately $12\overline{)25}$, the result must be approximately 2. Thus, the decimal point of the quotient should be located directly above the decimal point in the operation:

$$\begin{array}{r} 2.13 \\ 12\overline{)25\overset{\uparrow}{.}56} \\ \underline{24} \\ 1\ 5 \\ \underline{1\ 2} \\ 36 \\ \underline{36} \end{array}$$

Thus, you know exactly where the decimal point belongs if the divisor is a whole number. With this in mind, let us apply three familiar principles—namely, that a fraction is an indicated division, that any number over itself equal 1 $\left(\text{that is, } \frac{x}{x} = 1\right)$, and that multiplying any number by 1 equals that same number ($x \cdot 1 = x$). Now suppose that we have the division problem $3.2\overline{)9.28}$. This is the division indicated by the fraction $\frac{9.28}{3.2}$. Then

$$\frac{9.28}{3.2} = \frac{9.28}{3.2} \times 1$$

$$= \frac{9.28}{3.2} \times \frac{10}{10}$$

$$\frac{9.28}{3.2} = \frac{92.8}{32}$$

Thus, the divisions indicated by each fraction are equal, or

$$3.2\overline{)9.28} = 32\overline{)92.8}$$

But we know the correct location for the decimal point when dividing by a whole number:

$$\begin{array}{r} 2.9 \\ = 32\overline{)92\overset{\uparrow}{.}8} \\ \underline{64} \\ 28\ 8 \\ \underline{28\ 8} \end{array}$$

Therefore, the other quotient must also be 2.9.

```
       2.9           2.9
3.2)9.28  = 32)92.8
    6 4
    2 88
    2 88
```

Thus, we can always position the decimal easily by "making the divisor a whole number," so to speak. But when we "move the decimal point to the end of the divisor and move the decimal point in the dividend a corresponding number of places," we are, in effect, multiplying by a fraction equal to 1, which does not change the value. Stated differently, moving both decimals one place to the right is the same as multiplying by $\frac{10}{10}$; moving the decimals two places is equivalent to multiplying by $\frac{100}{100}$; moving the decimals three digits is equivalent to multiplying by $\frac{1{,}000}{1{,}000}$; and so forth.

Having determined why it is permissible to "move the decimal points," let us now consider some special cases. The decimal point in a dividend must often be moved more places to the right than there are digits. In this case, we must append zeros to the dividend in order to fill in the required number of places. (Suppose that the decimal point in the dividend 14.3 must be moved three places to the right. This is the same as multiplying 14.3 by 1,000; so $14_{x}3\_\,\_. = 14{,}300$.) If the dividend is a whole number, we obviously must first insert a decimal point before moving it to the right. (For example, if a divisor must be multiplied by 10 to become a whole number and the dividend is 17, then 17 must also be multiplied by 10. That is, the decimal point must be moved one place to the right: $17 \times 10 = 17. \times 10 = 17_{x}\_. = 170$.)

It has been pointed out that appending zeros to the right of a decimal fraction does not change its value. (That is, 0.52 = 0.520 = 0.5200 = 0.52000, and so forth.) When a division problem does not terminate (or end) within the number of digits contained in the dividend, the operation may be continued by appending zeros to the dividend. The process may be continued until the quotient terminates or until it contains as many decimal places as required.

Once the decimal point has been located in the quotient, each succeeding place must contain a digit. If normal division skips some of the first decimal places, these places must be held by zeros. (This procedure is similar to that in multiplication, when zeros must be inserted after the decimal point in order to complete the required number of decimal places.)

Example 7 shows how these procedures apply.

Example 7 (a) 3.02)15.402 (b) 4.725)37.8

```
(a)            5.1          (b)            8.
      3.02)15.40.2               4.725)37.800.
           15 10                       37 800
              30 2
              30 2
```

(c) $0.72\overline{)36.}$ (d) $6.4\overline{)0.3968}$

$$\begin{array}{r} 50. \\ 0.72\overline{)36.00.} \\ \underline{36\ 0} \\ 0 \\ \underline{0} \end{array} \qquad \begin{array}{r} .062 \\ 6.4\overline{)0.3.968} \\ \underline{3\ 84} \\ 128 \\ \underline{128} \end{array}$$

A special case develops when the divisor of a problem is a power of 10, as shown in Example 8.

Example 8 (a) $10\overline{)13.6}$ (b) $100\overline{)245.3}$

$$\begin{array}{r} 1.36 \\ 10\overline{)13.60} \\ \underline{10} \\ 3\ 6 \\ \underline{3\ 0} \\ 60 \\ \underline{60} \end{array} \qquad \begin{array}{r} 2.453 \\ 100\overline{)245.300} \\ \underline{200} \\ 45\ 3 \\ \underline{40\ 0} \\ 5\ 30 \\ \underline{5\ 00} \\ 300 \\ \underline{300} \end{array}$$

(c) $100\overline{)7.31}$ (d) $1{,}000\overline{)629}$

$$\begin{array}{r} .0731 \\ 100\overline{)7.3100} \\ \underline{7\ 00} \\ 310 \\ \underline{300} \\ 100 \\ \underline{100} \end{array} \qquad \begin{array}{r} .629 \\ 1{,}000\overline{)629.000} \\ \underline{600\ 0} \\ 29\ 00 \\ \underline{20\ 00} \\ 9\ 000 \\ \underline{9\ 000} \end{array}$$

In each operation in Example 8, the quotient contains the same digits as the dividend, but the decimal point is moved to the left the same number of places as there were zeros in the divisor. Knowing this, students should never actually have to perform a division by a power of 10. Rather, you can obtain the answer simply by moving the decimal point to the left. This is shown in Example 9.

Example 9 (a) $13.6 \div 10 = 1.3_{x}6 = 1.36$ (b) $245.3 \div 100 = 2.45_{x}3 = 2.453$

(c) $7.31 \div 100 = .\_7_{x}31 = 0.0731$ (d) $629 \div 1{,}000 = .629 = 0.629$

Hint. Students often confuse multiplication and division by powers of 10 and forget which way to move the decimal point. If this happens, you can quickly determine the correct direction by making up an example using small numbers. For example, $4 \times 10 = 40$ (that is, $4.0 \times 10 = 40$). The decimal point of the 4 moved to the right; therefore, the decimal point moves toward the right for *all* multiplication by powers of 10. Hence, the decimal point obviously moves to the left for division.

Note. Addition with decimal fractions is both commutative and associative; multiplication is similarly commutative and associative. Further, the distributive property of multiplication over addition also applies. You can verify these properties for yourself by selecting some decimal values and following the examples for whole numbers in Section 2 of this Appendix.

SECTION 4 PROBLEMS

1. In the number 6,432.85179, identify the place value associated with each of the following digits.

a. 1 b. 2 c. 3 d. 4 e. 5 f. 6 g. 7 h. 8 i. 9

2. Write in words the values of the following numbers.

a. 0.5 b. 0.06 c. 0.018 d. 0.32 e. 8.6 f. 17.44 g. 500.103 h. 9.0015

3. Convert the following common fractions (and mixed numbers) to their decimal fraction equivalents.

a. $\frac{2}{5}$ b. $\frac{1}{4}$ c. $\frac{6}{20}$ d. $\frac{21}{30}$ e. $\frac{16}{50}$ f. $\frac{15}{200}$ g. $\frac{1}{7}$ h. $\frac{4}{9}$ i. $\frac{5}{12}$

j. $\frac{6}{11}$ k. $\frac{4}{13}$ l. $\frac{7}{26}$ m. $\frac{18}{3}$ n. $\frac{163}{4}$ o. $\frac{108}{20}$ p. $5\frac{1}{9}$ q. $3\frac{5}{6}$ r. $9\frac{107}{250}$

4. Find the common fraction equivalent to each of the following decimals.

a. 0.06 b. 0.4 c. 0.25 d. 0.325 e. 0.155

f. 0.008 g. 0.036 h. $0.45\frac{1}{2}$ i. $0.04\frac{1}{4}$ j. $0.036\frac{4}{5}$

5. Multiply or divide, as indicated, by moving the decimal points.

a. $356.2 \div 10$ b. 2.645×10 c. $556.63 \div 100$ d. $104.88 \div 1{,}000$

e. $0.88452 \times 1{,}000$ f. $16 \div 10$ g. 26.5×10 h. 8.812×100

i. 2.8×100 j. $3.651 \div 1{,}000$ k. 525×100 l. $0.644 \div 100$

m. $54.6 \times 10{,}000$ n. $99.45 \div 10{,}000$

6. Add. [Arrange Parts (j)–(m) in columns before adding.]

a.	b.	c.	d.	e.
4.96	0.164	0.1887	12.713	175.024
6.01	0.917	0.5295	81.591	716.054
4.53	0.244	0.4897	20.307	505.149
+8.70	+0.415	+0.0972	+81.633	+347.664

f.	g.	h.	i.
18.131	1.02	6.3	52.7
4.01	62.6	341	149.66
215.8	7.482	8.98	62.435
0.145	138.5	12.455	9.15
42.5	16.933	0.138	317.8
+ 0.005	+541.87	+530.68	+274.511

j. 123.7 + 0.515 + 22.7 + 156.68 + 7.37+ 120.64

k. 54.88 + 6.83 + 328.6 + 330 + 5.81 + 146.9

l. 3.691 + 304.84 + 26.641 + 31.106 + 0.482 + 56.2

m. 94.171 + 135 + 8.06 + 26.71 + 125.7 + 206.728

7. Subtract. [Arrange Parts (n)–(q) in columns before subtracting.]

a.	b.	c.	d.	e.
135.86	5.224	38.267	385.002	56.1146
− 88.91	− 1.499	− 9.408	− 144.686	− 48.3036

f.	g.	h.	i.	j.
15.001	634.703	27.549	34.1006	265.4
− 11.156	− 75.881	− 14.62	− 5.091	− 123.056

k.	l.	m.
108.6	543.4	64
− 56.321	− 88.355	− 26.303

n. 54.2 − 6.18　　o. 314 − 28.88　　p. 105.5 − 36.606　　q. 5.4 − 0.8212

8. Multiply.

a.	b.	c.	d.	e.
6.24	18.3	6.06	352.8	0.124
× 6.4	× 5.4	× 0.48	× 0.015	× 5.25

f.	g.	h.	i.	j.
0.311	0.514	0.005	0.3551	0.418
× 0.47	× 0.82	× 0.029	× 2.64	× 0.033

9. Divide. (When quotients are not exact whole numbers, follow the procedure recommended in this section for changing fractions to decimals.)

a. $6.5\overline{)507}$　　b. $5.17\overline{)635.91}$　　c. $0.125\overline{)5.75}$　　d. $0.122\overline{)8.296}$

e. $5.62\overline{)764.32}$　　f. $1.742\overline{)5{,}696.34}$　　g. $0.64\overline{)0.8128}$　　h. $3.32\overline{)0.7968}$

i. $13.5\overline{)4.644}$　　j. $0.724\overline{)23.58792}$　　k. $5.6\overline{)4.368}$　　l. $28.44\overline{)1.83438}$

m. $25.8\overline{)11{,}352}$　　n. $5.62\overline{)230.42}$　　o. $0.285\overline{)91.485}$　　p. $4.132\overline{)1.15696}$

Appendix A Glossary

Addend. Any of the numbers being combined in addition. (Example: In $2 + 3 = 5$, the 2 and 3 are addends.)

Associative property. Pertains to the grouping of three numbers for addition or multiplication; states that the result will be the same regardless of which two numbers are added (or multiplied) first. That is, $(a + b) + c = a + (b + c)$ for addition; $(a \cdot b)c = a(b \cdot c)$ for multiplication.

Common fraction. An indicated division, using a horizontal line. (Examples: $\frac{3}{5}$ or $\frac{7}{4}$)

Commutative property. Pertains to the order of two numbers for addition or multiplication; states that the result will be the same regardless of which number is taken first. That is, $a + b = b + a$ for addition; $a \cdot b = b \cdot a$ for multiplication.

Counting numbers. (See "Natural numbers.")

Decimal fraction. An abbreviated fraction whose denominator is a certain power of 10 indicated by the location of a dot called the decimal point. (Example: 0.37 means $\frac{37}{100}$.)

Decimal notation. The system used to denote decimal fractions.

Denominator. The number below the division line of a common fraction. (Example: The denominator of $\frac{3}{5}$ is 5.)

Difference. The result in a subtraction operation.

Digit. Any of the 10 characters 0–9 with which decimal (or base 10) numbers are formed.

Distributive property. States that a factor times the sum of two numbers equals the sum of the individual products; that is, $a(b + c) = ab + ac$.

Dividend. In a division operation, the number that is being divided. (Dividend ÷ Divisor = Quotient)

Divisor. In a division operation, the number that divides into another.

$$4\overset{\displaystyle 3}{\overline{)12}}$$

(Example: In 4)12, the divisor is 4.)

Exponent. A superscript that indicates how many times another number, called the base, should be written for repeated multiplication times itself. (Example: The exponent 4 denotes that $10^4 = 10 \cdot 10 \cdot 10 \cdot 10 = 10{,}000$.)

Factor. Any number (multiplicand or multiplier) used in multiplication.

Fraction. (See "Common fraction.")

Higher terms. An equivalent fraction with a larger denominator. (Example: $\frac{3}{5}$ may be expressed in higher terms as $\frac{9}{15}$.)

Improper fraction. A common fraction whose numerator is equal to or larger than the denominator. (Examples: $\frac{7}{7}$ or $\frac{15}{8}$)

Integer. Any number, positive or negative, that can be expressed without a decimal or fraction. Any number from the set $\{\ldots, -3, -2, -1, 0, 1, 2, 3, \ldots\}$.)

Least (or lowest) common denominator. The smallest number (denominator) into which all given denominators will divide.

Minuend. In a subtraction operation, the number from which another number is subtracted. (Minuend − Subtrahend = Difference)

Mixed number. A number composed of a whole number and a fraction. (Example: $12\frac{3}{7}$)

Multiplicand. The first (or top) factor in a multiplication operation. (Multiplicand × Multiplier = Product)

Multiplier. The second (or bottom) factor in a multiplication operation. (Example: In 8×5, the multiplier is 5.)

Natural (or counting) number. Any number that may be used in counting objects. (Any number from the set $\{1, 2, 3, 4, \ldots\}$.)

Numerator. The number above the division line of a common fraction. (Example: The numerator of $\frac{4}{9}$ is 4.)

Operation. Any of the arithmetic processes of addition, subtraction, multiplication, and division.

Place value. The value associated with a digit because of its position in a whole number or decimal number.

Prime number. A number (larger than 1) that can be evenly divided only by itself and 1. (Examples: 2, 3, 5, 7, 11, . . .)

Product. The result in a multiplication operation.

Proper fraction. A fraction whose numerator is smaller than the denominator. (Example: $\frac{4}{5}$)

Quotient. The result in a division operation.

Rational number. Any number that can be expressed as a common fraction (the quotient of two integers).

Reciprocal (of a fraction). A fraction formed by interchanging the original numerator and denominator. (Example: The reciprocal of $\frac{5}{8}$ is $\frac{8}{5}$.)

Subtrahend. In a subtraction operation, the number being subtracted from another. (Example: In $10 - 7 = 3$, the subtrahend is 7.)

Sum. The result in an addition operation.

Whole number. Any number zero or larger that can be expressed without a decimal or fraction. (Any number from the set $\{0, 1, 2, 3, 4, \ldots\}$.)

APPENDIX

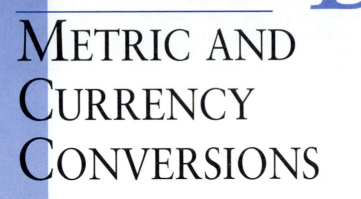

METRIC AND CURRENCY CONVERSIONS

OBJECTIVES

Upon completion of Appendix B, you will be able to:

1. Define and use correctly the terminology associated with each topic.
2. Given any multiple of a metric unit (Section 1: Examples 1, 2; Problems 1–4):
   - **a.** Name its single equivalent metric unit
   - **b.** Convert the given measurement to its equivalent measurement in another given metric unit.
3. Given a metric measurement, find its equivalent in a given U.S. unit (Section 1: Example 3; Problems 5, 6, 11, 12, 17, 18).
4. Given a U.S. measurement, determine its equivalent in a given metric unit (Section 1: Example 4; Problems 7, 8, 11–16).
5. Given temperature readings in either Celsius or Fahrenheit, convert to the other temperature scale (Section 1: Example 5; Problems 9, 10, 19, 20).
6. Given any multiple of a U.S. dollar, find its equivalent in the currency of another country (Section 2: Example 1; Problems 1, 2, 5–8, 13, 14).
7. Given any multiple of a foreign currency, determine its equivalent in U.S. dollars (Section 2: Example 2; Problems 3, 4, 9–12).

There is practically no individual, business, or industry that is not affected by international trade and commerce. Many of the employees who are hired by corporations and the customers who purchase goods come from culturally diverse backgrounds. Similarly, many products sold in the United States today come from El Salvador, Saudi Arabia, or Turkey, to name but a few countries of origination. U.S.-made products are sold worldwide; our economy is indeed global. To recognize the importance of the global marketplace, two international topics—metric conversions and currency conversions—are covered in this Appendix.

SECTION 1

METRIC CONVERSION

In 1971, a National Bureau of Standards report to Congress, entitled "A Metric America: A Decision Whose Time Has Come," described the United States as an isolated island in a metric world. The report recommended the passage of the Metric Conversion Act. Although the act was passed, it lacked clearly stated objectives and a timetable for implementation. In 1988, a metric usage provision was included in the Omnibus Trade and Competitiveness Act. The legislation designated the **metric system** as "the preferred system of weights and measurements for United States trade and commerce." It required federal agencies to use the metric system in procurements, grants, and other business-related activities. An executive order was signed by President Bush in 1991 that gave specific management authority to the Secretary of Commerce to coordinate implementation of the metric usage provisions of the Omnibus Trade Act.

The goal of these legislative and executive acts is to help U.S. industry and business compete more successfully in the global marketplace. The metric system (SI—International Systems of Units) is the international standard of measurement. The United States is the only developed country in the world that has not completely converted. Many U.S. industries, however, have converted to metrics. Today, we purchase photographic film in millimeters, medicines in milligrams, and soft drinks in liters.

Metric measurement is a decimal system, based on the number 10. The metric system was established with the **meter** (m) as the basic unit for length, the **gram** (g) as the basic unit for weight, and the **liter** (l) as the basic unit for liquid volume or capacity. Prefixes before each basic unit are used to name the other units of the system and to identify its multiple of the basic unit. The principal prefixes and indicated multiples of the basic unit are as follows:

METRIC PREFIX	MULTIPLE OF BASIC UNIT	METRIC PREFIX	MULTIPLE OF BASIC UNIT
deca (dk)	10	**deci** (d)	0.1 or 1/10
hecto (h)	100	**centi** (c)	0.01 or 1/100
kilo (k)	1,000	**milli** (m)	0.001 or 1/1000
mega (M)	1,000,000	**micro** (μ)	0.000001 or 1/1000000

Using the information in the table, we see that

1 *kilo*meter = 1000 meters

1 *deca*liter = 10 liters

1 *centi*gram = 0.01 gram

The following chart, which is similar to a place-value table for numbers, will help identify the power of 10 associated with each prefix.

Example 1 Identify the unit name for the given metric measurements.

(a) 10 meters = 1 ___ (b) 1000 liters = 1 ___
(c) 0.1 gram = 1 ___ (d) 100 grams = 1 ___
(e) 0.001 liter = 1 ___ (f) 0.01 meter = 1 ___

Using the basic units and prefix definitions above,

(a) 10 meters = 1 decameter (b) 1000 liters = 1 kiloliter
(c) 0.1 gram = 1 decigram (d) 100 grams = 1 hectogram
(e) 0.001 liter = 1 milliliter (f) 0.01 meter = 1 centimeter

Conversion to larger or smaller metric units is accomplished by first moving to the right or left in the chart of prefixes and then moving the decimal in the value the same number of places. To convert meters to decameters: Since decameters is 1 place to the left of meters, move the decimal 1 place to the left in the value in meters. Consider 40 meters as an example:

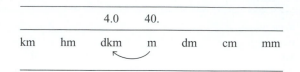

		4.0	40.			
km	hm	dkm	m	dm	cm	mm

So 40 meters = 4 decameters. To convert 40 meters to millimeters, move the decimal three places to the right in the value in meters, since millimeters is three places to the right of meters:

			40.			40 000.
km	hm	dkm	m	dm	cm	mm

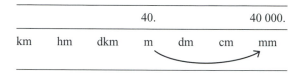

So 40 meters = 40 000 millimeters.

Example 2 (a) Convert 10 grams to centigrams. To convert grams to centigrams, move the decimal in the grams value to the right two places.

			10.		1 000.	
kg	hg	dkg	g	dg	cg	mg

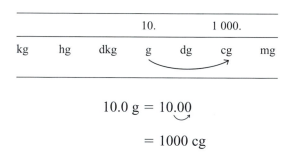

$$10.0 \text{ g} = 10.00$$
$$= 1000 \text{ cg}$$

(b) Convert 65 deciliters to hectoliters. To convert deciliters to hectoliters, move the decimal in the deciliters value to the left three places.

	0.065			65.		
kl	hl	dkl	l	dl	cl	ml

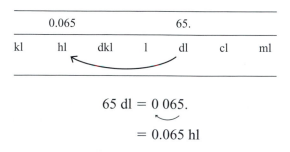

$$65 \text{ dl} = 0\,065.$$
$$= 0.065 \text{ hl}$$

The major challenge when anyone first uses the metric system is to develop a mental picture of the quantities that metric measures represent. By rough approximation, 1 meter is slightly longer than 1 yard; 1 liter is nearly equivalent to 1 quart. The gram is so small that it is used to weigh only small amounts, such as food items. Larger amounts are weighed using the kilogram, which is somewhat more than 2 pounds. Distances are expressed in kilometers, each of which is about ⅝ mile. As the following illustration shows, 1 inch is equivalent to 2.54 centimeters.

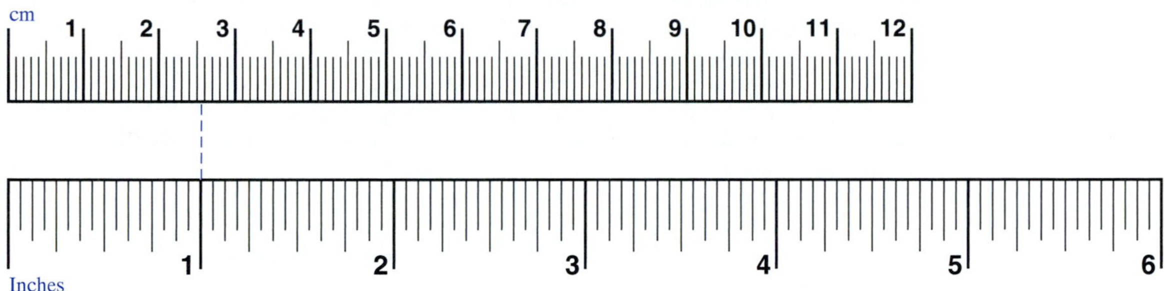

Example 3 Convert the following metric measurements to their U.S. equivalents, using the equivalents given in Table B-1.

(a) 10 meters = ? feet (b) 5 liters = ? quarts
(c) 200 grams = ? ounces (d) 500 kilograms = ? pounds

(a)
$$\begin{array}{rcr} 1\text{ meter} & = & 3.28\text{ feet} \\ \times 10 & & \times\ 10 \\ \hline 10\text{ meters} & = & 32.8\text{ feet} \end{array}$$

(b)
$$\begin{array}{rcr} 1\text{ liter} & = & 1.06\text{ quarts} \\ \times 5 & & \times\ 5 \\ \hline 5\text{ liters} & = & 5.3\text{ quarts} \end{array}$$

(c)
$$\begin{array}{rcr} 1\text{ gram} & = & 0.035\text{ ounce} \\ \times 200 & & \times\ 200 \\ \hline 200\text{ grams} & = & 7.0\text{ ounces} \end{array}$$

(d)
$$\begin{array}{rcr} 1\text{ kilogram} & = & 2.20\text{ pounds} \\ \times 500 & & \times\ 500 \\ \hline 500\text{ kilograms} & = & 1{,}100\text{ pounds} \end{array}$$

TABLE B-1 METRIC MEASURES WITH U.S. EQUIVALENTS

LENGTH	WEIGHT	LIQUID CAPACITY
1 millimeter = 0.039 inch	1 gram[a] = 0.035 ounce	1 liter[b] = 33.8 ounces
1 centimeter = 0.394 inch	1 kilogram = 2.20 pounds	= 2.11 pints
1 decimeter = 3.94 inches		= 1.06 quarts
1 meter = 39.37 inches	TEMPERATURE	= 0.264 gallon
= 3.28 feet		
= 1.09 yards	$C = \frac{5}{9}(F - 32)$	
1 kilometer = 0.621 mile	or $C = (F - 32) \div 1.8$	

[a]1 gram is defined as the weight of 1 cubic centimeter of pure water at 4° Celsius.
[b]1 liter is defined as the capacity of a cube that is 1 decimeter long on each edge.

Example 4 Use the equivalents given in Table B-2 to find the metric equivalents of the following U.S. measurements.

(a) 50 yards = ? meters (b) 4 gallons = ? liters
(c) 6 inches = ? centimeters (d) 150 pounds = ? kilograms

(a)
$$\begin{array}{rcr} 1 \text{ yard} & = & 0.914 \text{ meter} \\ \times\ 50 & & \times\ 50 \\ \hline 50 \text{ yards} & = & 45.7 \text{ meters} \end{array}$$

(b)
$$\begin{array}{rcr} 1 \text{ gallon} & = & 3.79 \text{ liters} \\ \times\ 4 & & \times\ 4 \\ \hline 4 \text{ gallons} & = & 15.16 \text{ liters} \end{array}$$

(c)
$$\begin{array}{rcr} 1 \text{ inch} & = & 2.54 \text{ centimeters} \\ \times\ 6 & & \times\ 6 \\ \hline 6 \text{ inches} & = & 15.24 \text{ centimeters} \end{array}$$

(d)
$$\begin{array}{rcr} 1 \text{ pound} & = & 0.454 \text{ kilogram} \\ \times\ 150 & & \times\ 150 \\ \hline 150 \text{ pounds} & = & 68.1 \text{ kilograms} \end{array}$$

Metric quantities will develop meaning to you as you use them, including the **Celsius** (or **centigrade**) temperature scale. You will learn, for example, that 10°C

TABLE B-2 U.S. MEASURES WITH METRIC EQUIVALENTS

LENGTH	WEIGHT	LIQUID CAPACITY
1 inch = 2.54 centimeters	1 ounce = 28.4 grams	1 ounce = 0.030 liter
1 foot = 0.305 meter	1 pound = 454 grams	1 pint = 0.473 liter
1 yard = 0.914 meter	= 0.454 kilogram	1 quart = 0.946 liter
1 mile = 1.61 kilometers		1 gallon = 3.79 liters
	TEMPERATURE	
	$F = \frac{9}{5}C + 32$	
	or $F = 1.8C + 32$	

United States Department of Commerce
Technology Administration
National Institute of Standards and Technology
Metric Program, Gaithersburg, MD 20899

All You Will Need to Know About Metric
(For Your Everyday Life)

Metric is based on the Decimal system

The metric system is simple to learn. For use in your everyday life you will need to know only ten units. You will also need to get used to a few new temperatures. Of course, there are other units which most persons will not need to learn. There are even some metric units with which you are already familiar; those for time and electricity are the same as you use now.

BASIC UNITS

METER: a little longer than a yard (about 1.1 yards)
LITER: a little larger than a quart (about 1.06 quarts)
GRAM: a little more than the weight of a paper clip

(comparative sizes are shown)

1 METER

1 YARD

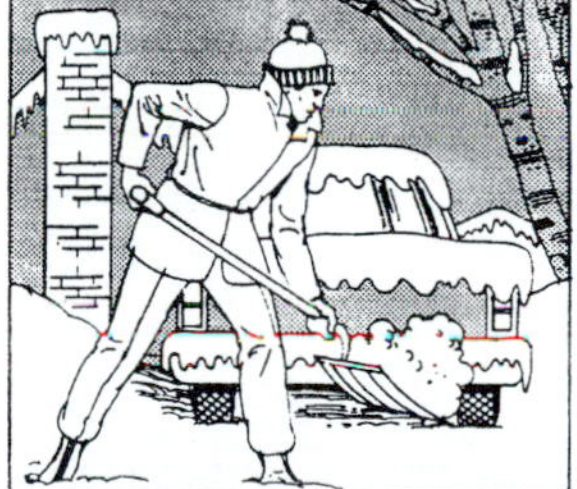

25 DEGREES FARENHEIT

COMMON PREFIXES
(to be used with basic units)

milli: one-thousandth (0.001)
centi: one-hundredth (0.01)
kilo: one-thousand times (1000)

For example
1000 millimeters = 1 meter
100 centimeters = 1 meter
1000 meters = 1 kilometer

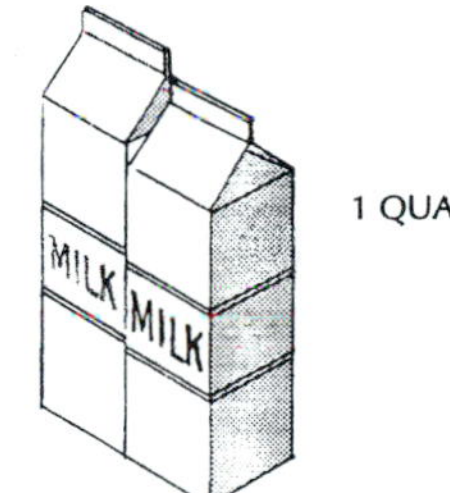

1 LITER

1 QUART

OTHER COMMONLY USED UNITS

millimeter:	0.001 meter	diameter of a paper clip wire
centimeter:	0.01 meter	a little more than the width of a paper clip (about 0.4 inch)
kilometer:	1000 meters	somewhat further than 1/2 mile (about 0.6 mile)
kilogram:	1000 grams	a little more than 2 pounds (about 2.2 pounds)
millimeter:	0.001 liter	five of them make a teaspoon

25 DEGREES CELSIUS

OTHER USEFUL UNITS
hectare: about 2 1/2 acres
metric ton: about one ton

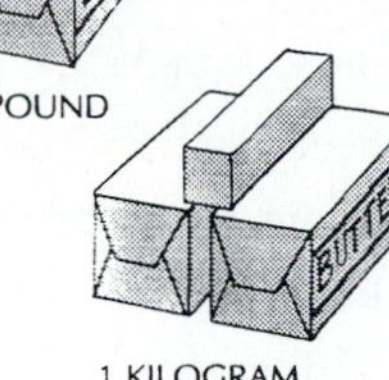

1 POUND

1 KILOGRAM

WEATHER UNITS:

FOR TEMPERATURE
degrees celsius

FOR PRESSURE
kilopascals are used
100 kilopascals = 29.5 inches of Hg (14.5 psi)

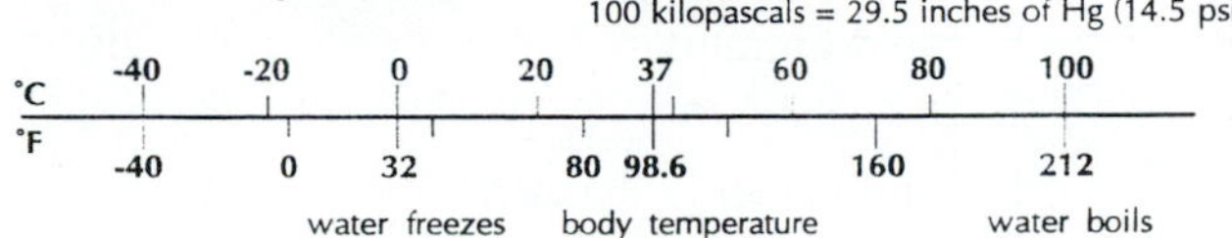

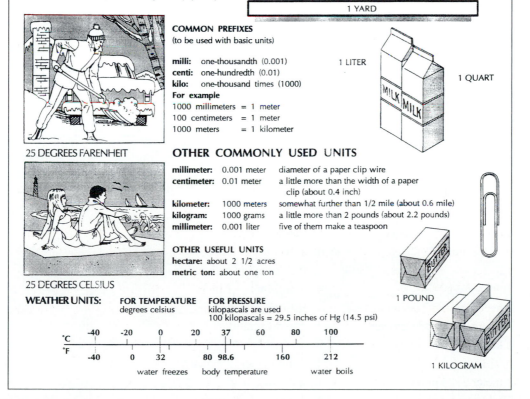

Courtesy of U.S. Department of Commerce Metric Program

is a cool day, that 20°–25°C is the comfortable indoor range, and that 35°C is a hot summer day.

Temperature conversions may be done using the formulas in Tables B-1 and B-2. Either the decimal or the fractional form of each formula may be used, as convenient.

Example 5 Find the following temperature equivalents.

(a) 32°F = ?°C (b) 35°C = ?°F
(c) 212°F = ?°C (d) 10°C = ?°F

(a)
$$\begin{aligned} C &= \tfrac{5}{9}(F - 32) \\ &= \tfrac{5}{9}(32 - 32) \\ &= \tfrac{5}{9}(0) \\ C &= 0^\circ \qquad (32^\circ F = 0^\circ C) \end{aligned}$$

(b)
$$\begin{aligned} F &= \tfrac{9}{5}C + 32 \\ &= \tfrac{9}{5}(35) + 32 \\ &= 63 + 32 \\ F &= 95^\circ \qquad (35^\circ C = 95^\circ F) \end{aligned}$$

(c)
$$\begin{aligned} C &= (F - 32) \div 1.8 \\ &= (212 - 32) \div 1.8 \\ &= 180 \div 1.8 \\ C &= 100^\circ \qquad (212^\circ F = 100^\circ C) \end{aligned}$$

(d)
$$\begin{aligned} F &= 1.8C + 32 \\ &= 1.8(10) + 32 \\ &= 18 + 32 \\ F &= 50^\circ \qquad (10^\circ C = 50^\circ F) \end{aligned}$$

If you once become accustomed to the metric system, the present U.S. system of weights and measures will seem as inefficient as any foreign monetary system that is not based on dollars and 100 cents.

SECTION 1 PROBLEMS

Name each metric unit or indicate the multiple of each given unit.

1. a. 1 decameter = ? meters
b. 100 liters = 1 __?__
c. 0.001 gram = 1 __?__
d. 1 deciliter = ? liter
e. 1 meter = ? centimeters
f. 0.1 decagram = 1 __?__
g. 1000 meters = 1 __?__
h. 1 milliliter = ? liter
i. 1 hectogram = ? decigrams
j. 0.1 centiliter = 1 __?__
k. 1 kilometer = ? decameters
l. 100 decigrams = 1 __?__

2. a. 1 kilogram = ? grams
b. 10 meters = 1 __?__
c. 0.01 liter = 1 __?__
d. 1 millimeter = ? meter
e. 1 meter = ? kilometer
f. 0.01 hectoliter = 1 __?__
g. 10 decigrams = 1 __?__
h. 1 liter = ? centiliters
i. 1 hectometer = ? decameters
j. 1000 centigrams = 1 __?__
k. 0.01 deciliter = 1 __?__
l. 1 hectoliter = ? kiloliter

Convert the following metric measurements as indicated.

3. a. 3.6 hectoliters = ? liters
b. 45 decimeters = ? meters
c. 2150 milligrams = ? grams
d. 0.8 kilometer = ? meters
e. 4.25 grams = ? centigrams
f. 188 liters = ? hectoliters
g. 30 meters = ? decimeters
h. 1200 grams = ? kilograms
i. 2.9 centigrams = ? milligrams
j. 76 meters = ? hectometers

4. a. 1.5 hectometers = ? meters
b. 125 centimeters = ? meters
c. 36 decagrams = ? grams
d. 2137 milliliters = ? liters
e. 1475 grams = ? kilograms
f. 17.4 grams = ? centigrams
g. 24 liters = ? decaliters
h. 50 meters = ? millimeters
i. 75 hectometers = ? kilometers
j. 362 grams = ? kilograms

Determine the U.S. equivalent of the following metric measurements.

5. a. 6 meters = ? yards
b. 50 kilograms = ? pounds
c. 10 liters = ? quarts
d. 80 kilometers = ? miles
e. 20 centimeters = ? inches
f. 0.5 liter = ? ounces
g. 5 meters = ? feet
h. 400 grams = ? ounces

6. a. 10 centimeters = ? inches
b. 3 meters = ? inches
c. 4 kilograms = ? pounds
d. 6 liters = ? ounces
e. 20 meters = ? yards
f. 5 decimeters = ? inches
g. 8 liters = ? quarts
h. 50 kilometers = ? miles

Find the metric equivalent of the following U.S. measurements.

7. a. 8 feet = ? meters
b. 100 pounds = ? kilograms
c. 4 quarts = ? liters
d. 20 pints = ? liters
e. 80 miles = ? kilometers
f. 6 inches = ? centimeters
g. 9 ounces = ? grams
h. 5 gallons = ? liters

8. a. 8 inches = ? centimeters
b. 10 miles = ? kilometers
c. 8 ounces = ? grams
d. 6 quarts = ? liters
e. 3 pounds = ? grams
f. 20 ounces = ? liters
g. 6 feet = ? meters
h. 50 gallons = ? liters

Convert each temperature measurement as indicated.

9. a. 77°F = ?°C
b. 95°F = ?°C
c. 59°F = ?°C
d. 14°F = ?°C
e. 20°C = ?°F
f. 80°C = ?°F
g. 55°C = ?°F
h. −5°C = ?°F

10. a. 104°F = ?°C
b. 41°F = ?°C
c. 68°F = ?°C
d. 5°F = ?°C
e. 35°C = ?°F
f. 45°C = ?°F
g. 10°C = ?°F
h. 0°C = ?°F

Find each measurement below as indicated.

11. An American-made car has a 120-inch wheelbase, and a European car has a 250-centimeter wheelbase.
a. How many centimeters is the wheelbase of the American car?
b. How many inches is the European car's wheelbase?
c. Which wheelbase is longer, and by how many centimeters?

12. Track sprinters run the 100-yard dash, while swimmers compete in 100-meter races.

a. How many meters is 100 yards?

b. How many yards is 100 meters?

c. Which race is longer, and by how many meters?

13. Elaine Stevens is 5 feet, 6 inches tall and weighs 120 pounds. Express her (a) height in centimeters and (b) weight in kilograms. (*Hint:* First convert her height to inches.)

14. Bill Gray is 5 feet, 10 inches tall and weighs 170 pounds. Express his (a) height in centimeters and (b) weight in kilograms.

15. The United States is approximately 3,000 miles wide. The earth is approximately 25,000 miles in circumference. Express each distance in kilometers.

16. Tasty Food Processors markets a pancake mix with a net weight of 40 ounces and a pancake syrup that contains 30 fluid (liquid) ounces. What metric equivalents should the labels contain for (a) the mix in grams and (b) the syrup in liters?

17. The Import House offers a gold pendant that weighs 20 grams and a silver tea service that weighs 1.75 kilograms.

a. How many ounces of gold are in the pendant?

b. How many pounds of silver are in the tea service?

18. The gas tank of a European car holds 50 liters, and 4 liters of oil are required in the motor. What is the U.S. equivalent of (a) the gas tank capacity in gallons and (b) the oil in quarts?

19. The high temperature in New York City was 95°F. The high in Madrid that same day was 30°C. Determine (a) the New York temperature in Celsius and (b) the Madrid temperature in Fahrenheit. (c) Which city was warmer, and by how many degrees Celsius?

20. Miami had a low temperature of 50°F on a day when the low in Madrid was 5°C. What were (a) the Miami temperature in Celsius and (b) the Madrid temperature in Fahrenheit? (c) Where was the air colder, and by how many degrees Celsius?

SECTION 2

FOREIGN CURRENCY CONVERSION

Although the economy is global, there is not a universal currency used worldwide. Before international transactions can be conducted, money must be converted or exchanged into the local currency. An **exchange rate** is the relative value of the currency of one country to that of another. For example, the exchange rate of U.S. dollars to Japanese yen is equal to the number of yen a dollar will purchase. Under the flexible (floating) exchange rate system, which has been in use since March 1973, the rate is determined by supply and demand in the marketplace for a particular country's currency. Just as prices will vary for any good or service, a currency will fluctuate as the supply and demand for it changes over time.

Foreign exchange rates are printed daily in newspapers. An illustration is shown in Table B-3, which includes conversions for the British pound (£), the Canadian dollar (C$), the French franc (₣), the German mark or Deutsche mark (DM), the Japanese yen (¥), and the Mexican (new) peso (N$).

According to the illustration in Table B-3, one U.S. dollar would be exchanged for 114.73 (or 115) yen. Likewise, a U.S. dollar could be exchanged for 9.48 Mexican pesos. When the dollar is strong, it will buy more foreign goods and investments (imports). When the dollar is weak, U.S. dollars will purchase less in foreign goods; conversely, U.S. products (exports) will be cheaper, relatively speaking, in other countries with stronger currencies. Thus, a weak dollar generally means fewer imports and more exports.

The question frequently arises as to rounding. It is common practice to convert American dollars to whole Japanese yen, whereas it is common to convert American dollars to other countries' currencies by rounding to two decimal places. When converting currencies to American dollars, round to two decimal places (or correct to cents).

In 1999, eleven European nations (Austria, Belgium, Finland, France, Germany, Ireland, Italy, Luxembourg, Netherlands, Portugal, and Spain) adopted a common currency, the euro. The euro notes and coins are currently used for noncash transactions ranging from government bond issues to credit-card and check purchases. In the year 2002, euro notes and coins will enter into general circulation. During the transitional years, businesses, banks, and individuals receive notice of transactions in each nation's currency along with the euro figures.

Example 1 Sam Perdue, on an exploratory business trip to Germany, exchanges $1,500 to German marks. How many marks does he receive?

Using the sample chart in Table B-3, U.S. $1 = DM1.8185. Thus, for $1,500,

$$\$1,500 \times \text{DM}1.8185 = \text{DM}2727.75$$

Or, from the chart, since DM1 = 0.5499,

$$\$1,500 \div 0.5499 = \text{DM}2727.78$$

Note. Most people use the first approach (multiplication) when converting U.S. dollars to a foreign currency.

TABLE B-3 SAMPLE FOREIGN CURRENCY EXCHANGE RATES

COUNTRY AND CURRENCY	$1 U.S. IN FOREIGN CURRENCY	FOREIGN CURRENCY IN U.S. DOLLARS
Britain (£)	0.6208	1.6108
Canada (C$)	1.5039	0.6649
France (₣)	6.0976	0.1640
Germany (DM)	1.8185	0.5499
Japan (¥)	114.73	0.008716
Mexico (N$)	9.4820	0.105463

Example 2 While in Japan, Terry Daniels paid 17,850 yen for dinner and 44,625 yen for a hotel room one night. How much did Ms. Daniels pay for the dinner and room in U.S. dollars?

From the sample chart in Table B-3, ¥1 = U.S. $0.008716. Thus,

¥17,850 × 0.008716 =	$155.58	for dinner
¥44,625 × 0.008716 =	$388.95	for the room
	$544.53	total

SECTION 2 PROBLEMS

For Problems 1–14, determine the currency equivalencies, using the sample chart in Table B-3. Round all answers to two decimal places except for answers in yen, which should be rounded to whole numbers.

	U.S. ($)	GERMAN (DM)	JAPANESE (¥)	MEXICAN (N$)
1. a.	$ 10			
b.	600			
c.	1,800			
2. a.	$ 20			
b.	500			
c.	2,000			
3. a.		2812	X	X
b.		X	53550	X
c.		X	X	3390
4. a.		3515	X	X
b.		X	12495	X
c.		X	X	1356

5. How much will Marcia Murphy receive when she exchanges $8,000 into Canadian dollars?

6. Phil Kelly exchanges $6,000 for Canadian dollars. How much will he receive?

7. During a trip to the United States, a French family purchased 4 tickets to a theme park for $80. How much did they pay for the tickets in French francs?

8. Before traveling to France, Sarah Delano exchanged $2,000 into French francs. How much did she receive in francs?

9. On a recent trip to Great Britain, Charles Ness purchased a sweater for £50.144. How many U.S. dollars did it cost?

10. While in Great Britain, Lucy and John Goodwin attended a stage production. The tickets cost £18.804 each. How much did they pay in U.S. dollars for two tickets?
11. While visiting Germany, Elizabeth Smith purchased a crystal necklace for DM158.18. How much had the necklace cost in U.S. dollars, rounded to the nearest dollar?
12. On a recent trip to Germany, Bob Glazer purchased gasoline for his car at DM3.60 a liter. What was the equivalent price per liter in U.S. dollars to the nearest cent?
13. If a Big Mac meal cost $3.99 in the United States, what would it cost in
    a. Germany
    b. Japan
    c. Canada
14. A Kentucky Fried Chicken dinner cost $4.29 in the United States. How much would the meal cost in
    a. Great Britain
    b. Mexico
    c. France

APPENDIX B GLOSSARY

Celsius (centigrade). The metric temperature scale, using 100 degrees for the range from the freezing point of water (0°C) to the boiling point (100°C).

Centi-. Prefix indicating 0.01 of a basic metric unit. (Example: 1 centimeter = 0.01 meter)

Centigrade. (See "Celsius.")

Deca-. Prefix indicating 10 times a basic metric unit. (Example: 1 decaliter = 10 liters)

Deci-. Prefix indicating 0.1 of a basic metric unit. (Example: 1 decigram = 0.1 gram)

Exchange rate. The relative value of one country's currency to that of another.

Gram. The basic metric unit of weight; defined as the weight of 1 cubic centimeter of pure water at 4°C.

Hecto-. Prefix indicating 100 times a basic metric unit. (Example: 1 hectometer = 100 meters)

Kilo-. Prefix indicating 1000 times a basic metric unit. (Example: 1 kilogram = 1000 grams)

Liter. The basic metric unit of liquid volume (or capacity); defined as the capacity of a cube that is 1 decimeter long on each edge.

Mega-. Prefix indicating 1 million times a basic metric unit.

Meter. The basic metric unit of length; originally defined (erroneously) as 1 ten-millionth of a quadrant $\left(\frac{1}{4}\right)$ of the earth's circumference; now officially defined as a multiple of a certain wavelength of krypton 86, a rare gas.

Metric system. The decimal system of weights and measures, where each unit is 10 times the size of the next smaller unit. Its basic units are the meter (length), the gram (weight), and the liter (capacity).

Micro-. Prefix indicating 1 millionth of a basic metric unit.

Milli-. Prefix indicating 0.001 of a basic metric unit. (Example: 1 milliliter = 0.001 liter)

METRIC TALK

When the United States goes to the metric system, perhaps some of our favorite sayings will have to be changed thus:

- Traffic was 2.54 centimetering along the freeway.
- It hit me like 907 kilograms of bricks.
- Cried 3.79 liters of tears.
- A miss is as good as 1.61 kilometers.
- A decigram of salt.
- Beat him within 2.54 centimeters of his life.
- All wool and 91.4 centimeters wide.
- Give her 2.54 centimeters and she'll take 1.61 kilometers.
- Give him 454 grams of flesh.
- Missed it by 1.61 country kilometers.

Answers to Odd-Numbered Problems

CHAPTER 1

SECTION 1

1. a. 2,490 b. 12,800 c. 109,800
d. 57,200 e. 424,800 f. 495,582
g. 714,825 h. 26,522,818 i. 452,610
j. 219.30 k. 94.90 l. 13.60
m. 173.7 n. 8.58 o. 625
p. 65,536 q. 343 r. 4.0401

3. a. 56 b. 16 c. 40
d. 30 e. 1.7 f. ($1,426)
g. ($175) h. ($17,500) i. ($525)

5. a. $\frac{26}{25}$ or $1\frac{1}{25}$ b. $\frac{307}{300}$ or $1\frac{7}{300}$ c. $\frac{199}{200}$
d. $\frac{49}{50}$ e. 5,400 f. 1,620
g. 976 h. 500 − 60g i. 300 + 27*b*
j. *jkl* + *jm* k. *w* − *wxy*

7.

a.	b.	c.
43.3	8.94	18.925
156.6	26.45	0.566
1,681.0	160.06	337.009

d. 5.1; 5.08; 5.085
23.7; 23,68; 23.675

9. a. 175.4; 175 b. 33.3; 33 c. 45.4; 45
d. 4.28; 4.28

11. a. $573.91 b. $760.81 c. $1,497.64
d. $381.63

SECTION 2

1. a. 3,009 b. 579 c. 171
d. 69 e. 150 f. 48
g. 120 h. 912

3. a. 29.92 b. 873 c. 25.11
d. 810

5. a. 3.92; 7.35; 22.05 b. 1.68; 3.5; 196 c. 5,600
d. 48 e. 11,680 f. 1,492
g. 742.50 h. 18

CHAPTER 2

SECTION 1

1.	65	**3.**	30	**5.**	28	**7.**	9	**9.**	3	**11.**	11
13.	23	**15.**	10	**17.**	8	**19.**	22	**21.**	5	**23.**	2
25.	7	**27.**	5	**29.**	128	**31.**	126	**33.**	150	**35.**	92

SECTION 2

1. a. $6x$ b. $A + O$ c. $n + 10$

d. $n - 18$ e. $\frac{2}{3}g$ f. $\frac{1}{4}P + 5$

g. $2(r + s)$ h. $g = h - \$4$ i. $d = 2(a + b)$

j. $b = 8.5f$ k. $m = \frac{1}{3}n - 9$ l. $\$10b$

3.	82	**5.**	$13.50	**7.**	$62
9.	$1,800	**11.**	3,200	**13.**	120
15.	$160	**17.**	24¢	**19.**	Staff: $100,000 Manager: $180,000
21.	P: 15 C: 20	**23.**	M: 810 I: 90 T: 105	**25.**	18
27.	S: $32 J: $80 F: $25	**29.**	L: 30 C: 24	**31.**	L: 16 C: 10
33.	F: 125 S: 150	**35.**	F: 65 L: 135		

SECTION 3

1. a. 8 to 21; 8 : 21; 8/21 b. 1 to 3; 1 : 3; 1/3 c. 2 to 5; 2 : 5; 2/5
d. 5 to 8; 5 : 8; 5/8 e. 9 to 4; 9 : 4; 9/4

3. a. 4 b. 6 c. 8
d. 6

5. 4 to 1 **7.** 11 to 2 or 5.5 to 1 **9.** $18,480

11. 220 **13.** 16 **15.** 96

17. 360 **19.** 36 **21.** 6,300

23. 40 **25.** 600 **27.** 10

CHAPTER 3

SECTION 1

1. $0.11; \frac{11}{100}$

3. $0.32; \frac{8}{25}$

5. $0.02; \frac{1}{50}$

7. $0.305; \frac{61}{200}$

9. $0.0625; \frac{1}{16}$

11. $1.74; \frac{87}{50}$ or $1\frac{37}{50}$

13. $1.284; \frac{321}{250}$ or $1\frac{71}{250}$

15. $0.008; \frac{1}{125}$

17. $0.0175; \frac{7}{400}$

19. $1.05; \frac{21}{20}$ or $1\frac{1}{20}$

21. $0.0075; \frac{3}{400}$

23. $0.004; \frac{1}{250}$

25. $0.00375; \frac{3}{800}$

27. $0.016; \frac{2}{125}$

29. $0.125; \frac{1}{8}$

31. 9%

33. 40%

35. 67.4%

37. 211%

39. 128%

41. 0.1%

43. 50%

45. 0.5%

47. 310%

49. 3%

51. 50%

53. 0.5%

55. $41.\overline{66}\%$ or $41\frac{2}{3}\%$

57. 175%	**59.** 480%		

SECTION 2

1. 56	**3.** 128	**5.** 19.5	**7.** 222
9. 70	**11.** 27	**13.** 35	**15.** $14.\overline{44}\%$ or $14\frac{4}{9}\%$
17. 35%	**19.** 3%	**21.** 25%	**23.** 320
25. 75	**27.** 600	**29.** 270	**31.** 8,000
33. 500	**35.** 9,600	**37.** 40%	**39.** 20%
41. 60%	**43.** 25%	**45.** 0.5%	**47.** 72
49. 48	**51.** 72	**53.** 225	**55.** 2,000

SECTION 3

1. 45%	**3.** 86%	**5.** $112,500	**7.** $6.80
9. $51,000	**11.** $300,000	**13.** $62,000	**15.** $40
17. 5.5%	**19.** 5.6%	**21.** $33\frac{1}{3}\%$	**23.** 6%
25. $12\frac{1}{2}\%$	**27.** 3,000	**29.** $45	**31.** 155
33. 3%	**35.** 400	**37.** 24%	**39.** $33\frac{1}{3}\%$
41. $104	**43.** $3,000	**45.** 5%	

CHAPTER 4

SECTION 1

1. a. 32 b. 140
30 137.5
22 154

3. a. 2.5 b. $2.\overline{66}$

5. a. \$600 b. \$515

7. a. \$104 b. \$99.54

9. \$4,800

11. \$133,000

13. \$82

15. a. 54.5 b. 54 c. 60

17. a. 1,620 b. 1,630 c. 1,640

SECTION 2

1. a. 43.6 b. 38.89 c. 20–39

3. a. $816.\overline{66}$ b. 825 c. 800–899

5. a. \$210 b. \$200 c. \$150–\$249

7. a. 9.69 b. 8 c. 6–8

SECTION 3

1. a. 50% b. 2.5% c. 47.5% d. 97.5%
e. 8 f. 42

3. a. 97.5% b. 50% c. 68% d. 16%
e. 4.96 or 5 f. 15.5 or 16 g. 21

5. a. ± 4 c. $\pm 1\sigma$: 3 (10, 12, 16); $\pm 2\sigma$: 5 (all 5)

7. a. ± 5 c. $\pm 1\sigma$: 4 (34, 35, 35, 40)
$\pm 2\sigma$: 6 (all 6)

9. a. ± 6 b. $\pm 1\ \sigma$: 5 (35, 41, 44, 44, 46)
$\pm 2\ \sigma$: 8 (all 8)

11. a. ± 5 c. $\pm 1\ \sigma$: 5 (have 4: 25, 31, 31, 34)
$\pm 2\ \sigma$: 8 (all 8)

13. a. ± 4 c. $\pm 1\ \sigma$: 7 (have 6: 16, 18, 20, 20, 23, 23)
$\pm 2\ \sigma$: 10 (all 10)

SECTION 4

1. a. $1,619.00 b. $164.40 c. $14.05
d. New York (176.6) e. Los Angeles (145.2) f. Dallas (142.3)
g. New York (255.3) h. New York (143.9) i. Atlanta (125.8)

	YEAR	INDEX
3.	1998	100
	1999	98
	2000	110
	2001	120
	2002	130
5.	1990	100
	1995	70
	1999	125
	2001	140

CHAPTER 5

SECTION 1

1. a. $0.63; $16.38 b. $5.28; $71.28 c. $32.50; $34.45
   d. $41.00; $43.05 e. $58.00; $61.48 f. $24.00; $1.68
   g. $36.00; $2.88 h. $87.00; $4.35

3. $30.74
5. $27.90; $337.90
7. a. $89 b. $94.34
9. a. $50 b. $57.50
11. a. $190 b. $8.55
13. a. $89.00 b. $4.45
15. a. $62.00 b. $4.03
17. a. $87.00 b. $6.96
19. a. $13.00 b. $0.65 c. $1.95
21. a. $70.00 b. $4.90 c. $9.10
23. a. $29,500 b. $1,500 c. $3,000

SECTION 2

1. a. 1.85% b. $1.85 per C c. $18.45 per M d. 19 mills
3. a. 1.93% b. $1.93 per C c. $19.24 per M d. 20 mills
5. a. 2.5% b. $2.45 per C c. $24.49 per M d. 25 mills
7. a. $2,632 b. $740 c. $930
   d. $50,000 e. $68,000 f. $65,000
   g. 1.4% h. $3.20 per C i. $9.60 per M

9. $630
11. $1,845
13. $22,000
15. $72,000
17. 2.4%
19. 15 mills
21. $2.20 per C
23. $600 increase
25. 1.6%

CHAPTER 6

SECTION 1

1. a. $1,960 + 505 = $2,466   b. $3,364 + 530 = $3,894

	PREMIUM	REFUND DUE
3.	a. $2,262	X
	b. 4,204	X
	c. 5,503	$ 971
	d. 3,900	7,800

		INSURANCE REQUIRED	INDEMNITY
5.	a.	$400,000	$125,000
	b.	352,000	90,000
	c.	540,000	350,000
	d.	720,000	600,000
	e.	800,000	700,000

		RATIO OF COVERAGE	COMPENSATION
7.	a.	$\frac{22}{53}$	$198,000
		$\frac{31}{53}$	279,000
	b.	$\frac{15}{28}$	$450,000
		$\frac{2}{7}$	240,000
		$\frac{5}{28}$	150,000

	Insurance required	Total indemnity	Co. ratio	Co. payment
9.	a. $480,000	$300,000	$\frac{5}{8}$	$187,500
			$\frac{3}{8}$	112,500
	b. $560,000	$660,000	$\frac{7}{12}$	$385,000
			$\frac{5}{12}$	275,000

11. $432

	Premium	Refund
13.	a. $3,402	$2,268
	b. $2,363	$3,307

15. a. $80,000 b. $500,000 c. $600,000

17. a. $28,000 b. $2,100,000 c. $3,500,000

19. a. $72,000 b. $168,000 c. $240,000

		a.	b.	c.
21.	A:	$12,000	$216,000	$ 240,000
	B:	17,000	306,000	340,000
	C:	21,000	378,000	420,000
		$50,000	$900,000	$1,000,000

		a.	b.	c.
23.	AA	$21,600	$180,000	$ 300,000
	BB	25,200	210,000	350,000
	CC	18,000	150,000	250,000
	DD	7,200	60,000	100,000
		$72,000	$600,000	$1,000,000

25. a. $1,500,000 b. $700,000; $800,000

27. a. $3,750,000 b. $1,125,000
$1,687,500
$937,500

SECTION 2

1. a. $301.20 b. $313.30 c. $523.05

3. a. $180.60 b. $336.30 c. $254.80

5. $436.90 **7.** $325.50 **9.** $455 **11.** $576.80

13. a. $90,000 b. $30,000

15. a. $20,000 b. 0

17. a. $41,500 b. $85,500

19. a. $217,750 b. $312,250

SECTION 3

1. a. $103.20 b. $1,689.50 c. $511.80 d. $2,217.00

3. a. $25,100 b. $10,900 c. 18 years, 91 days d. $6,600

5. a. $353.65 b. 20 years c. $656.00 d. $666.00

7. a. $2,385.10 b. $1,463.70 c. $47,702.00 d. $43,911.00

9. a. $355.00 b. $3,550.00 c. $8,610.00 d. $5,060.00
e. Term: 0; Whole life: $50,000

11. a. $320.25 b. $51,791.25 c. $52,177.50
d. 10 more years to build cash value plus insurance coverage for extra 10 years

13. a. $2,806 b. $18,700 c. $50,700
d. 28 years, 186 days

15. $12,760 **17.** a. $154,500 b. $37,650

19. a. $616.80 b. 10 years

21. a. $575 b. $555

23. a. $57,600 b. $41,184 c. Gained $16,416 and $7,600 over face value

25. a. $464.40 b. $94,737.60 c. $432
d. Gained $6,609.60 over the face value, but beneficiary lost $15,552

Chapter 7

Section 1

1.	$1,539.50	**3.**	$134.12	**5.**	$1,013.47
7.	$1,167.57	**9.**	$942.18	**11.**	$1,053.42

Section 2

1. Cash short, $0.90

3. Cash over, $0.25

5. Cash over, $0.56; cash short, $0.13; cash register, $5,626.17

CHAPTER 

SECTIONS 1 AND 2

		ANNUAL	MONTHLY	SEMI-MONTHLY	WEEKLY	BIWEEKLY
1.	a.	X	$4,000.00	$2,000.00	$ 923.08	$1,846.15
	b.	$72,000.00	X	3,000.00	1,384.62	2,769.23
	c.	24,000.00	2,000.00	X	461.54	923.08

3. a. $570 b. $7,200 c. 4%
d. $220; $420 e. $420; $6,000 f. $423

5. $429 **7.** $150,000

9. a. $726 b. $376

11. $3,600 **13.** $110; $285 **15.** $412.75

	EMPLOYEE	GROSS COMMISSION
17.	a. Dempsey	$156.00
	b. Gold	257.00
	c. Keller	221.40
	d. Miller	156.00
	e. Wright	203.00
	Total	$993.40

	EMPLOYEE	GROSS WAGES
19.	Bellis	$ 548.00
	Chambers	297.00
	Duggan	332.00
	Meyer	450.00
	Quader	460.00
	Total	$2,123.00

21. Gross proceeds, $1,714.80
Net proceeds, $1,413.51

23. Prime cost, $1,187.00
Gross cost, $1,397.35

SECTION 3

	EMPLOYEE	GROSS WAGES
1.	Becker	$ 234.00
	Doyle	270.00
	Margo	226.20
	Murray	237.50
	Richards	216.00
	Stark	284.00
	Wu	259.00
	Total	$1,726.70

	EMPLOYEE	TOTAL GROSS WAGES
3.	Agnew	$ 700.00
	Barski	709.50
	Clancy	592.00
	Ibar	817.00
	Nozek	756.50
	Total	$3,575.00

	EMPLOYEE	TOTAL GROSS WAGES
5.	Coffey	$ 364.00
	Hanson	442.00
	Yoder	434.30
	Total	$1,240.30

7. a. $399.00 b. $368.00 c. $465.50
9. a. $249.20 b. $365.75 c. $429.25

	EMPLOYEE		TOTAL GROSS WAGES
11.	Dinh		$ 286.00
	Horn		342.40
	Rook		368.65
	Soka		339.70
	Vern		418.00
		Total	$1,754.75

SECTION 4

1.	Gross production	748					
	Base wages	− 500	×	$0.52	=	$260.00	
	Premium wages	248	×	0.70	=	173.60	
	Total gross wages					$433.60	
3.	Net production	1,693					
	Base wages	− 1,000	×	$0.15	=	$150.00	
	Premium wages	693	×	0.24	=	166.32	
						$316.32	
	Dockings	14	×	0.08	=	− 1.12	
	Total gross wages					$315.20	

	EMPLOYEE		GROSS WAGES EARNED
5.	Abbott		$ 392.00
	Breck		402.73
	Dunn		311.08
	Jones		456.56
	Simon		415.14
	Woods		456.75
	Yoe		469.58
		Total	$2,903.84

	EMPLOYEE	TOTAL GROSS WAGES
7.	Chang	$ 226.20
	Evans	264.86
	Hall	338.80
	Long	334.11
	Thomas	310.76
	Total	$1,474.73

	EMPLOYEE	TOTAL GROSS WAGES
9.	Allen	$ 198.10
	Dill	210.84
	Edwards	207.70
	Jenks	228.75
	Strong	306.82
	Total	$1,152.21

SECTION 5

1.	Net Wages Due:		Column Totals:	
	Beasley	$ 446.31	Social Security:	$224.12
	Kirtley	685.21	Medicare:	52.41
	Mason	576.95	Fed. Inc. Tax:	391.00
	Musser	604.55	Insurance:	74.00
	Smith	560.45	Total Deductions:	$741.53
		$2,873.47		
3.	Net Wages Due:		Column Totals:	
	#44	$424.69	Social Security:	$177.66
	#45	455.74	Medicare:	41.56
	#46	458.66	Fed. Inc. Tax:	317.00
	#47	500.80	Union Dues:	43.00
	#48	448.39	Total Deductions:	$579.22
		$2,286.28		

5.

Net Wages Due:		Column Totals:	
#G-4	$ 426.98	Social Security:	$187.86
#G-5	388.34	Medicare:	43.94
#G-6	446.86	Fed. Inc. Tax:	344.00
#G-7	499.98	United Way:	303.00
#G-8	389.04	Total Deductions:	$878.80
	$2,151.20		

7.

Net Wages Due:		Column Totals:	
#54	$ 482.57	Social Security:	$202.49
#55	557.76	Medicare:	47.36
#56	532.51	Fed. Inc. Tax:	380.00
#57	497.27	Insurance:	80.00
#58	486.04	Total Deductions:	$709.85
	$2,556.15		

9.

Net Wages Due:		Column Totals:	
E	$ 594.69	Social Security:	$181.48
F	574.99	Medicare:	42.44
G	550.32	Fed. Inc. Tax:	390.00
H	553.08	Insurance:	40.00
	$2,273.08	Total Deductions:	$653.92

11.

Net Wages Due:		Column Totals:	
#20	$ 469.41	Social Security:	$160.49
#21	446.12	Medicare:	37.53
#22	479.88	Fed. Inc. Tax:	287.00
#23	587.83	Insurance:	129.44
	$1,974.24	Total Deductions:	$614.46

SECTION 6

LINE	PROBLEM NUMBERS 1	3	5	13	15
2	$15,000	$50,000	$29,500	$75,100	$76,500
3	2,250	7,500	4,200	10,014	10,155
5	2,250	7,500	4,200	10,014	10,155
6(a)	1,860	5,456	3,224	7,924	5,617
(b)	0	744	186	0	0
7	435	1,450	856	2,178	2,219
8	2,295	7,650	4,266	10,102	7,836
10	2,295	7,650	4,266	10,102	7,836
11	4,545	15,150	8,466	20,116	17,991
13	4,545	15,150	8,466	20,116	17,991
14	4,545	15,150	8,466	20,116	17,991
15	0	0	0	0	0

	EMPLOYEE	1ST QTR.	2ND QTR.	3RD QTR.	4TH QTR.
7.	G	$19,500	$19,000	$19,000	$18,200
	H	21,000	25,000	29,000	1,200
	I	26,000	26,000	24,200	0

	EMPLOYEE	SOCIAL SECURITY EARNINGS	MEDICARE EARNINGS
9.	1	$10,200	$18,000
	2	21,200	29,000
	3	0	28,000

11. b. $72,200; $76,300; $63,900; $45,300

13. and 15. See above (before Problem 7).

		EMPLOYEE	1ST QTR.	2ND QTR.	3RD AND 4TH QTR.
17.	a.	G	$ 7,000	0	0
		H	7,000	0	0
		I	7,000	0	0
	b. Totals		$21,000	0	0

	1st Qtr.	2nd Qtr.	3rd and 4th Qtr.
c. State	$756	0	0
Federal	$168	0	0

	Employee	1st Qtr.	2nd Qtr.	3rd and 4th Qtr.
19. a.	A	$ 6,800	$200	0
	B	7,000	0	0
	C	7,000	0	0
	D	7,000	0	0
b. Totals		$27,800	$200	0

	1st Qtr.	2nd Qtr.	3rd and 4th Qtr.
c. State	$1,001.00	$222.00	0
Federal	7.20	1.60	0

21. a. $4,724 b. $7,250 c. $56 d. $378

CHAPTER 9

QUICK PRACTICE—STRAIGHT LINE

1. a. $1,200 annually b. $600 annually c. $550 annually

QUICK PRACTICE—DECLINING BALANCE

1. a. $760; $456; $274: $164; $246
   b. $400; $200; $100; $100
   c. $600; $400; $267; $178; $118; $87

QUICK PRACTICE—UNITS OF PRODUCTION

1. a. $1,440; $1,600; $2,000; $1,520; $640
   b. $2,500; $2,000; $1,750; $1,250
   c. $420; $528; $480; $396; $336; $240

QUICK PRACTICE—MACRS

1. a. $1,000; $1,333; $444; $223
   b. $829; $1,420; $1,015; $725; $518; $518; $518; $257
   c. $900; $1,440; $864; $518; $518; $260

SECTION 1

	a.	b.	c.	d.
1.	$700 annually	$1,600	$467	$514
		$ 960	$622	$882
		$ 576	$207	$630
		$ 346	$104	$450
		$ 318		$321
				$321
				$321
				$161

3. $900 annually 5. $150 annually 7. $4,000 annually

9.	11.	13.
$4,000	$3,000	$200
$2,667	$1,500	$150
$1,778	$ 750	$113
$1,185	$ 750	$ 84
$ 790		$ 63
$ 780		$ 48
		$ 36
		$ 6

15. $1,240
$1,984
$1,190
$ 714
$ 714
$ 357

17. $24,338
$25,397
$25,397

		FIRST YEAR	SECOND YEAR	THIRD YEAR
19.		$20,909	$21,818	$21,818
		5,000	8,000	4,800
	Totals	$25,909	$29,818	$26,618

21. a. $675; $738; $684; $603
b. $1,260; $1,440; $1,584; $1,116
c. $1,176; $1,218; $1,260; $1,155; $840; $651

		STRAIGHT LINE	DECLINING BALANCE	MACRS
23.	First year	$800	$1,680	$ 840
	Second year	800	1,008	1,344
	Third year	800	605	806

SECTION 2

1. a. $1,800
$2,880
$1,728
$1,037
$1,037
$ 518

b. $300
$900
$900
$900
$900
$600

c. $975
$813
$406
$203
$203

3. $100
$600
$600
$600
$600
$600
$600
$500

5. $250
$433
$145
$ 22

SECTION 3

1. a. $ 2,400 / $ 3,000 / $ 3,600 / $ 4,500   b. $ 8,000 / $ 4,800 / $ 6,400 / $12,800

	a.	b.
1.	$ 2,400	$ 8,000
	$ 3,000	$ 4,800
	$ 3,600	$ 6,400
	$ 4,500	$12,800
3.	$24,000	$10,000
	$12,000	$11,000
	$11,200	$ 7,000
	$ 8,800	$ 8,000
5.	$ 7,200	$ 4,000
	$ 1,800	$ 8,000
	$ 5,400	$ 6,000
	$ 3,600	$12,000

7.		**9.**		**11.**	
	$ 8,000		$12,000		$21,000
	$12,000		$ 8,800		$24,000
	$16,000		$ 8,400		$15,000
	$24,000		$ 7,600		$18,000
			$ 3,200		$12,000

CHAPTER 10

SECTIONS 1 AND 2

1.

	AMOUNTS	PERCENTS
Sales		104.0%
Sales discounts		4.0
Net sales	$500,000	100.0%
Goods available for sale	232,000	
Cost of goods sold	149,000	29.8
Gross profit	$351,000	70.2
Salaries		40.0
Depreciation		6.4
Utilities		3.6
Maintenance		3.3
Advertising		2.0
Insurance		1.7
Office supplies		1.6
Miscellaneous		1.4
Total expenses	300,000	60.0
Net income from operations	$ 51,000	10.2
Net income	$ 37,000	7.4%

3.

	AMOUNTS	PERCENTS
Cash		4.7%
Accounts receivable		5.6
Notes receivable		3.5
Inventory		7.8
Total current assets	$124,000	21.6
Net building	270,000	47.0
Net truck	20,000	3.5
Total plant assets	450,000	78.4
Total assets	$574,000	100.0%
Accounts payable		5.9
Notes payable		8.0
Total current liabilities	$ 80,000	13.9
Mortgage		39.9
Total liabilities	$309,000	53.8
Owner's equity		46.2
Total liabilities and owner's equity	$574,000	100.0%

5.

			INCREASE OR DECREASE		PERCENT OF NET SALES	
	20X2	20X1	AMOUNT	%	20X2	20X1
Net sales			$25,000	8.3%	100.0%	100.0%
Invent., Jan. 1			8,000	9.8	27.2	27.3
Purchases			(15,000)	(10.0)	41.5	50.0
Gds. avail. for sale	$225,000	$232,000	(7,000)	(3.0)	69.2	77.3
Invent., Dec. 31			(9,000)	(10.0)	24.9	30.0
Cost of gds. sold	144,000	142,000	2,000	1.4	44.3	47.3
Gross profit	$181,000	$158,000	23,000	14.6	55.7	52.7
Salaries			30,000	50.0	27.7	20.0
Rent			1,500	3.5	13.5	14.2
Advertising			—	—	1.8	2.0
Depreciation			(500)	(9.1)	1.5	1.8
Utilities			200	7.1	0.9	0.9
Miscellaneous			2,800	127.3	1.5	0.7
Total expenses	153,000	119,000	34,000	28.6	47.1	39.7
Net income	$ 28,000	$ 39,000	($11,000)	(28.2%)	8.6%	13.0%

7.

			INCREASE OR DECREASE		PERCENT OF TOTAL ASSETS	
	20X2	20X1	AMOUNT	%	20X2	20X1
Cash			($6,000)	(18.2%)	5.8%	6.8%
Accounts receivable			4,000	8.3	11.3	9.9
Inventory			1,000	1.2	18.0	16.9
Total current assets	$162,000	$163,000	(1,000)	(0.6)	35.1	33.6
Total plant assets			(22,000)	(6.8)	64.9	66.4
Total assets	$462,000	$485,000	(23,000)	(4.7)	100.0%	100.0%
Current liabilities			(6,000)	(7.1)	17.1	17.5
Long-term liabilities			(22,000)	(12.8)	32.5	35.5
Total liabilities	$229,000	$257,000	(28,000)	(10.9)	49.6	53.0
Preferred stock			—	—	9.7	9.3
Common stock			2,000	2.6	17.3	16.1
Retained earnings			3,000	2.9	23.4	21.6
Total equity	233,000	228,000	5,000	2.2	50.4	47.0
Total liab. and equity	$462,000	$485,000	(23,000)	(4.7)	100.0%	100.0%

9. a. 1.6 to 1 b. 1.0 to 1 c. 6.8 to 1 d. $53

11. a. 2.1 to 1 b. 1.1 to 1 c. 16.1%
d. 1.1 to 1 e. 10 times f. 35.6 days

SECTION 3

1. $73,500 **3.** a. $103,600 b. $102,400

5. 3.8 times **7.** a. $44,100 b. 4 times

9. 3.2 times **11.** a. $73,650 b. 3.4 times

SECTION 4

1. a. (1) $78.00 b. (1) $300.80 c. (a) $1,207.30
(2) $78.75 (2) $304.50 (b) $1,260.00
(3) $77.00 (3) $297.50 (c) $1,160.00

5. a. $1,953.60 b. $1,800.00 c. $2,080.00

7.

	WEIGHTED AVERAGE	FIFO	LIFO
Gross profit	$17,061.30	$17,114.00	$17,014.00
% of gross profit	42.7%	42.8%	42.5%

CHAPTER 11

SECTION 1

1. a. $1.40/share b. $6/share; $0.80/share c. $5/share; $1/share

		Preferred dividend	Common dividend	Earnings per share
3.	a.	$7.00	$1.00	$16
	b.	4.00	7.00	12

5. $1.70/share

7. $7/share; $3.50/share

9. $4/share; $8/share

11. $5/share; 0/share

13. $7/share; $9.80/share

15. $12/share; $0.50/share

17. $10/share; $5/share

19. $3/share; $2.50/share

21. a. $2.50/share; $5.25/share b. $11.25/share

SECTION 2

1. a. K: $15,000
L: $45,000

b. A: $6,000
B: $7,200
C: $8,400

c. S: $6,280
T: $6,580
U: $3,140

d. V: $25,500
W: $9,800
X: $8,400

3. $25,000/each

5. a. Q: $14,000
R: $21,000
S: $7,000

b. Q: ($6,000)
R: ($9,000)
S: ($3,000)

7. B: $15,000
C: $16,000
D: $19,000

9. J: $30,000
G: $45,000

11. J: $30,000
G: $34,800

13. a. $18,000 b. T: $31,500
S: $45,500

15. a. $25,500 b. M: $14,895
P: $15,105

17. a. M: $9,080
K: $4,920

b. M: $5,580
K: $1,420

19. a. T: \$9,700 b. T: \$4,500
C: \$9,130 C: \$4,450
B: \$8,560 B: \$4,400

21. a. R: \$1,250 b. R: \$800
S: \$7,750 S: \$6,400
T: \$7,000 T: \$6,100

CHAPTER 12

SECTION 1

1. a. 80%; \$57.60 b. 75%; \$180 c. 60%; \$170 d. 33⅓%; \$108

e. 60%; \$174; 40% f. 0.7 or $\frac{7}{10}$; \$105; 30%

g. 0.4375 or $\frac{7}{16}$; \$160; 56.25% h. 0.336; \$500; 66.4%

3. a. \$59.95 b. \$325 c. \$25 d. \$16

5. \$316.50

7. a. (1) 0.64 (2) \$28.80 (3) 36% b. (1) 0.6375 (2) \$12.75 (3) 36.25% c. (1) 0.447 (2) \$17.88 (3) 55.3% d. (1) 0.51 (2) \$102 (3) 49%

9. Hodgkins: 55% **11.** \$3 **13.** \$50 **15.** 40%

17. a. 12.5% b. 30%

19. a. C: \$25.00 H: \$22.75 b. 9%

21. a. S: \$320 D: \$300 b. 6.25%

23. a. T: \$175 J: \$140 b. 20%

SECTION 2

1. a. 0 b. 3% c. 1% d. 2%
e. 3% f. 2%

3. a. \$824 b. \$588 c. \$318.50 d. \$780
e. \$441 f. \$834.20 g. \$528

5. a. \$490; \$1,500 b. \$297; \$560 c. \$450; \$220
d. \$400; \$390 e. \$360; \$345.60

7. \$250.48 **9.** \$184.79 **11.** \$2,447.60

13. a. \$294 b. \$891 c. \$1,185

15. a. \$435 b. \$435

CHAPTER 13

SECTION 1

1. a. (1) \$15 (2) \$12 (3) Profit, \$3
b. (1) \$18 (2) \$22 (3) Loss, \$4
c. (1) \$14 (2) \$15 (3) Loss, \$1
d. (1) \$7 (2) \$10 (3) Loss, \$3
e. (1) \$24 (2) \$14.40 (3) Profit, \$9.60

3. a. 1.4; \$126; \$36; 28.57% or $28\frac{4}{7}\%$

b. 1.2; \$72; \$12; $16.\overline{66}\%$ or $16\frac{2}{3}\%$

c. 1.125 or $\frac{9}{8}$; \$36; \$4; $11\frac{1}{9}\%$

d. 1.25; \$12; \$3; 20%

e. 1.6; \$40; \$24; 37.5%

f. $22\frac{2}{9}\%$; $\frac{11}{9}$ or $1.\overline{22}$; \$12; $18.\overline{18}$

g. $33\frac{1}{3}\%$; 1.33 or $\frac{4}{3}$; \$27; 25%

5. a. 1.42 b. \$18.46 c. \$5.46

7. a. 25% b. 20%

9. a. 1.375 or $\frac{11}{8}$

b. \$11; \$6.60; \$8.80; \$22.55

c. 27.27% or $27\frac{3}{11}\%$

11. a. \$60 b. \$68; 20%

13. a. \$35; \$52.50 b. 50% c. $33\frac{1}{3}\%$

15. a. \$11.25 b. 20%

17. a. \$38.40 b. 37.5%

SECTION 2

1. a. $1.\overline{33}$ or $\frac{4}{3}$; \$40; \$10; 33⅓% b. 1.6 or $\frac{8}{5}$; \$56; \$21; 60%

c. 1.5 or $\frac{3}{2}$; \$28; \$14; 50% d. 2; \$6; \$6; 100%

e. 10%; $1.\overline{11}$ or $\frac{10}{9}$; \$20, 11⅑%

3. a. 1.429 or $\frac{10}{7}$ b. \$20 c. \$6

5. a. \$40 b. \$25

7. a. 2.5 or $\frac{5}{2}$ b. \$12.50; \$16; \$18; \$21.25 c. 150%

9. a. \$443 b. 20% c. $16\frac{2}{3}$%

11. a. Markup on selling price, 2.0 vs. markup on cost, 1.5
b. Markup on selling price, $1.3\overline{8}$ vs. markup on cost, 1.2

13. a. 1.6 or $\frac{8}{5}$ b. \$120 c. 60%

15. a. \$221 b. 1.538 or $\frac{20}{13}$

17. \$50

SECTION 3

	TOTAL COST	REQUIRED SALES	AMOUNT TO SELL	SELLING PRICE
1. a.	\$ 42	\$ 50.40	63	\$0.80
b.	28	35.00	19	1.85
c.	200	225.00	48	4.69

	TOTAL COST	REQUIRED SALES	AMOUNT AT REGULAR PRICE	AMOUNT AT REDUCED PRICE	REGULAR SELLING
3. a.	\$32	\$ 38.40	76	4	\$0.49
b.	75	105.00	23	2	4.35

5. a. \$14 b. 33⅓% c. 25%

7. \$0.77 **9.** \$0.97 **11.** \$9.16 **13.** \$0.94 **15.** \$168

CHAPTER 14

SECTION 1

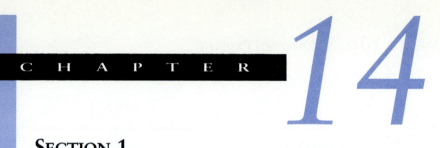

1. a. $50; $62.50 b. $48; $57.60
c. $40; $32; 12.5% d. $40; $36; 16⅔%
e. $27; 46%; $50 f. $148.75; 40.5%; $250
g. $240; 60%; $600; $320 h. $128; 60%; $320; $160

3. a. $60 b. $80

5. a. $1,280 b. $960 c. 16⅔%

7. a. $40 b. $30 c. 40%

9. a. $120 b. $54 c. 55%

11. a. $72 b. $50.40 c. 30%

13. a. $73.50 b. $110.25 c. $183.75

15. a. $35 b. $87.50 c. $100

SECTION 2

1.

	REGULAR SELLING PRICE	MARKDOWN PERCENT	MARKDOWN AMOUNT	SALE PRICE	WHOLESALE COST	OVERHEAD	THC	OPERATING PROFIT OR (LOSS)
a.			$20	$30			$26	$4
b.		30%		21			22	(1)
c.		$22\frac{2}{9}$	10		$23		31	
d.		$11\frac{1}{9}$	2	16	12			
e.	60	10	54		9			
f.	40		8		18		26	

3.

	REGULAR SELLING PRICE	MARKDOWN PERCENT	MARKDOWN AMOUNT	SALE PRICE	WHOLESALE COST	OVERHEAD	THC	OPERATING LOSS	GROSS LOSS AMOUNT	GROSS LOSS PERCENT
a.		$16\frac{2}{3}$%		$30			$35	$ 5	$2	6.25%
b.			$24	36		$6	46		4	10
c.	$40		8		$35			11	3	8.57
d.	90		27		70		74	11		10

5. ($15.50) **7.** ($1.20)

9. a. $16\frac{2}{3}$% b. $5.60, profit

11. a. $200 b. $80 c. $40; 28.57%

13. a. $0.60 b. $0.08 c. $13\frac{1}{3}$%

15. a. $6.30 b. $1.80 c. 28.57%

17. a. $560; $644; $670 b. Operating profit c. $26; 4.6%

19. a. $100; $130; $143 b. Operating profit c. $13; 13%

SECTION 3

1. a. $1,120 b. $548.80 c. $554.40 d. $1,103.20; $13.79
e. $19.31 f. $11.59 g. $1,313.20
h. Operating loss, $10.64 i. 1.0%

3. a. $8,245 b. $8,280; $82.80 c. $115.92
d. $57.96 e. $10,143 f. Operating loss, $207 g. 2.5%

CHAPTER 15

SECTION 1

1. a. $60; $660 b. $30; $630 c. $32; $1,232
   d. $17; $817 e. $162; $2,562
3. a. $660; $60 c. $1,232; $32 e. $2,562; $162
5. 8%
7. 10 months or 300 days
9. 10%
11. $8,000
13. 6.25 years

SECTION 2

1. a. 86 days b. 102 days c. 153 days
   d. 318 days e. 109 days f. 190 days
   g. 244 days h. 204 days
3. a. September 20 b. December 13
   c. October 16 d. September 17
   e. December 14 f. May 22
5. a. August 18; 181 days b. August 24; 92 days
   c. October 26, 2002; 273 days d. March 30, 2006; 122 days
   e. June 13, 2004; 152 days

SECTION 3

1. a. (1) $21.90
      (2) $21.60
   b. (1) $43.80
      (2) $43.20
3. (1) $379.56
   (2) $374.36
   (3) $373.33
5. a. $37.50 b. $75.00 c. $112.50 d. $56.25
7. a. $20 b. $40 c. $160 d. $360
9. a. $72 b. $45 c. $54
11. a. $20 b. $100 c. $80
13. a. $36.60 b. $18.30 c. $54.90
15. a. $57 b. $68.40 c. $102.60

SECTION 4

1. a. $5,000 b. William J. McDaniel c. Cardinal Bank
d. May 1, 20XX e. May 31, 20XX f. $5,000
g. 9% h. 30 days i. $37.50
j. $5,037.50

3. a. 270 days; $324; $5,124 b. June 21; $16.50; $916.50
c. November 21; $138; $4,638 d. $800; April 4; $820

5. $6,177.53; $177.53 **7.** $407.12 **9.** $2,150

11. 10% **13.** 90 days **15.** 9 months or 270 days

17. $1,500 **19.** $3,000

SECTION 5

1. a. $700 b. $2,400

3. a. $6,150; $6,014.67 b. $8,320; $7,923.81

5. a. $9,900; $9,428.57 b. $7,875; $7,720.59

7. $4,000 **9.** $6,000 **11.** $3,533.65

13. $5,980.77 **15.** $12,336.63 **17.** $2,559.01

19. Plan I: $30,000

CHAPTER 16

SECTION 1

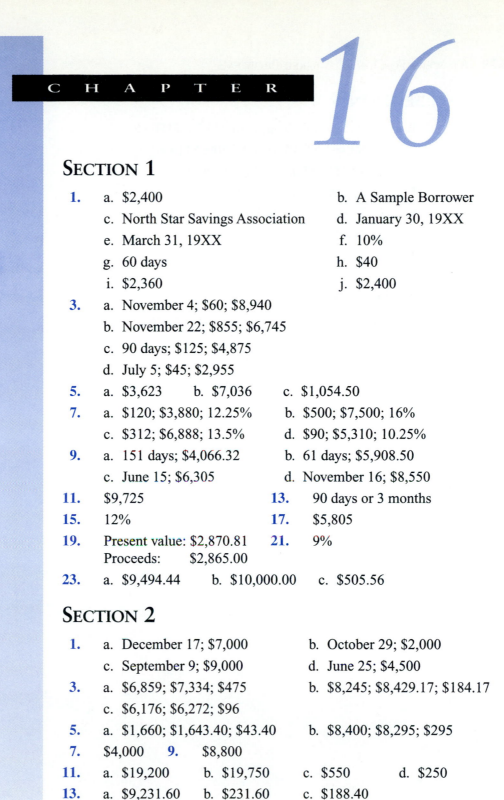

1. a. \$2,400 b. A Sample Borrower
 c. North Star Savings Association d. January 30, 19XX
 e. March 31, 19XX f. 10%
 g. 60 days h. \$40
 i. \$2,360 j. \$2,400
3. a. November 4; \$60; \$8,940
 b. November 22; \$855; \$6,745
 c. 90 days; \$125; \$4,875
 d. July 5; \$45; \$2,955
5. a. \$3,623 b. \$7,036 c. \$1,054.50
7. a. \$120; \$3,880; 12.25% b. \$500; \$7,500; 16%
 c. \$312; \$6,888; 13.5% d. \$90; \$5,310; 10.25%
9. a. 151 days; \$4,066.32 b. 61 days; \$5,908.50
 c. June 15; \$6,305 d. November 16; \$8,550
11. \$9,725
13. 90 days or 3 months
15. 12%
17. \$5,805
19. Present value: \$2,870.81
 Proceeds: \$2,865.00
21. 9%
23. a. \$9,494.44 b. \$10,000.00 c. \$505.56

SECTION 2

1. a. December 17; \$7,000 b. October 29; \$2,000
 c. September 9; \$9,000 d. June 25; \$4,500
3. a. \$6,859; \$7,334; \$475 b. \$8,245; \$8,429.17; \$184.17
 c. \$6,176; \$6,272; \$96
5. a. \$1,660; \$1,643.40; \$43.40 b. \$8,400; \$8,295; \$295
7. \$4,000
9. \$8,800
11. a. \$19,200 b. \$19,750 c. \$550 d. \$250
13. a. \$9,231.60 b. \$231.60 c. \$188.40
15. a. \$6,090 b. \$6,158.25 c. \$68.25 d. \$141.75
17. a. \$4,260.81 b. \$60.81 c. \$65.19

SECTION 3

1. a. $6,212.50; 20 days; $6,243.72; $160.92
b. $7,462.50; 15 days; $7,500; $114.80
c. $4,158; 45 days; $4,200; $42.86

3. a. $5,500 b. $113.40

5. a. $3,000 b. $73.45

7. a. $9,000 b. $280.36

9. a. $2,700 b. $33.30

SECTION 4

1. $68.75; $5,068.75 **3.** $44; $3,916

5.

Simple Interest	*Simple Discount*
a. $80	a. $80
b. $1,600	b. $1,520
c. $1,680	c. $1,600

7. a. $2,400 b. $2,400

9. a. $4,000 b. $4,000

11. 10.25%

13. Present value: $8,000
Proceeds: $7,980

15. $1,083.55 **17.** $5,326.82

19. a. $8,396.85 b. $396.85 c. $327.15

21. a. $5,817 b. $5,915 c. $98 d. $85

CHAPTER 17

SECTION 1

1. a. $2,640.20 b. $606.25 c. $3,320.00

3. $5,050 **5.** $5,062.50

SECTION 2

1. a. Total interest: $ 9.32 b. Total interest: $ 29.48
Total payments: $184.32 Total payments: $529.48

3. Total interest: $ 14.63
Final balance: $229.63

5. $10.97

7. a. $50.00 b. $2.98 c. $137.98

SECTION 3

1. a. $47/month b. $62.50/month c. $21.50/month

3. a. 38.5% b. 18% c. 24.5% d. 19.75%

5. Monthly interest: $ 40.14
Payment to principal: $799.86

7. $22/month **9.** $100/month

11. a. $39 b. $42.90/month c. $559 d. 21.25%

13. a. $400 b. 14.25%

15. a. $50 b. 10.5%

17. a. $1,544 b. $144 c. 11.25%

19. a. $1,510 b. $210 c. 12.25%

21. a. 25.0% b. 25.0% c. 24.5%

23. a. $108 b. $418/month c. 15.25% d. $107.87

25. a. $110 b. $462/month c. $109.81

SECTION 4

1. a. Interest saved: $ 33.75 b. Interest saved: $ 75.50
Balance due: $626.25 Balance due: $1,112.00

3. a. Interest saved: $ 30.80 b. Interest saved: $ 70.31
Balance due: $629.20 Balance due: $1,117.19

5. a. $560 b. $140/month c. 18.25%
d. $206.56 e. $1,753.44

7. a. $1,031.25 b. $568.75/month c. 19.75%
   d. $188.29 e. $3,224.21

9. a. $378 b. $442/month c. $126 d. $2,084

11. a. $180 b. $92/month c. $67.50 d. $760.50

CHAPTER 18

Section 1

		Number of periods	Rate per period
1.	a.	6	4.5%
	b.	24	0.75
	c.	10	3.75
	d.	32	1.5
	e.	18	2.5

3. a. $5,243.18; $1,243.18 b. $5,267.24; $1,267.24

5. a. $3,151.88; $151.88 b. $3,152.83; $152.83

7. a. $7,319.75; $319.75 b. $7,321.37; $321.37

9. More interest is earned when interest is compounded more often.

Section 2

1. a. $5,243.18; $1,243.18 b. $5,267.24; $1,267.24

3. a. $3,151.88; $151.88 b. $3,152.84; $152.84

5. a. $7,319.75; $319.75 b. $7,321.38; $321.38

7. a. $580.74; $80.74 b. $2,289.65; $589.65
c. $4,835.42; $1,635.42 d. $6,190.50; $1,690.50
e. $15,102.81; $7,002.81

9. a. $580.74; $80.74 b. $1,161.47; $161.47
c. $2,322.94; $322.94 d. Doubling the principal doubles the interest.

11. a. $1,082.86; $82.86 b. $1,171.66; $171.66
c. $1,368.57; $368.57 d. No

13. a. $1,125.51; $125.51 b. $1,266.77; $266.77
c. $1,604.71; $604.71 d. No

15. a. $1,020.10 b. $179.90 c. $1,248.72 d. $68.82

17. a. $1,012.55 b. $387.45 c. $1,485.91 d. $98.46

19. a. $1,025.26 b. $374.74 c. $1,608.43 d. $233.69

SECTION 3

1.	a.	(1) $7.28	b.	(1) $8.54	c.	(1) $7.51		
		(2) $1,007.28		(2) $808.54		(2) $4,007.51		
	d.	(1) $6.26						
		(2) $606.26						
3.	a.	(1) $84.85	b.	(1) $50.91				
		(2) $84.38		(2) $50.63				
5.	a.	(1) $15.28	b.	(1) $11.56				
		(2) $1,915.28		(2) $1,511.56				
7.	a.	(1) $106.08	b.	(1) $60.12				
		(2) $9,106.08		(2) $5,060.12				
9.	a.	(1) $61.01	b.	(1) $57.27	c.	(1) $59.79		
		(2) $5,161.01		(2) $5,757.27		(2) $4,559.79		

SECTION 4

1. a. $2,956.31; $243.69 b. $1,950.02; $549.98
c. $5,143.51; $1,656.49 d. $1,713.27; $86.73
e. $3,680.10; $719.90

3. a. $4,709.53 b. $290.47

5. a. $64,011.78 b. $10,988.22

7. a. $58,966.39 b. $41,033.61

9. a. $8,882.90 b. $7,275.73

11. a. $7,320.00 b. $6,366.29

13. a. $5,450.00 b. $5,133.38

CHAPTER 19

Section 1

1. a. $17,240.74; $14,000; $3,240.74
   b. $10,580.08; $9,600; $980.08
   c. $6,547.36; $4,800; $1,747.36
   d. $41,682.13; $36,000; $5,682.13
   e. $43,269.72; $39,600; $3,669.72

3. a. $4,253.77 b. $4,000.00 c. $253.77

5. a. $34,885.02 b. $30,000.00 c. $4,885.02

7. a. $30,000.00 b. $35,199.61 c. $5,199.61

9. a. $160,000.00 b. $184,989.33 c. $24,989.33

Section 2

1. a. $9,141.51; $16,000; $6,858.49
   b. $47,170.21; $54,000; $6,829.79
   c. $82,066.44; $120,000; $37,933.56
   d. $28,790.42; $40,800; $12,009.58
   e. $22,178.53; $26,000; $3,821.47

		Amount of annuity	Total interest	Present value of annuity	Total interest
3.	a.	$265,362.18	$73,362.18	$140,810.01	$51,189.99
	b.	10,999.96	3,799.96	4,817.51	2,382.49
	c.	265,079.37	29,879.37	208,643.55	26,556.45

5. a. $4,908.38 b. $5,400.00 c. $491.62

7. a. $4,540.87 b. $4,800.00 c. $259.13

9. a. $4,700 b. $2,100 c. $3,500

11. a. $14,900 b. $10,400 c. $6,400

13. a. $241,000.00 b. $186,935.99

15. a. $29,000.00 b. $25,829.56

CHAPTER 20

SECTION 1

1. a. $1,265.74; $25,314.80; $4,685.20
b. $3,327.61; $39,931.32; $10,068.68
c. $720.68; $60,537.12; $14,462.88
d. $5,684.67; $90,954.72; $9,045.28
e. $4,859.94; $174,957.84; $25,042.16

3. a. $1,010.00; Premium, $10; $62.50; 6.19%
b. $880.00; Discount, $120; $51.25; 5.82%
c. $1,000; Par, $97.50; 9.75%
d. $955; Discount, $45; $85; 8.9%
e. $1,062.50; Premium, $62.50; $100; 9.41%

5. a. $3,946.25 b. $63,140.00 c. $11,860.00

7. a. $5,931.28 b. $23,725.12 c. $1,274.88

9. a. $7,150.20 b. $686,419.20 c. $313,580.80

11. a. $1,040.00 b. Premium, $40 c. $520,000.00
d. $82.50 e. 7.93% f. $41,250.00

13. a. $932.08 b. $3,547.92

SECTION 2

1. a. $17,349.03; $520,470.90; $220,470.90
b. $5,726.61; $229,064.40; $79,064.40
c. $1,584.10; $95,046.00; $15,046.00
d. $50,511.95; $707,167.30; $207,167.30
e. $293; $28,128; $8,128

3. a. $1,468; $528,480; $328,480
b. $2,652; $636,480; $336,480
c. $514.20; $154,260; $94,260
d. $810; $194,400; $104,400

5. a. $954 b. $4,344

7. a. $2,819.38 b. $18,942.64

9. a. $7,338.80 b. $26,776 c. $171,776

11. a. $2,970.18 b. $28,210.80 c. $203,210.80

	20 YEARS	25 YEARS
13.	a. $ 2,145.90	$ 2,010.20
	b. 515,016.00	603,060.00
	c. 530,016.00	618,060.00
	d. $ 88,044.00	

	PERIODIC PAYMENT	TOTAL INTEREST	APPLIED TOWARD PRINCIPAL
15.	a. $1,219.45	$1,097.27	$4,999.98
	b. 1,144.57	867.43	5,999.99

SECTION 3

1.	a. $9,322.73	b. $4,322.73	
3.	a. $16,017.24	b. $13,982.76	
5.	a. $4,522.12	b. $144,707.84	c. $55,292.16
7.	a. $3,561.41	b. $28,491.28	c. $3,491.28
9.	a. $2,256,297.90	b. $2,000,000.00	c. $256,297.90
11.	a. 25,548.19	b. $3,251.81	
13.	a. $11,806.48	b. $3,193.52	
15.	a. $127,971.73	b. $15,971.73	
17.	a. $9,850.88	b. $9,105.28	
19.	a. $9,025.15	b. $9,600.00	c. $574.85
21.	a. $25,500.88	b. $510,017.60	c. $289,982.40
23.	a. $1,160.76	b. $160.76	

APPENDIX

SECTION 1

1. a. millions b. tens c. billions
   d. hundred thousands e. ones
3. a. Six hundred thirty-three million, five hundred twenty thousand, four hundred eighty-one
   b. Twenty-five million, five hundred forty-three thousand, one hundred twenty-eight
   c. One hundred fifty million, two hundred eighty-six thousand, four hundred thirteen
   d. Six million, forty-six thousand, one hundred twenty-five
   e. Eight hundred twelve billion, three hundred forty-four million, six hundred one thousand, twenty-two

SECTION 2

1. a. 20 b. 19 c. 27
   d. 25 e. 27 f. 20
   g. 19 h. 225 i. 206
3. a. 22 b. 24 c. 28
   d. 37 e. 284 f. 221
   g. 213 h. 315 i. 268
5. a. 206 b. 262 c. 1,412
7. a. 8 b. 664 c. 1,921
   d. 323
9. a. 4,660 b. 6,400 c. 55,300
   d. 11,200 e. 82,800 f. 2,075,000
11. a. 1,824 b. 37,842 c. 381,351
    d. 1,750,060
13. a. 598 b. 746 c. 18

d. 26 e. 291

15.

Monday:	$ 2,457.47	Dept. #1	$ 2,404.35
Tuesday:	2,315.53	Dept. #2	3,002.09
Wednesday:	2,189.82	Dept. #3	2,651.89
Thursday:	2,361.51	Dept. #4	1,991.98
Friday:	2,732.17	Dept. #5	1,962.61
Saturday:	2,627.90	Dept. #6	2,671.48
Daily Totals:	$14,684.40 =	Dept. Totals	$14,684.40

SECTION 3

1. a. $\frac{2}{3}$ b. $2\frac{1}{3}$ c. $\frac{7}{9}$ d. $\frac{4}{7}$

e. $\frac{2}{3}$ f. $1\frac{2}{3}$ g. $\frac{3}{4}$ h. $1\frac{1}{5}$

i. $\frac{63}{107}$ j. $\frac{7}{12}$

3. a. $\frac{5}{14}, \frac{3}{7}, \frac{1}{2}, \frac{5}{8}, \frac{5}{7}, \frac{3}{4}$ b. $\frac{13}{24}, \frac{7}{12}, \frac{5}{8}, \frac{3}{4}, \frac{7}{9}, \frac{5}{6}$

5. a. $2\frac{5}{6}$ b. $9\frac{1}{5}$ c. $8\frac{1}{2}$ d. $3\frac{3}{7}$

e. $16\frac{1}{3}$ f. $3\frac{1}{12}$ g. $15\frac{1}{3}$ h. $13\frac{1}{2}$

i. $5\frac{2}{3}$ j. $11\frac{1}{5}$ k. $11\frac{1}{3}$ l. $9\frac{1}{2}$

7. a. $\frac{1}{2}$ b. $\frac{11}{20}$ c. $\frac{25}{56}$ d. $\frac{1}{12}$

e. $\frac{1}{15}$ f. $11\frac{11}{12}$ g. $18\frac{5}{12}$ h. $20\frac{5}{6}$

i. $15\frac{2}{7}$ j. $20\frac{1}{2}$

9. a. $\frac{1}{40}$ b. $\frac{7}{32}$ c. $24\frac{1}{2}$ d. 28

SECTION 4

1. a. thousandths b. ones c. tens

d. hundreds e. hundredths f. thousands

g. ten-thousandths h. tenths i. hundred-thousandths

3.

a. 0.4	b. 0.25	c. 0.3
d. 0.7	e. 0.32	f. 0.075
g. $0.14\frac{2}{7}$ or 0.142857	h. $0.44\frac{4}{9}$ or $0.44\overline{4}$	i. $0.41\frac{2}{3}$ or $0.41\overline{6}$
j. $0.54\frac{6}{11}$ or $0.54\overline{54}$	k. $0.30\frac{10}{13}$ or 0.307692	l. $0.26\frac{12}{13}$ or 0.269231
m. 6.0	n. $40\frac{3}{4}$ or 40.75	o. 5.4
p. $5.11\frac{1}{9}$ or $5.11\overline{1}$	q. $3.83\frac{1}{3}$ or $3.83\overline{3}$	r. 9.428

5.

a. 35.62	b. 26.45	c. 5.5663
d. 0.10488	e. 884.52	f. 1.6
g. 265	h. 881.2	i. 280
j. 0.003651	k. 52,500	l. 0.00644
m. 546,000	n. 0.009945	

7.

a. 46.95	b. 3.725	c. 28.859
d. 240.316	e. 7.811	f. 3.845
g. 558.822	h. 12.929	i. 29.0096
j. 142.344	k. 52.279	l. 455.045
m. 37.697	n. 48.02	o. 285.12
p. 68.894	q. 4.5788	

9.

a. 78	b. 123	c. 46
d. 68	e. 136	f. 3,270
g. 1.27	h. 0.24	i. 0.344
j. 32.58	k. 0.78	l. 0.0645
m. 440	n. 41	o. 321
p. 0.28		

APPENDIX B

SECTION 1

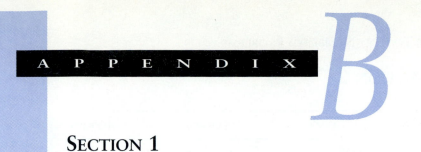

1. a. 10 b. hectoliter c. milligram d. 0.1
e. 100 f. gram g. kilometer h. 0.001
i. 1 000 j. milliliter k. 100 l. decagram

3. a. 360 b. 4.5 c. 2.15 d. 800
e. 425 f. 1.88 g. 300 h. 1.2
i. 29 j. 0.76

5. a. 6.54 b. 110 c. 10.6 d. 49.68
e. 7.88 f. 16.9 g. 16.4 h. 14

7. a. 2.44 b. 45.4 c. 3.784 d. 9.46
e. 128.8 f. 15.24 g. 255.6 h. 18.95

9. a. 25° b. 35° c. 15° d. −10°
e. 68° f. 176° g. 131° h. 23°

11. a. 304.8 centimeters b. 98.5 inches
c. American is longer by 54.8 centimeters.

13. a. 167.64 centimeters b. 54.48 kilograms

15. 4 830 kilometers; 40 250 kilometers

17. a. 0.7 ounce b. 3.85 pounds

19. a. 35°C b. 86°F
c. New York (by 5°C)

Section 2

		German	Japanese	Mexican
1.	a.	DM 18.19	¥ 1,147	$ 94.82
	b.	1,091.10	68,838	5,689.20
	c.	3,273.30	206,514	17,067.60

3. a. $1,546.32 b. $466.74 c. $357.52

5. C$12,031.20 7. Fr487.81 9. $80.77 11. $87.00

13. a. DM7.26 b. ¥458 c. C$6.00

INDEX

Selective page number references that appear in ***boldface*** *in the index refer the student to glossary definitions of important terms and concepts.*

A

D

K

L

M

N

Q

R

S

T

U

V

W